The Role of Air-Sea Exchange
in Geochemical Cycling

NATO ASI Series

Advanced Science Institutes Series

A series presenting the results of activities sponsored by the NATO Science Committee, which aims at the dissemination of advanced scientific and technological knowledge, with a view to strengthening links between scientific communities.

The series is published by an international board of publishers in conjunction with the NATO Scientific Affairs Division

A Life Sciences	Plenum Publishing Corporation
B Physics	London and New York
C Mathematical and Physical Sciences	D. Reidel Publishing Company Dordrecht, Boston, Lancaster and Tokyo
D Behavioural and Social Sciences E Engineering and Materials Sciences	Martinus Nijhoff Publishers The Hague, Boston and Lancaster
F Computer and Systems Sciences G Ecological Sciences	Springer-Verlag Berlin, Heidelberg, New York and Tokyo

Series C: Mathematical and Physical Sciences Vol. 185

The Role of Air-Sea Exchange in Geochemical Cycling

edited by

Patrick Buat-Ménard

Centre des Faibles Radioactivités,
Laboratoire Mixte CNRS–CEA,
Gif-sur-Yvette, France

D. Reidel Publishing Company

Dordrecht / Boston / Lancaster / Tokyo

Published in cooperation with NATO Scientific Affairs Division

Proceedings of the NATO Advanced Study Institute on
The Role of Air-Sea Exchange in Geochemical Cycling
Bombannes, France
16-27 September 1985

Library of Congress Cataloging in Publication Data

NATO Advanced Study Institute on the Role of Air-Sea Exchange in Geochemical Cycling
  (1985 : Bombannes, France)
  The role of air-sea exchange in geochemical cycling.
  (NATO ASI series. Series C, Mathematical and physical sciences; vol. 185)
  "Proceedings of the NATO Advanced Study Institute on the Role of Air-Sea Exchange
in Geochemical Cycling, Bombannes, France, 16–27 September, 1985"—T.p. verso.
  Companion vol. to: Air-sea exchange of gases and particles / edited by Peter S. Liss
and W. George N. Slinn. Dordrecht ; Boston : D. Reidel Pub. Co., c1983.
  "Published in cooperation with NATO Scientific Affairs Division."
  Includes bibliographies and index.
  1. Ocean-atmosphere   interaction—Congresses.   2. Sea-water—Composition—
Congresses. 3. Geochemistry—Congresses. I. Buat-Ménard, Patrick, 1946–   .
II. North Atlantic Treaty Organization. Scientific Affairs Division. III. Title. IV. Series:
NATO ASI series. Series C, Mathematical and physical sciences; vol. 185.
GC190.2.N16   1985   551.46′.01   86–17678

ISBN-13: 978-94-010-8606-6   e-ISBN-13: 978-94-009-4738-2
DOI: 10.1007/978-94-009-4738-2

Published by D. Reidel Publishing Company
P.O. Box 17, 3300 AA Dordrecht, Holland

Sold and distributed in the U.S.A. and Canada
by Kluwer Academic Publishers,
101 Philip Drive, Assinippi Park, Norwell, MA 02061, U.S.A.

In all other countries, sold and distributed
by Kluwer Academic Publishers Group,
P.O. Box 322, 3300 AH Dordrecht, Holland

D. Reidel Publishing Company is a member of the Kluwer Academic Publishers Group

---

CONTENTS

PREFACE - P. Buat-Ménard   xiii

ACKNOWLEDGEMENTS   xv

BASIC CONCEPTS IN GEOCHEMICAL MODELLING - R. Wollast   1

   1. Introduction   1
   2. First Order Models   3
      2.1. First order decay reaction   3
      2.2. Instantaneous perturbation in a first order decay
           model   8
      2.3. First order production model   9
   3. Heterogeneous Catalysis and Enzymatic Type Reactions   10
   4. Reversible Reactions   12
   5. Model with Coupled Components in the Reservoir   19
   6. Second Order Reactions   25
   7. Periodic Fluctuations   27
   8. Coupling of Reservoirs   30
   9. Conclusions   33

ATMOSPHERIC PATHWAYS TO THE OCEANS - J. T. Merrill   35

   1. Introduction   35
   2. Atmospheric Structure and Transports   36
      2.1. Boundary layer   37
      2.2. Cloud scale transport   41
      2.3. Storms and midlatitude circulation   42
      2.4. Global scale exchange   47
   3. Variability and Representativeness   49
      3.1. Seasonal and interannual variability   50
      3.2. Representativeness of observations   52
   4. Modeling of Atmospheric Transport   53
      4.1. Source identification models   53
      4.2. Mechanistic models   56
      4.3. Tropospheric chemistry system models   57

MODELING OCEANIC TRANSPORT OF DISSOLVED CONSTITUENTS -
   J.L. Sarmiento   65

   1. Introduction   65
   2. Box Models   66
   3. Advection-Diffusion Models   71
   4. Equations of Motion   78
   5. Conclusion   80

VERTICAL TRANSPORT OF PARTICLES WITHIN THE OCEAN –
   J.C. Brun-Cottan   83

1. Introduction   83
2. Determination of the Lognormal Coefficients L, σ
    and of N   86
    2.1. Particle size data collection   86
    2.2. Calculation of the coefficients L, σ and N
          from data   87
3. Determination of Suspended Particulate Matter Physical
    Properties using Lognormal coefficients   90
    3.1. Surface area concentration   90
    3.2. Mass concentration   90
    3.3. Vertical fluxes   92
    3.4. Residence time   96
    3.5. Application to the open sea   97
4. Suspended Particulate Matter Sedimentation with
    Dissolution Process   99
    4.1. Sedimentation at steady state   99
    4.2. Sedimentation at non steady state   102
5. Conclusion   105
6. Appendix   106
    6.1. Specific properties of the lognormal law   106
    6.2. Evaluation of the lognormal coefficients   107

AIR-SEA GAS EXCHANGE RATES: INTRODUCTION AND SYNTHESIS –
   P.S. Liss and L. Merlivat   113

1. Introduction   113
2. Basic Principles   114
3. Models   115
    3.1. Film model   115
    3.2. Surface renewal models   115
    3.3. Boundary-layer models   116
4. Laboratory (Wind Tunnel) Studies   116
    4.1. Smooth surface regime   118
    4.2. Rough surface regime   118
    4.3. Breaking wave (bubble) regime   118
5. Field Measurements   118
    5.1. Box method   119
    5.2. Dissolved gas balance method   119
    5.3. Micrometeorological techniques   119
    5.4. Natural and bomb-produced ^{14}C 120
    5.5. The radon deficiency method   120
    5.6. Sulphur hexafluoride   121
    5.7. Summary   121
6. Synthesis   122
    6.1. Comparison with field data   125

THE OCEAN AS A SOURCE FOR ATMOSPHERIC PARTICLES - E.C. Monahan   129

    1. Introduction   129
    2. The $\overset{\bullet}{W}$ $\partial E/\partial r$ Model   130
    3. The $\overset{\bullet}{W}$ $\partial^2 E/\partial t$ ∂r Model   135
    4. Comparison of $\overset{\bullet}{W}$ $\partial E/\partial r$ and W $\partial^2 E/\partial t$ ∂r Models   139
    5. Oceanic Whitecap Coverage   146
    6. Global Sea-to-Air Salt Flux   152
    7. Toward a Comprehensive Marine Aerosol Generation Model   155

THE OCEAN AS A SINK FOR ATMOSPHERIC PARTICLES - P. Buat-Ménard   165

    1. Overview   165
    2. Assessement of Wet Deposition   167
    3. Field Approach to Dry Deposition   172
    4. Accurate Deposition Measurements do not Guarantee
        Accurate Net Air to Sea Transfer Rates   177
    5. Relative Importance of Wet and Dry Removal Rates   179
    6. Conclusion   181

ATMOSPHERIC, OCEANIC, AND INTERFACIAL PHOTOCHEMISTRY AS FACTORS
INFLUENCING AIR-SEA EXCHANGE FLUXES AND PROCESSES - O.C. Zafiriou   185

    1. Introduction   185
    2. Environmental Photochemistry   186
        2.1. Stratospheric photochemistry   186
        2.2. Homogeneous tropospheric photochemistry   187
        2.3. Heterogeneous tropospheric photochemistry   191
        2.4. Seawater photochemistry   194
        2.5. Soil photochemistry   198
    3. Interaction of Photochemistry with Air-Sea Exchange
        Processes   198
        3.1. Air-sea gas exchange   198
        3.2. Rainout-washout deposition processes   201
        3.3. Dry deposition   202
        3.4. Marine aerosol generation   203
    4. Summary   203

CARBON DIOXIDE: ITS NATURAL CYCLE AND ANTHROPOGENIC PERTURBATION -
    U. Siegenthaler   209

    1. Introduction   209
    2. The Natural Cycle of Carbon Dioxide   210
        2.1. Reservoirs, fluxes, residence times   210
        2.2. Air-sea exchange of CO_2 213
        2.3. Regional variability of air-sea fluxes   215
        2.4. Marine carbonate chemistry   219
        2.5. The oceanic carbon cycle   220

    2.6. The cycle of oxygen 221
  3. Anthropogenic Increase of Atmospheric CO_2 223
    3.1. Observations and airborne fraction 223
    3.2. Modelling the oceanic response to carbon cycle
          perturbations 225
    3.3. CO_2 release from the terrestrial biosphere and the
          "missing CO_2 sink" 229
    3.4. Scenarios for future CO_2 concentrations 231
    3.5. Carbone isotope perturbations 233
  4. Climatic Effects of CO_2 Increase 234
  5. Natural CO_2 Variations 239
    5.1. Seasonal variations 239
    5.2. Correlation with El Niño 240
    5.3. Glacial/interglacial changes 242

CO_2 AIR-SEA EXCHANGE DURING GLACIAL TIMES: IMPORTANCE OF DEEP
    SEA CIRCULATION CHANGES - J.C. Duplessy 249

  1. Introduction 249
  2. Evidence from Polar Ice Cores 249
    2.1. Data 249
    2.2. Discussion: is the ocean able to absorb the
          missing CO_2? 250
  3. Evidence from Deep Sea Sediments 250
    3.1. Data 250
    3.2. Various hypotheses explaining the sedimentary
          record 251
    3.3. Cadmium as a proxy-indicator for past phosphate 252
  4. Broecker's two box Model for the CO_2 Cycle 253
  5. Evidence for Deep Water Circulation during the Last
      Climatic Cycle 253
    5.1. Geochemical basis 253
    5.2. Glacial to interglacial contrasts 258
    5.3. Disappearance of North Atlantic Deep Water during
          the glacial to interglacial transition 263
    5.4. Enhanced North Atlantic Deep Water formation
          during the inception of the glaciation 263
  6. Conclusion 265

EXCHANGE OF CO AND H_2 BETWEEN OCEAN AND ATMOSPHERE - R. Conrad
    and W. Seiler 269

  1. Introduction 269
  2. Determination of the Supersaturation Factors of
      CO and H_2 270
  3. Spatial and Temporal Changes of dissolved CO and H_2 272
  4. Processes Sustaining CO and H_2 Concentrations
      in Surface Water 274
    4.1. Production processes 274

     4.2.   Consumption processes   276
     4.3.   Transport processes   277
  5. Calculation of Fluxes by the "Laminar Film Model"   278
  6. Role of Oceans in the budget of atmospheric CO and H$_2$ 280

THE AIR-SEA EXCHANGE OF LOW MOLECULAR WEIGHT HALOCARBON GASES -
     P.S. Liss   283

  1. Introduction   283
  2. Gases for which the Oceans are a net Source for the
     Atmosphere   284
     2.1.   Alkyl (mainly Methyl) halides   284
     2.2.   Haloforms   287
     2.3.   Other organo-halides   289
  3. Gases for which the Oceans are a net Sink for the
     Atmosphere   289
  4. Summary   290

SEA-AIR EXCHANGE OF HIGH-MOLECULAR WEIGHT SYNTHETIC ORGANIC
     COMPOUNDS - E. Atlas and C.S. Giam   295

  1. Introduction   295
  2. Compounds of Interest   296
  3. Sampling/Analytical Aspects of Trace Organics   296
  4. Distribution of High Molecular Weight Organics in
     the Marine Environment   297
     4.1.   Water and organisms   297
     4.2.   Atmospheric concentrations   300
     4.3.   Atmospheric deposition   303
  5. Air-Sea Exchange Mechanisms for Synthetic Organics   303
     5.1.   Dry deposition   305
     5.2.   Wet deposition   310
     5.3.   Adsorption and partitioning in surface waters   311
  6. Air-Sea Fluxes in the North Pacific   314
  7. Relative Importance of Atmospheric Deposition to the
     CHC Cycle   319
  8. Summary and Conclusions   321

THE OCEAN AS A SOURCE OF ATMOSPHERIC SULFUR COMPOUNDS -
     M.O. Andreae   331

  1. Sources of Sulfur to the Atmosphere: an Overview   331
  2. Seaspray and the Production of Aerosol Sulfate   332
  3. Sulfate Reduction by Geological and Biological Processes 334
  4. Assimilatory Sulfate Reduction   336
  5. Biosynthesis of Dimethylsulfide   338
  6. Marine Chemistry and Distribution of Dimethylsulfide   338
  7. Estimating the Air/Sea Flux of Dimethylsulfide   343

8. Chemical Reactions and Transformations of Dimethylsulfide
    in the Marine Atmosphere   346
9. A Model of the Cycle of Biogenic Sulfur over the Oceans   349
10. Carbonyl Sulfide   352
    10.1. Photochemical production of COS   352
    10.2. Air/Sea exchange of COS   353
11. Formation and Emission of other Sulfur Species: Hydrogen
    Sulfide, Carbon Disulfide, Methylmercaptan,
    Dimethyldisulfide etc...   354
    11.1. Hydrogen sulfide   354
    11.2. Carbon disulfide   355
    11.3. Methylmercaptan, dimethyldisulfide and other
          sulfur compounds   356
12. Conclusion   356

CYCLING OF MERCURY BETWEEN THE ATMOSPHERE AND OCEANS -
    W.F. Fitzgerald   363

1. Introduction   363
2. Global Models   364
3. Physico-Chemical Models   367
4. Atmospheric Hg Determinations   370
    4.1. Total gaseous Hg   370
    4.2. Volatile Hg species   371
5. Hg Analysis in Seawater and Rainwater   371
    5.1. Reactive and Total Hg   372
    5.2. Volatile Hg   372
    5.3. Determinations of Hg in rain   372
6. Air-Sea Exchange of Hg   373
    6.1. Preliminary studies   373
    6.2. Present status   375
    6.3. Summary   382
7. Ocean Sources of Hg   382
    7.1. Hg evasion from the Equatorial Pacific Ocean: 1980   382
    7.2. Hg evasion from the Equatorial Pacific Ocean: 1984   386
8. Hg Deposition to the Sea Surface   393
    8.1. Precipitation   393
    8.2. Dry depositional Hg flux to the Equatorial
          Pacific Ocean   396
    8.3. Air-sea exchange in the Equatorial Pacific Ocean   396
    8.4. Physico-chemical aspects   397
9. Atmospheric Cycling of Hg over the Oceans:
    Global Perspectives   397

THE AIR-SEA EXCHANGE OF PARTICULATE ORGANIC MATTER: THE SOURCES AND
    LONG-RANGE TRANSPORT OF LIPIDS IN AEROSOLS - R.B. Gagosian   409

1. Introduction   409
    1.1. Background   409

2.   Sampling and Analytical Methodology   413
3.   Source and Long-Range Transport Studies   415
     3.1.   Introduction   415
     3.2.   North Pacific Trades: Enewetak   416
     3.3.   South Pacific Westerlies: New Zealand   424
     3.4.   Short-Range transport: Coastal Peru   429
4.   Conclusions   436

THE MARINE MINERAL AEROSOL - R. Chester   443

1.   Introduction   443
2.   The Concept of the Marine Dust Veil   443
3.   Sources of Material to the Marine Atmosphere   445
4.   The Distribution of Material in the Marine
     Dust Veil   446
     4.1.   Introduction   446
     4.2.   The Atlantic Ocean and surrounding waters   448
     4.3.   The Mediterranean   450
     4.4.   The Pacific ocean   451
     4.5.   The Indian ocean   452
     4.6.   Summary   454
5.   The Composition of Material in the Marine
     Dust Veil   455
     5.1.   Introduction   455
     5.2.   The mineral composition of the marine dust veil   455
     5.3.   Chemical chacacteristics of the mineral aerosol   459
6.   The Influence of the Marine Dust Veil on Oceanic
     Cycles   464
     6.1.   The water column   464
     6.2.   The sediment column   467
     6.3.   Summary   471

AIR TO SEA TRANSFER OF ANTHROPOGENIC TRACE METALS -
     P. Buat-Ménard   477

1.   Introduction   477
2.   The Fate of Atmospheric Trace Metals in Ocean Waters   478
     2.1.   Physical and chemical forms of metals in the
            atmosphere   478
     2.2.   Biogeochemical cycling of trace metals
            in the ocean   480
3.   Geographical Variability of Metal Fluxes from the
     Atmosphere to the Ocean   483
4.   Effects of Eolian Inputs of Anthropogenic Origin on the
     Distribution of Trace Metals in Surface and Deep
     Marine Waters   485
5.   Conclusions   491

THE IMPACT OF ATMOSPHERIC NITROGEN, PHOSPHORUS, AND IRON SPECIES ON
   MARINE BIOLOGICAL PRODUCTIVITY – R.A. Duce   497

      1.   Introduction   497
           1.1.   Atmospheric deposition: geochemical and
                  biological impact   497
           1.2.   Nutrients and primary productivity   499
           1.3.   A simple model of vertical transport to the photic
                  zone   500
      2.   Nitrate and other Nitrogen Species   502
           2.1.   Calculations of nitrogen fluxes in open
                  ocean regions   502
           2.2.   Atmospheric nitrogen input to coastal waters   509
      3.   Phosphorus   510
           3.1.   Atmospheric phosphorus and the phosphorus cycle   510
           3.2.   Calculations of phosphorus fluxes in open
                  ocean regions   511
      4.   Iron   516
           4.1.   Introduction   516
           4.2.   Solubility of atmospheric iron   517
           4.3.   Calculations of iron fluxes in open ocean regions   518
      5.   General Discussion and Conclusions   521

LIST OF PARTICIPANTS   531

INDEX   537

PREFACE

This book arises from a NATO-sponsored Advanced Study Institute on 'The Role of Air-Sea Exchange in Geochemical Cycling' held at Bombannes, near Bordeaux, France, from 16 to 27 September 1985. The chapters of the book are the written versions of the lectures given at the Institute. The aim of the book is to give a comprehensive up-to-date coverage of the subject, presented in a teaching mode. The chapters contain much recent research material and attempt to give the reader an understanding of how the role of air-sea exchange in geochemical cycling can be quantitatively assessed.

In the last decade, major advances in the fields of marine and atmospheric chemistry have underlined the role of physical, chemical and biological processes at and near the air-sea interface in a number of geochemical cycles (C, S, N, metals etc...). Further, there is strong concern over the anthropogenic perturbation of these cycles on both regional and global scales.

The first part of the book (Chapters 1 to 8) provides a review of topics fundamental to such studies. These topics include concepts in geochemical modelling, assessment of atmospheric transport from sources to the oceans, description of mixing and transport processes within the ocean for both dissolved and particulate materials, quantification of air-sea fluxes for both gases and particles, photochemical transformations in the atmospheric and oceanic boundary layers.

In the second part (Chapters 9 to 19), various case studies are presented which illustrate the theme of the book. The general approach here is to provide the reader with an overview of the biogeochemical cycles of the elements or compounds of interest, to focus on present knowledge of its air-sea exchange and to assess the effect of such exchange on atmospheric and/or oceanic chemistry. The case studies have been selected on one or more of the following grounds: i) to highlight and illustrate important processes of air-sea exchange (high-molecular weight hydrocarbons, mercury); ii) to focus on substances of global environmental concern (CO_2, sulphur compounds, halocarbons); iii) to include elements whose global cycles have been significantly perturbed by human activity (heavy metals); iv) to emphasize several elements important in marine biological systems (N, P, Fe); v) to illustrate air-sea transfer as a major pathway in some geochemical cycles (S, C, mineral dust etc...). In addition, it seemed appropriate to review the available evidence of past natural perturbations of the CO_2 cycle, especially during glacial times, in order to properly assess the strengths and the effects of geochemical perturbations due to human activities.

Clearly, the book does not pretend to exhaust all the geochemical aspects of air-sea exchange. Nevertheless, it is hoped that the reader will be given an entry into the various interdisciplinary approaches required in this field of research. Although some overlap between chapters has been eliminated following the presentation at the Institute, it still exists. However, this allows individual chapters to stand by themselves. Also, the book is not devoid of underlap. In this regard, the reader should be aware that this publication is building on that arising from a previous N.A.T.O. A.S.I (Air-Sea Exchange of gases and particles, P.S. Liss and W.G.N. Slinn eds., D. Reidel Publishing Company, 1983). It is hoped that the two volumes taken together will be considered as 'companion' textbooks in a rapidly expanding field of research.

April 1986   Patrick Buat-Ménard
                                        Gif sur Yvette, France.

ACKNOWLEDGEMENTS

The idea of holding the Institute originated from the N.A.T.O Science Committee's Special Programme Panel on Global Transport Mechanisms in the Geosciences. I wish to thank D. Whelpdale and R.A. Duce for their advice in the early stage of planning of this Institute, P.S. Liss and L. Merlivat for their continuous help as members of the Organizing Committee.

Major financial support came from the N.A.T.O Advanced Study Institute Programme. I also wish to acknowledge the financial assistance of the following French Institutions: The Centre National de la Recherche Scientifique, through the Programme Interdisciplinaire de Recherches en Océanographie (PIROCEAN) and the Programme Interdisciplinaire de Recherches sur l'Environnement (PIREN), The Ministère de l'Environnement, Service de la Recherche, des Etudes et du Traitement de l'Information sur l'Environnement (S.R.E.T.I.E.) as well as my own laboratory.

I would like to thank all the staff at Village Les Dunes, Bombannes, were the A.S.I was held, for making our stay there such a pleasant one. Marc Ewald of the University of Bordeaux helped in several ways with the local organization.

Special thanks are due to Joëlle Buat-Ménard, Dominique Mertens and Pat Fitzgerald for assisting the participants with many secretarial, social and travel activities. Claude Lalou and Peter Liss aided in preparing the book chapters for publication. Finally, the participants in the A.S.I greatly contributed to its success. They provided the lecturers with many useful suggestions both oral and written and this input has .considerably improved the quality of the chapters presented here. In addition, 52 poster presentations were given by the participants from which stemmed much stimulating discussions. On behalf of all the lecturers, I would like to thank the participants for contributing to the enthusiastic two-way interchange which so characterized the Institute.

BASIC CONCEPTS IN GEOCHEMICAL MODELLING

Roland WOLLAST
University of Brussels
Laboratory of Oceanography
50 av. F.D. Roosevelt
B-1050 Brussels (Belgium)

1. INTRODUCTION

During this Nato-Advanced Course devoted to the Role of Air-Sea Exchange
in Geochemical Cycling, basic concepts like steady-state, residence time
and response time, stability or instability of the system, were often
evoked by several speakers. It was therefore important that the participants became rapidly acquainted with these concepts. In addition, it
should be noted that in the literature these terms are not always well
defined. Authors often neglect to consider the fundamental properties
of their systems in terms of stationary state or stability before they
develop their procedures for calculation and draw their conclusions.
The aim of this chapter is to present simplified systems where these
concepts appear clearly and their various basic properties can be easily
shown. The more complicated geochemical cycles frequently exhibit be-
haviours similar to those of simplified systems which will be considered
later. It is difficult, however, in complex systems to formulate theoretically a model without considerable simplifications and to distinguish
then between the various factors influencing the response of the geochemical cycles.

The first attempts to elaborate geochemical cycles, mainly considered on a global scale, were initially limited to the estimation of
the amount of the components of interest in a number of convenient
reservoirs and of the fluxes of these compounds between the reservoirs
(Garrels and Mackenzie, 1971; Mackenzie and Wollast, 1977; Holland, 1978).
Because steady-states were assumed for these cycles, the only constraints
in the calculations were to satisfy the mass balance for each reservoir
and for the entire system. The early developments of these models were
essentially devoted to the increase of the number of reservoirs, and to
the introduction of compartments in the reservoirs (like soils, biota,
...) and of speciation of the components (dissolved, particulate, organic, living, isotopes, etc...).

These unsophisticated steady-state models were nevertheless extremely powerful since the quantification of the fluxes between the
reservoirs (1) helped one to visualize the dynamical aspects of the present-day geochemical cycles, (2) raised many unanswered questions about the

1

P. Buat-Ménard (ed.), The Role of Air-Sea Exchange in Geochemical Cycling, 1–34.
© *1986 by D. Reidel Publishing Company.*

origin of some fluxes, and (3) allowed one to estimate the influence of man's impact and interference on the natural cycles.

However, the weakness of this steady-state approach is that the reservoirs are considered as black boxes where the processes responsible for the fluxes are ignored. The only kinetic concept derived from these models was the residence time which was also considered a rough estimate of the response time of the systems. It is thus not possible to use them to predict quantitatively how they would react to perturbations (such as those due to man's activities) or to describe what could have been the slow evolution of the cycle of an element on geological time scales.

The first progress in the development of kinetic geochemical cycles was obtained by assuming that the output flux of the component of interest from one reservoir is proportional to the amount or concentration of the component in that reservoir (Broecker, 1971; Holland, 1978; Lerman, 1978). This assumption of first order dependence can be easily justified in many cases, for example in a well mixed reservoir where the output flux is due to an advective process. In many instances first order kinetics are reasonable approximations of the decay of a component in a reservoir owing to physical, chemical and biological processes.

More recently coupling between components in one reservoir and between fluxes among reservoirs have been introduced in the mathematical description of geochemical cycles, which leads to powerful extensions of useful concepts such as response time, feedback between cycles and the possible instability behaviour of the system (Lasaga, 1980, 1981; Berner, Lasaga and Garrels, 1983).

The introduction of kinetic expressions in the geochemical models unfortunately complicates the mathematical solution of the problem. Answers to simple questions such as the existence of a steady-state for the system considered are not always obvious. It is therefore preferable to start with very simple cases where various concepts may be discussed separately.

We will present here mainly a system of one reservoir with an input and an output, and with various processes obeying different kinetic laws, occurring inside the reservoir. The reservoir will be considered as well-mixed and the concentrations of the components in the reservoir are thus homogeneous. In this case, if the flux out of the reservoir is due to an advective process, then the concentration in the advective agent is the same as that in the reservoir. The component of interest is furthermore submitted to a process which may be physical, chemical or biological and which produces or consumes that component in the reservoir.

We will first consider the classical first order decay reaction in the system and use this example to define some basic concepts and calculation procedures. We will then introduce cases of more complicated kinetics including second order reactions, reversible reactions and the reaction of the considered component with another component in the system (coupled reactions). Finally we will briefly discuss the effect of coupling of two reservoirs without reactions occurring inside the reservoirs.

For all the systems described in this chapter, we have illustrated their behaviour graphically by selecting small and simple numerical values, with no units given explicitly. However, in many cases, we have considered the behaviour of dissolved silica and opal in the ocean as a representative example of a real system. The numbers used in these models are personal "best guesses" and should be considered only for the purpose of illustration. Most of the data have been selected from previous estimates made by the author (Wollast, 1974, 1981; Wollast and Mackenzie, 1983) and from additional data found in Calvert (1983), Hurd (1983) and Spencer (1983). Also the bibliography has been intentionally limited to a restricted number of selected references where more complete literature can be found.

2. FIRST ORDER MODELS

2.1. First Order Decay Reaction

We have mentioned in the introduction that the first attempts to introduce kinetic constraints in geochemical models were simply to assume that the rate of output of a component from a reservoir was directly proportional to its amount in the reservoir. In this simplified approach, the kinetic constants can be estimated directly from the existing steady-state models by dividing the output flux of the component of interest by its amount in the reservoir.

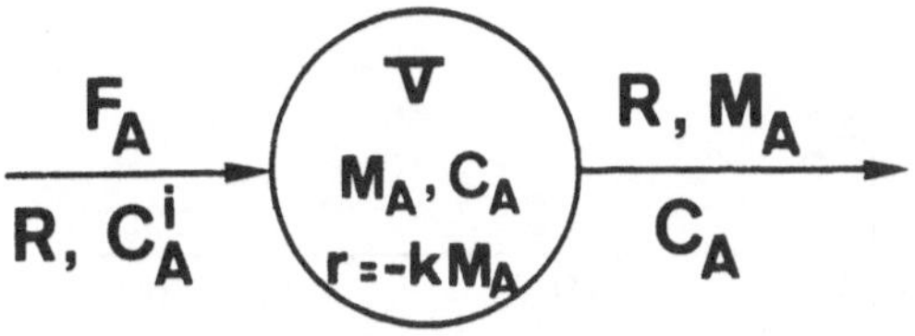

Figure 1. One-reservoir system with a first order decay reaction.

To start we will consider here a slightly more generalized model, presented schematically in figure 1, where the component A is removed from the reservoir by two first order processes: advection with a kinetic constant R and reaction with a kinetic constant k. The mathematical formulation of the geochemical model expresses simply the mass balance of the component in the system. If M_A represents the amount of A in the reservoir and F_A its flux to the reservoir, then the fluctuation of M_A in the system is given by

$$\frac{dM_A}{dt} = F_A - kM_A - RM_A = F_A - M_A (k + R) \tag{1}$$

This equation can be easily integrated if k, R and F_A remain constant. If at time $t = 0$, $M_A = M_A^0$, then the solution is:

$$M_A = \frac{F_A}{k + R} - \left(\frac{F_A}{k + R} - M_A^0 \right) e^{-(k + R)t} \tag{2}$$

A typical example of the evolution of M_A as a function of t is given in figure 2. As t increases, M_A approaches a constant value M_A^∞ which is given by equation (2) when $t \to \infty$. In other words, this system may reach a steady-state, and in this case only one steady-state, where the mass of A in the reservoir is given by relation (3):

$$M_A^\infty = \frac{F_A}{k + R} \tag{3}$$

The value of M_A^∞ can also be deduced immediately from equation (1) without integrating it since at steady-state $\frac{dM_A}{dt} = 0$.

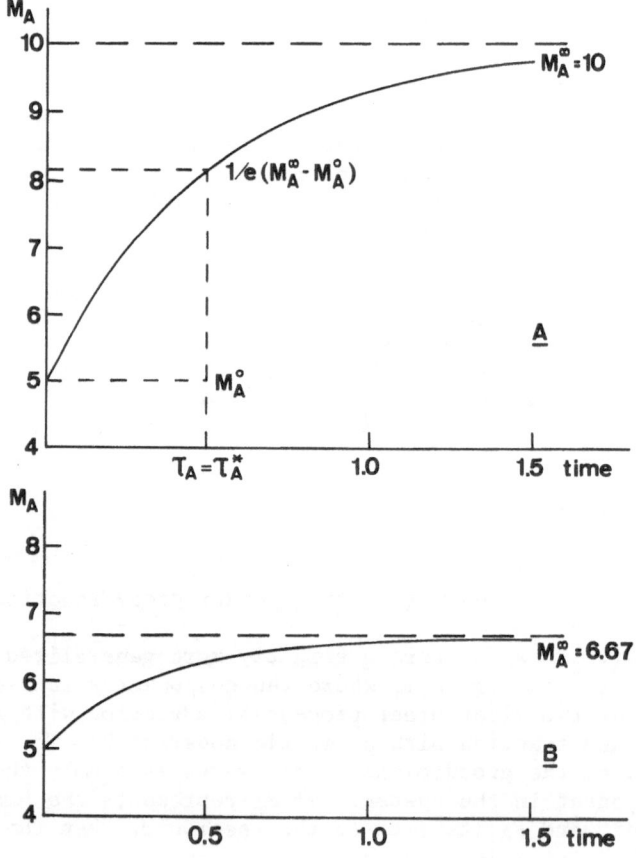

Figure 2. Evolution of M_A as a function of time in a system with a first order decay reaction   a) after doubling F_A
                                b) after doubling R and assuming that
$F_A = RM_A^i$ (see text).

The existence of a steady-state allows furthermore the definition of a mean residence time for component A in the reservoir. The residence time τ_A is defined as:

$$\tau_A = \frac{M_A^\infty}{f_A^+} = \frac{M_A^\infty}{f_A^-} \tag{4}$$

In this relation f_A^+ represents the total input and production of A and f_A^- the total output and consumption of A at steady-state where by definition $f_A^+ = f_A^-$. In the case of a system with a first order decay, according to equation (1):

$$f_A^+ = F_A \text{ and } f_A^- = M_A^\infty (k + R)$$

Therefore,

$$\tau_A = \frac{M_A^\infty}{F_A} = \frac{1}{k + R} \tag{5}$$

In addition, one may also define the response time, which represents the time necessary for the system to approach a new steady-state after the conditions such as amount, input flux and rate constants are changed. It may also be defined as the time required to decay substantially a perturbation of the system away from its original steady-state. In the case considered here, combining equations (2) and (3) results in:

$$M_A = M_A^\infty - (M_A^\infty - M_A^0) e^{-(k + R)t} \tag{6}$$

where $(M_A^\infty - M_A^0)$ represents the initial departure from steady-state.

Since in many cases the solution of the rate equation is an exponential function of time, such as the one shown in equation (6), the response time has often been chosen by convenience as the time required to reduce the initial perturbation from steady-state to e^{-1} ($= 0.37$), or in other words to reduce the perturbation by 63%.

In the case of the first order kinetic considered in this section, the response time τ_A^* is, according to equation (6):

$$\tau_A^* = \frac{1}{k + R} \tag{7}$$

Thus, here the response time is equal to the residence time of the component in the system as illustrated in figure 2(a).

In some recent geochemical models, the main interest concerns the evolution with time of the distribution of components among reservoirs. These models simply postulate that the fluxes out of the reservoirs are proportional to the amounts in the reservoirs. The kinetic constant used in these models is equal to the sum of k and R in our case and does not therefore distinguish between the advective and the reaction term.

Let us take, for example, the case of dissolved silica in the oceans. According to the values selected previously by the author, the

total mass of dissolved silica in the ocean is roughly equal to 1.8 x
10^{17} moles and the river input to the ocean 7 x 10^{12} moles.y^{-1}. If
steady-state is assumed, then the river input must be compensated by an
equal flux of withdrawal of dissolved silica from the oceanic water
column. It is well known that this removal is accomplished by planktonic
organisms, mainly diatoms and radiolarians in order to build their siliceous skeletons. Assuming first order kinetics for the precipitation
of dissolved silica with further removal of the skeletons by sedimentation, one obtains for the rate constant k = 7 x 10^{12} moles.y^{-1}/
1.8 x 10^{17} moles = 4 x $10^{-5} y^{-1}$. The mean residence time of dissolved
silica in the ocean with respect to the river input is then given by
1/4 x $10^{-5} y^{-1}$ = 26 x 10^3 y.

It has been argued that the eutrophication of rivers due to man's
activities has led to an increase in the rate of uptake of dissolved
silica by freshwater diatoms, which in turn has decreased, to a certain
extent, the flux of this component from the rivers to the oceans. According to the above theory if this perturbation persists, it will take
about 26 x 10^3 years before the total amount of dissolved silica in the
ocean is significantly reduced.

This model is however oversimplified and provides only limited
useful information. On the other hand, the derivation of a more sophisticated model requires the identification of the rate-controlling
processes and especially the evaluation of the rate laws and the rate
constants. A considerable effort is needed in this field at the present
time for many geochemical cycles.

In the case of silica, the behaviour of this component in the
ocean may be better described if the biological precipitation and the
sedimentation of particulate silica are explicitly taken into account.
Therefore, it is interesting to divide the ocean into two compartments:
> - the surface water (< 200m depth) where most of the biologi-
>   cal activity related to silica uptake is confined.

> - the intermediate and deep waters where the siliceous skele-
>   tons of dead organisms are sinking and gradually dissolving.

In such a system, the transfer of particulate Si from the surface
water to the deep water is compensated by an upward flux of dissolved
silica due to upwelling and vertical turbulent diffusion. This system,
taking into account existing estimated fluxes, is represented schematically in figure 3. The data concerning the concentration of opal
skeletons in the deep ocean is however very scarce. The total mass of
particulate silica in the reservoir was evaluated here indirectly by
considering mainly experimental data regarding the rate of dissolution
of opal in sea-water. Thus for a dissolution flux of 153 x 10^{12} moles
y^{-1} and a rate constant close to 1 y^{-1}, the total mass of opal in the
deep water is estimated to be 150 x 10^{12} moles, which corresponds also
to a mean concentration in the deep waters of 7 μg SiO_2 /l. The first
order rate constant related to the sedimentation process is then given
by 7 x 10^{12} moles y^{-1}/150 x 10^{12} moles = 5 x 10^{-2} y^{-1}.

The mean residence time and thus the response time for particulate
silica in the deep water is thus approximately 1 year. Since dissolved
silica is essentially present below 200 m, the mean residence time for

dissolved silica in deep water with respect to the dissolution influx
or upwelling outflux is given by 1.8×10^{17} moles/1.5×10^{14} moles = 1.2×10^{3}y. This time is close to the time necessary to renew the deep water
by surface water, known also as the ventilation time, estimated to be
around one thousand years.

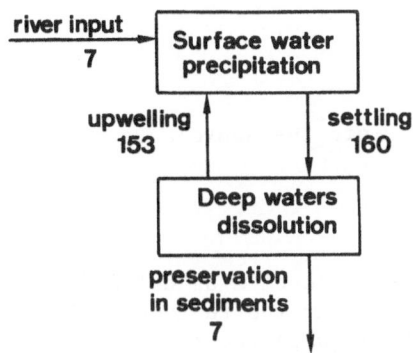

Figure 3.Schematic representation of the fluxes of silica in the ocean
induced by biological activity (fluxes in 10^{12} moles y^{-1}).

It is sometimes also useful to develop models in term of concentration instead of the amount of the component considered. An interesting particular case is given for example by a lake of volume V where
the input and output fluxes are due to river water flow and where the
component A may be an organic compound carried by the river and submitt
ed to a first order decay in the lake. If the river discharge into and
out of the lake are the same and equal to Q, then the mass balance may
be rewritten as:

$$V \frac{dC_A}{dt} = QC_A^i - VkC_A - QC_A \qquad (8)$$

where C_A and C_A^i are the concentrations in the lake and in the incoming
river respectively, and with $RM_A = QC_A$.

Since the residence time of water in the lake is in fact given by
$\tau_H = \frac{V}{Q}$, which is also equal to 1/R, equation (8) can be rewritten:

$$\frac{dC_A}{dt} = \frac{1}{\tau_H} C_A^i - C_A (k + \frac{1}{\tau_H}) \qquad (9)$$

Integration of equation (9) leads to:

$$C_A = \frac{C_A^i}{1 + k\tau_H} - \left(\frac{C_A^i}{1 + k\tau_H} - C_A^0 \right) e^{-(\frac{1}{\tau_H} + k)t} \qquad (10)$$

and at steady-state:

$$C_A^\infty = \frac{C_A^i}{1 + k\tau_H} \tag{11}$$

As one may expect, the concentration of A in the lake will reach at steady-state the concentration of the river water C_A^i if there is no decay reaction ($k = 0$). The residence time of A is given by dividing its content in the lake by the input:

$$\tau_A = \frac{VC_A^\infty}{QC_A^i}$$

Substituting C_A^∞ by equation (11) into the above equation and replacing V/Q by τ_H, one obtains:

$$\tau_A = \frac{\tau_H}{1 + k\tau_H} = \frac{1}{k + R} \tag{12}$$

which represents also the response time of A as defined earlier (equation 7). Equation (12) shows thus that the residence time of A in the lake will be maximum and equal to τ_H when the rate of degradation becomes very small. Increasing the rate of decay will decrease the residence time and thus also the response time. Figure 2(a) shows the evolution of the mass of A in a hypothetical lake when the input is doubled by doubling the concentration in the influent (C_A^i). Figure 2(b) shows how the system evolves if the input is doubled by doubling the river discharge (Q). The difference between the two cases is due to the fact that changing Q affects also the output of A from the lake.

2.2. Instantaneous Perturbation in a First Order Decay Model

In the model that we have discussed here above it is easy to integrate the differential equation, but this is often not the case when more elaborated geochemical cycles are considered. It is nevertheless possible to evaluate the response time of the system in this case by applying the method of perturbation. The principles of the method will be outlined here in the case of the first order kinetic model.

Let us impose at time $t = 0$ a small instantaneous perturbation α_0 of the amount of A in the reservoir, initially at steady-state. We will assume that the evolution of the amount of A in the reservoir has the following form:

$$M_A(t) = M_A^\infty + \alpha(t) \tag{13}$$

with

$$\frac{dM_A}{dt} = \frac{d\alpha}{dt} = \alpha'$$

According to equation (1) the mass balance of the perturbated system is given by:

$$\alpha' = F_A - k(M_A^\infty + \alpha) - R(M_A^\infty + \alpha) \tag{14}$$

or

$$\alpha' = F_A - kM_A^\infty - k\alpha - RM_A^\infty - R\alpha \tag{15}$$

Since

$$F_A - kM_A^\infty - RM_A^\infty = 0 \tag{16}$$

equation (15) reduces to:

$$\alpha' = -\alpha(k + R) \tag{17}$$

which can be easily integrated:

$$\alpha = \alpha_0\, e^{-(k + R)t} \tag{18}$$

The perturbation is thus decaying exponentially with time and the system is considered to be stable since it goes back to its initial steady-state after the imposed small perturbation. The response time of the system to the perturbation is identical to that previously derived in equation (7).

In more complicated systems it is sometimes avantageous to assume that the evolution of the perturbation follows an exponential function:

$$\alpha = \alpha_0\, e^{\xi t} \text{ with } \alpha' = \alpha_0 \xi\, e^{\xi t} \tag{19}$$

Applying these relations to equation (1), one obtains:

$$\alpha_0 \xi e^{\xi t} = F_A - kM_A^\infty - k\alpha_0 e^{\xi t} - RM_A^\infty - R\alpha_0 e^{\xi t} \tag{20}$$

which reduces, according to the steady-state condition given by equation (16), to

$$\xi = -(k + R) \tag{21}$$

This last method is often used because it is much faster and apparently more efficient. However, it presupposes that the mathematical solution of the evolution of the perturbation in the system is reduced to a simple exponential function. It will be shown later that this may not always be the case.

2.3. First Order Production Model

We will consider here that component A is produced at a rate proportional to its amount. This case is often encountered in biological systems where the rate of growth of a population is proportional to its size.

Let us take again the example of silica in the ocean. If we accept that the rate of silica uptake by organisms in the surface waters is 400×10^{12} moles y^{-1} and that the mean life time of plankton in the ocean is 0.05 y or $k = 20y^{-1}$, then according to the first order production kinetics, the mass of living siliceous organisms in the surface waters of the ocean is equal to 20×10^{12} moles. More generally first order production kinetics are named autocatalytic processes. The mass balance for the system in this case becomes:

$$\frac{dM_A}{dt} = F_A + kM_A - RM_A \tag{22}$$

with the following solutions:

$$M_A^\infty = \frac{F_A}{R - k} \tag{23}$$

and

$$M_A = M_A^\infty - (M_A^\infty - M_A^0)\, e^{(k - R)t} \tag{24}$$

Note that M_A^∞ has positive values only if $R > k$ which is a necessary condition for reaching a steady-state.

If $k > R$, M_A increases exponentially with time and such a system will never reach a stable state with a finite amount of component A. In nature the situation is more complex, however, and when the amount of A increases, other kinetic processes may become involved to limit the amount of A in the reservoir. First order kinetics are always simplifying assumptions either because the reaction is more complex or simply reversible. We will consider a few examples of such cases here below.

3. HETEROGENEOUS CATALYSIS AND ENZYMATIC TYPE REACTIONS

The kinetics developed in the following section represent a typical case of a more complex rate law often encountered in natural systems, which can be reduced to simple first order kinetics under certain circumstances. We will consider here that component A is adsorbed on an active site and that the reaction rate of A is proportional to the amount of A bound to the active sites. The adsorption of A on the active site S_0 may be described by:

$$S_0 + A \rightleftharpoons S_A \tag{25}$$

We will further assume that the adsorption is at equilibrium in the system and is of the Langmuir type with:

$$S_0 + S_A = S_{tot} = \text{constant} \tag{26}$$

The equilibrium condition is given by:

$$K_{eq} = \frac{S_A}{S_0 \cdot A} = \frac{S_A}{(S_{tot} - S_A)A} \tag{27}$$

and the rate equation becomes:

$$\text{rate} = kS_A = \frac{kS_{tot}\, A}{\left(\frac{1}{K_{eq}} + A\right)} = \frac{k_m A}{(K + A)} \tag{28}$$

where k_m and K are equal to kS_{tot} and $\frac{1}{K_{eq}}$ respectively.

This rate equation describes very satisfactorily a large number of different natural processes. It is a classical expression, known as the Menten-Michaelis relation, used by the biologists to describe organic processes like the uptake of a substrate by a biological population, which in turn governs its growth. It has been successfully used to simulate, for example, the rate of degradation of organic matter by bac-

teria or the uptake of nutrients by phytoplankton. It may also be used
to represent the adsorption of the component A onto a solid S which is
then eliminated from the reservoir by settling.

If the amount of A is small with respect to K, the rate law reduces to a first order kinetic law, and for A much larger than K, to a
zero order rate law. Note also that k_m is the maximum rate that the
process can reach at high values of A.

As pointed out earlier, the same kinetic law can be used to describe, for example, the uptake of a nutrient by a biological population
or the removal of a dissolved compound A by adsorption on a settling
solid. In this case, the mass balance equation for the one-reservoir
system becomes:

$$\frac{dM_A}{dt} = F_A - \frac{k_m M_A}{K + M_A} - RM_A \tag{29}$$

The steady-state value for M_A is given by $\frac{dM_A}{dt} = 0$ or from equation (29):

$$RM_A^2 - M_A(F_A - KR - k_m) - KF_A = 0 \tag{30}$$

with

$$M_A^\infty = \frac{(F_A - KR - k_m) \pm \sqrt{(F_A - KR - k_m)^2 + 4\ KRF_A}}{2\ R} \tag{31}$$

Since the square root is necessarily greater than $(F_A - KR - k_m)$ only
the positive root leads to positive values of M_A^∞. There is thus only
one possible steady-state for this system and it exists for any finite
values of the constants F_A, K, k_m and R.

Going back to the uptake of dissolved silica in surface waters
discussed in the previous section, we may assume that the removal of
silica by plankton (400×10^{12} moles y^{-1}) obeys Menten-Michaelis kinetics.
If the mean concentration of dissolved silica in surface waters is taken
as 3 μmoles 1^{-1}, then $M_A = 200 \times 10^{12}$ moles. There are however two unknown
constants k_m and K in the kinetic equation. Field observations indicate
that the populations which adapt and develop have generally a Michaelis
constant K whose value is close to the mean concentration of the substrate in the environment considered. Therefore, to cover roughly the
observed productivity of diatoms in the oceans one can use for k_m and
K 1.2×10^{15} moles y^{-1} and 400×10^{12} moles respectively. First order
kinetics may be used as a more crude approximation with a first order
uptake constant of $k = 2\ y^{-1}$. Note that the residence time of dissolved silica in the photic zone is, according to this model, only
0.5 y.

It is also often assumed that the removal process of dissolved
trace metals in the ocean is governed by adsorption processes on
particles that are removed from the water column by sedimentation. In
the case of the ocean there is no advection term R for the dissolved
components and the model is then simplified and represented by two terms:
the input rate of the dissolved element F_A and the removal reaction. Thus

$$\frac{dM_A}{dt} = F_A - \frac{k_m M_A}{K + M_A} \tag{32}$$

with

$$M_A^\infty = \frac{F_A K}{k_m - F_A} \tag{33}$$

and

$$\tau_A = \frac{M_A^\infty}{F_A} = \frac{K}{k_m - F_A} \tag{34}$$

It is interesting to point out that M_A^∞ grows faster than F_A due to the increased occupation of adsorption sites when F_A increases. Relation (34) shows furthermore that the residence time of an element is directly proportional to K, which is in turn inversely proportional to the distribution coefficient of the element between water and particulate matter (K_{eq}).

4.   REVERSIBLE REACTIONS

The case of an open system in which a reversible reaction occurs is interesting from a fundamental point of view because it demonstrates the basic difference between a thermodynamical model and a kinetic model for a stationary state. Let us consider that the first order reaction occurring in our system is reversible.

$$A \underset{k_b}{\overset{k_f}{\rightleftharpoons}} B$$

and is characterized by the equilibrium constant:

$$K_{eq} = k_f/k_b \tag{35}$$

To simplify we will assume that the reaction occurs only in the reservoir and that there is no input of B (figure 4).

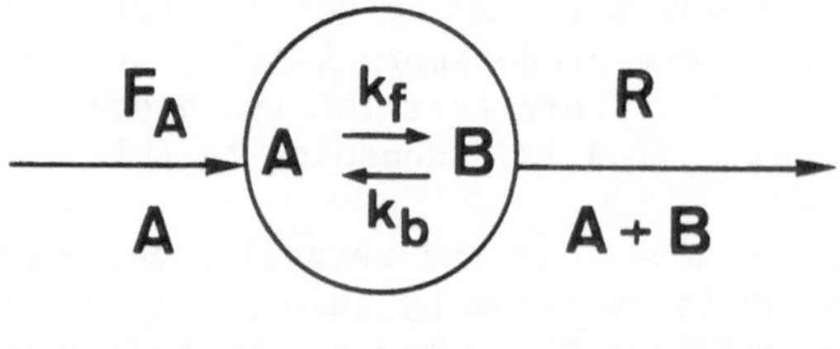

Figure 4.   One-reservoir system with a reversible reaction.

We must now write a mass balance for each of the components in the system which gives the following set of differential equations:

$$\frac{dM_A}{dt} = F_A - k_f M_A + k_b M_B - R M_A \tag{36}$$

$$\frac{dM_B}{dt} = k_f M_A - k_b M_B - R M_B \tag{37}$$

At steady-state, $dM_A/dt = dM_B/dt = 0$ and

$$k_b M_B^{\infty} = M_A^{\infty} (k_f + R) - F_A \tag{38}$$

$$k_f M_A^{\infty} = M_B^{\infty} (k_b + R) \tag{39}$$

Combining (38) and (39) gives:

$$R (M_A^{\infty} + M_B^{\infty}) = F_A \tag{40}$$

This equation simply states that the input flux of A must be equal at steady-state to the output of A plus B. If the input flux is due to advection then:

$$F_A = R M_A^i \quad \text{and} \quad M_A^i = M_A^{\infty} + M_B^{\infty} \tag{41}$$

where M_A^i is the amount of A in the input flux. This property is very useful in more complicated systems. Substituting M_A^{∞} from (39) into (38) and rearranging the terms one obtains:

$$M_A^{\infty} = \frac{(k_b + R) F_A}{R (k_f + k_b + R)} \tag{42}$$

and

$$M_B^{\infty} = \frac{k_f F_A}{R (k_f + k_b + R)} \tag{43}$$

This system will thus reach a steady-state for any finite value of the constants k_b, k_f, F_A and R.

If A and B are in a system reaching thermodynamical equilibrium, then their distribution must obey the following constraint:

$$\frac{M_B^{eq}}{M_A^{eq}} = K_{eq} = \frac{k_f}{k_b} \tag{44}$$

In our steady-state system, we can see from (42) and (43) that:

$$\frac{M_B^{\infty}}{M_A^{\infty}} = \frac{k_f}{k_b + R} \tag{45}$$

Thus the stable value for M_B at steady-state is always smaller than its equilibrium value and will approach it if the advection term R is small with respect to the back reaction constant k_b. Thermodynamic equilibrium requires that R = 0 or in other words that the system be closed.

Bearing in mind that the residence time is given by the amount of the component in the reservoir divided by the total positive or negative fluxes appearing in the kinetic equation one has for component A and B respectively:

$$f_A^- = k_f M_A^\infty + R M_A^\infty \qquad \text{and} \qquad (46)$$

$$\tau_A = \frac{1}{R + k_f} \qquad (7)$$

$$f_B^- = k_b M_B^\infty + R M_B^\infty \qquad \text{and} \qquad (48)$$

$$\tau_B = \frac{1}{R + k_b} \qquad (49)$$

where f^- designates the negative flux of the subscript species.
These expressions are very similar to those obtained for the first order reactions. However, it should be noted that the time necessary to renew A in the reservoir by considering only the input flux F_A is longer than τ_A and is according to equation (42):

$$t_{renew} = \frac{k_b + R}{R(k_b + k_f + R)} \qquad (50)$$

The occurrence of a backward reaction may modify significantly the response time of the system for the components. The response time after a perturbation of the amount of the component A from the steady-state can be evaluated using the perturbation method described earlier. Let us assume that the perturbations of A and B from the steady-state are respectively α and β, such that:

$$M_A = M_A^\infty + \alpha \qquad (51)$$

$$M_B = M_B^\infty + \beta \qquad (52)$$

with the initial conditions that at time t = 0, $\alpha = \alpha_0$ and $\beta = \beta_0 = 0$. The differential equations (36) and (37) become then:

$$\frac{dM_A}{dt} = F_A - k_f(M_A^\infty + \alpha) + k_b(M_B^\infty + \beta) - R(M_A^\infty + \alpha) \qquad (53)$$

$$\frac{dM_B}{dt} = k_f(M_A^\infty + \alpha) - k_b(M_B^\infty + \beta) - R(M_B^\infty + \beta) \qquad (54)$$

As shown previously, the introduction of the steady-state condition allows these equations to be reduced to:

$$\alpha' = - k_f \alpha + k_b \beta - R\alpha \qquad (55)$$

$$\beta' = k_f \alpha - k_b \beta - R\beta \qquad (56)$$

From equation (55) one obtains

$$\beta = \frac{\alpha' + k_f \alpha + R\alpha}{k_b} \tag{57}$$

and differentiation of the same equation (55) gives:

$$\alpha'' = - k_f \alpha' + k_b \beta' - R\alpha' \tag{58}$$

Substitution of β' and β from equations (56) and (57) into equation (58) and rearrangment of the terms result in:

$$\alpha'' + \alpha'(k_f + k_b + 2R) + \alpha R(k_f + k_b + R) = 0 \tag{59}$$

The solution of this second order differential equation exhibits two real roots and is of the form:

$$\alpha = \alpha_1 e^{-RT} + \alpha_2 e^{-(k_f + k_b + R)t} \tag{60}$$

The solution for β is then easily obtained by differentiating equation (60) and introducing α and α' in equation (57) gives:

$$\beta = \alpha_1 \frac{k_f}{k_b} e^{-Rt} - \alpha_2 e^{-(k_f + k_b + R)t} \tag{61}$$

The initial conditions applied to (60) and (61) give at t = 0.

$$\alpha_0 = \alpha_1 + \alpha_2$$

and

$$\alpha_2 = \alpha_1 \frac{k_f}{k_b} \quad or \quad \frac{\alpha_2}{\alpha_1} = K_{eq} \tag{62}$$

The perturbations α and β will thus evolve according to the following equations:

$$\frac{\alpha}{\alpha_0} = \frac{e^{-RT}}{1 + K_{eq}} + \frac{1}{1 + \frac{1}{K_{eq}}} e^{-(k_f + k_b + R)t} \tag{63}$$

and

$$\frac{\beta}{\alpha_0} = \frac{e^{-RT}}{1 + \frac{1}{K_{eq}}} - \frac{1}{1 + \frac{1}{K_{eq}}} e^{-(k_f + k_b + R)t} \tag{64}$$

The evolution of the perturbation of A as a function of time and the effect of the reverse reaction on the response time are well demonstrated in figure 5. In this figure the decay of a small perturbation from the same steady-state is shown for various values of the backward reaction rate. Increase in k_b results in a large increase of the response time especially if R is small compared to the reaction rates. Thus the

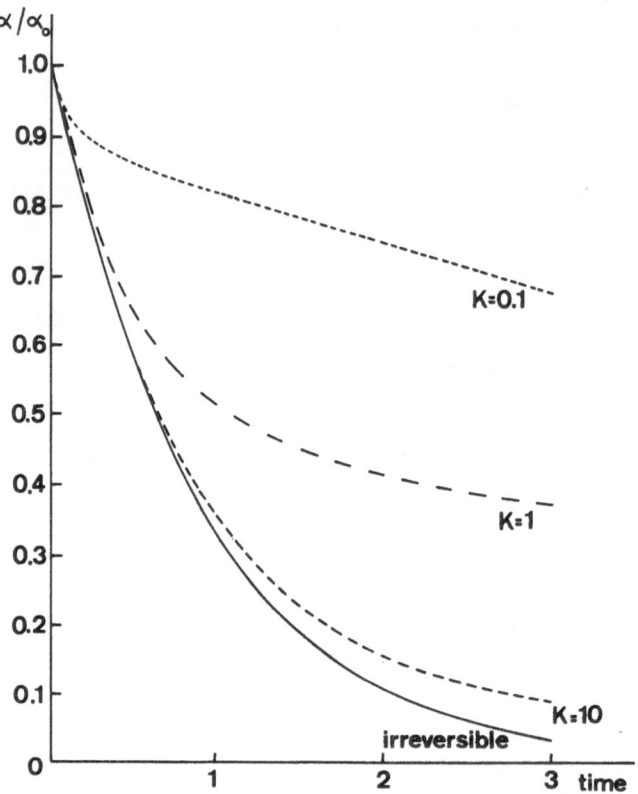

Figure 5.   Evolution of the perturbation of component A involved in a
            reversible reaction with component B, for various values of
            the equilibrium constant.
            (k_f = 1, R = 0.1).

existence of a reversible reaction tends to destabilize the system,
especially when the equilibrium constant is in favor of the component
considered.

     To compare to the example of simple first order kinetics, illus-
trated previously in figure 2, we have represented in figures 6 and 7
the influence of doubling the input flux F_A and the advection rate R in
the case of a system with a reversible reaction. The initial conditions
are similar to those used in the example of figure 2.

     The response of component A in the case represented by figure 6
has a similar shape to the one shown in figure 2 except that the response time of A compared to its residence time is longer. Doubling
the input flux F_A also results in doubling the amount of B at steady-
state. The response time of B is, however, relatively large and is

almost three times the residence time of this component in the reservoir. In fact, the theoretical equations show that both the maximum perturbation of B and the response time increase as K_{eq} increases.

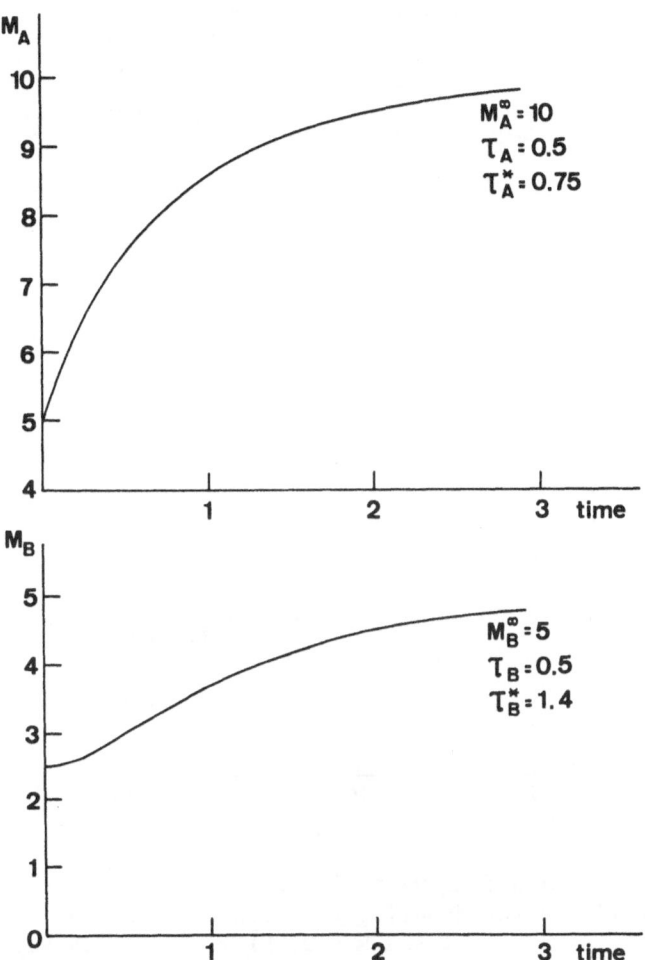

Figure 6.   Influence of doubling the input flux of component A in a
            reservoir with a reversible reaction on the evolution of
            M_A and M_B . (k_f = 1, k_b = 1, R = 1, F_A^0 = 7.5 and F_A = 15).

Figure 7 shows the case where the input flux depending on an advective process is doubled, which in turn also doubles the output flux. As a matter of fact, the perturbation is rapidly damped because of the increased outflux. Both the relative amplitude of the perturbation and the response time are small compared to the previous case.

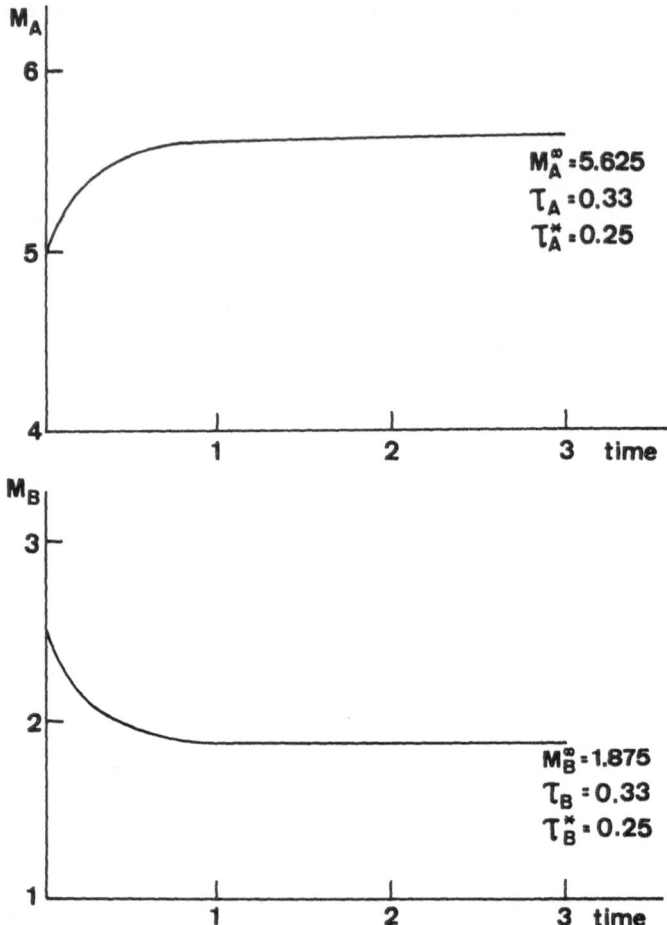

Figure 7. Evolution of M_A and M_B with time in a reservoir with a reversible reaction after doubling the advective term R
($k_f = 1$, $k_b = 1$, $R^0 = 1$ and $R = 2$, $F_A^0 = R^0 M_A^1 = 7.5$ and
$F_A^f = R M_A^1 = 15$).

The reversible reaction may be used to describe in a more detailed manner the behaviour of silica in the photic zone. Dissolved
silica which is consumed to produce particulate opal is also produced
in the surface waters by dissolution of the skeletons after the death
of the organisms and the removal of a protective organic coating. On
the other hand, opal is also removed from the surface water by settling.
The system may thus be described by the following equations:

$$\frac{dM_S}{dt} = F_S - k_f M_S + k_b M_p \qquad (65)$$

$$\frac{dM_p}{dt} = k_f M_S - k_b M_p - RM_p \qquad (66)$$

where the subscripts S and p refer to the dissolved and particulate
species of SiO_2 respectively. At steady-state, equations (65) and (66)
give

$$M_p^\infty = F_S/R \qquad \text{and}$$

$$M_S^\infty = \frac{F_S}{k_f} \left(1 + \frac{k_b}{R}\right)$$

Taking the same value for the dissolution rate of opal as previously ($k_b = 1\ y^{-1}$) and a first order constant for the production term
($k_f = 2\ y^{-1}$), the steady-state condition allows one to evaluate M_p and
R from equations (65) and (66). This gives $M_p = 240 \times 10^{12}$ moles for the
reservoir content of particulate silica and $R = 0.67\ y^{-1}$ for the settling constant.

Let us now assume that the input flux of dissolved silica to the
surface waters F_S is doubled due to a modification of the water circulation in the ocean. We can predict from the model that the system will
reach a new steady-state where the reservoir content of dissolved and
particulate silica, as well as the sedimentation and dissolution fluxes,
will be multiplied by a factor of two. Although the residence times of
dissolved and particulate silica are respectively equal to 0.5 y and
0.6 y, the response times predicted by the model are respectively 2 y
and 2.75 y for the dissolved and particulate forms of silica.

5. MODEL WITH COUPLED COMPONENTS IN THE RESERVOIR

We will now consider the case where component A reacts with another
component B in the reservoir to produce C. The system is represented in
figure 8.

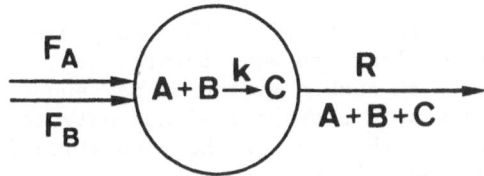

Figure 8.   One-reservoir system with a coupled reaction between components A and B.

Since we are mainly interested in the effect of coupling component A with component B we will consider only the mass balance for these two species:

$$\frac{dM_A}{dt} = F_A - kM_A M_B - RM_A \tag{67}$$

$$\frac{dM_B}{dt} = F_B - kM_A M_B - RM_B \tag{68}$$

At steady-state, equations (67) and (68) give:

$$M_A^\infty = \frac{F_A}{kM_B^\infty + R} \tag{69}$$

$$M_B^\infty = \frac{F_A}{kM_A^\infty + R} \tag{70}$$

and substitution of M_B^∞ in M_A^∞ results in:

$$M_A^{\infty 2} + M_A^\infty \left(\frac{F_B}{R} - \frac{F_A}{R} + \frac{R}{k} \right) - \frac{F_A}{k} = 0 \tag{71}$$

and

$$M_A^\infty = \frac{\left(\frac{F_A}{R} - \frac{F_B}{R} - \frac{R}{k} \right) \pm \sqrt{\left(\frac{F_A}{R} - \frac{F_B}{R} - \frac{R}{k} \right)^2 + 4\frac{F_A}{k}}}{2} \tag{72}$$

Since the square root is always larger than $\frac{F_A}{R} - \frac{F_B}{R} - \frac{R}{k}$, only the positive root gives a possible real solution $(M_A^\infty > 0)$. A similar equation is obtained for M_B^∞:

$$M_B^\infty = \frac{\left(\frac{F_B}{R} - \frac{F_A}{R} - \frac{R}{k} \right) + \sqrt{\left(\frac{F_B}{R} - \frac{F_A}{R} - \frac{R}{k} \right)^2 + \frac{4F_B}{k}}}{2} \tag{73}$$

Thus there exists a steady-state and only one for each component in this system.

The application of the case of coupled components to the silica problem discussed earlier leads to some interesting conclusions regarding biological systems. In such a system we may suppose that the rate of growth of the organisms is proportional not only to the amount of substrate (M_S) but also to the size of the population (M_p). The rate equations (65) and (66) used in the preceding section for dissolved and particulate silica then become:

$$\frac{dM_S}{dt} = F_S - k_2 M_S M_p + k_b M_p \tag{74}$$

$$\frac{dM_p}{dt} = k_2 M_S M_p - k_b M_p - R M_p \qquad (75)$$

and the second order rate constant k_2 can be evaluated at steady-state by the condition $k_2 M_p = k_f$ and thus $k_2 = 8 \times 10^{-15} \text{moles}^{-1} y^{-1}$, according to the numerical data selected previously. At steady-state, the masses of dissolved and particulate silica are:

$$M_S = \frac{R + k_b}{k_2} \qquad (76)$$

$$M_p = \frac{F_S}{R} \qquad (77)$$

This result indicates that if we change the input flux F_S of nutrients in such a system, the mass of organisms M_p will change proportionally to F_S. On the contrary, the mass or concentration of nutrients reached at steady-state in the system is independent of F_S and depends only on R, k_b and k_2. In other words the biomass in the system adapts itself to the input of nutrients and reduces the dissolved silica content to a constant limiting value. Such a system acts as a chemostat. The same conclusions are obtained if Menten-Michaelis kinetics, instead of the first order law used here, are applied to the uptake of the substrate.

The residence time for the two components may be obtained in this system by dividing M_A^∞ and M_B^∞ by F_A and F_B, respectively. Note that if M_B is very large, it will not be affected much by its reaction with A. The value of M_B may then be considered as constant and the reaction term $(- k M_B M_A)$ can be approximated by $(- k' M_A)$. We then return to a first order decay reaction discussed earlier. It is thus more interesting to discuss the effect of coupling A with B when the amount of B is comparable to the amount of A. Let us assume again that at time $t = 0$ we impose on M_A^∞ a perturbation α. As shown previously, the mass balance equations (67) and (68) become:

$$\frac{dM_A}{dt} = F_A - k (M_A^\infty + \alpha)(M_B^\infty + \beta) - R (M_A^\infty + \alpha) \qquad (78)$$

$$\frac{dM_B}{dt} = F_B - k (M_A^\infty + \alpha)(M_B^\infty + \beta) - R (M_B^\infty + \beta) \qquad (79)$$

By introducing the steady-state condition and neglecting the second order terms of the perturbation, equations (78) and (79) can be reduced to:

$$\alpha' = - k M_A^\infty \beta - k M_B^\infty \alpha - R\alpha \qquad (80)$$

$$\beta' = - k M_A^\infty \beta - k M_B^\infty \alpha - R\beta \qquad (81)$$

As previously, differentiation of α' gives:

$$\alpha'' = - k M_A^\infty \beta' - k M_B^\infty \alpha' - R\alpha' \qquad (82)$$

and substitution of α' and β' in equation (82) by equations (80) and (81) gives the following second order differential equation:

$$\alpha'' + \alpha' (kM_A^\infty + kM_B^\infty + 2R) + \alpha(kM_A^\infty R + kM_B^\infty R + R^2) = 0 \tag{83}$$

Here again there are two real roots and we have finally, after imposing the initial condition at time t = 0: $\alpha = \alpha_0$ and $\beta = \beta_0 = 0$.

$$\frac{\alpha}{\alpha_0} = \frac{M_A^\infty}{M_A^\infty + M_B^\infty} e^{-RT} + \frac{M_B^\infty}{M_A^\infty + M_B^\infty} e^{-(kM_A^\infty + kM_B^\infty + R)t} \tag{84}$$

and

$$\frac{\beta}{\alpha_0} = \frac{- M_B^\infty}{M_A^\infty + M_B^\infty} e^{-Rt} + \frac{M_B^\infty}{M_A^\infty + M_B^\infty} e^{-(kM_A^\infty + kM_B^\infty + R)t} \tag{85}$$

The solution is very similar to the previous case except for the expressions of the preexponential term. Both exponential terms decrease with time and the system is thus stable for all values of the kinetic constants k and R. However, it is more difficult in this case to demonstrate explicitly the influence of the reaction of A with B on the stability of the system since M_A^∞ and M_B^∞ are both present in equation (84) and (85).

To illustrate the evolution of a perturbation in this case we have represented nevertheless in figure 9, α and β as a function of time, as well as the contribution of the various terms in equations(84) and (85). As one may expect, a positive perturbation of A induces a negative perturbation of B. The response time of A ($\tau_A^* = 0.67$) is slightly larger than its residence time ($\tau_A = 0.5$). The influence of the advection term becomes rapidly the predominant term responsible for removal of the perturbation of both α and β. The perturbation of component B is maximum at a time equal to its residence time.

The influence of the amount of B on the response time of A is shown in figure 10, where the relative value (α / α_0) of the perturbation remaining at a time equal to the residence time is expressed as a function of the ratio M_B^∞ / M_A^∞. The response time is close to the residence time ($\alpha / \alpha_0 = 1 / e \stackrel{\sim}{=} 0.37$) as long as M_B^∞ considerably exceeds M_A^∞. The response time increases drastically if the system becomes depleted in B as one may expect.

Finally, figure 11 shows the evolution of a perturbation in a system with a very small output (R = 0.01) and in the presence of an excess of B. The perturbation of A disappears with a response time close to the residence time of A and is progressively transmitted to component B.

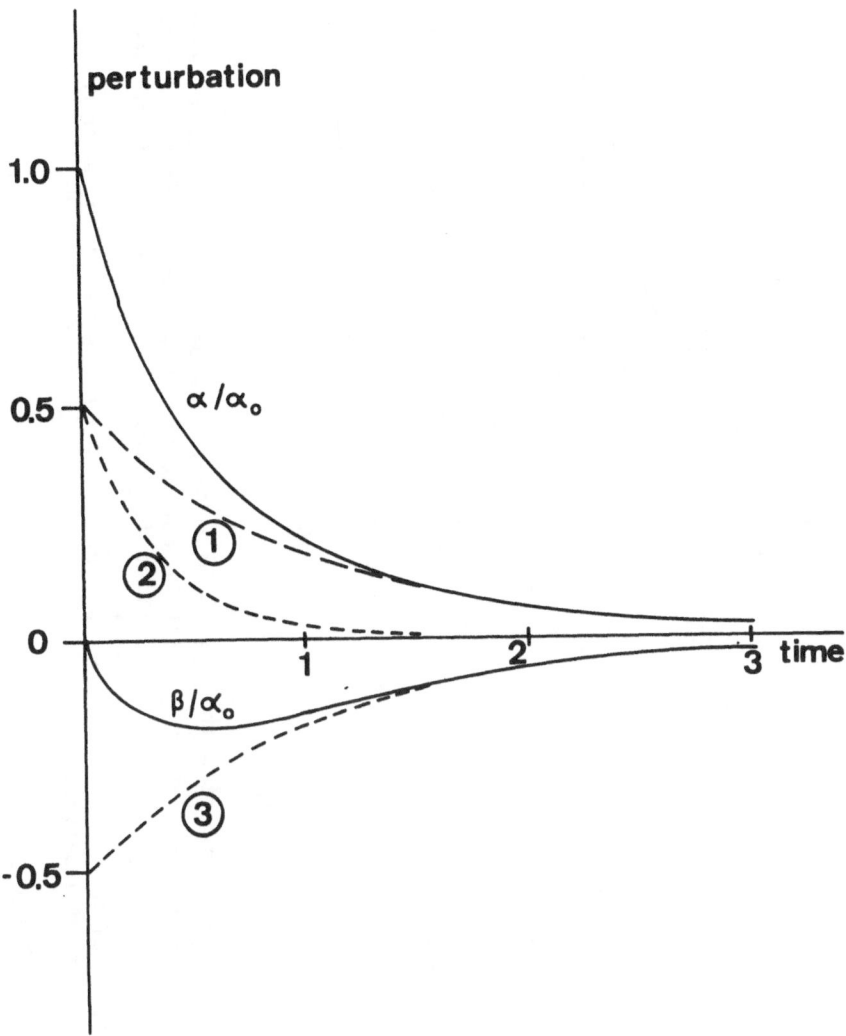

Figure 9. Evolution of the perturbations of A and B in a system where
the two components are coupled through a reaction (A + B → C).
Curves 1 and 2 represent the contributions of the first and
second terms in equation (84) respectively to α/α_0.
Curve 3 represents the contribution of the first term in
equation (85) to β/α_0. The second term in equation (85)
is also given by curve number 2.
($k = 0.1$, $R = 1$, $M_A^\infty = M_B^\infty = 10$, $\tau_A = \tau_B = 0.5$).

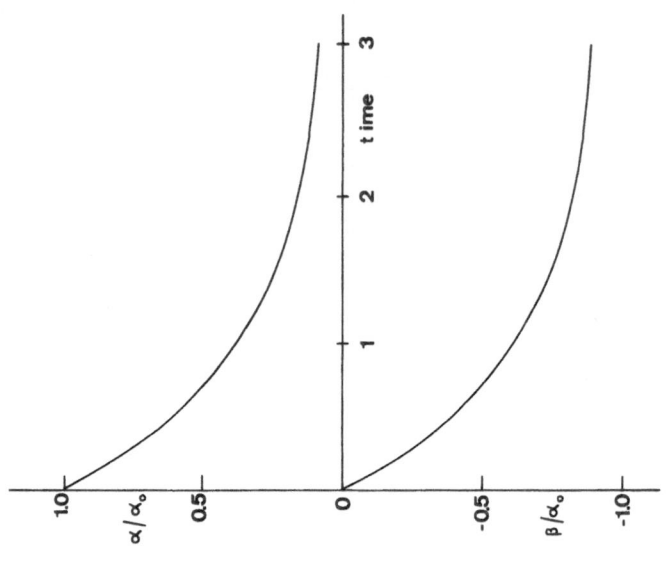

Figure 11. Perturbation of A and B in a reservoir with an output flux negligible compared to the reaction flux and in the presence of an excess of B.
($k = 0.005$, $R = 0.01$, $M_A^\infty = 10$, $M_B^\infty = 200$).

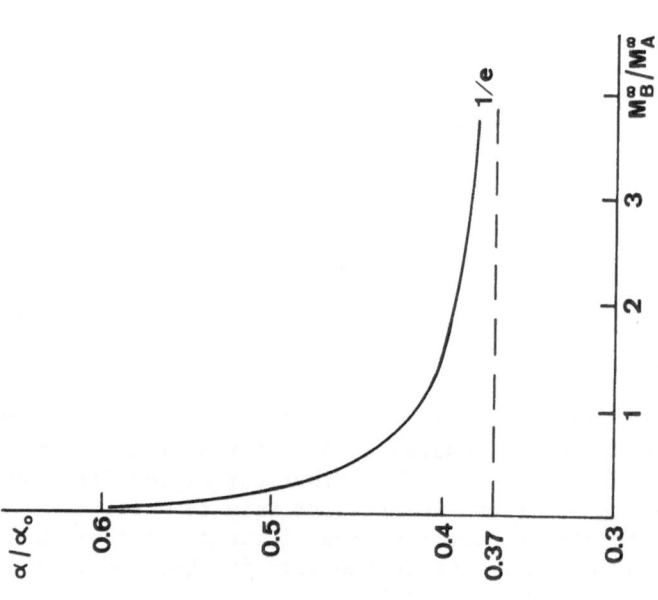

Figure 10. Influence of the relative amount of B to A at steady-state on the perturbation of A. The values of α/α_0 represent those attained by the system at a time equal to the residence time of A in the reservoir. At large values of M_B^∞/M_A^∞, α/α_0 approaches $1/e$ since $\tau_A = \tau_A^*$ for a pseudo-first order reaction.
($R = 1$, $M_A^\infty = 10$ and the product kM_B^∞ has been kept at 10 to maintain a constant contribution of the reaction flux).

6. SECOND ORDER REACTIONS

Processes with a rate dependence higher than the first order with respect to one component are less common in nature. However, apparent
higher order kinetics may be observed in complex systems due to interactions of several reactions or processes. Such systems are very interesting because they may exhibit several steady-states. We will consider here a theoretical example of a second order production rate in a
one-reservoir model. The mass balance in this case is given by:

$$\frac{dM_A}{dt} = F_A + kM_A^2 - RM_A \tag{86}$$

The steady-state value for M_A is thus:

$$kM_A^{\infty\,2} - RM_A^{\infty} + F_A = 0 \tag{87}$$

or

$$M_A^{\infty} = \frac{R \pm \sqrt{R^2 - 4F_A k}}{2k} \tag{88}$$

This system may reach a steady-state only if $R^2 > 4F_A k$. However,
if this condition is obeyed, then the two roots of equation (88) correspond to $dM_A/dt = 0$. This condition is in fact not sufficient to define
a stable steady-state. It is easy to see that the second derivative
d^2M_A/dt^2 must be smaller than zero when the system approaches a stable
steady-state, indicating that the rate of change of M_A is decreasing.
In the case considered here this implies that:

$$\frac{d^2M_A}{dt^2} = 2kM_A^{\infty} - R < 0 \tag{89}$$

or

$$M_A^{\infty} < \frac{R}{2k}$$

If for instance $F_A = 8, R = 6$ and $k = 1$, $(M_A^{\infty})_+ = 4$ and $(M_A^{\infty})_- = 2$,
and only $M_A^{\infty} = 2$ corresponds to a stable steady-state.
It is possible to integrate the differential equation (86) but it
is rather complicated to deduce useful information from the analytical
solution. The perturbation method allows one to approximate more simply
and rapidly the behaviour of the system. If we introduce a perturbation
α of M_A, then equation (86) becomes:

$$\frac{dM_A}{dt} = F_A + k (M_A^{\infty} + \alpha)^2 - R (M_A^{\infty} + \alpha) \tag{90}$$

Introducing the steady-state condition and neglecting the second
order terms, one obtains:

$$\alpha' = 2 kM_A^{\infty}\alpha - R\alpha \tag{91}$$

and the solution is:

$$\alpha = \alpha_0 \, e^{(2 \, kM_A^\infty - R)t} \tag{92}$$

Taking into account the value of M_A^∞ given by equation (88), α may have two values defined by:

$$\alpha = \alpha_0 \, e^{\pm \left(\sqrt{R^2 - 4 \, F_A k} \right) t} \tag{93}$$

The exponent may be positive or negative depending on the value of M_A^∞. In other words the perturbation may increase or decrease depending on the steady-state considered and only the decreasing one may lead to a stable steady-state.

The evolution of the system under various circumstances is shown in figure 12. If we start from an empty reservoir it reaches rapidly

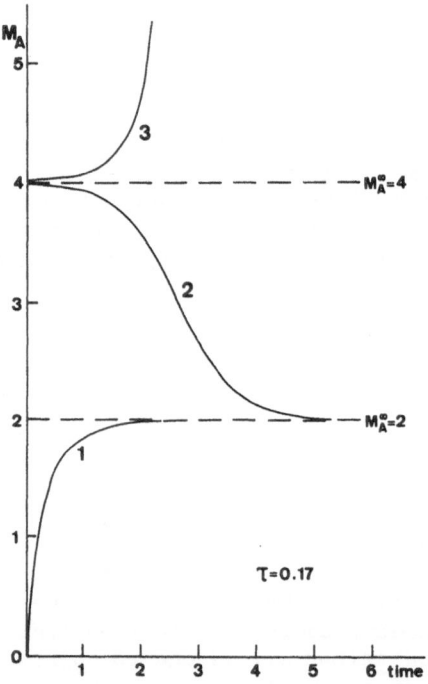

Figure 12. Evolution of the amount of component A in a system with a second order reaction. In case 1, the initial system is started in the absence of A in the reservoir and the system evolves towards the stable steady-state where $M_A^\infty = 2$. Case 2 corresponds to the evolution of the system if a small negative perturbation is imposed on the unstable steady-state $M_A^\infty = 4$. Case 3 corresponds to a small positive perturbation on the same unstable steady-state.

the stable steady-state value for M_A^{00}. On the contrary when the perturbations are close to the unstable steady-state the changes of M_A are
extremely slow. If the initial perturbation is negative the system is
evolving towards the stable steady-state. If the initial perturbation
is positive, the system evolves again first very slowly, but then suddenly M_A goes to an infinite value.

For higher order reactions it is theoretically possible to reach
several stable steady-states.

7. PERIODIC FLUCTUATIONS

The input of a component in a reservoir is often variable and it is important to evaluate the effect of these fluctuations on the behaviour
of the component in the reservoir and more precisely its impact on the
ability of the system to reach a steady-state. It is rather easy to
treat the case of a periodic fluctuation of the input (Holland, 1978),
which can be represented by:

$$F_A = f_a + b \sin \omega t \tag{95}$$

where b is the amplitude of the fluctuation and ω its frequency. If
we combine this input with a first order removal process by advection
the mass balance is given by:

$$\frac{dM_A}{dt} = (f_A + b \sin \omega t) - RM_A \tag{96}$$

Note that the same equation is obtained if we consider a periodic fluctuation of the production of A inside the reservoir. A typical case
of such a fluctuation is the production of organic matter by plants.
The rate fluctuates periodically at least with two distinct frequencies:
diurnally and seasonally. One may also consider long term climatic
changes with periods of several thousand years.

Integration of equation (96) gives:

$$M_A = \frac{f_A}{R} - \left(\frac{f_A}{R} - M_A^0 - \frac{b\omega}{R^2 + \omega^2} \right) e^{-Rt} + \frac{b}{R^2 + \omega^2} (R \sin\omega t - \omega \cos\omega t) \tag{97}$$

if $M_A = M_A^0$ at time $t = 0$. For large values of t, the exponential term
e^{-Rt} approaches zero and equation (97) is reduced to:

$$M_A \simeq \frac{f_A}{R} + \frac{b}{R^2 + \omega^2} (R \sin\omega t - \omega \cos \omega t) \tag{98}$$

We have compared in figure 13 the evolution of M_A given by this
equation to the fluctuation of the input function. Equation (98) and
figure 13 show that the system fluctuates also periodically around a
"steady-state" value. The mass of A in the system oscillates with the
same frequency as that of the input flux. However, the relative amplitude of the fluctuation of M_A is attenuated with respect to that of F_A,
and has a different phase. The phase lag between F_A and M_A is given
by:

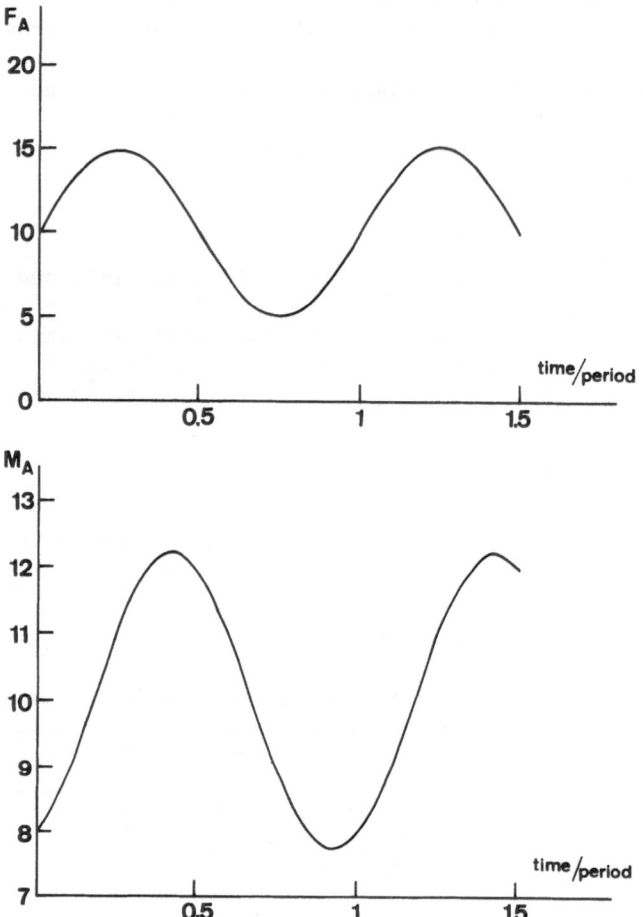

Figure 13.   Comparison of the periodic fluctuation of the input flux
             (or the production term) and that of the amount of component
             A in the reservoir (f_A = 10, b = 5, ω = 2, R = 1).

$$\delta = \cos^{-1}\left(\frac{R}{\sqrt{R^2 + \omega^2}}\right) \tag{99}$$

and the amplitude of the fluctuation of M_A is $b/(R^2 + \omega^2)$.

One may consider other interesting limiting cases for this system.
If ω is small for the time scale considered, $\sin \omega t$ may become negligible and the differential equation is reduced to the classical first
model. This is a rather trivial case since it means simply that the
fluctuation of the input is small over the period of time considered and

thus may be taken constant as a first approximation.

If on the other hand ω becomes very large compared to R, then the amplitude of the fluctuation of M_A decreases and may be neglected as a first approximation. At steady-state, $M_A^\infty \simeq f_A/R$ which is again the solution for a simple first order decay system, with a constant input. Hence, when the fluctuations of the input are occurring at high frequency, the fluctuations of the content of the reservoir are leveled off. Figure 14 shows an example of such a case.

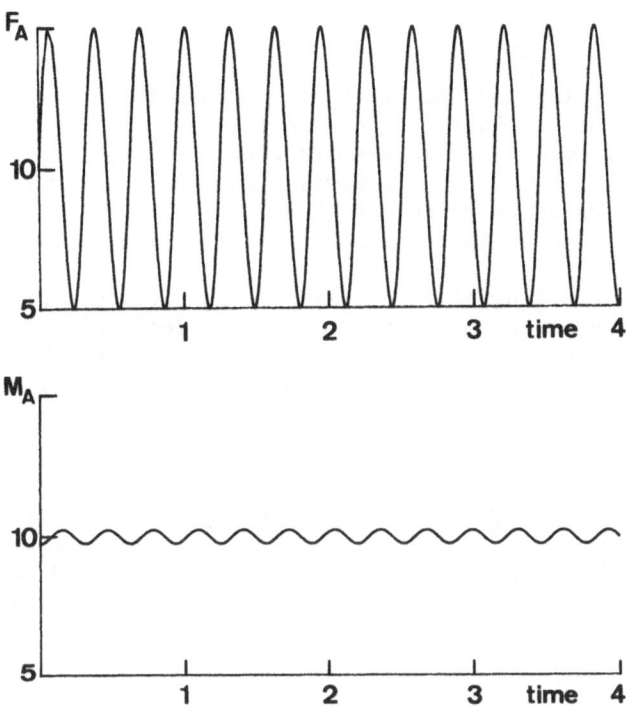

Figure 14.   Comparison of the periodic fluctuation of the input flux and that of the amount of component A in the case of a high frequency fluctuation of the input function. (f_A = 10, b = 5, ω = 20, R = 1).

Taking again the example of the cycle of silica in the oceans we will estimate what is the influence of the seasonal and diurnal fluctuation of the planktonic activity on the mass of siliceous skeletons in the surface water. According to our previous estimates the mean annual production rate of skeletons (f_A) is 400 x 10^{12} moles y^{-1} and the first order removal rate constant (R) by dissolution and settling is 1.67 y^{-1}. For the seasonal fluctuation of the planktonic activity, ω is equal to 1 x $2\pi y^{-1}$ and if the amplitude b is taken similar to f_A (b $\simeq f_A$) then the fluctuation of M_A around its mean value (f_A/R = 240 x 10^{12} moles) will have an amplitude of $\sim$ 10 x 10^{12} moles with a phase lag of 76 days. For the diurnal fluctuation of the planktonic activity, $\omega = 365 \times 2\pi \ y^{-1}$ and the amplitude of the relative fluctuation of the mass M_A would be less than one part per million.

In conclusion of the discussion of these limiting cases it appears that periodic fluctuations of kinetic factors must be explicitly included in the models only if their frequency induces significant changes at time scales comparable to those due to other kinetic processes occurring in the system. Such a system acts thus as a very powerful filter which attenuates efficiently the periodic fluctuations imposed on the input fluxes.

8.   COUPLING OF RESERVOIRS

We have considered until now the effect of various parameters on the behaviour of a system consisting of only one reservoir. Geochemical cycles involve usually several reservoirs connected among them by multidirectional fluxes of components. Sometimes natural reservoirs are divided into subreservoirs where matter is exchanged,for example, the troposphere and the stratosphere in the atmosphere or surface water and deep water in the ocean.

In fact, the coupling of reservoirs has considerable effect on the stability of the system and may introduce strong feedback effects.

To illustrate this concept, we will simply consider the case of two reservoirs exchanging component A at a rate proportional to the amount of A in each reservoir (figure 15).

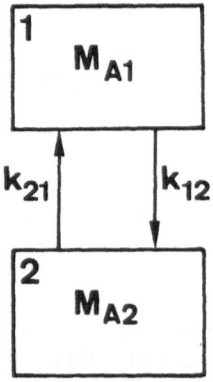

Figure 15.   Two-reservoirs system with coupled fluxes of one component.

The mass balance for each reservoir is:

$$\frac{dM_{A_1}}{dt} = k_{21}M_{A_2} - k_{12}M_{A_1} \tag{100}$$

$$\frac{dM_{A_2}}{dt} = k_{12}M_{A_1} - k_{21}M_{A_2} \tag{101}$$

At steady-state we have:

$$\frac{M^{\infty}_{A_1}}{M^{\infty}_{A_2}} = \frac{k_{21}}{k_{12}} \tag{102}$$

and $M^{\infty}_{A_1} + M^{\infty}_{A_2} = M^{0}_{A_1} + M^{0}_{A_2}$ = constant $\tag{103}$

since A is neither produced nor consumed in this system. The residence times of A in each reservoir are given by:

$$\tau_{A_1} = \frac{M^{\infty}_{A_1}}{k_{12}M^{\infty}_{A_1}} = \frac{1}{k_{12}} \tag{104}$$

and

$$\tau_{A_2} = \frac{M^{\infty}_{A_2}}{k_{21}M^{\infty}_{A_2}} = \frac{1}{k_{21}} \tag{105}$$

In order to solve the set of differential equations describing the often intricate coupling between reservoirs, Lasaga (1980, 1981) has recently proposed the use of matrix algebra. It is obvious that this powerful method is particularly suitable for solving such complicated systems. The use of matrices in the case considered here is explained in great detail by Lasaga (1981). He shows that the solutions of equations (100) and (101) are:

$$M_{A_1} = \frac{k_{21}(M^{0}_{A_1} + M^{0}_{A_2})}{k_{12} + k_{21}} + \frac{k_{12}M^{0}_{A_1} - k_{21}M^{0}_{A_2}}{k_{12} + k_{21}} e^{-(k_{12} + k_{21})t} \tag{106}$$

$$M_{A_2} = \frac{k_{12}(M^{0}_{A_1} + M^{0}_{A_2})}{k_{12} + k_{21}} - \frac{k_{12}M^{0}_{A_1} - k_{21}M^{0}_{A_2}}{k_{12} + k_{21}} e^{-(k_{12} + k_{21})t} \tag{107}$$

The evolution of M_{A_1} and M_{A_2} is illustrated with one example in figure 16.

The response time needed to remove a perturbation is given by the exponential decay term. Reduction of the perturbation to $\frac{1}{e}$ will require a time:

$$\tau^{*}_{A_1} = \tau^{*}_{A_2} = \frac{1}{k_{12} + k_{21}} \tag{108}$$

We see that the response time is always necessarily smaller than the residence time in the two reservoirs. If the residence time of A in reservoir 1 is much shorter than in the other reservoir, then the response time in both reservoirs approaches the residence time of A in reservoir 1. In other words the coupling of the reservoirs has increased the stability of the system and it is important to underline that it is the fastest process which controls the response time of the entire system in such a case.

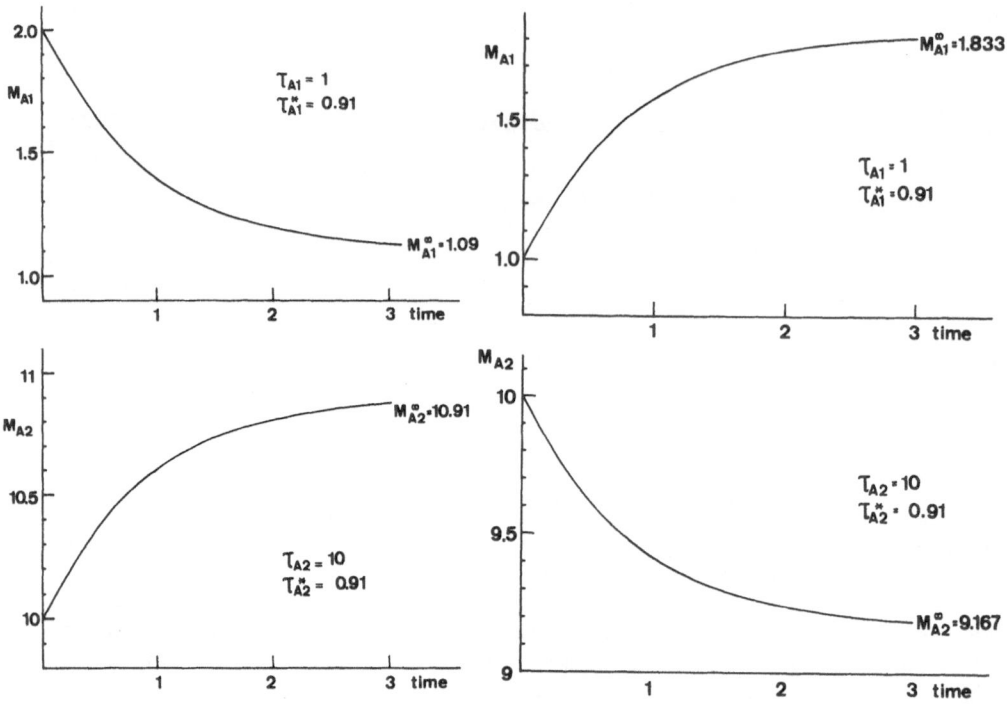

Figure 16.   Evolution of the mass of component A in two reservoirs
              after   a) doubling the amount of A in reservoir 1 and
                      b) doubling the rate of transfer k_{21}.
              (Initial conditions: $k_{12} = 1$, $k_{21} = 0.1$, $M_{A_1}^0 = 1$, $M_{A_2}^0 = 10$)

        Let us take again the example of the silica cycle in the ocean.
We have considered previously the effect of doubling the upwelling flux
on the processes occurring in the surface water. This case can be described by a system with two boxes representing respectively the surface waters and the deep waters with an upward flux of dissolved silica
compensated by a downward flux of particulate silica between these two
reservoirs. According to the numerical values used previously, $k_{12} =$
$0.67 y^{-1}$ and $k_{21} = 8.9 \times 10^{-4} y^{-1}$ in the present situation with steady-
state mass of skeletons in the surface waters equal to 240×10^{12} moles
and that of dissolved silica in the deep waters of 1.8×10^{17} moles.
If the upwelling rate k_{21} is doubled, the system will reach a new
steady-state where the amount of opal in surface water will also be
doubled. The content of dissolved silica in the deep waters will be
decreased, however, by a negligible amount. According to equation (108)
the response time of the two reservoirs is then given by
$\tau_{A_1}^* = \tau_{A_2}^* = 1/(0.67 + 1.8 \times 10^{-3}) = 1.5$ y, which is equal to the residence
time of the particulate silica in the surface water but is much shorter
than the residence time of dissolved silica in the deep water close to

1100 y. Other examples of more complicated coupling between reservoirs can be found in the recent publications of Lasaga and his co-workers.

9. CONCLUSIONS

This chapter has presented some elementary concepts concerning modelling of geochemical cycles. The aim was to show that even in simple systems, basic and fundamental notions such as residence time, response time and stability of the system may become conceptually complex. We have shown in several examples that the response time may be very different from the residence time even in systems restricted to only one reservoir. The use of the residence time of a component in the reservoir as an approximation of its response time may be sometimes misleading. It is thus important to evaluate carefully the properties of the systems in terms of their stability behaviour. Mathematical methods like the theory of perturbation and matrix algebra serve as powerful tools for this purpose and one can foresee a rapid development in kinetic modelling of geochemical cycles in the near future. However, one should be aware that false conclusions can be easily drawn if simplified kinetics and unreasonable approximations are applied.

ACKNOWLEDGEMENTS

The author is deeply grateful to J.P. Vanderborght and L. Chou for their helpful suggestions and comments. The discussions with many participants of the NATO Air-Sea Exchange Course, especially,L.S. Austin, K.A. Hunter, and J. Merrill, are greatly appreciated.

REFERENCES

Berner, R.A., Lasaga, A.C. and Garrels, R.M. (1983). -'The carbonate-silicate geochemical cycle and its effect on atmospheric carbon dioxide over the past 100 million years.' *A.J.S.* Vol. 283. Sept. pp 641-683.

Broecker, W.S. (1971). - 'A kinetic model for the chemical composition of sea water.' *Quaternary Res.*, 1, 188-207.

Calvert, S.E. (1983). - 'Sedimentary geochemistry of silicon'. In *Silicon Geochemistry and Biogeochemistry*. S.R. Aston (ed.). Academic Press, London.

Garrels, R.M. and Mackenzie, F.T. (1971). - 'Evolution of sedimentary rocks.' Norton and Co. New York, 397 pp.

Holland, H.D. (1978). - 'The chemistry of the atmosphere and oceans.' John Wiley & Sons, New York, 350 pp.

Hurd, D.C. (1983). - 'Physical and chemical properties of siliceous skeletons'. In *Silicon Geochemistry and Biogeochemistry*. S.R. Aston (ed.), Academic Press, London.

Lasaga, A.C. (1980). - 'The kinetic treatment of geochemical cycles.' *Geochim. Cosmochim. Acta*, 44, 815-828.

Lasaga, A.C. (1981). - 'Dynamic treatment of geochemical cycles: global kinetics.' In *Kinetics of Geochemical Processes*. A.C. Lasaga and R.J. Kirkpatrick (eds.), Mineralogical Society of America. Reviews in Mineralogy, Vol. 9, pp 69-110.

Lerman, A. (1979). - 'Geochemical processes, water and sediment environ-
    ments.' 481 pp, Wiley & Sons, New York
Mackenzie, F.T. and Wollast, R. (1977). - 'Sedimentary cycling models
    of global processes.' In *The Sea: Ideas and Observations in
    Progress in the Study of the Seas*. Goldberg, E.D. (ed.), pp
    739-785. John Wiley & Sons, Inc.
Spencer, C.P. (1983). - 'Marine biogeochemistry of silicon.' In *Silicon
    Geochemistry and Biogeochemistry*. S.R. Aston (ed.), Academic Press,
    London.
Wollast, R. (1974). - 'The silica problem.', In *The Sea*, vol. 5, E.D.
    Goldberg (ed.), Wiley-Interscience, New York.
Wollast, R. (1981). - 'Interactions between major biogeochemical cycles:
    marine cycles.' In *Some Perspectives in the Major Biogeochemical
    Cycles*. G.E. Likens (ed.), SCOPE 17, Wiley, Chichester, New York,
    Brisbane, Toronto.
Wollast, R. and Mackenzie, F.T. (1983). - 'The global cycle of silica.'
    In *Silicon Geochemistry and Biogeochemistry*. S.R. Aston (ed.),
    Academic Press, London.

ATMOSPHERIC PATHWAYS TO THE OCEANS

John T. Merrill
Center for Atmospheric Chemistry Studies
Graduate School of Oceanography
University of Rhode Island
Kingston, Rhode Island 02881   USA

1.   INTRODUCTION

The meteorological factors contributing to geochemical cycling in
general and air-sea exchange in particular include processes as small
in scale and transitory as the gust of air over a sand dune and as
large in scale and as drawn out as the Southern Oscillation. Many of
the smaller scale phenomena are so closely linked with specific source
mechanisms or exchange processes that they are appropriately covered
in later sections. Here I will discuss those aspects of meteorology
which may contribute to an understanding of atmospheric transport
processes at intermediate and larger scales. We have at our disposal
the full armamentarium of atmospheric science - the deductions of
physics applied to the wind and rain.
     Clear accounts of the structure and dynamics of the atmosphere can
be found in several books. In Wallace and Hobbs (1977) a wide range
of subjects is covered at a fundamental level. Pasquill and Smith
(1983) provide an up to date treatment of diffusion, including both
theory and observation. A comprehensive (but now somewhat dated)
discussion of large scale motion systems is in Palmen and Newton
(1969). Reiter (1972,1978) is a useful review of atmospheric
transport, and the chapter by Hasse (1983) covers most of the
prerequisite material.
     Many practical people would at this juncture abandon all hope of
learning anything about atmospheric transport, realizing from the
above that such understanding may depend upon fathoming something as
imponderable as the general circulation of the atmosphere. Now, I
could provide assurances that learning about transport is easier than
understanding the general circulation. After all, we do not need to
know "why" the wind blows as it does, just "how" it moves materials
around. Lest we become cocksure of ourselves, however, I note again
that we must deal with motions spanning the range 1 to 10^4 km and
10^3 to 10^8 seconds. Also, it is quite easy to so miscast thought
experiments on atmospheric transport as to get correct but bewildering
amd anti-intuitive results regarding everyday phenomena.

P. Buat-Ménard (ed.), The Role of Air-Sea Exchange in Geochemical Cycling, 35–63.
© *1986 by D. Reidel Publishing Company.*

The quantitative and qualitative aspects of transport can both be discussed using a simplified form of the mass conservation equation:

$$\frac{dC}{dt} = \frac{\partial C}{\partial t} + u\frac{\partial C}{\partial x} + v\frac{\partial C}{\partial y} + w\frac{\partial C}{\partial z} = \text{Sources} - \text{Sinks} \tag{1}$$

The substantive derivative of the mass concentration is made up of the local tendency and the advective terms. Since mass is conserved this net tendency must balance the difference between all production and loss of the species under discussion. The advective terms constitute the scalar product of the velocity and the gradient of the concentration and are expressed in (1) in the usual Cartesian coordinate system with u (the velocity component along x) positive to the east, v positive to the north and w positive vertically upwards. The amtospheric circulation is very "flat", i.e. h/ℓ << 1, where h and ℓ are the height (vertical scale) and horizontal scale of the flow. Also, away from the boundary layer, the flow is statically stable, and vertical exchange takes place at the expense of energy. Thus it is traditional to treat the horizontal and vertical motions separately. In part of what follows this separation will be set aside in favor of an alternate view which takes account of this motion implicitly as part of the overall three dimensional pattern.

There are five important classes of transport that exist in the atmosphere: boundary layer flow, boundary layer-free troposphere exchange, cloud transport, synoptic-scale transport by storm systems, and global-scale transport by mean winds and waves. The discussion that follows centers first on boundary layer phenomena, then cloud transport, then storms. At larger scales I deal first with the midlatitude circulation systems and then global-scale processes.

2. ATMOSPHERIC STRUCTURE AND TRANSPORTS

The processes important in transport are different in the various environments encountered around the world. Within a given environment different processes may dominate at different spatial and temporal scales. Also, even at the same scale different mechanisms may be more important for one state of matter than for another. Because the specifics of individual chemical species are covered by other authors in the later sections of this book, the last point will be glossed over here. Where no particular state of matter is mentioned it is correct to assume that the transport of both gases and particles is being discussed. In each of the following sections the atmospheric structures and then the dominant modes of transport at various scales are discussed.

Most substances of importance in geochemical cycling have their sources at or near the surface, within the atmospheric boundary layer. Also, obviously, the same is true of all air-sea exchange. Therefore, the following discussion begins with this critical region of the atmosphere. The chapters by Panofsky (1985) and Wyngaard (1983) and the references therein provide further details on the observed and modeled structure of the boundary layer.

2.1 Boundary Layer

The atmospheric boundary layer is that part of the atmosphere in which the influence of the surface is felt directly. Of course this definition, to be applied, requires a specification of what "influence" is of concern. The most fundamental choices are these: molecular viscous influence, thermal influence and momentum (frictional) influence. The corresponding labels are: the viscous sublayer (thickness δ), the mixed layer (z_M) and the Ekman layer (z_E). The viscous sublayer is generally only millimeters thick and is not considered here.

The structure of the boundary layer is time dependent and heterogeneous, dependent as it is upon the surface condition and the overlying synoptic scale flow. Nevertheless, the following general statements can be made about the height of the mixed layer (summary in Table 1). Over land the structure alternates diurnally between convective and inversion, with short-lived passages through neutral stability at sunrise and sunset. Over the sea the stratification is neutral or near-neutral except under very light winds or strong temperature gradients.

The height of the boundary layer is variable under stable stratification, but is limited by the lowest strong inversion in convective cases. The factor of 4 in the formula for the height of the neutral boundary layer is heuristic; in nature the height is quite variable, but typically does scale with the friction velocity and the Coriolis parameter.

TABLE 1. Boundary layer characteristics in various environments.

Area	Boundary Layer Type	Height
Land: Day	Convective	z_I
Land: Night	Inversion	Variable
Land: Transition	Neutral	$u*/4f$
Sea	Neutral	$u*/4f$

The concentration profile within and above the boundary layer is dependent upon a host of factors, even if we assume that the source of the substance of interest is at the surface and that gravitational settling effects are unimportant. As a first approximation one may assume that in neutral or convective conditions the concentration is nearly independent of height above the viscous sublayer. In a steady state situation the flux of the substance decreases linearly from the

near-surface value to its value at the top of the boundary layer. In
micrometeorology there is much discussion of the "constant flux"
layer, lying above the viscous sublayer, up to 10% of the boundary
layer height. The flux referred to is the momentum flux, i.e. stress;
and it is not constant, but decreases by 10% in that distance.
While one sometimes hears "constant flux" layers mentioned in
discussions of material exchange, Slinn (1983, Section 6.3) shows that
caution is in order. Under homogeneous conditions one can imagine
boundary layer transport as simple advection of air with a given
concentration of the substance of interest. Because the wind speed
increases markedly with height it is appropriate to use winds averaged
over the depth of the boundary layer.

An important type of boundary layer transport is the cyclic land
breeze-sea breeze system caused by the temperature contrast between
the land and the sea. Solar heating during the day and radiational
cooling at night heat and cool the land and the resulting pressure
gradient results in onshore and offshore winds alternately with

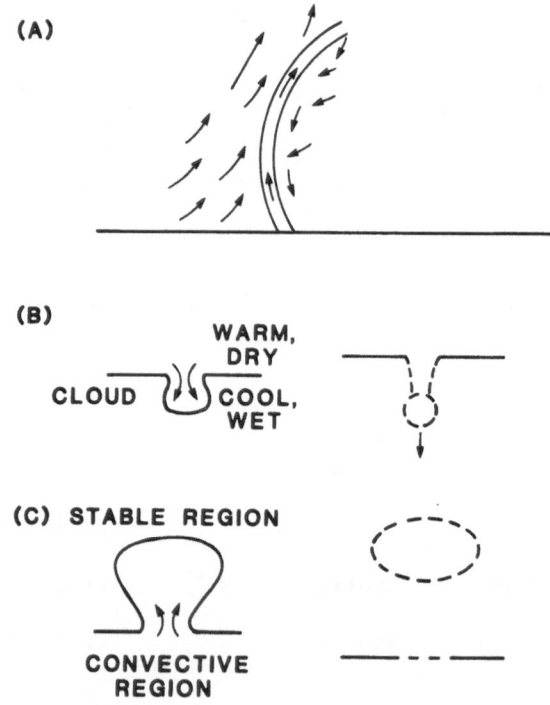

Figure 1. Sketch of exchange mechanisms between the boundary
layer and the free atmosphere. a) Cold frontal uplifting. Vectors
are the motion relative to the frontal surface, which advances to the
left. b) Cloud-top entrainment. The warm, dry air in the inversion
layer above the cloud is incorporated into the cloudy boundary layer.
c) Penetrative convection. The air in the stable region is mixed into
the penetrating updraft.

relative calms (See Hasse, 1983, Section 4). This system is typical
of any coastal location, particularly in the summer, unless the
synoptic scale flow is adverse or clouds block the sun. There must be
a sufficient area of land and of water: small lakes and narrow
peninsulas are exempt. The theory of these circulations is quite
straightforward, but its direct application to practical problems is
limited. Inclusion of such features as coastal topography or varying
inversion height cause the investigator to resort to numerical
modeling almost immediately.

Because coastal areas are typically heavily populated and
consequently have their share of pollution, the land breeze-sea breeze
cycle may result in an exchange between continental and ocean areas.
The first thing that comes to mind is the deposition of airborne
pollutants over the coastal sea. This process is certainly
important. Accumulation of pollutants while air parcels make multiple
passes through such a system has been documented (Cass and Shair,
1984). While total sulfur oxides air quality at an urban coastal site
was dominated by fresh emissions from nearby sources, sulfate
concentrations, which reached a peak of 60 μg m^{-3} and averaged
30 μg m^{-3}, were due largely to the oldest material in the atmosphere
in their observations. The "reverse" process of onshore transport of
pollutants or significant natural substances can also take place, as
the nearshore waters are, in general, contaminated by municipal
outfalls, by riverine or estuarine runoff and by air pollution. The
presence of significant coastal upwelling offshore can make for a
particularly rich mixture of substances.

Boundary layer-free troposphere exchange is a subject that has not
received adequate attention. I believe that this is in part becaused
it is a very great challenge to learn anything using either
observational or theoretical techniques. The processes we will
discuss are frontal uplifting, entrainment and penetrative convection.

When a front advances it displaces the air in advance of it both
horizontally and vertically. Thinking of the front as a material
surface as a first approximation we would expect the vertical
displacement to be

$$\delta z = \frac{\partial z}{\partial y} \delta y \tag{2}$$

where the slope of the front multiplies δy, the distance it advances.
The slope in (2) is predictable from theory (see Holton, 1979, Section
9.4) and is usually in the range $(0.2 - 5.0) \times 10^{-2}$ (Palmen and
Newton, 1969). In observational analyses a much larger uplifting is
observed (Sanders, 1955), corresponding to significant convergence
before the front. At the mesoscale this uplifting is gentle
(1 cm s^{-1} vertical velocity), but at the microscale large vertical
velocities are observed (6 m s^{-1}, Shapiro (1984)). The clouds and
precipitation associated with fronts are caused by this uplifting.
Since boundary layer air is lifted into the free troposphere in this
process, it is an important exchange mechanism.

The sketch in Figure 1a shows a hypothetical cold front as a pair

of solid lines. The vectors show the air motion relative to the
frontal surface, which is moving to the left. The warm air overlies
the cold air except near the ground, where frictional drag retards the
motion. In a warm frontal situation the motion of the frontal surface
would be to the right.

In the context of mixing across an interface (as in boundary
layer-free troposphere exchange) entrainment refers to movement of air
from the laminar to the turbulent side, i.e. into the boundary layer.
This entrainment will tend to make the boundary layer grow in height
and will dilute any tracers concentrated there. On the other hand,
penetrative convection will have the opposite effect. Buoyant plumes
or bubbles from the convective layer impinge upon and distend the
boundary. If there is sufficient excess buoyancy the bubble will
penetrate. The subsequent detrainment enriches the free troposphere
with boundary layer air.

The sketch in Figure 1b illustrates entrainment. The process can
be initiated by radiational cooling of the air above the boundary
layer, and may be accelerated if the boundary layer is cloudy
(Deardorff, 1980). In that case the parcel is cooled and sinks
because cloud droplets mixed into it by turbulence evaporate. Figure
1c shows deep penetrative convection. Note that the decrease in the
boundary layer thickness is highly exaggerated and may be negligible
owing to entrainment coincident with the convection. Also common is
shallow penetrative convection in which the bulge in the base of the
stable region has insufficient buoyancy and collapses into a weakly
turbulent extension of the boundary layer.

The following example is indicative of the problems that can occur
when, by choice or necessity, time averaging is done. The air above a
flat, featureless plain is heated by its contact with the ground,
which is warmed by sunlight. Convection develops and a large heat
flux, upward, is established, about 300 W m^{-2}; meanwhile the
temperature gradient is reduced to a small, negative value, 0.5 °C per
100 m. After nightfall the ground cools rapidly and the air
temperature near the ground drops. The stable temperature inversion
suppresses vertical exchange and, while the temperature gradient grows
larger and larger reaching 5 °C per 100 m, the heat flux (which
reversed to downward after sunset) remains small, about -30 W m^{-2}.
The turbulent mixing coefficient ("eddy heat viscosity") satisfies

$$F = -K\frac{\partial T}{\partial z} \tag{3}$$

Now the average of F over the entire day is positive and large,
about 270 W m^{-2}. The full day average of $\partial T/\partial z$ is also positive,
+ 4.5 °C per 100 m. Thus the mixing coefficient must be negative. On
the average, in this thought experiment, the heat flux is up gradient,
from cooler to warmer areas.

While negative viscosity values can be instructive (Starr, 1968),
this is a case of inappropriate use of averaging. Taking the day and
night periods separately yields positive mixing coefficients and
downgradient flux. When averaging over times or places dominated by
different processes we must keep our wits about us.

2.2 Cloud Scale Transport

Clouds are important contributors to atmospheric transport. Also, but
not covered further here, clouds are important in the transformation
of many species (by heterogeneous chemical and photochemical
processes) and in removal by precipitation.

The global average cloudiness is 50%, but clouds are not
responsible for 50% of the vertical transport. First, a substantial
fraction of the cloud cover is due to high and middle level clouds
that are poorly coupled with the major high-concentration reservoirs
of the geochemical cycles and have limited vertical motion. Also,
clouds are not long-lived, dense aggregations of condensed droplets or
ice crystals; they are commonly transient, inhomogeneous volumes of
air with a sparse admixture of droplets and particles. They are
treated separately here because they are an additional identifiable
transport agent at a scale smaller than that of synoptic systems.

The models discussed here are specific to subtropical cumulus
clouds, but the same ideas could be applied to cloud systems in other
environments.

Vigorous convection takes place sporadically in the subtropics and
is intermittent even in the humid tropics. Thus the effects of
convective motions are generally not included explicitly in transport
models. The most that has been done is to "enhance" the vertical
diffusivity of the model in environments where clouds are common.
However, since clouds move lower level (primarily boundary layer) air
higher into the troposphere, where it is deposited or detrained, the
flux of any tracer high in concentration in the lower levels is not
proportional to the concentration gradient, and thus not treatable as
a diffusive process. Gidel (1983) used a model of cloud mass fluxes
in which a wide range of entrainment rates are included, each
associated with a given mass flux and present at a frequency specified
as the fractional areal coverage. He finds that <u>increasing</u> mixing
ratios with height can result for substances with photochemical
lifetimes varying in the vertical. Consequently, models that
essentially ignore cloud transport may significantly underestimate the
amount of some species in the upper troposphere. The results are
somewhat controversial. Chatfield and Crutzen (1984) use a
mathematically less formal (read "heuristic") model based on a two
dimensional geometry to estimate the cloud transport. Their results
also have relatively high concentrations of reactive species in the
high troposphere, even when removal of soluble species by rainfall is
taken into account. Since the reactive lifetimes are much longer
there, the substances can make important contributions to the cycles,
i.e., they can be transported further. These models explicitly
account for the vertical mixing associated with evaporating detrained
cloud droplets. Also, the downward vertical transport in the
downdraft surrounding clouds is taken into account. As discussed by
Slinn (1983, Section 4) downward motion in rainshafts is important as
well. Cloud ensemble models will be needed to determine these effects
accurately in different enviroments around the world.

2.3 Storms and midlatitude circulation

The occurrence of storms is categorized by the frequency of their
formation, cyclogenesis, by their passage and by their intensity. In
early marine atlases the storm statistics were thought to be subject
to an important "fair weather" bias: the most severe storms would be
undersampled because prudent mariners avoid such storms when
possible. Satellite meteorology has helped mariners to avoid severe

Figure 2. Simplified indication of mean jet position and storm
tracks in the northern hemisphere winter. The axis of mean maximum
wind (heavy line) is shown, and maximum speeds are indicated. The
maximum occurrence of cyclones (short arrows) and anticyclones
(double-shafted arrows) are also shown. Adapted from Palmen and
Newton (1969).

weather, but has also contributed to our knowledge of the storm tracks and the fate of storms. In part because of this there are now atlases that are known to give adequate coverage except in the far southern hemisphere. The subject of this section is the structure of and transport associated with the life cycle of storms. The emphasis is on midlatitude cyclonic storms because they are the most prevalent type, especially over populated areas.

The extratropical cyclones form in the baroclinic zone (the region of strong latitudunal temperature gradient) below the polar front jet stream. The mean jet axis in the winter lies between ~20°N to ~50°N, with equatorward jogs over the eastern oceans and poleward trends over the continents (Figure 2). In the eastern Atlantic meandering of the jet is common, and the mean axes shown are the places where it is most steady. Individual synoptic maps show intermediate positions as well. The two major cyclone tracks lead from the southwestern to the northeastern parts of the Atlantic and Pacific oceans, in approximate alignment with the major ocean currents. Other important cyclone tracks are present over the continents and are influenced strongly by topography. The merging storm tracks over the North Pacific and North Atlantic in Figure 2 mark the positions of the Aleutian and Icelandic lows. The Aleutian and Icelandic lows are important sites of frontogenesis despite their being poleward of the mean maximum wind axis. These are two of the important "semi-permanent centers of action," the other two being the marine subtropical highs. These anticyclone centers have their maximum frequency of occurrence along the double-shafted arrows.

The general picture is that storms form on the warm air side of the jet, move along the front beneath the jet and move to the left of the jet as they reach mature form. The structure of these storms, particularly their air mass structure, is discussed by Hasse (1983, Section 3). The surface pressure fall in cyclogenesis is caused by upper level divergence, which moves air out of the vertical column faster than air is accumulated through low level convergence.

The overall structure of the atmosphere in the winter hemisphere is shown in Figure 3. The principal air masses are shown, as are the fronts, tropopauses and jets, and their relationship to the surface wind systems. Note that the polar front is drawn dashed near the ground; this indicates that cross-frontal mixing is not always prevented there. This is because frontolysis (frontal dissipation) can allow exchange. Note also that the arctic and subtropical fronts are not always present and that they typically do not span the entire depth of the troposphere.

At the scale of storms, synoptic scales, the vertical and meridional motions are related. Embedded within the storm are small scale convective systems, but at synoptic scales the motion is quasi-adiabatic. That is, the flow involves no exchange of heat between individual parcels of air and their surroundings, either by mixing, divergence of radiational fluxes or latent heat exchange. Since there is little mixing in the atmosphere above the boundary layer this approximation is a good one away from deep convection (where both latent heating and mixing are strong). Under these circumstances the potential temperature, Θ, is constant, even though a parcel may migrate vertically, experiencing changes in pressure and temperature. Recall that Θ is the temperature a parcel of air would have if it were moved dry adiabatically to the reference pressure, p_0:

$$\Theta = T\left[\frac{p_0}{p}\right]^{R/c_p} \tag{4}$$

where R and c_p are the gas constant and specific heat at constant pressure of air. Because the free atmosphere is statically stable Θ increases monotonically with height and simply connected two dimensional surfaces of constant Θ can be imagined. The height of the surface varies, increasing with increasing latitude and undulating with the large-scale waves. Since the temporal vertical excursions of Θ surface are small, the vertical motion of an air parcel is related to its horizontal motion: equatorward motions (which are of colder air) are coupled with descent, while poleward movements (of warmer air) are coupled with ascent.

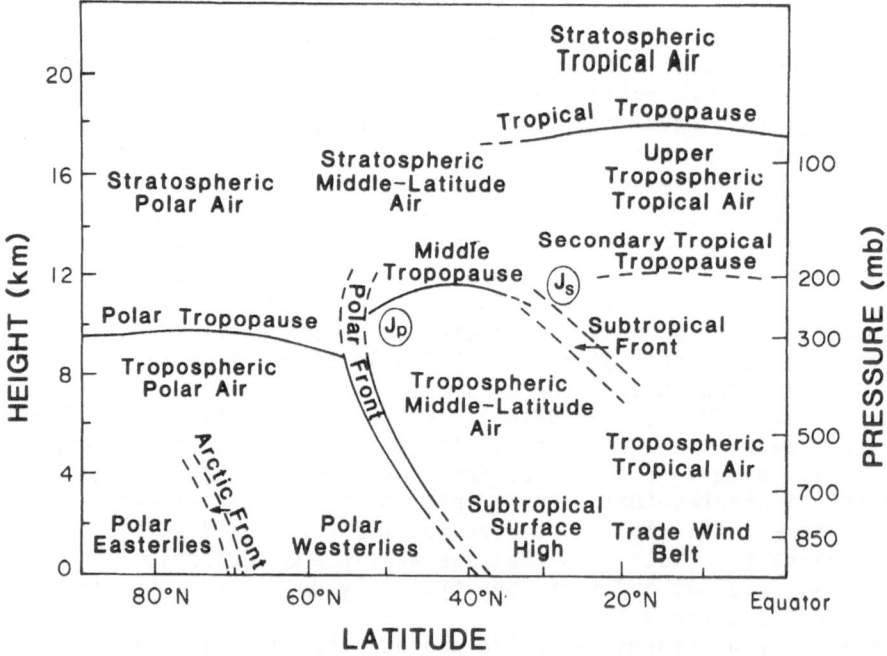

    Figure 3. Simplified characterization of the principal air masses, fronts, tropopauses, and jet streams for the northern hemisphere winter. Positions are typical but approximate. Adapted from Palmen and Newton (1969).

    Utilizing the adiabatic flow approximation it is possible to visualize the circulation of a growing storm. The movements of individual air parcels on the 295 K isentropic surfaces are shown in Figure 4. The lows and highs indicated are of the streamfunction on the 295 K surface, and the chart can be interpreted as a weather map on a sloping surface which lies at 600-1000 mb, above and within the planetary boundary layer. Six circles are shown at the initial time. The left-most three circles started out southwest of the surface cyclone, and the rightmost three to the southeast. After 48 hours the circles have been deformed into narrow filaments. Downward motion of 50-150 mb accompanies the movement towards the cold front, while there is upward motion of 100 to 250 mb for the parcels moving toward the warm front and occluded front (Merrill, Bleck and Boudra, 1986).

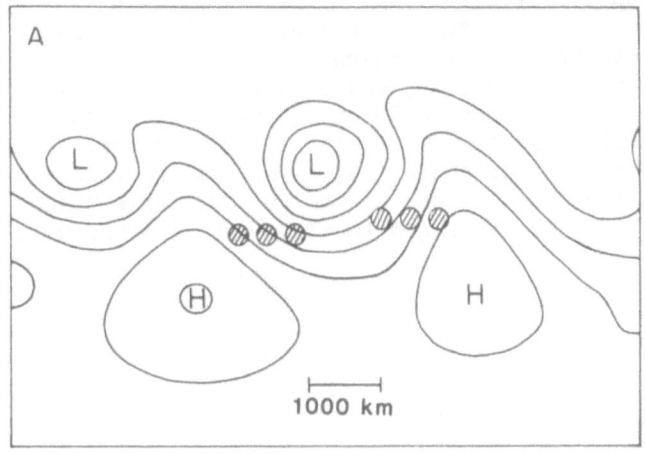

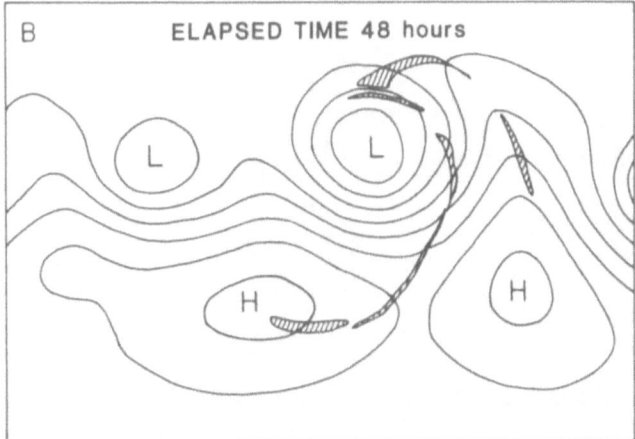

Figure 4. Isentropic deformations for a midlatitude storm. The air initially in the six circles at 295 K is deformed into the thin lines at the fronts. The time elapsed is 48 hours. The streamlines of the isentropic flow are shown. At the initial time the center of the surface low is about 600 km east of the low indicated at 295 K.

This example is meant to show that the motion in midlatitudes is complex but comprehensible. The coupling of vertical and horizontal motions means that latitudinal heat transfer (the energetic "goal" of the storm) is accompanied by vertical excursions which may exchange air from within the boundary layer with air from the free atmosphere. The horizontal "scale" of the exchange is large, 1200-2500 km. When initially compact parcels of air have been deformed into narrow filaments small scale mixing becomes dominant and the exchange process is complete.

It is questionable whether conventional climatological wind data can be combined with concentration data for trace substances to estimate the net atmospheric transport. In an attempt at such an estimate Galloway et al. (1984) use an "advection climatology" prepared by Whelpdale et al. (1984). This is a set of onshore and offshore wind estimates at several heights in the atmosphere for eight North American eastern coastal segments whose boundaries were chosen to be homogeneous and climatologically distinct. There is little doubt that the air moves generally as indicated by Whelpdale et al. But Galloway et al. must combine these values with vertical and horizontal profiles of the concentration of the substances of interest (compounds of S and N). The use of the few available height profiles of these concentrations, averaged together and combined with

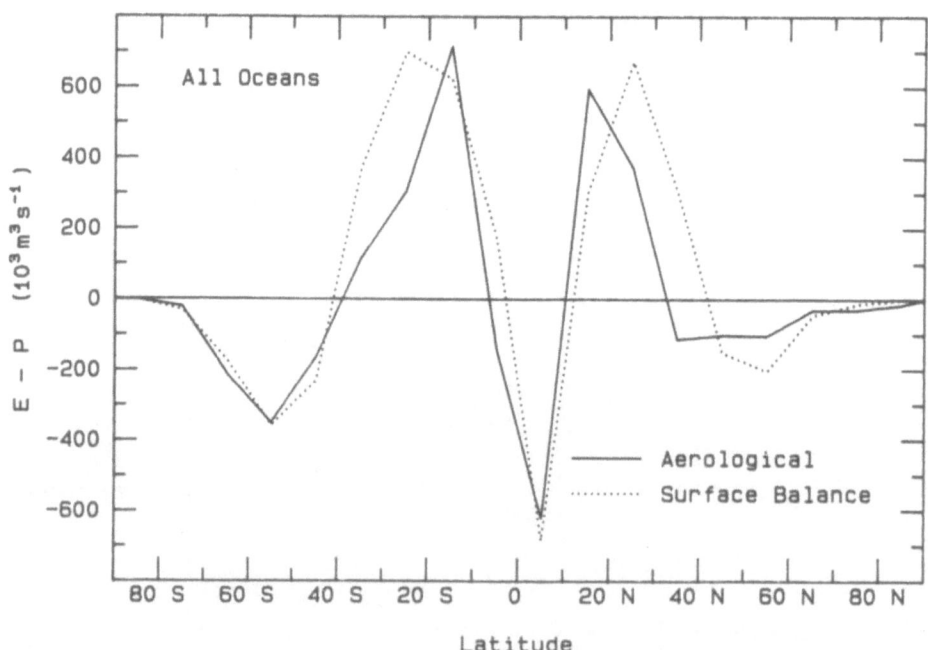

Figure 5. Meridional profiles of the annual mean evaporation minus precipitation over all the oceans combined in 10° zonal bands. Aerological data are from Bryan and Oort (1984), while the surface water balance data are from Baumgartner and Reichel (1975).

climatological winds, produces a very uncertain result. Galloway et
al. estimate that their layer average flux values are uncertain by
25-100%. Although they estimate an additional 10-20% uncertainty
due to the neglect of eddy effects, I believe that this is an
underestimate. The correlation between latitudinal and vertical
exchange discussed above in the context of adiabatic transport cannot
be accounted for when monthly average data are used. While the net
flux may not be misestimated badly I believe that the vertical
distribution of flux may be far wrong. Homogeneous estimates of
concentration and wind which can be used to estimate fluxes (but so
far only of water) are discussed in the next section.

2.4 Global scale exchange

The indvidual storms and systems of storms discussed above cannot
account for the largest scale of transport.   At the global scale we
must consider such factors as the meandering of the Inter Tropical
Convergence Zone (ITCZ) and the waxing and waning of the monsoons to
account for phenomena like interhemispheric exchange.
     Using in situ measurements (as opposed to remote sensing) places
severe limits on our ability to understand the budgets of trace
substances with global sources and sinks. Water is an excellent
example because good data for its height distribution are available,
colocated with the wind data from weather balloons. Thus, an analysis
of the transport and cycling of water is indicative of the highest
expectations we should have about budget studies. Furthermore, an
independent set of surface measurements exists that can be used to
check the atmospheric branch of the hydrological cycle; the absence of
adequate precipitation measurements over the oceans is a serious
problem. There are differences between water vapor and other tracers:
it is confined to the lower and middle troposphere, and it condenses
and falls out of the atmosphere if lifted. Put another way, water is
a tracer which is weighted by the temperature.
     The aerological method and the atmospheric circulation data set
prepared for global budget studies are discussed by Oort (1983). The
use of these data in studying atmospheric water vapor and its relation
to the other branchs of the hydrological cycle are discussed by
Peixoto and Oort (1983) and by Bryan and Oort (1984). Briefly, the
method uses wind and specific humidity measurements to estimate the
local tendency and horizontal flux vector of the vertically integrated
water vapor content, i.e. the left hand side of (1). The tendency and
flux of condensed-phase water are known to be negligible on this scale
(two orders of magnitude smaller) and are ignored. The sum of the
local tendency and the advection is the evaporation minus
precipitation, E-P, as these are the only source and sink.
     The meridional profiles of the annual mean oceanic E-P values
calculated by Bryan and Oort (1984) are shown in Figure 5. The values
are volume fluxes of water over 10° zonal bands. The tropical region
of net precipitation is centered north of the equator. The source for
this water is the subtropical oceans, where the evaporation exceeds
the precipitation. The transports required are discussed below. In

the midlatitudes of both hemispheres precipitation associated with cyclonic storms causes the difference E-P to be negative.

There are independent estimates of the term E-P. The data of Baumgartner and Reichel (1975) are also shown in Figure 5. These are based on a balanced global water budget; there were obvious incompatibilities in the surface water balance, and subjective adjustments were applied by Baumgartner and Reichel. Despite these difficulties the initital impression is of adequate overall agreement. The aerological results have the midlatitude region of excess precipitation extending further equatorward than the surface-based estimates, particularly in the northern hemisphere. There are more serious discrepancies in the budgets over the continents.

The rate of interhemispheric exchange of water vapor can be obtained from these data. There is an excess of precipitation in the northern hemisphere; the global excess value is larger by a factor of ~8 than the oceanic value discernable in Figure 5. As discussed below (Section 3.1) there is a strong seasonal cycle in the flux of water vapor across the equator evaluated by the aerological method, but the net flux is toward the northern hemisphere, consistent with the E-P estimates in both sign and magnitude. It is senseless to attempt to compute a cross-equatorial "piston velocity" from those flux values. Rather we should consider the longitudinal variations which produce them.

Another method of assessing the interhemispheric exchange is by the use of radionuclides. The tracers of greatest use here are anthropogenic; natural tracers generally show lower hemispheric asymetry and thus less "signal". For example, Weiss et al. (1983) report on ^{85}Kr measurements in air taken during transects of the Atlantic. They find a significant decrease as they cross the latitude of the ITCZ from north to south. Also, southern hemisphere concentrations are uniform, while the northen hemisphere concentrations increase with increasing latitude. The data are consistent with known northern hemisphere sources (^{85}Kr is a waste product of nuclear fuel reprocessing) and an interhemispheric exchange time of 1.0 - 1.7 years. The abrupt decrease in ^{85}Kr concentration south of the ITCZ seems to indicate that the ITCZ is a significant barrier to interhemispheric exchange.

This is because the ITCZ represents the boundary between the trade wind circulations of the hemispheres. The barrier is not like an impervious material surface, however. The water vapor flux maps (see Peixoto and Oort, 1983) indicate that the cross-equatorial flux is divided about evenly between exchange through the ITCZ and monsoonal exchange. Because the monsoonal flux is dominated by a low level jet the half of the transport associated with the monsoon takes place very rapidly. Using data for the concentration of fission products from nuclear bomb tests in the South Pacific, Rangarajan and Eapen (1981) showed that the transit time from ~ 15°S to ~15°N is only 3-5 days during the summer monsoon.

I wish to emphasize again the importance of structured flow in atmospheric transport. The monosoon is not a collection of eddies acting independently (although there are mesocale events and fluctuations). Because it can draw air and tracers from one area and move them to another without drastic dilution it can act as an antidiffusive process. Note, for instance, that it carries water vapor into the northern hemisphere from the south, against the gradient. There is a diffusive component to interhemispheric exchange, but more than half of the low-level flux is carried by jets and other identifiable structured circulations. Thus both the rate and "style" of such exchange may depend in part upon the distribution with height of the substance of interest.

3. VARIABILITY AND REPRESENTATIVENESS

The temporal and spatial variability discussed up to this point is that associated with "weather". At time scales longer than one year we characterize fluctuations or trends as aspects of the climate. We now recognize that there are natural variations in the climate at many different time scales. There must be corresponding variability in geochemical cycling, and one of the topics discussed here is the determination of the representativeness of results.

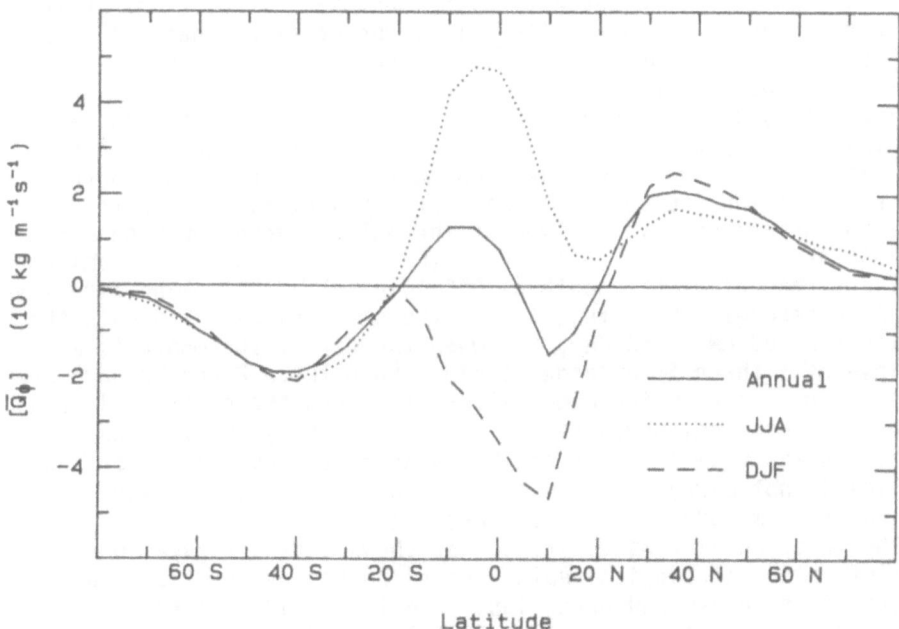

Figure 6. Meridional profiles of the vertically integrated meridional water vapor flux across latitudinal walls in the atmosphere. Positive values are towards the north. The square bracket denotes an average over longitude in this case. Annual and two seasonal profiles are shown. Adapted from Peixoto and Oort (1983).

3.1 Seasonal and interannual variability.

The seasonal dependence of atmospheric transport has been mentioned
briefly above in the context of the spatial structure of the global
circulation. Here we emphasize the temporal variation and
variability. The variation can be seen in Figure 6, where the total
meridional water vapor flux in the atmosphere is shown as a function
of latitude. This is the amount of water passing through a vertical
wall at each latitude, per meter of length of the wall, per second.
Notice that the flux is predominantly poleward in midlatitudes. In
that region the flux is dominated by the transient and standing
eddies. That is, by midlatitude storms and features like the Aleutian
and Icelandic lows. There is a large seasonal variation in the lower
latitudes. The flux into the northern hemisphere during June, July
and August is ~19 x 10^8 kg s^{-1} while the outflow in December,
January and February is ~-14 x 10^8 kg s^{-1}. For the entire year
there is a net influx into the northern hemisphere of
~3 x 10^8 kg s^{-1}; this cross-equatorial flow balances the larger
negative evaporation minus precipitation there. The net cross
equatorial flow is also consistent with the mean position of the ITCZ,
north of the equator. However, when we examine the global maps of
this flux (see Peixoto and Oort, 1983) it is apparent that much of the
seasonal variation is associated with monsoonal circulations in the
Indian Ocean and Western Pacific. Thus the moisture that falls as
rain in the northern hemisphere summer monsoon originates as vapor in
the southern hemisphere.

There is a seasonal dependence of the near-equatorial flow away
from the monsoons as well. In Figure 7 trajectories to hypothetical
sites at arbitrarily chosen positions are shown for the northern
hemisphere spring and summer. These are trajectories backward in
time, so the air parcels arrived at the points along 160°W on March
20, 1984 and on July 28, 1981. Note that air from northern hemisphere
locations reaches as far south as 5°S in March. Air coming directly
from Asia reaches 5°N at this time. The small circles indicate the
positions at 00 GMT each day, and when the day of the month is a
multiple of 5 there is a large circle. So between 8 and 12 days are
required for air from the midlatitudes to reach the equatorial areas.
In the northern hemisphere summer (i.e. late July) the ITCZ has moved
north, and air from the southern hemisphere reaches these locations.
The ITCZ is not always visible as a cloud band, but generally
oscillates from ~5°N to ~12°N and back each year.

The interannual variability is not shown on the figures but is
apparent in the aerological data. Both the net evaporation minus
precipitation (Bryan and Oort, 1984) and the meridional fluxes
discussed above (Peixoto and Oort, 1983) extend over 10 years and have
standard relative errors of ~20%. While the overall water balance
is maintained, this high level of variability is significant.

Recent work has shown that interannual variability extends to such
fundamental quantities as the mass of the atmosphere. There is a
substantial redistribution of the atmospheric mass between the
northern hemisphere and southern hemisphere (Christy and Trenberth,

1985 and the references therein). The redistribution is presumed to
be forced by large thermal anomalies (e.g. the Southern Oscillation),
as the mean annual cycle of pressure exhibits interhemispherec mass
exchange in response to the cooling of one hemisphere relative the
other. Since hemispheric average anomalies of 1 mb can occur in one
month, it should not be surprising that fluctuations exist: this is
0.1% of the atmospheric mass, 2.5×10^{15} kg of air.

 Along with the mass redistribution there may be abrupt transitions
in the usual patterns of flow. During the now well-known El Nino –

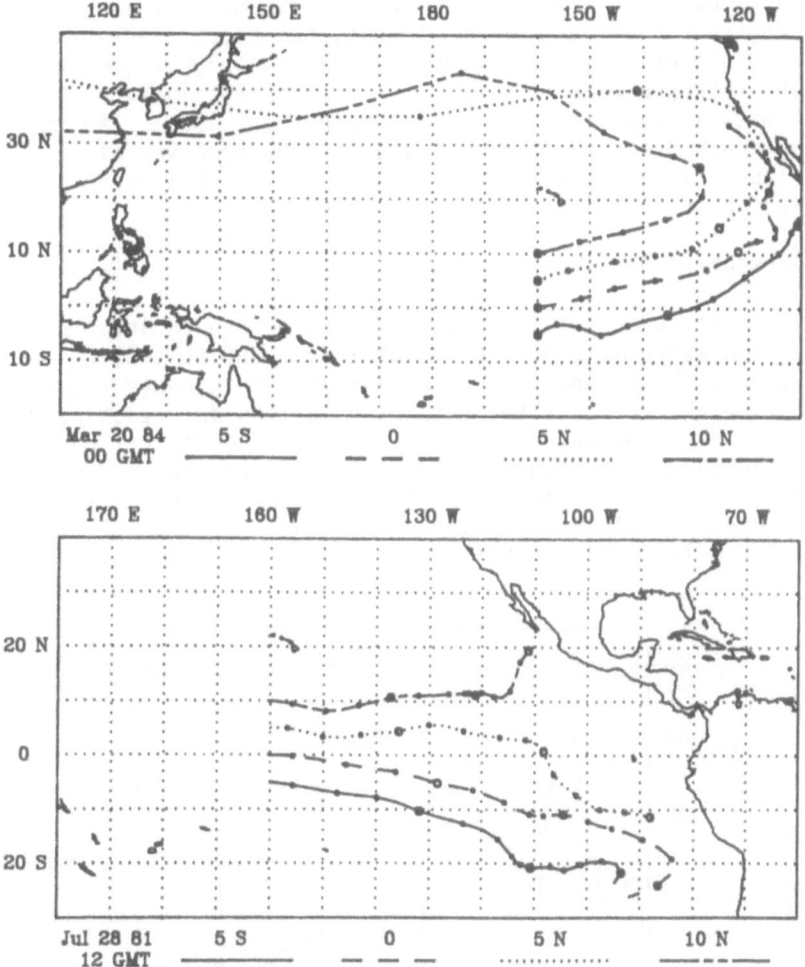

Figure 7. Interhemispheric trajectories. Isentropic trajectories
at 305 K, backward in time from March 20, 1984 and from July 28,
1984. The trajectories "begin" at intervals of 5° of latitude. Note
that the March trajectories remain in the northern hemisphere while
those in July cross into the southern hemisphere.

Southern Oscillation events the trade wind field over vast regions of
the central and western Pacific Ocean weakens, and westerly winds
reach as far as the equator at surface levels (Rasmussen and
Carpenter, 1982; Philander and Rasmussen, 1985). There are coincident
changes in the precipitation pattern, with the heavy rainfall
characteristic of the far western Pacific shifting into the central
basin. While the details are not yet known it is clear that there are
important shifts in the large scale atmospheric transport patterns
during such events. Oceanic sources with a dependence on productivity
will be altered by shifts in the pattern of upwelling. Transformation
and removal processes dependent upon clouds and rain will have their
geographical distribution changed. An experiment over the Pacific
lasting one year might avoid such events, but because they recur every
3 years or so a multi-year experiment will probably include one or
more. The return period of El Nino - Southern Oscillation events
varies from 2-10 years (Rasmussen and Carpenter, 1982). No matter
what the length of an experiment there will be some variation. If we
wish to take nature as we find it we must accept that such variations
are inseparabale from the average conditions.

3.2 Representativeness of observations.

It is difficult to accept how easily we can be fooled into thinking
that representative data have been obtained. The following anecdote
may illustrate this point. The SEAREX (Sea-Air Exchange Program) has
had as on one of its goals the estimation of concentrations and fluxes
of various trace substances at remote sites. Experiments were to be
conducted in the major wind regimes, and the North Pacifc trades were
the clear first choice. To encompass the expected annual variation
one experiment was planned for the dry and one for the wet season
(January and July). The results (obtained later) indicated that

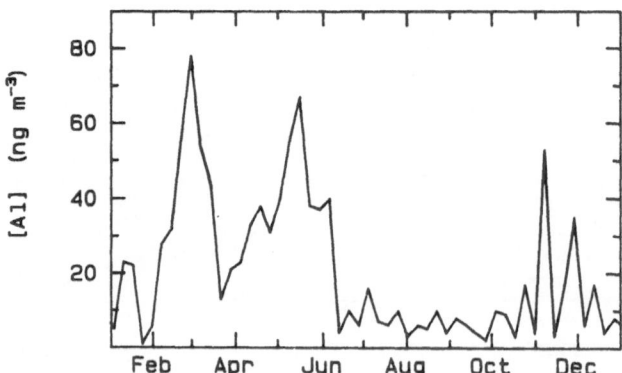

Figure 8. Weekly average aerosol aluminum concentrations at
Enewetak. Note low values in January and July and high values in
April. Unpublished data from the North Pacific Aerosol Transport and
Deposition Network, 1981-1985.

mineral aerosol was present and that the flux is large. There is
indeed an annual cycle, but the highest concentrations are in March -
May, and the values in July and January are typically <u>both</u> low (see
Figure 8). The experiment planned for January, 1979 was delayed
because of equipment damage and logistical problems brought on by the
passage of Typhoon Alice directly over the site at Enewetak Atoll. If
the storm had veered and the experiment had gone ahead in January we
would most likely have missed the most important "signal". The
experiment began in April, and Duce et al. (1980) were able to report
a striking seasonal variation. The result would have been found by
others, but at Hawaii, closer to the westerly wind belt, not in the
deep subtropics (Shaw, 1980; Bodhaine et al., 1981; Parrington, Zoller
and Arras, 1983).

4. MODELING OF ATMOSPHERIC TRANSPORT

The means used in modeling are diverse, as are the goals. This
discussion begins with the most fundamental models for geochemical
cycles, source identification models. Next, for use in understanding
processes and determining which factors may dominate, there are
mechanistic models. Finally, for use in global budget studies there
are tropospheric chemistry system models. While it is not possible to
give an exhaustive treatment, a brief discussion will be given of the
strengths and weaknesses of the various approaches. Since the theme
of the ASI is geochemical cycling, the discussion of models does not
emphasize air quality or pollution models.

4.1 Source identification models

Data on the concentration and deposition of a trace substance are
obtained at a remote site. Where does this material come from?
This is the question addressed by source identification models. We
are concerned here only with the meteorological aspect, that is, what
can be learned about the source using meteorological data?
     Quite often both the source and the observation are within the
boundary layer. If the transport includes free tropospheric travel
there must be a mechanism for exchange.
     Of crucial importance is that the calculations discussed here
provide a meteorologically consistent source region, an area from
which the "tagged" air probably came. Of course the air parcel has a
complex history before passing the source area, and the meteorological
data cannot rule out the possibility that the tracer source may be
further upwind. Obviously the source identification process may not
be simple. Furthermore, these calculations usually give either little
or no information whatever on the time variation of the concentration,
and the interpretation can be tricky.
     Fields of meteorological data can be used to estimate the
displacement of hypothetical air parcels. The type of trajectory
analysis that results depends upon both the method of analysis used to
obtain the meteorological fields and the assumptions made in

calculating the trajectories. The fields can be subjectively prepared (that is, hand analyses, either routine or specially constructed) based on upper air observations, or can be objectively analyzed fields (that is, machine products). The objectivity and reproducibility inherent in the computer-generated analyses is no guarantee of superiority. The human eye-brain system is wonderfully adept at pattern recognition, continuity checking and subjective smoothing. However, the accumulated advances in objective analysis and forecasting have resulted in generally high quality analyses from computers. Forecasting enters into the picture because the first guess analysis is forecast from the earlier analyses in the objective system. This is discussed by Bourke, Seaman and Puri (1985).

With a set of analyzed fields available several types of trajectory analysis can be undertaken. In isobaric trajectory analysis the motion along constant pressure surfaces is tracked. This method is useful only for very short times (<1 day), as the air does not remain on isobaric surfaces. Averaging winds over the depth of the boundary layer is a justifiable approach for tracking air that remains in that layer. Because the simplest nondispersive calculation is purely antisymetirc in time the trajectories can be calculated forward or backward. Available operational models include diffussion and deposition as well (Heffter, 1981).

For transport above the boundary layer, relaxation of the isobaric assumption is essential. Although several different techniques are available (and are discussed by Danielsen, 1974) the most widely useful is the isentropic method of Danielsen (1961). Although developed for manual use on hand-analyzed charts, the technique can be automated and used on gridded objective analyses (the interested reader may read Danielsen and Bleck (1967) for a detailed discussion). The use of data on surfaces of constant potential temperature (discussed above in Section 2.3) is an essential step. Even when the best techniques are used there is an uncertainty in the end point of any trajectory, and the uncertainty grows with time. The errors introduced by the interpolation process can be minimized, but two sources of error are unavoidable: the sparsity of basic meteorological data in some areas and the 12 hour time interval between routine upper air observations and analyses. Adequately accurate results can be obtained for periods up to 5 days (Merrill, Bleck and Boudra, 1986), except in regions of rapid change or strong deformation. (Unfortunately these are often the most interesting locations). Less encouraging results, particularly for isobaric trajectories, have been found in other studies (Kuo, et al., 1985). While the calculation can be continued indefinitely, the uncertainties soon dominate. In part this is because individual air parcels do not retain their identity indefinitley. As discussed above and shown in Figure 4, compact parcels are stretched into thin filaments by the deformation inherent in synoptic scale flow. As the stretching continues the concentration gradient at the edge of the hypothetical air parcel increases until small-scale mixing processes set in. Subsequently the parcel loses its identity. Thus, while the trajectory techniques can be extended for very long times the results

can only be used in a qualitative way. The mineral aerosol transport
to the North Pacific subtropics provides a simple example of this
qualitative limitation. Asia is the source of the dust there in
spring; trajectories back from the site show this (Merrill, Bleck and
Avila, 1985). However, most of the mineral aerosol flux from Asia is
confined to the region beneath the westerlies. Being "connected" to
the source doesn't mean you are in the "main plume". A minute
fraction of the aeolian dust flux from Asia may reach the equator:
even this minute fraction is the dominant source in such a "clean"
area.

The trajectory methods provide only qualitative information on
transport. Even though the calculations may be extended to account
for diffusion and removal, the Lagrangian techniques are generally not
suited to quantitative evaluation of fluxes. However, isobaric
trajectory calculations combined with simple source and deposition
parameterizations have been shown to predict concentrations of sulfur
dioxide and particulate sulfate in Europe; the results for
concentrations in rain were much less encouraging (Eliassen and
Saltbones, 1983). The latter is attributable to factors not
accurately predicted by the calculations; perhaps concentrations above
the boundary layer or subgridscale removal processes.

It has been assumed that Eulerian models, following the evolution
of fields of concentration, could not be used in source
identification. This is because even simple initial distributions of
substances result in complex fields of concentration after short
times, and the approach to a well-mixed state is often slow. The
deformation processes described above start this mixing at the largest
scales, and diffusion and ultimately molecular diffusion complete it.

In an Eulerian model one accounts for the evolution of the
concentration using (1), approximated by finite differences on a
three-dimensional grid. In the simplest case the field of motion is

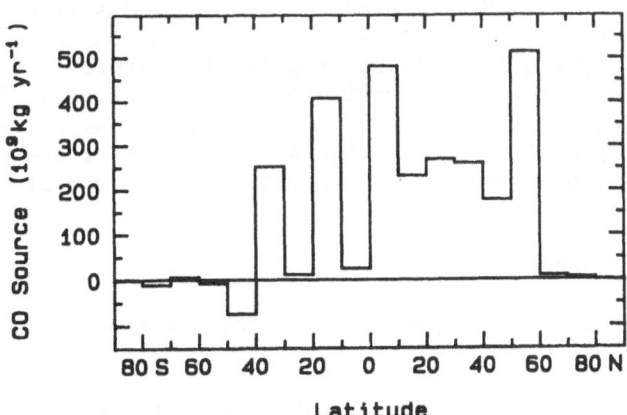

Figure 9. Net non-methane source strength distribution for CO.
Data from Hameed and Stewart (1979), based on an annual average, one
dimensional diffusion model.

determined separately in an earlier model run. That is, the u, v and w fields on the left hand side are archived, along with the related thermodynamic variables. This is possible if the tracers to be studied are passive, i.e. do not affect the motion in any way, no matter what the concentration (e.g., passive tracers are not radiatively active). At each time step the concentration in each grid volume is changed to account for the amount added or removed by advection, sources and sinks. Since these processes interact in a nonlinear way, it is usually impossible to discern which process or branch of the flow is responsible for which maximum, say, in the concentration.

Something can be learned using Eulerian models and an insightful trick developed by Julius Chang of the National Center for Atmospheric Research (Acid Deposition Modeling Project, 1985, Section 4.4). A high frequency oscillation modulates each source: the frequencies are kept far apart and the amplitude small. Simple Fourier analysis of the time series of concentration at a place of interest reveals a mixture of frequencies, with the amplitude of each frequency representing the strength of the corresponding source. The frequencies must be high enough for the signal to "pass through" the system response (see Wollast, this volume), and diffusion inexorably reduces the highest frequency components preferentially. Also, the sums and differences of the frequencies will appear in the results and the highest frequency must not exceed the reciprocal of the time step of the calculation, so only a few frequencies may be available. The technique can be applied sequentially to each of several critical substances in a model, perhaps including chemical reactions, as a means of determining which reactions are controlled by which substances.

Because source identification calculations deal with specific events they require the most of routine meteorological observations. In remote areas where such observations are sparse the analysis will be taxed to its limits. Remote areas may be generally free of local sources, and the distances to the important sources may be very large. Then the synergistic effect of multiple chemical tracers (e.g., substances having different temporal dependencies in their sources, or having different removal mechanisms) can be crucial to unambiguous interpretation. In continental areas there are usually enough sources to go around, and the problem becomes one of apportionment.

4.2 Mechanistic models

In these models one or two aspects of a transport problem are treated accurately while all other factors are parameterized simply. The role of the carefully-treated factors can be examined in this way.

For example, the latitudinal distribution of the source of CO was determined by Hameed and Stewart (1979) using a vertically and zonally averaged model of the troposphere. The formalism used is similar to the two box (hemispheric) system described by Czeplak and Junge (1974). The meridional transport was taken to be purely diffusive,

and the latitudinal distribution of diffusivity was based on the
annual average of the variance of the meridional wind. The continuity
equation (1) is thus reduced to a second order ordinary differential
equation. In this calculation Hameed and Stewart took the CO
concentration (which had been measured independently) as given. The
net source strength is the sum of the explicit loss terms, the
difference between photocehmical loss and production, and the net loss
due to transport. The results are shown in Figure 9. The
most-poleward positive peaks in each hemisphere are caused by
convergence of the transport flux, and transport contributes
significantly in the tropical area as well. The midlatitude sources
could be anthropogenic, but the tropical source must be natural.
While these results seem to support the idea that large natural,
non-methane, sources are needed to balance the global budget of CO,
the use of a more realistic tracer model leads to very different
conclusions regarding the role of transport (see the next section).

There are also two dimensional models, most of them time
dependent. Because inclusion of numerous chemical interactions with
time dependence in three dimensional models is infeasible in most
circumstances, it is often necessary to employ such simplified
models. Their usefulness in understanding observed distributions of
some substances is clear; this is particularly true for
photochemically active species. However, it is also clear that the
use of eddy diffusion in these models puts limits on their
applicability (Crutzen and Gidel, 1983). Examples (including clear
expositions on the equations and the meteorological data required) are
in Hyson, Fraser and Pearman (1980) and Gidel, Crutzen and Fishman
(1983). Since the two dimensions are usually height and latitude,
these models tell us little or nothing about zonal transport. This is
less important for long-lived species, but for substances with shorter
residence times zonal transport is crucial, particularly for most
continent-ocean combinations.

4.3 Tropospheric chemistry system models

These are numerical models of the concentration of substances using
the meteorological results of a global atmospheric general circulation
model. The idea is to specify a series of sources and sinks for the
substances of interest and to account for their interactions
accurately enough to follow the evolution of their distribution in a
self-consistent way. The emphasis here is again on the
transport-related aspects, as this part alone presents a sufficient
challenge that few research groups have attempted credible efforts
along these lines. There are at this time no models that meet the
criterion of self consistency; in all cases certain things must be
specified ab initio.

Let us first address the question of passive vs. active
substances. As discussed above, passive substances do not affect the
field of motion. This can be used as an adequate first approximation
in many circumstances: e.g. water, a manifestly active substance, can
be ignored or carried as a source for rain (but not of latent heat) in

short-term modeling of midlatitude storms. However, the latent heat
release does strengthen the circulation, and a better model would
carry the water explicitly. In smaller scale systems the water plays
an even more crucial role and must be accounted for in a prognostic
mode. Water vapor is usually carried in circulation models as an
active variable. Ozone does not affect the circulation in the
troposphere enough to be deemed active; however, it is dominant in
parts of the stratosphere where it absorbs solar ultraviolet light and
alters the temperature and wind distribution. Thus ozone can be
modeled as a passive tracer in the troposphere (Levy, Mahlman, Moxim
and Liu, 1985; Mahlman, Levy and Moxim, 1980).

So, to the extent that all relevant active substances are
included, the general circulation model might be expected to yield
winds and rain in agreement with observations (with only fundamental
constants specified). This expectation is not usually fulfilled.
There are many processes that are not adequately treated owing, in
part, to the large grid size: clouds, noted above, terrain-induced
effects and boundary layer-free troposphere exchange are examples.
However, several models now exist that agree adequately with
observations of the average circulation and seasonal variations.

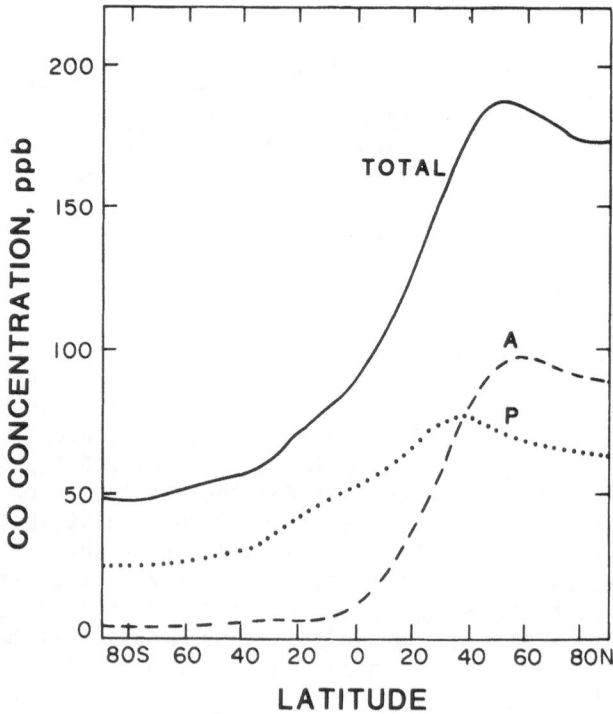

Figure 10. Total CO concentration vs. latitude and contributions
due to athropogenic (A, dashed line) and plant sources (P, dot-dash
line) in a three dimensional model. Adapted from Pinto et al. (1983).

Currently these have no dynamic interaction with a variable ocean; i.e. the sea surface temperature distribution is fixed in a seasonally varying cycle. Also, most models have severe approximations in the radiative heat transfer calculations. Usually the circulation model history is archived for a "final year" and (if they extend over more than one year) the tracer calculations suffer a noticeable discontinuity as the second year begins; although the second year is a repeat of the first it does not end exactly the way it began. Gradually more and more of these limitations are being removed. Then the substances that can be treated as passive variables can be included in a model run which takes the circulation (and thus the clouds, rain, sea surface temperature, gross aerosol distribution, etc.) as given. Initial distributions of sources are used to specifiy the starting conditions. The sources may be specified as areal and temporal distributions specific to the tracer; perhaps the source strength depends upon regional energy utilization or upon biomass burning. If a fully self-consistent treatment of the steady state behavior is desired, the sources are turned on and the transport model integrated until the transients decay. This can take a long while: long lived substances such as N_2O could require 1000 years of model integration to reach equilibrium. Reinitialization techniques can sometimes be used to reduce this to a feasible value, 6 years in the case of N_2O (Levy, Mahlman and Moxim, 1982). The distributions of interest are those calculated after all initializations are complete. Of course it is possible to study time series at a given station, average values and variations, height vs. latitude cross sections or any other type of statistic desired.

For example, Pinto et al. (1983) used the meteorological fields from a general circulation model of limited spatial resolution in a tracer simulation of the carbon monoxide cycle. Their results for the latitudinal distribution of the contribution of two source types in their most successful experiment are shown in Figure 10. Methane oxidation is not shown, as it is nearly uniform at 20-25 ppb. The "plant" source is the sum of non-methane hydrocarbon oxidation and vegetative burning (forest fires and slash and burn clearing of the land). While the strength of the natural non-methane source required in this calculation is not significantly different from that used in earlier studies, the transport pattern is entirely different. Although transport plays a role, the model simulations indicate that local sources and sinks dominate the latitudinal variation in the CO concentration. As can be seen in Figure 10, the non-methane natural source is imporant even at middle and high latitudes. Also, the anthropogenic CO does not have a significant interhemispheric flux. The more detailed treatment of transport in this model is an advance over the purely diffusive treatment of Hameed and Stewart (1979; see Figure 9, above). The absence of interhemispheric exchange seems consitent with the limited lifetime (2 months) of CO in the troposphere. However, the limited horizontal resolution of the parent general circulation model would be likely to reduce the role of transport in both the middle latitude and cross-equatorial areas, so further work is needed.

System models can be used to perform experiments in which source strengths are varied, or in which proposed measurement networks are simulated, for example. As more processes are simulated using techniques based on first principles the models become more powerful tools. It is clear that continuous interaction between those making observations and those using models is essential.

ACKNOWLEDGEMENTS

My work on atmospheric transport has been supported by the National Science Foundation, most recently by grants ATM 83-11694, OCE 84-05608 and OCE 85-00739. My colleagues have provided stimulating discussion and collaboration, particularly Dr. Rainer Bleck and Dr. Robert Duce. The participants at the ASI contributed significantly by asking probing questions. Dr. Mitsuo Uematsu provided the unpublished data in Figure 8. The computer facilities of the National Center for Atmospheric Research (also supported by NSF) were used in the calculations. The computer-drawn figures were prepared by Ms. Ruth Platner. It is a pleasure to thank all of these people for their help.

REFERENCES

Acid Deposition Modeling Project, 1985: 'The NCAR Eulerian regional acid deposition model'. Report ADMP 85-3. Available from the National Center for Atmospheric Research as TN-256+STR.
Baumgartner, A. and E. Reichel, 1975: The World Water Balance - Mean Annual Global, Continental and Maritime Precipitation, Evaporation and Runoff. Elsevier, Amsterdam, 179 pp.
Bodhaine, B. A., B. G. Mendoca, J. M. Miller and J. M. Harris, 1981: 'Seasonal variation in aerosols and atmospheric transmission at Mauna Loa Observatory'. Journal of Geophysical Research, 86, 7395-7398.
Bourke, W., R. Seaman, and K. Puri, 1985: 'Data assimilation'. In: Advances in Geophysics, 28B (S. Manabe, ed.), Academic Press, New York, 123-155.
Bryan, F. and A. H. Oort, 1984: 'Seasonal variation of the global water balance based on aerological data'. Journal of Geophysical Research, 89, 1,717-1,730.
Cass, R. G. and F. Shair, 1984: 'Sulfate accumulation in a sea breeze/land breeze circulation system'. Journal of Geophysical Research, 89, 1429-1438.
Chatfield, R. B. and P. J. Crutzen, 1984: 'Sulfur dioxide in remote oceanic air: cloud transport of reactive precursors'. Journal of Geophysical Research, 89, 7111-7132.
Christy, J.R. and K.E. Trenberth, 1985: 'Hemispheric interannual fluctuations in the distributions of atmospheric mass'. Journal of Geophysical Research, 90, 8053-8065.

Crutzen, P.J. and L.T. Gidel, 1983: 'A two-dimensional photochemical
    model of the atmosphere, 2: Tropospheric budgets of anthropogenic
    chlorocarbons, CO, CH_4, CH_3Cl and the effect of various NO_x
    sources on tropospheric ozone'. Journal of Geophysical Research,
    88, 6641-6661.
Czeplak, G. and C. Junge, 1974: 'Studies of interhemispheric exchange
    in the troposphere by a diffusion model'. In: Advances in
    Geophysics, 18B (F.N. Frenkiel and R.E. Munn, eds.), Academic
    Press, New York, 57-72.
Danielsen, E. F., 1974: 'Review of trajectory methods'. In: Advances
    in Geophysics, 18B (F.N. Frenkiel and R.E. Munn, eds.), Academic
    Press, New York, 73-94.
Danielsen, E.F., 1961: 'Trajectories: isobaric, isentropic and
    actual'. Journal of Meteorology, 18, 479-486.
Danielson, E.F. and R. Bleck, 1967: 'Research in four dimensional
    diagnosis of cyclonic storm cloud systems'. AFCRL Report
    67-0617. Available from the National Technical Information
    Service as AD 670 847.
Duce, R.A., C. K. Unni, B. J. Ray, J. M. Prospero and J. T. Merrill,
    1980: 'Long-range atmospheric transport of soil dust from Asia to
    the tropical North Pacific: Temporal variability'. Science, 209,
    1522-1524.
Eliassen, A. and J. Saltbones: 'Modelling of long-range transport of
    sulphur over Europe: a two-year model run and some model
    experiments'. Atmospheric Enviroment, 17, 1457-1473.
Galloway, J. N., D. M. Whelpdale and G. T. Wolff, 1984: 'The flux of
    S and N eastward from North America'. Atmospheric Enviornment,
    18, 2595-2607.
Gidel, L. T., 1983: 'Cumulus cloud transport of transient tracers'.
    Journal of Geophysical Research, 88, 6587-6594.
Gidel, L. T., P. J. Crutzen and J. Fishman, 1983: 'A two-dimensional
    photochemical model of the atmosphere, 1: Chlorocarbon emissions
    and their effect on stratospheric ozone'. Journal of Geohpysical
    Research, 88, 6622-6640.
Hameed, S. and R. W. Stewart, 1979: 'Latitudinal distribution of the
    sources of carbon monoxide in the troposphere'. Geophysical
    Research Letters, 6, 841-844.
Hasse, L., 1983: 'Introductory meteorology and fluid mechanics'. In:
    Air-Sea Exchange of Gases and Particles (P.S. Liss and W.G.N.
    Slinn, eds.), D. Reidel, Dodrecht, 1-51.
Heffter, J. L., 1981: 'Air Resources Laboratories atmopsheric
    transport and dispersion model'. NOAA Technical Memorandum ERL
    ARL-81.
Holton, J.R., 1979: An Introduction to Dynamic Meteorology. Academic
    Press, New York, 391 pp.
Hyson, P., P. J. Fraser and G. I. Pearman, 1980: 'A two-dimensional
    transport simulation model for trace atmospheric constituents'.
    Journal of Geophysical Research, 85, 4443-4455.
Kuo, Y.-H., M. Skumanich, P. L. Haagenson and J. S. Chang, 1985: 'The
    accuracy of trajectory models as revealed by observing system
    simulation experiments'. Monthly Weather Review, 113, 1852-1867.

Levy, H., II, J. D. Mahlman and W. J. Moxim, 1982: 'Tropospheric N$_2$O
    variability'. Journal of Geophysical Research, 87, 3061-3080.
Levy, H., II, J. D. Mahlman, W. J. Moxim and S. C. Liu, 1985:
    'Tropospheric ozone: The role of transport'. Journal of
    Geophysical Research, 90, 3753-3772.
Mahlman, J. D., H. Levy, II, and W. J. Moxim, 1980: 'Three dimensional
    tracer structure as simulated in two ozone precursor
    experiments'. Journal of the Atmospheric Sciences, 35, 1340-1374.
Merrill, J. T., R. Bleck and L. Avila, 1985: 'Modeling atmospheric
    transport to the Marshall Islands'. Journal of Geophysical
    Research, 12,927-12,936.
Merrill, J. T., R. Bleck and D. B. Boudra, 1986: 'Techniques of
    Lagrangian trajectory analysis in isentropic coordinates'.
    Monthly Weather Review, in press.
Oort, A.H., 1983: Global Atmospheric Circulation Statistics,
    1958-1973. NOAA Professional Paper 14, Washington, D.C.
    Available from U. S. Government Printing Office as 003-017-00510-1.
Palmen, E. and C. W. Newton, 1969: Atmospheric Circulation Systems,
    Their Structure and Physical Interpretation. Academic Press, New
    York, 603 pp.
Panofsky, H. A., 1985: 'The planetary boundary layer'. In: Advances
    in Geophysics, 28B (S. Manabe, ed.), Academic Press, New York,
    359-385.
Pasquill, F. and F. B. Smith, 1983: Atmospheric Diffusion, John Wiley
    and Sons, New York, 437 pp.
Parrington, J., W. Zoller, and N. K. Arras, 1983: 'Asian Dust:
    Seasonal transport to the Hawaiian Islands'. Science, 220,
    195-197.
Peixoto, J. P. and A. H. Oort, 1983: 'The atmospheric branch of the
    hydrological cycle and climate'. In: Variations in the Global
    Water Budget (A. Street-Perrott, M. Beran and R. Ratcliffe, eds.),
    D. Reidel, Dordrecht, 5-65.
Philander, S. G. and E. M. Rasmussen, 1985: 'The Southern Oscillation
    and El Nino'. In: Advances in Geophysics, 28A (S. Manabe, ed.),
    Academic Press, New York, 197-215.
Pinto, J. P., Y. L. Yung, D. Rind, G. L. Russell, J. A. Lerner, J. E.
    Hansen and S. Hameed, 1983: 'A general circulation model study of
    atmospheric carbon monoxide'. Journal of Geophysical Research,
    88, 3691-3702.
Rangararjan, C. and C. D. Eapen, 1981: 'Estimates of interhemispheric
    transport of radioactive debris by the East African low-level jet
    stream'. Journal of Geophysical Research, 86, 12153-12154.
Rasmussen, E. M. and T. H. Carpenter, 1982: 'Variations in tropical
    sea surface temperature and surface wind fields associated with
    the Southern Oscillation/El Nino'. Monthly Weather Review, 110,
    354-384.
Reiter, E. R., 1972: Atmospheric Transport Processes, Part 3:
    Hydrodynamic Tracers. Available from the National Technical
    Information Service as TID 25731.

Reiter, E. R., 1978: Atmospheric Transport Processes, Part 4:
    Radioactive Tracers. Available from the National Technical
    Information Service as TID 27114.
Sanders, F., 1955: An investigation of the structure and dynamics of
    an intense surface frontal zone. Journal of Meteorology, 12,
    542-552.
Shapiro, M. A., 1984: 'Meteorological tower measurements of a surface
    cold front'. Monthly Weather Review, 112, 1634-1639.
Slinn, W. G. N., 1983: 'Air-to-sea transfer of particles'. In:
    Air-Sea Exchange of Gases and Particles (P. S. Liss and W. G. N.
    Slinn, eds.), D. Reidel, Dordrecht, 299-405.
Starr, V. P., 1968: The Physics of Negative Viscosity Phenomena.
    McGraw-Hill, New York. 256 pp.
Whelpdale, D. M., T. B. Low and R. J. Kolomeychuck, 1984: 'Advection
    climatology for the east coast of North America'. Atmospheric
    Environment, 18, 1311-1327.
Wallace, J. M. and P. V. Hobbs, 1977: Atmospheric Science, An
    Introductory Survey. Academic Press, New York, 467 pp.
Weiss, W., A. Sittkus, H. Stockburger and H. Sartorius, 1983:
    'Large-scale atmospheric mixing derived from meridional profiles
    of Krypton 85'. Journal of Geophysical Research, 88, 8574-8578.
Wyngaard, J. C., 1983: 'Lectures on the planetary boundary layer'.
    In: Mesoscale Meteorology - Theories, Observations, and Models (D.
    K. Lilly and T. Gal-Chen, eds.), D. Reidel, Dordrecht, 603-650.

MODELING OCEANIC TRANSPORT OF DISSOLVED CONSTITUENTS

J. L. Sarmiento
Geophysical Fluid Dynamics Program
Princeton University
Princeton, New Jersey 08542

1. INTRODUCTION

The development of geochemical cycling models for the oceans requires a consideration of the physical as well as chemical and biological processes affecting the constituent of interest. The subject of this chapter is how one deals with the physical processes. The approach is to focus on the process of building ocean models by going through a hierarchy of models, from simple to complex; and thinking, along the way, through a series of examples, about the assumptions that are being made and how these assumptions determine the results that one obtains.

It is difficult to overemphasize the importance of beginning ones excursions into model development with some understanding of the underlying physical processes involved. The pitfalls of ignoring those processes are illustrated with a number of examples of where this leads to serious misconceptions and obviously unrealistic results. It is not the main objective of this chapter to discuss the physics, however, which shall have to be obtained from other sources, including any of a large number of standard textbooks.

Modelers often draw a useful distinction between different types of modeling approaches. The distinction is obvious, but is usually couched in jargon that obscures the meaning. Let us suppose we have an object moving according to the familiar equation $x = vt + x^{\circ}$, where x is the position of the object at time t, v is velocity (assumed constant with time), t is time, and x° is position at t = 0. Given the parameters of the model, velocity and x°, we can readily predict the position of our object at any time t. This is a _prognostic_ model. Often, however, we do not know what the parameters of our model are, nor even if the form of the equation we choose to work with is correct. If we have a large number of measurements of the position x as a function of time, we can use _diagnostic_ techniques, or a diagnostic model (in this case probably a least-squares fitting technique) to estimate what the slope (velocity) and intercept (x°) are.

Notice that we have used one and the same equation for both the prognostic and diagnostic approaches. The only real difference is that in one case, the prognostic, we have enough constraints that we can actually use our equation to tell us what we might expect to find if we went out and made measurements. In the other case, the diagnostic,

P. Buat-Ménard (ed.), The Role of Air-Sea Exchange in Geochemical Cycling, 65–82.

we presume a certain ignorance about our model parameters (we can
even assume ignorance about the form of the equation) and go out and
make the measurements and see what constraints those can put on our
model. If we are lucky, we can then turn around and use the results
obtained from the diagnostic model to predict what will happen in a
new situation.

I shall show below that the equations necessary to solve for
oceanic flow and temperature and salinity distributions, given
appropriate boundary conditions at the air–sea interface and ocean
floor–sea interface, are known. The difficulty is that we are unable to
solve them and must therefore always make approximations in order to
develop a model. The critical lesson to learn is to think very carefully
about the approximations, and to use as much as possible all the data
available to us to examine the reasonableness of our models. Most of
us are interested in developing a predictive ability, but the best way
to do this, generally, is to do some diagnostic work first.

Geochemists are accustomed to working with the conservation
equation to construct box models of the type discussed by Wollast in
this volume. The integral form of the conservation equation, discussed
in the following section, helps one to understand very clearly the
types of assumptions that are being made in constructing such models.

2. BOX MODELS

The basic building block for geochemical box models of the oceans is
the integral form of the conservation equation for a substance C:

$$\int (\frac{\partial C}{\partial t} - SMS)dVol = \oint (-VC + K \cdot \nabla C)dS \tag{1}$$

where SMS are all chemical and biological sources minus sinks, V is
velocity, K is the diffusion tensor; the left hand side (lhs) is a volume
integral, the right hand side (rhs) an integral around the entire
surface enclosing the volume of ocean under consideration.

Consider a simple two-box model of the world ocean with a surface
box 100 meters thick representing the upper ocean, and a deep box
3900 meters thick representing the deep ocean (Figure 1). The lhs of
the integral form of the conservation equation would give, for the deep
ocean, the time rate of change of tracer C in the deep ocean, summed
to the total tracer sources minus utilization or other sinks such as
radioactive decay. The rhs integral is difficult to perform because we
do not have the measurements needed. We would need to know the
vertical velocity and mixing, and the concentration and its vertical
gradient at every point along the interface between the upper and
lower boxes. We must therefore make some assumptions about how to
deal with these terms of the equation.

What would normally be done in such a situation is to define the
exchange in terms of the average concentration in each box:

$$\oint (-VC + K \cdot \nabla C) dS = \nu'(C_s - C_d) A + K' \frac{(C_s - C_d)}{\Delta z} A \tag{2}$$

where A is the area of the interface. Mixing and advection are often combined to arrive at the typical box model equation:

$$\frac{\partial C_d}{\partial t} = \nu(C_s - C_d) + SMS \tag{3}$$

where $\nu = (\nu' + K'/\Delta z) A/Vol$. A common misconception is that a box model assumes that the concentration in a given box is uniform. This is clearly not necessary.... we have made the assumption here that the interface concentration is equal to the average interior concentration, but we could as easily make any other assumption, or actually use measurements of the interface concentrations if they were available.

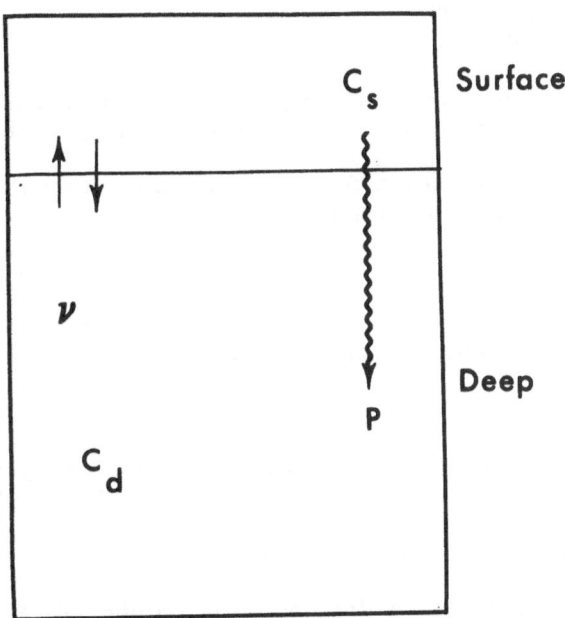

Figure 1. A two box model of the oceans. ν represents physical exchange between the surface and deep box, C_s and C_d represent average surface and deep concentration, and P represents particulate rain of phosphate in organic matter.

Notice that the surface integral in equation (2) is actually the integral of the velocity weighted tracer concentration, and the diffusivity weighted gradient of the tracer along that surface. In other words, the properties of the surface in the regions where the velocity and diffusivity are greatest are the ones that are most important in determining the exchange across the surface. For example, deep water formation regions, with their high sinking velocities, should receive stronger weight than other regions. The failure to do this can lead to difficulties as shall be shown below.

The above treatment illustrates the major problems that must be dealt with in geochemical model building: (1) what are the SMS terms, and (2) how does one deal with exchange across the boundaries of the boxes? Assumptions and simplifications must always be made, but it is important to be aware of them.

This model has been used with considerable success for obtaining valuable insights into the cycling of chemicals within the oceans (e.g., Broecker and Peng, 1982). Let us carry our example one step further, however, by considering a tracer for which the model does not work: the oxygen balance. If we assume a steady state distribution, we obtain for oxygen the equation:

$$0 = \nu \, (O_{2_s} - O_{2_d}) - rP \tag{4}$$

P is the particle rain of phosphate in organic matter, which is metabolized in the deep ocean, and r is the ratio O_2/PO_4 in the organic matter.

From the conservation equation for PO_4, one can obtain $P = -\nu$ $(PO_{4s} - PO_{4d})$. Substituting into (4) gives:

$$O_{2_d} = O_{2_s} + r \, (PO_4 - PO_{4_d}) \tag{5}$$

from which one can find, using global average values for the oxygen and phosphate:

$$O_{2_d} = 240 \ \frac{\mu mol}{kg} + 171 \ (0 - 2.2 \ \mu M/kg) = -136 \ \mu M/kg$$

The observed O_{2_d} is 166 $\mu M/kg$. What went wrong?

The answer is very simple: we did not do a proper job of setting up our boundary conditions. Most of the water sinking into the deep ocean comes from high latitudes, which have two properties very different from the global average properties we used in the above model: (1) they have much colder temperatures and therefore much higher oxygen concentrations; (2) most importantly, a high fraction of the nutrients are not used up by organisms. Nutrients that are used by organisms get exported into the deep ocean in reduced form, thus creating an oxygen demand. High latitude regions send down lots of oxygen without the oxygen demand that would accompany this process if high latitude organisms were more efficient.

How can we improve our model to take the high latitude processes into consideration? A sensible approach is a 3-box model of the oceans which includes a high latitude surface box (see figure 2). It is not difficult to follow an argument similar to the above and show that:

$$O_{2_d} = O_{2_h} + r\ (PO_{4_h} - PO_{4_d}) \tag{6}$$

$$\sim 340\ \mu M/kg + 171\ (1.2 - 2.2)\ \mu M/kg$$

$$= 169\ \mu M/kg,$$

a far more satisfactory result.

The three-box model has another interesting feature not possessed by the two-box model: it allows for changes in atmospheric pCO_2 simply by changing the preformed nutrient level, PO_{4_h}, which in turn, depends on the rate of nutrient supply by circulation, relative to the rate of nutrient removal by high latitude productivity, P_h. One can readily see how this occurs by substituting total carbon and alkalinity, on

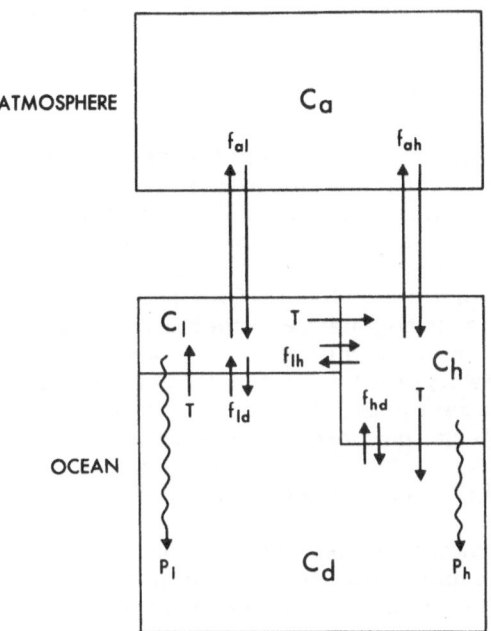

Figure 2. A three-box model of the oceans similar to the two box model in Figure 1, only the surface ocean has now been separated into a high latitude and low latitude component indicated by the subscripts h and ℓ, respectively. f is exchange between various boxes as indicated by the subscripts. $f_{\ell d}$ and $f_{\ell h}$ are assumed to be negligible. T represents thermohaline overturning. Taken from Sarmiento and Toggweiler (1984).

which the pCO_2 depends, in place of oxygen in equation (6). The consequences of this for the ice ages, during which time lower atmospheric CO_2 levels existed (e.g., Neftel, et al., 1982), have been fully explored by Sarmiento and Toggweiler, 1984; Siegenthaler and Wenk, 1984; Knox and McElroy, 1984; Toggweiler and Sarmiento, 1985; Wenk and Siegenthaler, 1985; and Ennever and McElroy, 1985.

In the two- and three-box models discussed above we bypassed the problem of having to know anything about the actual rate of ocean circulation or particle rain because ν, f and T, and P cancelled out of our equations. How can we actually find something out about these terms?

Consider first the two-box model. Equation 4 is a simple linear equation in two unknowns, ν and P. If we know the oxygen distribution and the Redfield ratio, we can solve for the ratio of ν to P (ignoring for the moment that we have already shown this model does not work for oxygen):

$$\nu/P = 171 \; / \; (74 \; \mu mol/kg) = 2.3 \; kg/\mu mol$$

We can also use PO_4 as well as many other tracers to obtain a solution, e.g.:

$$\nu/P = -1 \; / \; (-2.2 \; \mu mol/kg) = 0.45 \; kg/\mu mol$$

(The difference between the oxygen and phosphate calculations arises from the fact that our model is inconsistent with the observations, as shown above.) We can keep adding tracers, and then average all the estimates together to get a best value. This is essentially a diagnostic model.

Notice, however, that we are only able to find the ratio of ν to P. Unfortunately, our equation is homogeneous, which means that no matter how many tracers we have measurements of, we can only solve for this ratio. We need a new type of tracer with a behavior that does not lead to a homogeneous equation to allow us to separate ν from P. Radiocarbon, with its decay term, can provide the constraint we need:

$$0 = \nu \; (C\text{-}14_s - C\text{-}14_d) + rP - \lambda C\text{-}14_d \qquad (7)$$

We divide through by P, substitute in our values for ν/P and C-14, then solve for λ/P. Since we know the decay constant λ, we can obtain P, and then ν. Broecker and Peng (1982) give solutions for the case where total carbon and radiocarbon are used.

The addition of more boxes to simple exchange models such as the above complicates matters greatly because of the increased number of unknowns. Toggweiler and Sarmiento (1985) show how one can obtain solutions for the three-box model using atmospheric CO_2, oxygen, phosphate, and radiocarbon as constraints. They have to ignore some of the exchange terms in the model in order to obtain a unique solution. Broecker and Li (1970) have obtained an interesting solution for a model which separates the Pacific, Antarctic, and Atlantic, from each other; but the paper might serve as well as an illustration of the convolutions one must go through in order to obtain a solution for models with more than 3 boxes. The difficulty with adding more boxes using the basic box modeling approach discussed above, is that one quickly adds many more unknowns than constraints.

How can one progress? Clearly the models discussed above are not adequate to represent most oceanic phenomena we are interested in. The basic problem we have is one of placing additional constraints on the physical exchange processes. One way of doing this is simply to try and obtain direct measurements of the physical exchange processes. This option is usually impractical, however. Another approach, discussed in the following section, is to assume that the exchanges between different boxes are related to each other in some way, thus reducing the number of unknowns. There are ways of doing this that are not inconsistent with what we think we know about the ocean. A third approach, discussed in part 4, is to add further constraints to our model by using the equations of motion.

3. ADVECTION-DIFFUSION MODELS

One of the problems of considerable importance to chemical oceanographers is to correctly predict the uptake of fossil fuel CO_2 by the oceans. Given the limitations of our knowledge of the ocean circulation and mixing, the approach that has generally been taken is to carefully calibrate simple models such as those discussed above, with a variety of radiogenic, cosmogenic, and anthropogenic tracers that provide information about the exchange of the surface ocean with the deeper ocean. Broecker and Peng (1982) give an excellent discussion of the types of tracers one can use and of some of the modeling approaches that one can take.

Keeling (1973) showed very near the beginning of this effort that models with only a few boxes were unable to predict tracers with different time scales without significant adjustments to the model parameters in going from one tracer to another. Oeschger, et al (1975), showed for a two-box model that this problem arose from the fact that the large deep reservoir dilutes tracers entering it instantaneously, thus creating a large gradient between it and the smaller reservoir above, a gradient which is not very sensitive to how long the tracer has been in the ocean. In the real ocean, the process of "filling in" the deeper layers from the upper ocean is very complex, but as a generalization, one can say that it is much more gradual, with upper ocean layers interacting with small layers below them. A better way of simulating this process is to have a vertical stack of a large number of layers, but if one attempts to develop a model of this stack based on the principles developed in section 2, one ends out with a severely underdetermined problem.

Oeschger, et al.'s solution to this dilemna was to use the horizontal integral of the conservation equation. If the equation is integrated over the whole ocean basin, the horizontal advection and diffusion terms drop out, as does the vertical advection (there is no net vertical advection in the ocean as a whole). Dividing through by area gives:

$$\frac{\partial C}{\partial t} = \frac{\partial}{\partial z} \left(K \frac{\partial C}{\partial z} \right) + SMS \qquad\qquad (8)$$

where C is now the horizontal average. If we assume that K is constant everywhere, this one dimensional "diffusion" model has only one unknown. Oeschger, et al., showed that it does a much better job of predicting the various tracers than the two-box model does.

A one dimensional advection-diffusion model similar to the above has also been used for studying a variety of local phenomena, such as the oxygen minimum, and the temperature, salinity, and radiocarbon distribution in the Pacific (e.g., Wyrtki, 1962; Munk, 1966, and later Craig, 1969). The equation for these locally applied one dimensional advection-diffusion models is obtained by assuming that the horizontal contribution of mixing and advection are negligible. It has the same form as equation (8) except for including an additional term: $-w \, \partial C/\partial z$ on the rhs, where w is vertical advection.

The one dimensional diffusion and advection-diffusion models have had a particulary strong attraction for geochemists because of their simplicity. Properly used, they can provide considerable insight, as the papers referred to and many others have shown. However, as the variety and quantity of tracer data have increased, and as geochemists working with these tracers have become more sophisticated in their understanding of oceanic processes, it has become clear that these models have important drawbacks.

For example, Siegenthaler (1983) used improved data sets to show that the Oeschger, et al., model in fact gives significantly different predictions of fossil fuel CO_2 uptake, depending on the tracer that is used to calibrate the model. Sarmiento (1983b) used a tritium simulation with a three-dimensional primitive equation model such as the one discussed in section 4 to estimate the vertical diffusivity (K) that would be needed to fit the results with a model of the type given in equation (8). This calculation showed that the vertical diffusivity, far from being constant with depth, varied over several orders of magnitude as a function of depth. Tritium being a transient tracer, the diffusivity at a given depth also varied over as much as several orders of magnitude as a function of time. The reasons for these variations are that different processes are acting on the tracer at different times and depths, including surface convection, Ekman pumping, mixing and advection along surfaces of constant potential density, deep water formation events, etc. This also explains why the different tracers in the Siegenthaler study give different results.

A particularly interesting example of how the penetration of a tracer varies with time on a seasonal time scale is given by a unique data set of ^{90}Sr and ^{139}Cs measurements that was obtained by V. Bowen at several Weathership Stations in the North Atlantic during the period of maximum input of these bomb produced isotopes to the oceans. Figure 3 shows the surface concentrations measured at Weathership Station C. There is a large seasonal variation in the ^{90}Sr concentration in the surface ocean which is greatest during the time of maximum entry of the tracer, but then decreases with time. Also, the

⁹⁰Sr concentration in the oceans increases rapidly to its peak in ˜1965, then decreases gradually to a value which remains relatively constant for the remainder of the record after ˜1967.

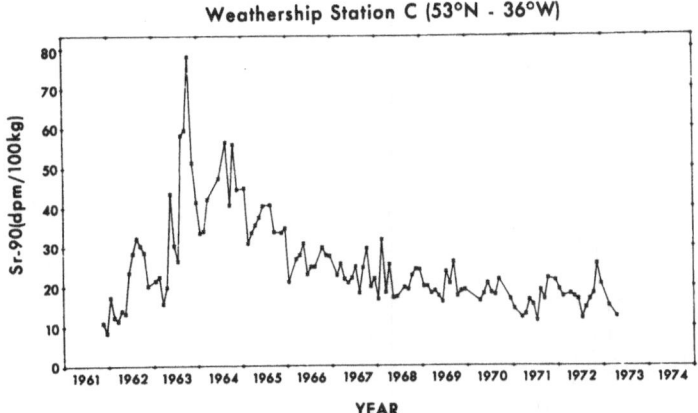

Figure 3. Surface ocean concentrations of ^{90}Sr made at Weathership Station C in the North Atlantic by V. Bowen's laboratory at Woods Hole Oceanographic Institution. Note the large seasonal fluctuations during the period of maximum input in the early 1960's, and the nearly flat distribution after ˜1967.

What is the mechanism for the seasonal variation in the surface concentrations? A first thought is that it might be due to the seasonality in the flux from the atmosphere. It is well known that the flux of bomb produced tracers from the stratosphere into the troposphere is maximum in the springtime. Figure 4 shows the concentration of ⁹⁰Sr in rainfall at 51°N, which gives some idea of how the fallout to the oceans has varied with time. The springtime maximum is clearly evident in these data. Once in the troposphere, the isotope is scavenged on a time scale of ˜one month. Another possible explanation is that the seasonal variation of the surface concentration is due to the seasonal variation in the thickness of the mixed layer. The relative importance of the seasonal signal in the flux and the seasonal variations in mixed layer thickness can be investigated by a simple one dimensional advection-diffusion model which has been modified to include variations in the thickness of the mixed layer with time.

First, however, we need an input function of Sr-90. Gwinn and Sarmiento (1984) and Sarmiento and Gwinn (in press), have devised a fallout model based on observations of ^{90}Sr concentrations in air and

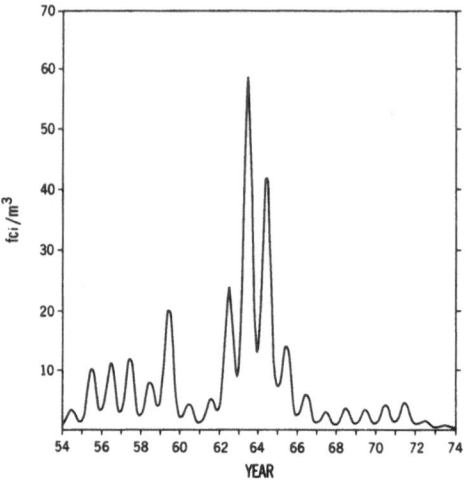

Figure 4   Concentration of ^{90}Sr in rainfall at 51°N from Gwinn and Sarmiento (1984) and Sarmiento and Gwinn (in press). Note the large peak which occurs in the springtime.

in rainfall. The air concentration at any given latitude is predicted from the equation:

$$C (t,\phi) = \bar{C}_{ref}(t) \ R \ (\phi) \ [1 + g(m,\phi)]$$

where $\bar{C}_{ref}$ is the deseasonalized trend at a reference latitude (51°N), R is the ratio of the observations at latitude ϕ to the reference latitude, and $(1 + g)$ represents the seasonal trend in the air concentration, and m is the month. This air concentration is then used in the equation:

$$F (\lambda,\phi,t) = C (t,\phi) \ [\nu_d \ (\phi) + \nu_w \ (\lambda,\phi,t)]$$

$$\nu_w \ (\lambda,\phi,t) = a \ (\phi) \ [P(\lambda,\phi,t) \ / \ P_o]^{b(\phi)}$$

to predict the fallout. λ is longitude, ν_d is the dry removal velocity, ν_w is the wet removal velocity, P is rainfall with $P_o = 1$ cm/month, and a and b are constants determined along with ν_d by fitting land based observations. Figure 5 shows the wet removal velocity as a function of latitude. The variations with latitude of the scavenging velocities are probably related to differences in the way in which clouds are formed in different regions. For comparison, ν_d has an average value of 0.17 ± 0.02 cm s^{-1}.

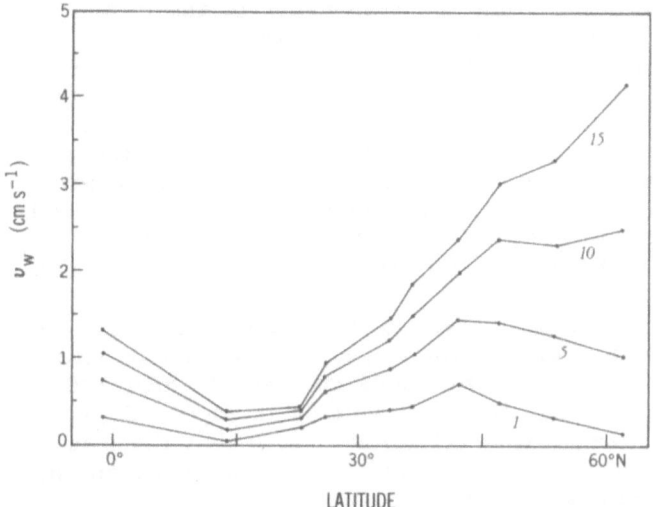

Figure 5 Wet deposition velocity, ν_W, as a function of latitude and rainfall in cm/month, which is shown next to each curve. Taken from Sarmiento and Gwinn (in press). The dry deposition velocity, ν_d, is 0.17 ± 0.02 cm s^{-1}.

Figure 6 gives the results of various simulations of the surface ^{90}Sr concentration for a location typical of Weathership Station C. It

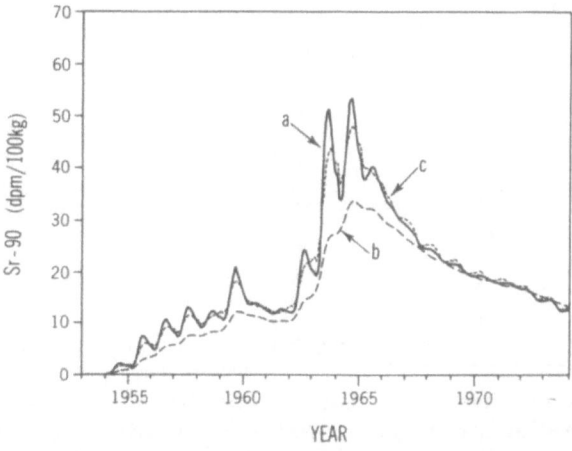

Figure 6 Simulations of surface ocean ^{90}Sr concentrations with a one dimensional advection-diffusion model. Curve "a" includes seasonal effects in input and mixed layer variations. Curve "c", almost identical to "a", has the seasonality in the input removed. Curve "b" has the seasonality in the mixed layer removed. The seasonality in the mixed layer thickness is the dominant effect in predicting the surface concentration.

is clear from this simple study that the seasonality of the input
function plays a minor role in the seasonality of the surface
concentration. The main factor is the change in the thickness of the
mixed layer. It is also clear that it would not be possible to simulate
the true surface concentration during the times of maximum fallout with
a model that had a constant mixed layer thickness. This statement
would apply to any tracer which has an annual input to the ocean
which is large relative to the amount of tracer present in the mixed
layer at any given time. Notice that with ^{90}Sr the seasonality
decreases dramatically with time as the amount of ^{90}Sr in the ocean
increases and the input goes down.

The one dimensional ^{90}Sr model with a time varying mixed layer
thickness gives useful insights into the tracer dynamics, but notice
that in the model the surface concentrations continue to drop off
exponentially with time, whereas the observations in Figure 3 remain
constant after 1967. A likely explanation for this is that there is
significant lateral input of ^{90}Sr to the surface, a process which the
one dimensional model cannot reproduce.

Another drawback of the one dimensional model is that it leaves out
high latitude deep water formation processes. This does not seem to
have a very strong effect on the transient tracer simulations, but it
does make it impossible to properly simulate the nutrient and oxygen
distributions for reasons similar to those discussed in the previous
sections. This can be readily demonstrated. Figure 7 shows a one
dimensional advection-diffusion pipe model which allows water to be
removed at the surface and put in at the bottom. This model has been

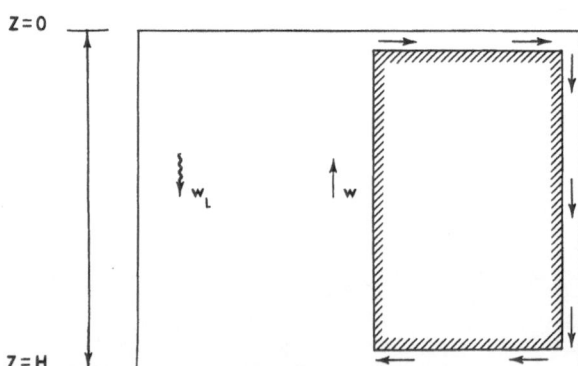

Figure 7. Schematic of a one dimensional advection-diffusion pipe model
from Sarmiento and Toggweiler (in press). Water rises within the main
body of the ocean as the velocity w, and is carried directly from the
surface to the bottom of the ocean by the pipe. w_L represents sinking
of particles.

used to predict the oceanic PO$_4$ distribution using a simple model of
total removal of PO$_4$ from the surface by organisms, and a regeneration
of this PO$_4$ which decreases exponentially with increasing depth

(Sarmiento and Toggweiler, in press). Sediment trap observations have suggested that the e-folding length scale of regeneration is of the order of 300 meters. The equation governing this model is:

$$PO_4 = PO_{4_s} + \frac{(\overline{PO}_4 - PO_{4_s})\, H\, (e^{wz/K} - e^{-az})}{[\frac{K}{w}(e^{wH/K}-1) + \frac{1}{a}(e^{-aH}-1)]}$$

$$\overline{PO_4} = \left[\int_0^H PO_4\ dz \right] / H$$

where H is the depth of the ocean and 1/a is a length scale of regeneration of phosphate by biological processes.

Figure 8 shows several calculations based on the model described above. The critical parameters are the regeneration length scale, 1/a, and the length scale K/w. The regeneration length scale controls

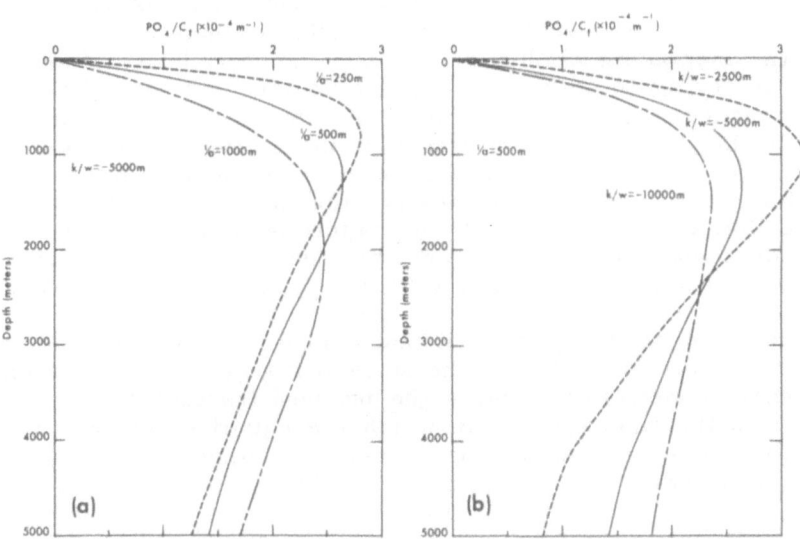

Figure 8. Simulations of oceanic PO_4 concentrations normalized by C_t = $\int_0^H PO_4$ dz. These simulations were done with the one dimensional advection-diffusion pipe model shown in Figure 7. (a) shows how the depth of the nutrient maximum depends on the regeneration length scale 1/a. (b) shows how the difference between the nutrient concentrations at the nutrient maximum and ocean floor depend on the advection-diffusion length scale K/w.

mainly the depth of the nutrient maximum. One can obtain reasonable
depths for the nutrient maximum with regeneration length scales which
are close to the value of 300 meters which the sediment trap data
suggest. The K/w length scale controls mainly the difference between
the concentration of PO_4 at the maximum, and the concentration at the
bottom of the ocean. The work of Munk (1966) and Craig (1969)
suggest that values of the order of 1000 meters for this length scale
are reasonable. However, the PO_4 observations, which show that the
near bottom values are fairly close to the nutrient maximum values,
require a K/w of the order of 10000 m. The difficulty is that the
waters entering the bottom of the ocean in this model are coming from
nutrient depleted surface ocean, whereas, in the ocean they come from
high latitude nutrient rich waters, as discussed in the previous section.

Similarly, Jenkins (1980) was able to show that a one dimensional
advection-diffusion model applied locally to tritium/helium-3
observations in the Sargasso Sea is unable to explain the observations,
since most of the exchange with the upper ocean is along sloping
surfaces of constant potential density, rather than vertically with the
waters directly above. He used a very simple box modeling approach,
with boxes oriented along the sloping surfaces of constant potential
density to model the observations. Sarmiento (1983a) used a similar
approach with a more extensive data set to elucidate some of the
physical mechanisms of the surface to thermocline exchange.

Clearly, the increasing quality of the data and complexity of the
problems one is trying to address with them require more sophisticated
modeling approaches. The terms in the conservation equation that we
have been ignoring, namely lateral exchange and advection processes,
need to be included in our models. Geochemists are increasingly
turning to two- and three-dimensional advection-diffusion models with
variable advection and diffusion and with outcrops from the deep ocean
to high latitude regions to study various problems of exchange within
the oceans (e.g., Broecker and Peng, 1982; Ku et al., 1980; Siegenthaler,
1983; Martin, et al., 1985).

The introduction of all these additional degrees of freedom lead to
the same problem encountered with the many box model: how does one
provide sufficient constraints to make the problem tractable? This
question leads to the final section, in which the equations of motion
and inverse diagnostic techniques for dealing with underdetermined
systems are discussed.

4. EQUATIONS OF MOTION

The simplest form of the equations of motion which has been used to
obtain velocity estimates from hydrographic observations is the
geostrophic approximation, first introduced by Sandström and
Helland-Hansen (1903). This technique only gives relative velocities at
a given location, i.e., the vertical velocity shear; and one must
therefore specify the velocity at some level in order to obtain absolute
velocities. Wunsch (1978) has shown that the geostrophic method, even
when coupled with mass conservation, has essentially an infinite

number of possible solutions for any given data set, but always within a given range that can be reduced with additional constraints (Wunsch, 1984). Inverse or diagnostic techniques have been developed by Wunsch and others to formalize the procedure for obtaining velocity fields by the geostrophic technique coupled with conservation constraints for mass and other tracers such as radiocarbon (e.g., Wunsch and Minster, 1982). The velocity fields obtained in this manner, or by the more traditional techniques of subjectively choosing a level where the motion would be expected to be near 0, can in principle be used in geochemical models (e.g., Thiele et al.,   in press).

Other techniques for obtaining velocity fields from hydrographic measurements without the ambiguity of the reference level problem have also been introduced recently (e.g., the β-spiral technique, Stommel and Schott, 1977). These techniques involve assumptions that do not seem always to be justified by the observations, but there does seem to have been some success in obtaining reasonable velocity fields in the carefully studied β-spiral region (Armi and Stommel, 1983).

One disadvantage of the geostrophic technique from the point of view of modeling chemical distributions is that it gives velocity fields consistent with the water column density structure at the time the measurements were made. For most geochemical modeling problems we are interested in a time averaged flow which may be quite different from the instantaneous flow estimated from the density structure, due to the great variability that characterizes most regions of the ocean. In recognition of this problem oceanographers are turning more and more to the use of chemical tracers, whose distribution reflects the average of processes occurring over longer time scales, to obtain average flow fields (e.g., Wunsch and Minster, 1982).

The geostrophic method makes use of only a small part of the equations of motion. In principle, one can obtain a full solution to the ocean circulation by solving a set of seven equations in seven unknowns subject to a specification of the boundary conditions of heat exchange, precipitation, evaporation, and wind forcing at the surface, and friction at the ocean floor. This is the technique employed in the primitive equation model developed by Bryan (1969). The seven equations are the equations of motion for two horizontal directions, the hydrostatic equation, continuity, the equation of state, and the conservation of heat and salt. The seven unknowns are: three advection components, pressure, density, temperature, and salinity.

Primitive equation models are solved by finite difference techniques, and give simulations that still differ in many respects from the observed ocean circulation due to (1) the fact that we do not know the boundary conditions to very great accuracy; (2) we are unable, due to limited computational resources, to develop models with high enough resolution to simulate important features of the ocean circulation such as mesoscale eddies; and (3) we must approximate all phenomena occurring on spatial scales smaller than the resolution of our model by some arbitrary specification of mixing which is not well known. Nevertheless, these models have been used to study the dispersion of tracers within the ocean with some success (e.g., Holland, 1971, and Sarmiento, 1983a), and plans exist for further studies along these lines.

5. CONCLUSION

The models of choice for most geochemists will probably continue to be box models and advection-diffusion models. This chapter has attempted to give a brief discussion of alternative approaches that will likely come to dominate the field in longer run, but the focus has been primarily on these two. The most important points I have attempted to make are (1) what it is that we are doing when we set up a box model, particularly regarding the boundary conditions; (2) how our knowledge of important oceanographic phenomena, such as the process of deep water formation and the tendency for advection and diffusion to move along surfaces of constant potential density, need to be taken into careful consideration in the construction of models; and (3) how we can use data and other contraints to make our models less arbitrary.

ACKNOWLEDGMENTS

The author wishes to thank Marty Jackson and Johann Callan for assistance in the preparation of the manuscript. Support was provided by DOE/Martin Marietta Grant #19X 27405C.

REFERENCES

Armi, L., and H. Stommel, 'Four views of a portion of the North Atlantic subtropiccal gyre', *J. Phys. Oceanogr.*, 13, 828-857, 1983.

Broecker, W. and Y.-H. Li, 'Interchange of water between the major oceans', *J. Geophys. Res.*, 75, 3545-3552, 1970.

Broecker, W.S., and T.-H. Peng, *Tracers in the Sea*, Lamont-Doherty Geological Observatory, Palisades, NY, 690pp., 1982.

Bryan, K., 'A numerical method for the study of the circulation of the world ocean', *J. Comput. Phys.*, 4, 343-376, 1969.

Craig, H., 'Abyssal carbon and radiocarbon in the Pacific', *J. Geophys. Res.*, 74, 5491-5506, 1969.

Ennever, F.K. and M.B. McElroy, 'Change in atmospheric CO_2: Factors regulating the glacial to interglacial transition', in *The Carbon Cycle and Atmospheric CO_2: Natural Variation Archean to Present*, *Geophys. Monogr. Ser.*, 32, E.T. Sundquist and W.S. Broecker, eds., pp 154-162, AGU, Washington, DC, 1985.

Gwinn, E. and J.L. Sarmiento, 'A model for predicting strontium-90 fallout in the northern hemisphere (1954-1974)', *Ocean Tracers Laboratory, Technical Report #3*, Department of Geological and Geophysical Sciences, Princeton University, Princeton, NJ., 1984.

Holland, W.R., 'Ocean tracer distributions. Part I. A preliminary numerical experiment', *Tellus*, 23, 371-392, 1971.

Jenkins, W.J., 'Tritium and He-3 in the Sargasso Sea', *J. Mar. Res.*, 38, 533-569, 1980.

Ku, T.L., C.A. Huh, and P.S. Chen, 'Meridional distribution of [226]Ra in the eastern Pacific along GEOSECS cruise tracks', *Earth Planet. Sci. Lett.*, **49**, 293-308, 1980.

Keeling, C.D., 'The carbon dioxide cycle: Reservoir models to depict the exchange of atmospheric carbon dioxide with the oceans and land plants', in *Chemistry of the Lower Atmosphere*, S.I. Rasool, ed., Plenum Press, New York, pp. 251-329, 1973.

Knox, F., and M.B. McElroy, 'Changes in atmospheric CO_2: Influence of the marine biota at high latitudes', *J. Geophys. Res.*, **89**, 1405-1427, 1984.

Martin, J.H., G.A. Knauer and W.W. Broenkow, 'VERTEX: the lateral transport of manganese in the northeast Pacific', *Deep-Sea Res.*, **32**, 1405-1427, 1985.

Munk, W., 'Abyssal recipes', *Deep-Sea Res.*, **13**, 707-730, 1966.

Neftel, A., H. Oeschger, J. Swander, B. Stauffer, and F. Zumbrunn, 'Ice core sample measurements give atmosphere CO_2 content during the past 40,000 years', *Nature, Lond.*, **295**, 220-223, 1982.

Oeschger, H., V. Siegenthaler, U. Schlotterer, and A. Gugelman, 'A box diffusion model to study the carbon dioxide exchange in nature', *Tellus*, 2, 168-192, 1975.

Sandström, J.W., and B. Helland-Hansen, 'Ueber die Berechnung von Meeresströmungen', *Report on Norwegian Fishery- and Marine-Investigations 2 (1902)*, **4**, 43pp., 1903.

Sarmiento, J.L., 'A tritium box model of the North Atlantic thermocline', *J. Phys. Oceanogr.*, **13**, 1269-1274, 1983a.

Sarmiento, J.L., 'A simulation of bomb tritium entry into the Atlantic ocean', *J. Phys. Oceanogr.*, **13**, 1924-1939, 1983b.

Sarmiento, J.L. and E. Gwinn, '[90]Sr Fallout Prediction', *J. Geophys. Res.*, in press, 1986.

Sarmiento, J.L., and J.R. Toggweiler, 'A new model for the role of the oceans in determining atmospheric P_{CO_2}', *Nature*, **308**, 621-624, 1984.

Sarmiento, J.L., and J.R. Toggweiler, A preliminary model of the role of upper ocean chemical dynamics in determining oceanic O_2 and atmospheric CO_2 levels, in *Proceedings, NATO ARI on Dynamic Processes in the Chemistry of the Upper Ocean, NATO Conference Series IV: Marine Sciences*, Plenum Press, ed. J.D. Burton, in press, 1986.

Siegenthaler, U., 'Uptake of excess CO_2 by an outcrop-diffusion model of the ocean', *J. Geophys. Res.*, **88**, 3599-3608, 1983.

Siegenthaler, U., and T. Wenk, 'Rapid atmospheric CO_2 variations and ocean circulation', *Nature*, **308**, 624-626, 1984.

Stommel, H., and F. Schott, 'The Beta spiral and the determination of the absolute velocity field from hydrographic station data', *Deep-Sea Res.*, **24**, 325-329, 1977.

Thiele, G., W. Roether, P. Schlosser and R. Kuntz, 'Baroclinic flow and transient-tracer fields in the Canary-Cape-Verde basin', *J. Phys. Oceanogr.*, in press, 1986.

Toggweiler, J.R., and J.L. Sarmiento, Glacial to interglacial changes in atmospheric carbon dioxide. The critical role of ocean surface water in high latitudes, in *The Carbon Cycle and Atmospheric CO₂: Natural Variations Archean to Present, Geophys. Monogr. Ser.*, 32, E.T. Sundquist and W.S. Broecker, eds., Chapman Conference, AGU, Washington, D.C., pp. 163-184, 1985.

Wenk, T., and U. Siegenthaler, The high-latitude ocean as a control of atmopsheric CO₂: in *The Carbon Cycle and Atmospheric CO₂: Natural Variations Archean to Present, Geophys. Monogr. Ser.*, 32, E.T. E.T. Sundquist and W.S. Broecker, eds., Chapman Conference, AGU, Washington, D.C., pp. 185-194, 1985.

Wunsch, C., 'The North Atlantic general circulation west of 50°N determined by inverse methods', *Rev. Geophys. Space Phys.*, 16, 583-620, 1978.

Wunsch, C., 'An estimate of the upwelling rate in the equatorial Atlantic based on the distribution of bomb radiocarbon and quasi-geostrophic dynamics', *J. Geophys. Res.*, 89, 7971-7978, 1984.

Wunsch, C., and J.-F. Minster, 'Methods for box models and ocean circulation tracers: Mathematical programing and nonlinear inverse theory', *J. Geophys. Res.*, 87, 5647-5662, 1982.

Wyrtki, K., 'The oxygen minima in relation to ocean circulation', *Deep-Sea Res.*, 9, 11-23, 1962.

VERTICAL TRANSPORT OF PARTICLES WITHIN THE OCEAN

Jean Claude Brun-Cottan

Laboratoire de Physique et Chimie Marines
Université Pierre et Marie Curie, Tour 24
4 place Jussieu, 75230 Paris Cedex 05, France

1 INTRODUCTION

The geochemical cycle of major elements in the Ocean, strongly depends on their behaviour in the particulate phase. For example, this is the case for the carbon cycle and for the CO_2 exchange between the atmosphere and the upper layer of the sea water, as these phenomena are related to the exit, from the surface water, of organic particulate matter and calcium carbonate particles. The horizontal transport of the suspended matter (SPM) is tied to the motion of the water masses. The vertical transport of the SPM, however, mainly depends on the gravity and vertical mixing effects. We intend here to describe these effects and how to use them to compute some useful physical and geochemical parameters. The vertical transport of the particles in the atmosphere, and their particle size distribution, follow in most cases the same basic laws which apply to marine particles (Slinn, 1983).

The settling of the suspended matter in the Ocean depends to a large extent on the physical parameters which determine the dynamic properties such as the sedimentation flux, the residence time in the water masses or the exchange capacity between the particles and the surrounding water. These physical parameters concerning the suspended matter are mainly the particle size distribution and the particle density.

This paper shows the way in which these physical parameters can be determined and how they can be used to model the dynamical properties of the suspended matter.

Early models describing the particle size distribution function (PSD) for marine suspended matter, used the power law function, also known as the Jungian function (Junge, 1963; Bader, 1970; Brun-Cottan, 1971, Carder et al., 1971; Harris, 1972; Sheldon et al, 1972):

$$\frac{dN}{Ndl} = \frac{1}{l_o} \left| \frac{1}{l_o} \right|^{-b} \tag{1}$$

83

P. Buat-Ménard (ed.), The Role of Air-Sea Exchange in Geochemical Cycling, 83–111.
© *1986 by D. Reidel Publishing Company.*

where l is the equivalent diameter of the particle. This means that l is
the diameter of a spherical particle having the same volume as the
considered one, whatever its shape. l_0 is a reference size; in this
paper l_0=1cm. N is the number of particles per unit volume having a size
greater than l_A and b is the parameter determining the curvature of the
function.

Knowing the density ρ_P of the particles and considering l_A and l_B
as the respective sizes of the smallest and the largest particles taken
into account, then the mass concentration of the particles whose sizes
lie between l_A and l_B is:

$$M = \int_{l_A}^{l_B} \rho_P N(\pi/6) l_0^{b-1} l^{-b+3} \, dl \tag{2}$$

It is obvious in (2) that
for b < 4: $(l_B \to \infty) \to (M \to \infty)$
and   for b > 4: $(l_A \to 0) \to (M \to \infty)$

Equation 1 has a real meaning only if the limits l_A and l_B are
strictly defined. The values, such as the mean size $l_N = \int (dN/N) dl$ or
the mean mass $\underline{m}$ = M/N vary very rapidly if these limits are changed.
This effect is dramatic with the change of l_A. These mean values, which
are the most important parameters, have no signification using the power
law. In addition the total concentration, in terms of number of particles or total mass, cannot be extrapolated outside of the given limits
l_A and l_B. Moreover the mathematical properties of the power law function cannot be applied to random processes, such as the ones which
generate disaggregation or coagulation of the suspended matter. The
power law function is often used because it is very easy to compute the
parameter b, but the above constraints restrict its numerical application, particularly for modelling the dynamical properties of the suspended particulate matter.

During recent years, the threshold detection has decreased with the
use of the electron microscope (Lambert et al.,1981) to reach 0.1 μm,
and it is now clear that below 1 μm, the particle concentration values
decreases with the particle size. Here, the all PSD values can no longer
be described by a power law function, but are well described by a
lognormal distribution function $\Lambda(L, \sigma^2)$:

$$d\Lambda = \frac{dN}{N} = \frac{1}{\sqrt{2\pi} \, l\sigma} \exp\left| -\frac{1}{2\sigma^2}(\ln l - L)^2 \right| dl \tag{3}$$

or in the case of shallow waters by the sum of several lognormal distributions (Brun-Cottan,1967,1975,1977; Lambert et al.,1981). The central
coefficient is L and the dispersion or quadratic coefficient is σ. The
lognormal distribution signify that the logarithm of the particle size
follows the normal law $\psi(L, \sigma^2)$:

$$d\psi = \frac{dN}{N} = \frac{1}{\sqrt{2\pi} \, \sigma} \exp\left| -\frac{1}{2\sigma^2}(\ln l - L)^2 \right| d\ln(l) \tag{4}$$

When we consider the normal law of the logarithm lnl of the sizes

of the particles, L is the mode (the most probable value) and also the
mean and the median value of the total number of particles. When we
consider the lognormal distribution of the linear particle sizes, L is
the logarithm of the median size l_N, of the total number of particles.
Because we are mainly interested in the mean, as the natural representative of a statistical population, we also often use the mean size $\underline{l}$. The
median value lies between the mode and the mean values; roughly twice as
close to the mean as to the mode (Calot, 1964). For the distribution of
the number of suspended particles, these three central parameters of the
lognormal law are:

Mode $= \exp(L-\sigma^2)$ (5)

Median $l_N = \exp(L)$ (6)

Mean $\underline{l}_N = \exp(L+(1/2)\sigma^2)$ (7)

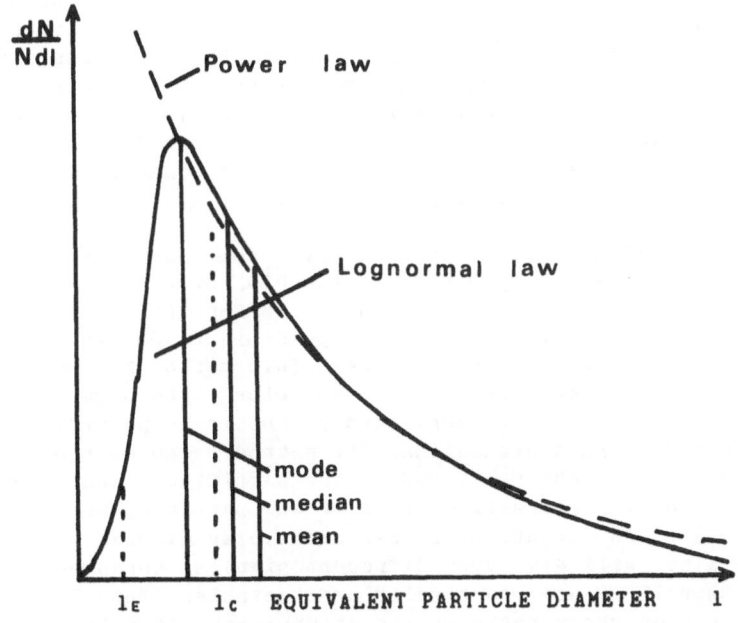

Figure 1. Representation, in linear coordinates, of the differential lognormal and power law functions of the sizes of a
given population of suspended particles. One can see, when the
size 1→0, that the lognormal function tends to 0 when the
power law tends to ∞. l_E is the electron microscope threshold
and l_C, the Coulter counter one.

The mathematical properties of the lognormal law imply that the
size distribution is due to many effects, each one having small action

and these actions being multiplied together (Calot,1964).   Additive effects of these actions would be a characteristic of the normal law.   The
values of l$_N$ obtained from the electron microscope usually lie between
0.6 and 2 µm, and are mainly arround 1 µm.

All the particles of which the size is greater than the mode size
are as well described by the lognormal distribution as by the power law
(Figure 1).

The following chapters describe the way in which the coefficients
of the best lognormal PSD are determined and how these two coefficients
are employed in modelling some geochemical properties of the SPM.

2 DETERMINATION OF THE LOGNORMAL COEFFICIENTS L, σ AND OF N

2.1 Particle size data collection

Many methods and technologies are used to determine the PSD of the
SPM. Some use the global effects of the whole population of particles,
such as the measurement of the scattering or of the attenuation of the
light (Jerlov, 1968; Morel, 1973). However these methods are not useful
for marine studies, because they are not sensitive enough to detect the
relative variations of the number of particles in the size range of the
order of several micrometers. The useful methods are those able to
detect the particles one by one and three are commonly used: the microscope, the Hyiac and the Coulter counter.

The technology based on microscopy, specially with electron beam,
allows one to obtain very precisely the PSD, with a low threshold value,
of 1 µm for the optical method and 0.1 µm for the electron one.   Moreover, with electron microscopy it is possible to obtain the chemical and
mineral composition of each particle. This method has some major disad-
vantages; it is very expensive, very slow, the organic matter is not
well detected and it is very hard to obtain a particle count large
enough to obtain good statistics. The method based on the measure of the
intensity of the light scattered by the particles, one by one, is very
sensitive to the refractive index of the particles in the sea water
(Morel,1973). An organic particle and a clay mineral particle, having
the same size, will give very different signals. Moreover, the particles
larger than 15 µm are practically undetectable. The Hyac technology,
which is set on the measure of the attenuation of a thin light beam due
to individual particles, is also sensitive to the refractive index, but
it is, nevertheless, usable for particles larger than 10 µm if it is
assumed that those particles are mainly organic in the open Ocean and
then have a quasi constant refractive index. The Coulter counter technology (Parson and Sheldon, 1967; Bader, 1970; Brun-Cottan, 1971) is based
on the measurement of an electrical pulse due to the passing of a
particle through a probe. The height strength of the pulse is proportional to the volume of the found particle and thus access directly the
total volume of the SPM, whatever the uncertainties introduced when

computing the equivalent diameter l. These pulses can be stored in the
memories of a multichannel analyser which allows one to obtain very
quickly a large count of particles and then to produce good statistics.
This system gives no information on the chemical or mineral composition
of the particles and requires great care to obtain accurate data. The
concentration in number of particles must be much below the maximum
values supplied by the manufactories, because they compute the coinci-
dence probability, using narrow size distributions, of two particles
being in the probe at the same time. This is not the case in nature.

2.2 Calculation of the coefficients L, σ and N from data

In order to compute the characteristic lognormal coefficients, and
to apply them in geochemical studies, some specific properties of the
lognormal law are needed. The main one is that the multiplication of a
lognormal distribution by a power law gives another lognormal distribution. These properties are described in the appendix (Paragraph 6.1).

It is then possible to compute the fraction of the total volume of
particles whose diameters lie between two given sizes. In place of the
concentration N of the number of particles (now N could be the concentration of the particles of which the size lie between 0 to ∞), one can
consider the differential concentration in volume of the particles
having size l. This concentration is $dV = dN(\pi/6)l^3$. Applying (81), one
obtains:

$$V/N = al_N^3 \exp([1/2]b^2\sigma^2) = \underline{v} \tag{8}$$

where $\underline{v}$ is the volume arithmetic mean of the PSD, then:

$$dV/N\underline{v} = dV/V = d\wedge(L+(1/2)b\sigma^2, \sigma^2) \tag{9}$$

The mass flux concentration, or the total surface area concentration of the SPM, can be obtained by the same way.

The data usually given by the Coulter technology or by microscope
allow one to obtain the number ΔN_i of particles whose sizes lies between
l_{i-1} and l_i.

L and σ are determined by the best fit, in terms of least squares,
between $\wedge$ and the experimental data. The range of the L values generally
found in sea water corresponds to mean diameters between 0.6 and 2 μm.
For the same samples, the corresponding median values of L' of the
volume variable ($\Delta V = \Delta N(\pi/6)l^3$) lie between 3 and 10 μm. These last
values are best centered in the scale of the sizes usually given by data
(1 to 40 μm for the Coulter counter) in which case the boundary effects
during computing are minimized if we use the variable $v=(\pi/6)l^3$. The
best way is then to determine L' and σ' and to compute L or l_N and σ
following:

$$L = (L' - \ln\frac{\pi}{6}\sigma'^2)/3 \; ; \quad \sigma = \frac{\sigma'}{3} \; ; \quad N = \frac{6 V}{\pi l_N^3 \exp(9/2)\sigma^2} \tag{10}$$

The simplest way to determine L' and σ' is to consider the normal function $\psi(L', \sigma'^2)$ corresponding to the lognormal function $\wedge(L', \sigma'^2)$.

$$\frac{d\psi}{du} = \frac{dV}{Vdu} = \frac{1}{\sqrt{2\pi}\sigma'}exp-\frac{1}{2\sigma'^2}(u-L')^2 \qquad (11)$$

where $v = (\pi/6)l^3$ and $u = lnv$ \hfill (12)

The experimental points are defined by the pairs (u_i, e_i) and the computed points by the pairs $(u_i, f(u_i))$,

$$u_i = (lnv_i+lnv_{i-1})/2 \; ; \qquad \Delta u_i = lnv_i-lnv_{i-1} = ln\left|\frac{v_i}{v_{i-1}}\right| \qquad (13)$$

with $\qquad v_i = \frac{\pi}{6}l_i^3 \qquad$ and $\qquad e_i = \frac{\Delta V_i}{\Delta u_i} = \frac{\Delta N}{\Delta u_i}v_i \qquad$ (14)

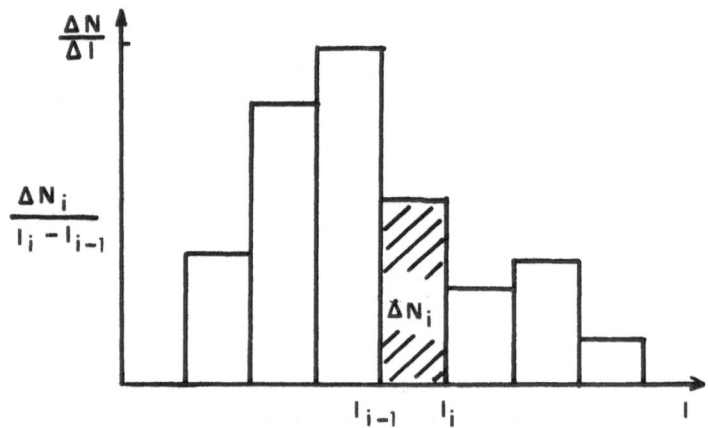

Figure 2. The particle sizes data are generally stored in number of particles per channel. Each channel is defined by a minimum size l_{i-1} or $ln(l_{i-1})$ and a maximum size l_i or $ln(l_i)$, depending on whether it is a linear or logarithm multichannel analyser. A logarithm multichannel analyser is the most convenient for marine research.

Some methods are avalaible to evaluate L', σ' and V, and we give those which look the most useful for the data given by the standard methods of measuring the PSD.

2.2.1 Narrow and well centered distribution

For data showing a median well centered on the u-axis and with small Δu_i, applying the basic properties of the normal distribution, we get:

$\qquad$ L' = $\Sigma e_i u_i / \Sigma e_i \qquad$ and with the König theorem: $\qquad$ (15)

$$\sigma'^2 = (\Sigma e_i u_i^2 / \Sigma e_i) - L'^2 \tag{16}$$

The proportion Pr_A of V corresponding to the volume of the particles smaller than the threshold value l_A is:

$$Pr_A = \int_{-\infty}^{u_A} d\psi_{(L', \sigma'^2)} \tag{17}$$

The proportion Pr_B of V corresponding to the volume of the particles greater than the maximum value l_B measured is:

$$Pr_B = \int_{u_B}^{\infty} d\psi_{(L', \sigma'^2)} \tag{18}$$

Pr_A and Pr_B are easily computed by numerical integration. Setting
$y_A = (u_A - L')/\sigma'$ and $y_B = (u_B - L')/\sigma'$, standard tables give the integration
of the normal error function $\psi_{(0,1|y)}$ and then:

$$Pr_A = \psi(0,1|y_A) \quad \text{and} \quad Pr_B = 1 - \psi(0,1|y_B) \tag{19}$$

With these values we have immediately:

$$V = \Sigma \Delta v_i . (1 - |Pr_A + Pr_B|) \tag{20}$$

This very simple method requires 50 to 100 size classes of data,
the considered sizes lying on a geometric or logarithmic progression.
The data must have a small dispersion around the theoretical curve.

2.2.2 Good statistical distribution

If each of the proportions Pr_A and Pr_B are estimated to be less
than 5 to 8%, one can assume that $V = \Sigma \Delta v_i$ and then use LogXProb paper
which then gives a graphic representation (Henri straight line) similar
to the straight line in Figure 5. Each pair $(l_i, (1/V)\Sigma \Delta v_i)$ is plotted
and the best straight line drawn through the points; the boundary points
are not plotted as Pr_A and Pr_B are neglected. The value on the lnl-axis
corresponding to the proportion 0.5 is $L' = \ln l_M$, σ' is the slope of this
line. This method was the one used by Lambert et al. (1981); it gives
good results with more than 10 size classes.

2.2.3 Large number of size classes

With a number of size classes greater than 50 and using a computer.
The coefficients of the best lognormal fit to the data can be found
using either the differential function $d\psi(0,1)$ or the normal function
$\psi_{(0,1)}$. The interest of using the differential function is that the lack
of knowledge of Pr_A and Pr_B does not affect the results, provided there
is only a small dispersion of the data arround the theoretical curve and
that the maximum of the function is clearly marked. The interest of
using the function $\psi_{(0,1)}$ is that the integration operation smooths the
dispersion of the points and does not need a good definition of the
maximum, but this requires that the proportions Pr_A and Pr_B are known,
or that each is less than 15% .

We have seen different possible methods of computing the 2 coefficients L and σ of the lognormal function describing the PSD and the
total concentration in volume and in number. Depending of the instrumen-
tation and computing possibilities, some of these methods are the more
useful. With the data from our counting system (Brun-Cottan, 1976), all
of these methods usually give results within a range of 5% for the mean
L and 10% for the dispersion σ. However, if the count is poor for the
largest sizes, which is generally the case, we prefer the last method,
it being the safest.

3 DETERMINATION OF SPM PHYSICAL PROPERTIES USING LOGNORMAL COEFFICIENTS

Total surface area, mass concentration (wet), mass flux, mean
settling velocity and residence time are the most important physical and
geochemical properties which can be determined, knowing the PSD and some
other physical properties of the suspended particles. For the time
dependant parameters, such as the flux or residence time, we assume in
this chapter that the system is at steady state.

3.1 Surface area concentration

The total surface area of the SPM is an important parameters for
the processes which involve surface absorption, such as the metallic,
stable or radioactive, elements. The elementary surface area concentration per unit volume of sea water of the particles whose the sizes lie
between l and l+dl, is:

$$dS = \pi l^2 dN \tag{21}$$

dS is a lognormal distribution multiplied by a power law function. Thus,
applying the specific lognormal property given by (81), the total
surface area concentration is:

$$S = N \pi l_N^2 \exp 2\sigma^2 \quad cm^2/cm^3 \tag{22}$$

If the total number N is unknown, but if the total mass concentration is measured and if an assumption can be made for the value of the
parameters L and σ, we can estimate the total surface area concentration
by computing N from (25).

Because of the importance of absorption/disorbption and dissolution
processes, which depend on surface properties, we must systematically
compute a value of the total surface area related to the mass concentration, when we handle PSD data. We propose here that this value must
be the ratio $\rho^2 S^3/M^2$, which is a dimensionless parameter.

3.2 Mass concentration

To compute the particle mass concentration, knowing the volume
concentration, we must know the absolute density of the particles. The

density taken into account here is the wet density, because it is this which has physical meaning for the Archimedeus force, and thus for the Stokes' parameter. The particle absolute density depends on the mineral and chemical composition, of the proportion of water in the structure and on the temperature. Except for some special cases where the studied particles are well known mineral or chemical species (clay minerals or calcium carbonates particles), the particle density varies, from particle to particle and with the particle size. Based on some studies on this subject, it appears that the particle density decreases when the size increases (Mac Cave, 1975). In the first general approximation, the particle absolute density ρ_p decreases following a power law which is:

$$\rho_p = \rho(1 + a(1/1_0)^{-\alpha}) \ gr/cm^3 \qquad (23)$$

where ρ is the absolute density of the sea water. The best estimates based on the values reported by Mac Cave are: $\alpha=0.5$ and $a=5.3 \times 10^{-3}$ for $1_0=1$ cm. In this article the results are given in C.G.S. units. It is obvious that the parameters of this law depend on the particular waters studied, but this approximation is certainly a better approach than taking a mean ρ_p value, often 1.5, for all the particles.

The elementary mass concentration of the particles, whose sizes lie between 1 and 1+d1 and whose density is given by (23), is:

$$dM = \rho\{1+a(1/1_0)^{-\alpha}\}(\pi/6)1^3 dN$$

$$\rho(\pi/6)1^3 dN + \rho(\pi/6)a1_0^{\alpha}1^{3-\alpha}dN \qquad (24)$$

dM is the sum of two lognormal distributions multiplied by a power law function. Thus, applying the specific lognormal property given by (81), the total mass concentration of the SPM is:

$$M = N\rho(\pi/6)1_N^3(B_1+a1_0^{\alpha}1_N^{-\alpha}B_2) \ gr/cm^3 \qquad (25)$$

with $\qquad B_1 = exp(9/2)\sigma^2 \qquad\qquad\qquad\qquad\qquad (26)$

and $\qquad B_2 = exp\{(3-\alpha)^2/2\}\sigma^2 \qquad\qquad\qquad (27)$

One can see that when $1\rightarrow0$, $\rho_p\rightarrow\infty$, but applying (79), we found that this only effects the particles which have a density greater than 2, which corresponds to the particles having a size less than 0.2 µm. The corresponding mass can be neglected with regard to the total mass concentration.

For the particles having a single ρ_p value, the mass concentration is:

$$M = N\rho_p(\pi/6)1_N^3 exp(9/2)\sigma^2 \qquad (28)$$

3.3 Vertical fluxes

3.3.1 Drag force and settling velocity

The drag force applied on a given particle depend on its velocity and thus on its size, its shape and relative density. For spherical particles having a diameter l and a velocity w, in laminar flow, Stokes showed that the drag force Fs is proportionnal to the velocity and is:

$$Fs = 3\pi\nu\rho_p lw \qquad (29)$$

When the gravity force is equal to the drag force, this gives the vertical Stokes velocity:

$$w = cl^2(\rho_p-\rho)/\rho \text{ cm/s} \qquad (30)$$

$c = g/(18\nu)$ is the Stokes drag coefficient
with g the gravity acceleration
and ν the kinematic viscosity coefficient.
The value of c, for sea water at 10°, is 3.6×10^3 cm^{-1}s^{-1}.

SETTLING VELOCITY in cm/s

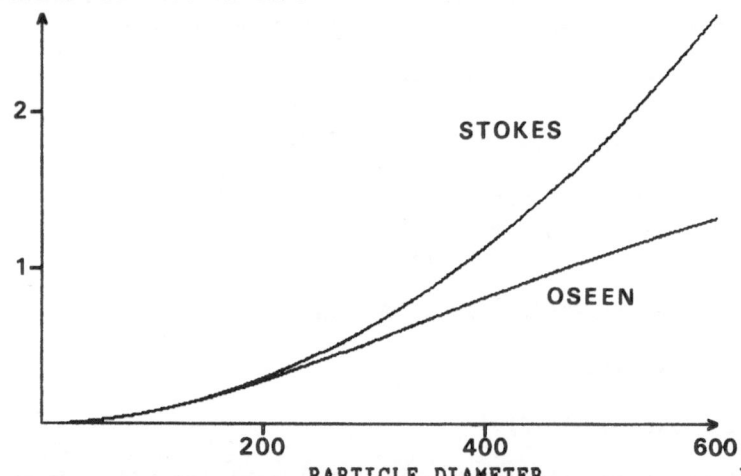

Figure 3. Stokes and Oseen settling velocity for spherical particles having a relative density of 0.2.

For the particles larger than a few micrometers and for particle density greater than 1.1, the flow arround the particle is not strictly laminar but pseudo laminar. In this case there is a quadratic term in the drag force equation, established by Oseen (Fortier, 1967; Brun-Cottan, 1975). The drag force becomes:

$$Fo = Fs(1+w(3l/8\nu)) \qquad (31)$$

This gives the general settling velocity for the SPM, including the fecal pellets and the large agglomerates:

$$w = \frac{8\nu}{3l}\left| -1 + \sqrt{1 + \frac{(\rho_p - \rho)\, l^3}{24\ \rho\ \nu^2 g}}\ \right| \tag{32}$$

Figure 3 gives, for a given density, the difference between the Stokes and the Oseen settling velocities.

3.3.2 Particle shape and settling velocity

The Stokes or Oseen settling velocity was established for spherical particles, but in nature particles can have any shape, except spherical. Generally, using optical diffusivity or Coulter technology, we do not have information about the particle shape and the reason for assuming sphericity is simplicity. In some particular cases, the particle shape is approximatly known. For example this is the case when studying clay minerals with an electron microscope (Lambert et al.,1980) and then it is possible to take into account the shape for computing the settling velocity. Several studies have been done on the subject, mainly theoritical (Lerman et al.,1974), but a few experiments have been performed at the length scale of marine particles (Boido,1957). From Boido experiments on small particles, some first order estimate of settling velocity can be found for simple particle shapes.

- Flat plate particle shape

This particle shape is often that of clay mineral particles. If the thickness "e" is small with respect to the mean square size l, the settling velocity is:

$$w = elg(\rho_p - \rho)/(7\nu\rho) \tag{33}$$

- Needle particle shape

This particle shape is often encountered in biological debris; these particles settle in a horizontal position. For a radius "r" small with respect to the length l, the settling velocity is:

$$w = \frac{gr^2(\rho_p - \rho)}{\rho\nu(0.4 + 5.4(r/l))} \tag{34}$$

3.3.3 Influence of the eddy diffusivity on the vertical SPM flux

In vertical one dimensional space, with a constant diffusivity coefficient Kz, the vertical flux of an SPM property having a concentration C carried by the particles having a settling velocity w, is:

$$\phi = Kz(dC/dz) - wC \tag{35}$$

In deep well stratified waters, the value of Kz usually lies between 1 to 100 cm^2/s, but it is generally very difficult to get an

estimate of the vertical particle concentration gradient. If this gradient exists, its value is generally below the background noise and the
random natural particle concentration variations. Thus it is difficult
to estimate the fraction of the vertical particle flux coming mainly
from the vertical settling velocity and the one comming mainly from the
vertical mixing. In fact, the a question is to know for which particle
sizes the main vertical transport process is sedimentation or mixing,
knowing that the largest particles settle with high velocity and thus
are not affected by turbulence.

We propose here a crude, but simple way to get an estimate of which
is the main transport process of particles in a given water column of
height h. During a time interval, we can estimate the square of the
settling depth of particles having a settling velocity w, divided by the
square of the length of a cloud. This cloud is defined by the volume
containing 1 σ'' of an initial quantity of particles released at the
center of the cloud at t=0. This length is $\sqrt{2tK_z}$. Applying this to the
water column of height h, we obtain a kind of settling coefficient C_s:

$$C_s = w. h/K_z \qquad\qquad (36)$$

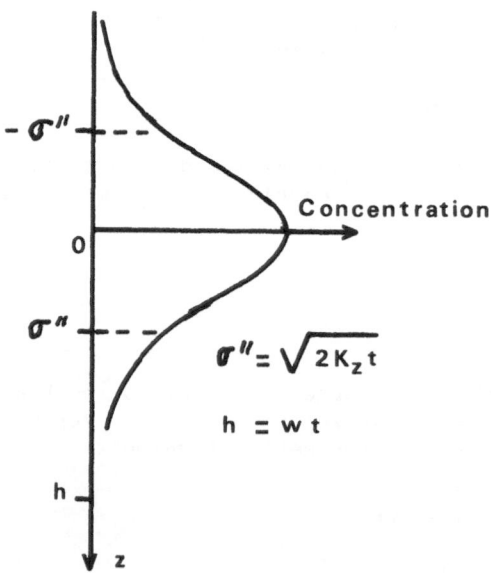

Figure 4. Representation of the settling height h, during a
given time t, compared with the length, during the same time,
of a cloud of particles containing σ'' (68%) of an initial
event.

If C_s is >>1, we can assume that the vertical transport is mainly
due to the Stokes or Oseen settling velocity, if C_s is <<1, we can

suggest that the vertical transport is mainly due to turbulent mixing. For example, with a water column thickness of 1000 m, a particle size of 10 μm having a relative density $\rho_p - \rho = 0.2$, for $K_z = 1$ cm^2/s, $C_s = 70$, which implies the dominant effect of the Stokes/Oseen vertical transport; for $K_z = 100$ cm^2/s, to get the same C_s value the particles must have a minimum size of 100 μm.

3.3.4 Vertical flux of the total particle surface area

From equation 21, given the elementary surface area concentration and applying the fact that the elementary surface area flux of the particles whose sizes lie between l and l+dl is: $d\phi_s = dS.w$, where $w = a(l/l_o)^{-\alpha} cl^2$ is the settling velocity, we have:

$$d\phi_s = a(l/l_o)^{-\alpha} c\pi l^2 dN \qquad (37)$$

One can see in equation 37 the major effect of the uncertainties of the particle density values. This effect is now proportional to the difference $\Delta\rho$ between the particle density and the water density, as it is proportional to ρ_p when computing the total mass concentration. Applying the specific lognormal property as it was done for the concentration properties, the vertical flux of the total particle surface area of the SPM is:

$$\phi_s = N\pi ac l_o^{\alpha} l_N^{4-\alpha} \exp\{(4-\alpha)^2/2\}\sigma^2 \quad cm^2/cm^2/s \qquad (38)$$

The mean settling velocity w_s of the SPM total surface area is:

$$w_s = ac l_o^{\alpha} l_N^{2-\alpha} \exp\{(6-\alpha)(2-\alpha)/2\}\sigma^2 \quad cm/s \qquad (39)$$

From the definition of the flux, w_s is the mean settling velocity of the SPM total surface area; this is available only for a steady state.

For the particles having a single ρ_p value, equations 38 and 39 simplify to give:

$$\phi_s = N\pi\{(\rho_p - \rho)/\rho\}cl_N^4 \exp 8\sigma^2 \qquad (40)$$

$$w_s = \{(\rho_p - \rho)/\rho\}cl_N^2 \exp 6\sigma^2 \qquad (41)$$

As in paragraph 3.1, if the total number N is unknown, but if the total mass concentration is measured and an assumption can be made for the parameters L and σ, we can compute N from (25), then S from (22) and estimates the total surface area flux, applying $\phi_s = S.w_s$.

3.3.5 Vertical flux of the total particle mass

The elementary mass flux of the particles whose sizes lie between l and l+dl is, applying the same procedure as for the surface area flux:

$$d\phi_M = \rho\{1+a(l/l_o)^{-\alpha}\}(\pi/6)l^3 a(l/l_o)^{-\alpha} cl^2 dN \qquad (42)$$

As for dM, $d\phi_M$ is the sum of 2 lognormal distributions multiplied by a power law function. From equation 24, given the elementary mass concentration and applying the fact that the elementary mass flux of the particles whose sizes lie between l and $l+dl$ is: $d\phi_M=dM.w$, where $w=a(l/l_0)^{-\alpha}cl^2$ remains the settling velocity, the total mass flux of the SPM is:

$$\phi_M = Np(\pi/6)acl_0{}^{\alpha}l_N{}^{5-\alpha}(B_3+al_0{}^{\alpha}l_N{}^{-\alpha}B_4) \ gr/cm^2/s \qquad (43)$$

with $$B_3 = \exp\{(5-\alpha)^2/2\}\sigma^2 \qquad (44)$$

and $$B_4 = \exp\{(5-2\alpha)^2/2\}\sigma^2 \qquad (45)$$

By definition, $\phi_M=M.w_M$, from (25) and (43), one obtains:

$$w_M = acl_0{}^{\alpha}l_N{}^{2-\alpha}\left|\frac{B_3+al_0{}^{\alpha}l_N{}^{-\alpha}B_4}{B_1+al_0{}^{\alpha}l_N{}^{-\alpha}B_2}\right| \qquad (46)$$

From the definition of the flux, w_M is the mean settling velocity of the SPM; this is true only for a steady state.

For the particles having a single ρ_p value, equations 43 and 46 simplify to give:

$$\phi_M = N\rho_p[(\rho_p-\rho)/\rho](\pi/6)cl_N{}^5\exp(25/2)\sigma^2 \qquad (47)$$

$$w_M = [(\rho_p-\rho)/\rho]cl_N{}^2\exp8\sigma^2 \qquad (48)$$

3.4 Residence time

One of the most popular parameters in marine geochemistry (Martin et al.,1985) is the residence time of chemical or mineral species or SPM in the water mass or in the water column.

The _true_ residence time of a single particle settling in a water column is simply the _total_ time this particle spends in this column. In place of a single particle, we can consider a kind of rain of identical particles produced at a constant rate by the surface waters . The residence time of each of these particle is identical and equal to that of the group of particles which represents the element of particle concentration, whatever the moment of observation chosen. If the production rate from the surface waters varies (Bishop et al., 1984), what we said above holds, except for the particles considered as a concentration element, since this concentration varies with time; for them the mean residence time loses its usual signification. Thus the mean residence time must be applied only in case of steady state and is given by:

$$\frac{\text{Concentration}}{\text{Flux}} \times h = \frac{\Sigma n_i c_i}{\Sigma n_i c_i w_i} \times h \qquad (49)$$

If one takes into account all the particles constituting the SPM, many logical and coherent definitions of their residence time in a water column of height h can be made. The one which appears the most conve-

nient for geochemical studies , if w if known using the PSD, is the
inverse of w multiplied by h, or the ratio of the concentration versus
the vertical flux multiplied by h. For instance for the mass parameter,
the mass concentration could be known by filtration and the vertical
flux by sediment-trap experiments. We want to point out the fact that
the residence time depends on the considered property taking. For a
given SPM, applying (49) with the inverse of (39) and (48), the residence time of the total surface of the SPM is greater than that of the
total mass. This might, at first, appear surprising, but it must be
remembered that this residence time is not tied to a given lot of
particles, but to sources of different size classes of particles. These
different residence times, must be taken into account when modelling the
scavenging effect of settling particles (Bacon, 1985), because it is a
surface area process and the measurements concern the mass concentration
of elements.

3.5 Application to the open sea

The physical properties defined above may be computed and applied
to the natural environment. Usually the particle size data comes from
Coulter-counter technology and Figure 5. shows how the results may be
arranged.

In the entire world Open Ocean, the values obtained for L are in
the range -9.4 to -8, which corresponds to l_N between 0.8 to 3 µm and σ
values lying between 0.4 to 1.2. Typical values are: L=-9.2, which
corresponds to a median diameter of 1 µm, σ=0.6, and the Stokes drag
coefficient $c=3.6X10^3$ $cm^{-1}s^{-1}$. For those who do not have a system measuring the particle size distribution, or want to compute global estimates, these typical values can be used. With these values and those of
a and α given in paragraph 3.2, the mean mass settling velocity of the
SPM given by (48) is $w=1.4X10^{-4}cm/s=43$ m/year. This implies for a standard mass concentration of 10 µg/l, a global wet vertical flux of
$1.4 \ 10^{-12}gr/cm^2/s=43$ $mg/cm^2/1000year$. Obviously these values are very
small with respect to the known rate of sedimentation and this implies
that the major part of the floor deposition comes from other sources
than the deep water SPM (Riley, 1963; Brun-Cottan, 1976; Mac Cave, 1975;
Honjo, 1978; Wiebe et al.,1976; Turner, 1977 Silver et al.,1978; Bishop
et al. 1984). If one of these other sources is the sink of large particles, agglomerates or fecal pellets, this means that these particles are
not included in the standard background SPM. This is quite understanda-
ble, since they have such a large settling velocity that they do not
have time, during their fall, to strongly interact with the background
SPM. The low velocity of 43 m/y shows that the major part of the SPM
does not sediment. Its transport is mainly driven by advection. Applying
(82) and (6)in this example shows that the mass flux is supported by the
particles having a mean size of 50 µm (the mass concentration being
supported by the particles having a mean size of 10 µm). This implies an
important removal of the SPM by the particles coming from the surface
waters.

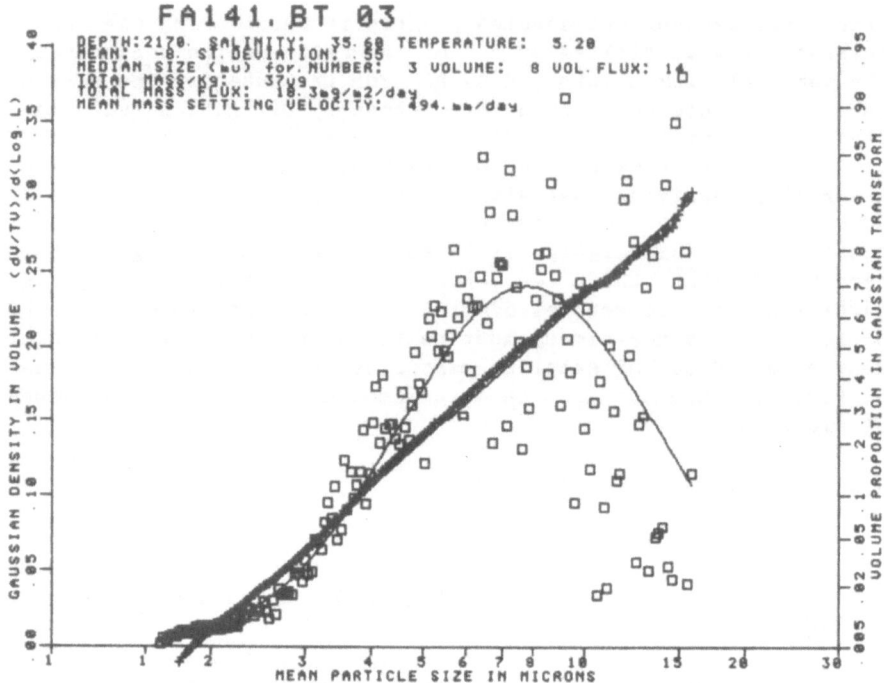

Figure 5.   Results computed from a sample of the Fluxatlante
cruise (may 1985).   The square represent the data coming from
the multichannel analyser, the + markers represent the integration of these values.   The curved line is the best normal
associated law fitting the data, the straight line is the
Henri representation of the lognormal corresponding function.
As usual the integration operation smooths the noise comming
from the low count of the largest particles.   The program
estimates that 10% of the fraction of the total volume is not
measured (right vertical axis).   The major particle parameters
described in this article are shown.   Here the median size
shifts from 3 µm for number concentration to 8 µm for the
volume flux (close to the one of the mass flux).   The program
is avalaible in Fortran 5, from the author (the graphic plot
above is done by a microcomputer using GSX system).

     The uncertainties on the particle density do not introduce a large
error in the estimates of the mass concentration M, because we know that
in marine water, the wet density usually is comprised between 1 and 2.
Then, the uncertainty on computing M does not excess 50 to 100%. Unfortunately, when computing the vertical mass flux, the density uncertainties apply two times, on the mass determination and on the settling
velocity. The effect on the settling velocity is proportional to Δ, then
much more important than the effect on the mass computation. We can say
that the error on the vertical mass flux value can exceed 2 or 3 times

the theoritical estimation.

In the coastal zone, or in the euphotic zone during the planktonic bloom, the PSD often shows more than one population. Only If these populations are statistically well separated, the properties defined above can apply to each of them.

An interesting choice of known mineral species is the clay particles, these particles are almost the only source of particulate aluminium in sea water, moreover their total mass concentration can be easily deduced from the direct measurement of total aluminium. From detailed electron microscope studies of clay particles in the Atlantic and Pacific oceans (Lambert et al., 1981), it appears that the PSD of these particles follows a very good lognormal law with a mean diameter of 1.7 μm, $\sigma=0.55$ and a density $\rho_P=1.6$ g/cm^3 (wet absolute density). Applying (48), one obtains a mean settling velocity of $w_M=7\times10^{-4}$ cm/s=200 m/y.

4 SPM SEDIMENTATION WITH DISSOLUTION PROCESS

4.1 Sedimentation at steady state

An important part of the SPM , mainly the organic matter and the calcium carbonates, is subject to the dissolution process during its fall inside the water masses. A principal parameter used to model geochemical cycles is the kinetic dissolution coefficient of these particles, either for all the SPM, or for specific mineral particles. It is also important to be able to compute the total amount of the material dissolved in the water column, knowing the regeneration rate.

We assume that sedimentation is constant with respect to depth in the considered water column and that the dissolution is proportional to the particle surface area (Lerman et al.,1975; Brun-Cottan, 1975) This last assumption apparently implies that the PSD varies with depth; effectively in a given water thickness the small particles stay longer than the large ones and the proportion of their mass which dissolves is then greater than that of the largest particles. Thus, one must observe a strong deformation of the PSD, mainly a lack of small particles, in the direction of deep waters. However, this is incorrect and the hundreds of measurements done in the shallow waters of the world Ocean show that the PSD is generally very well described by a lognormal law whose coefficients change very little. We think that the best way to explain this apparent contradiction is to accept the fact that most of the particles do not stay as individual bodies during their fall, but continually disaggregate or coagulate. The mechanisms driving these processes are not simple. The theoretical studies on the subject (Hunt,1980; Mac Cave,1984) do not give with certitude an applicable method.

We consider that the mechanisms of coagulation or agglomeration are unknown, but we assume that they constrain the PSD to remain unchanged. We shall see that this constraint allows us to compute, in the homoge-

neous case, the dissolution coefficient and the regeneration rate coefficient. The fact that the system remains invariable with respect to
depth, implies that the proportion of SPM which is dissolved in a water
thickness dz at depth z is the same for any considered depth. This
implies an exponential decrease of SPM versus depth. Because the PSD
remains constant, we can write:

$$\frac{N_{(z+dz)}}{N_z} = \frac{M_{(z+dz)}}{M_z} = \frac{\Phi_{(z+dz)}}{\Phi_z} = \text{constant} \qquad (50)$$

The total vertical mass flux at the depth z+dz is:

$$\Phi_{(z+dz)} = N_{(z+dz)}.\underline{\Phi} = (N_z+dN)\underline{\Phi} \qquad (51)$$

then:

$$\Phi_{(z+dz)}-\Phi_z = \underline{\Phi}dN \qquad (52)$$

where $\underline{\Phi}$ is the mean flux value. This difference of flux is the flux of
SPM which dissolves between z and z+dz, i.e. the regeneration rate. This
flux has to be determined with respect to the kinetic dissolution coefficient and the PSD coefficients.

Applying a dissolution process proportional to the surface area of
a particle having a size l, one can compute the mass this particle
looses during its fall between z and dz. Integrating this mass for all
the particles of the SPM gives the total mass lost by the SPM between z
and z+dz, thus one can identify this quantity with ΦdN and obtain at
depth z the SPM concentrations of particles coming from depth z_o:

$$\frac{N_z}{N_0} = \frac{M_z}{M_0} = \frac{\Phi_z}{\Phi_0} = \exp-(E\frac{\xi}{c}z) \qquad (53)$$

where

$$E = l_N^{-3}\frac{B_3+al_o{}^{\alpha}l_N{}^{-\alpha}B_6}{B_3+al_o{}^{\alpha}l_N{}^{-\alpha}B_4} \qquad (54)$$

with

$$B_5 = \exp(2-\alpha)^2\sigma^2 \qquad (55)$$

$$B_6 = (1-\{\alpha/3\})\exp2(1-\alpha)^2\sigma^2 \qquad (56)$$

The logic of this approach is given in Figure 6.

Knowing the decrease of mass concentration versus depth, one can
compute the kinetic dissolution coefficient:

$$\xi = (c/Ez)\ln(M_0/M_z) \qquad (57)$$

ξ is expressed in cm/s. It is not possible to express the global dissolution coefficient in gr/s, because this depends on the size distribution, but ξ , which is the velocity of the particle size decrease, is a
pure property of the particle's body.

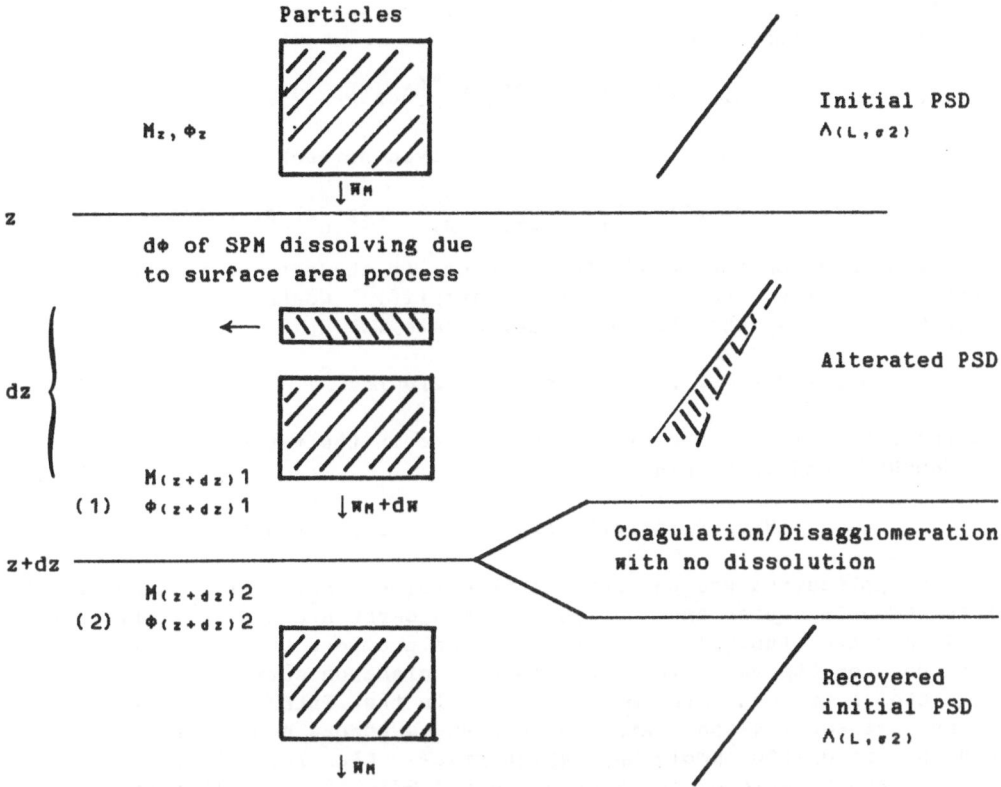

Figure 6. Sketch plan of the settling/dissolution model
The logic of what happens between z and z+dz is:
- At depth z, the particles have a lognormal PSD, a mass
concentration M_z, a mass flux ϕ_z and a mean mass settling
velocity w_M.
- Between z and z+dz, the particles dissolve as single bodies
subject to a dissolution process proportional to their surface
area (Brun-Cottan, 1976). This implies an alteration of the
PSD with a loss of small particles.
- Just before crossing the z+dz surface, the particles have a
mass concentration $M_{(z+dz)1}$, a mass flux $\phi_{(z+dz)1}$ and a mean
settling velocity w_M+dw.
- As they cross the z+dz surface, the particles recombine by
the coagulation or the disaggregation process to recover the
initial lognormal PSD .
- Just below z+dz surface, the particles have the initial
lognormal PSD, and thus the initial mean settling velocity.
The mass concentration is $M_{(z+dz)2}$ and the mass flux is
$\phi_{(z+dz)2}$. By definition there is no loss or gain of SPM at
z+dz, thus $\phi_{(z+dz)1}=\phi_{(z+dz)2}$. The fact that above and below
z+dz, the mass fluxes are equal and that the PSD's are different, implies that the mass concentrations are different,
$M_{(z+dz)2}<M_{(z+dz)1}$.

The flux of SPM which is dissolved is called the regeneration rate $dJ_z=-(E\xi/c)\phi_z$; its value at depth z is:

$$dJ_z = M_z\xi al_o{}^\alpha l_N{}^{(-1-\alpha)}\frac{B_3+al_o{}^\alpha l_N{}^{-\alpha}B_6}{B_1+al_o{}^\alpha l_N{}^{-\alpha}B_2}dz \ gr/cm^3/s \tag{58}$$

The total flux of SPM which is dissolved between 0 and z is:

$$J_{0\rightarrow z} = M_o cal_o{}^\alpha l_N{}^{2-\alpha}\frac{B_3+al_o{}^\alpha l_N{}^{-\alpha}B_6}{B_1+al_o{}^\alpha l_N{}^{-\alpha}B_2}(1-exp-\frac{\xi}{c}z) \ gr/cm^2/s \tag{59}$$

As we saw for the mass determination, in the case of constant known absolute density ρ_P, the equations simplify. Coefficient E becomes: $E=exp-(3L+(21/2)\sigma^2$ and the regeneration rate is:

$$dJ_z = M_z\xi([\rho_P-\rho]/\rho)l_N{}^{-1}exp(-[5/2]\sigma^2)dz \tag{60}$$

and the total flux of SPM wich is dissolved in the water column, between the depths 0 and z, becomes:

$$J_{0\rightarrow z} = M_o c([\rho_P-\rho]/\rho)l_N{}^2exp(8\sigma^2).(1-exp-E[\xi/c]z) \tag{61}$$

An application was performed with calcium carbonates at station 67 of the GEOSECS cruise and the measurements carried out at CFR laboratory at Gif/Yvette (Aubey,1983). The data was provided by the electron microscope for the particle sizes and with electron activation for measuring the suspended calcium carbonate. In the Intermediate Antarctic waters, between 1.4 and 2 km, the calcium carbonate concentration varies from 0.4 to $0.1X10^{-2}\mu mole/kg$, which gives $-(E\xi/c)\phi_z=10^{-18}gr.cm^{-3}.s^{-1}$, and then the kinetic dissolution value $\xi=2.5X10^{-12}cm/s$. This value can be compared with the one computed with the data given by Berger (1969); $\xi=3X10^{-12}cm/s$. At the same station, in the Deep Antarctic waters, between 3.5 and 4.6 km, the calcium concentrations varies from 1.25 to $0.3X10^{-2}\mu mole/kg$, which gives a dissolution coefficient value of $\xi=10^{-12}$ cm/s. The computed and measured dissolution coefficients are in the same range, for this exemple, but obviously this does not necessarily mean that the proposed model fits the general problem of SPM dissolution, but it looks a coherent approach to the problem. This model can be applied to the dissolution process of other componants, such as silicates or SPM organic matter.

4.2 Sedimentation at non steady state

We give here a brief introduction to non steady state effects on the SPM behaviour. Generally, theoretical geochemical studies are done at steady state, but this is an exeptional situation in the natural environment. The problem with the non steady state situation, is that we do not have enough data or enough knowledge of the main parameters which drive the system, to be able to solve it. Here we take a simple example where the production rate of SPM in the surface waters, varies in an exponential way at the depth 0. In the case of a decrease we can write:

$$\phi_{(0,t)} = \phi_{(0,0)}exp-\beta t \tag{62}$$

In the case of an increase to a maximum value when t→∞,

$$\Phi_{(0,t)} = \Phi_{(0,0)}(1-exp-\beta t) \qquad (63)$$

The solution for an increase is similar to that for a decrease, thus we discuss only the case of a decrease, see Figure 7.

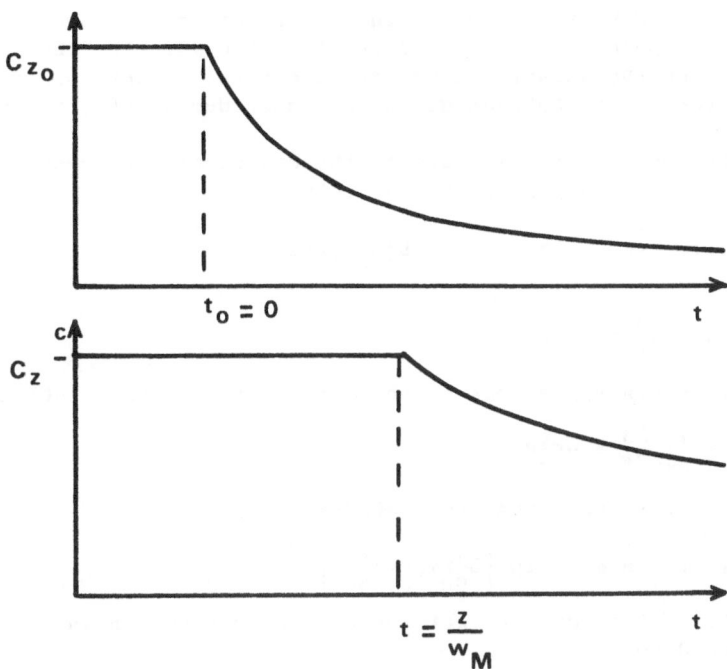

Figure 7. At $z_0=0$, the all particle parameters (number, mass, vertical fluxes, etc...), decrease at the exponential rate ß, at depth z, these parameters decrease with a lag time from the surface z_0, with an other exponential rate.

If the particles are considered as single stable bodies, a decrease of the flux of SPM coming from the surface waters must imply an alteration of the PSD. At a given depth, one would firstly see a decrease in the number of the largest particles leading to a lack of large particles. As far as we know, such an alteration of the PSD has never been seen for the global SPM and we can assume that a large part of the particle size measurements in nature were done at non steady state. As for the dissolution process, this implies also that the largest fraction of the SPM coagulates and disaggregates continuously. This assumption allows us to maintain the local concept of mean settling velocity in the case of non steady state sedimentation.

In the case of sedimentation with no dissolution, it is possible to compute easily the mass concentration at depth z and time t, knowing the concentration at time O and at depth O. For time $t<(z/w_M)$, the mass concentration remains constant, for time $t>(z/w_M)$, the mass concentration is:

$$\frac{M_{(z,t)}}{M_{(o,o)}} = \exp-\beta(t-\frac{z}{w_M}) \qquad (64)$$

In the case of sedimentation with dissolution, at non steady state it is impossible to compute the kinetic dissolution coefficient from the PSD and from the SPM data. But if this coefficient is known by another means, we can compute the mass concentration at any depth, the mean concentration (then the standing crop) and the total amount of dissolved SPM in the water column between depths O and z, during the time O to t.

Since the loss of material due to the dissolution process during Δt is that given at steady state, one can set:

$$M_{(z+dz,t+dt)} - M_{(z,t)} = -E\frac{\xi}{c}M_{(z,t)}dz \qquad (65)$$

then

$$\frac{\partial M}{\partial z}dz + \frac{\partial M}{\partial t}dt = -E\frac{\xi}{c}Mdz \qquad (66)$$

The constancy of the PSD allows us to write $dz=w_M.dt$, then (66) becomes:

$$\frac{\partial M}{\partial z} + \frac{1}{w_M}\frac{\partial M}{\partial t} = -E\frac{\xi}{c}M \qquad (67)$$

The solution of this equation, for $t>(z/w_M)$ is:

$$M_{(z,t)} = M_{(o,o)}\exp-\left| E\frac{\xi}{c}z+\beta(t-\frac{z}{w_M}) \right| \qquad (68)$$

From (68) it is easy to get the mean concentration in the water column between O to z:

$$\underline{M} = \frac{M_{(o,o)}}{z(E\frac{\xi}{c}-\frac{\beta}{w_M})} \exp-\left| E\frac{\xi}{c}z+\beta(t-\frac{z}{w_M}) \right| \qquad (69)$$

The standing crop in the water column is $\underline{M}$ multiplied by the height z of the given water column.

As we said before, the concept of regeneration rate is inconsistant at non steady state, but it is possible to compute the quantity $Q_{s(z,t)}$ of the SPM which dissolves between the depths O to z during the elapsed time O to t. This quantity is:

$$Qs_{z,t} = (R_0-R_z) + (Q_0-Q_t) \qquad (70)$$

R_0 is the total amount of SPM entered in the water column at depth O.

R_z is the total amount of SPM which has left the water column at depth z.

Q_0 is the SPM standing crop at time O.

Q_t is the SPM standing crop at t time. These parameters are, for

$t>(z/w_M)$:

$$R_0 = M_{(0,0)}\frac{w_M}{\beta}(1-\exp\beta\frac{z}{w_M}) \qquad (71)$$

$$R_z = M_{(0,0)}w_M\{\left|\frac{z}{w_M}+\frac{1}{\beta}\exp-(E\frac{\xi}{c})\right|\frac{1}{\beta}\exp-\left|(E\frac{\xi}{c}\frac{\beta}{w_M})z-\beta t\right|\} \qquad (72)$$

$$Q_0 = \frac{M_{(0,0)}c}{E\xi}\left|1-\exp-(E\frac{\xi}{c}z)\right| \qquad (73)$$

$$Q_t = \frac{M_{(0,0)}}{(\frac{\beta}{w_M}-E\frac{\xi}{c})}\left|\exp-(E\frac{\xi}{c}z)-\exp-\beta t\right| \qquad (74)$$

With equations 71 to 74 and for an elapsed time $t>(z/w_M)$, the total amount of SPM dissolved in the water column between 0 to z comes from the following equation 75:

$$\frac{Q_{S(z,t)}}{M_{(0,0)}} = \frac{w_M}{\beta}\frac{1}{E\frac{\xi}{c}}-\left|\frac{w_M}{\beta}\frac{1}{E\frac{\xi}{c}}+z\right|\exp-(E\frac{\xi}{c}z)-\left|\frac{w_M}{\beta}+\frac{1}{E\frac{\xi}{c}\frac{\beta}{w_M}}\right|\{\exp-(\beta t)-\exp-\left|(E\frac{\xi}{c}\frac{\beta}{w_M})z-\beta t\right|\}$$

If we want to compute $Q_{S(z,t)}$ for an elapsed time $t<(z/w_M)$, it is only necessary to divide the problem into 2 parts. For the upper fraction of the water column of which the depth z' is defined by z'=w_M.t, we apply equation 75. For the lower fraction z''=z-z', we apply equation 61; this fraction does not see the beginning of the decrease of SPM in the surface waters and we can thus use the steady state equations.

5 CONCLUSION

The use of the lognormal law allows us to describe the PSD of the total particle population of the SPM with two coefficients. These coefficients and the knowledge of the absolute density of the particles allows us to compute the value of the most important physical parameters concerning the geochemical properties of the SPM. These parameters are the wet mass concentration, the sedimentation flux, the mean settling velocity, and the residence time of the SPM. It appears that the mean settling velocity, and thus the residence time, depend on the considered SPM properties. For example, the mean settling velocity value of the total SPM surface area is greater than that of the total SPM mass. These concepts permit a global evaluation of some important SPM geochemical properties, they could be a useful tools for work with numerical models, as they represent the total SPM with very few parameters, but they must be used with care, for example the mean mass settling velocity of the total SPM has no real meaning in describing the vertical flux of aggregates or fecal pelets.

The computed mean settling velocity of the total SPM mass, in the open Ocean, which is generally less than 150 m/year, clearly shows that the transport of the SPM is mainly due to horizontal advection. This is not in contradiction with a very important vertical flux of large particles settling through the water column (Mac Cave, 1975; Brun-Cottan, 1975), which has no significant mass concentration.

Taking into account the fact that the PSD does not significantly vary inside the water mass, more sophisticated parameters can be obtained, such as the kinetic dissolution coefficient, the proportion of the SPM dissolved during its fall and thus, the input flux of dissolved material into the water column. The usefulness of these parameters increases markedly if the mineral and chemical structure of the SPM is known precisely. Thus, the global dissolution coefficient is not as interesting as that concerning the calcites or silicates or organic particles. The computed value for calcite dissolution is close to that which can be obtained from Berger measurements (1969).

Both the dissolution process and the non steady state imply that the particles cannot be considered as single bodies quietly settling in the water column, but must be considered mainly as small fractions of the SPM, continuously interacting by coagulation and disaggregation processes.

6 APPENDIX

6.1 Specific properties of the lognormal law

6.1.1 - If the sizes l of a population of particles are distributed following the lognormal law (3), any other property which is a power of l, follows a lognormal law (Aitchison and Brown, 1975, th. 2.1). Then if $p = a.l^b$, p follows the lognormal law:

$$\frac{dN}{Ndp} = \frac{d\Lambda}{dp} \; (\ln a + bL, \; b^2\sigma^2) \tag{76}$$

Thus the particle equivalent volumes $v = (\pi)l^3$ of the given population follow the lognormal law:

$$\frac{dN}{Ndv} = \frac{d\Lambda}{dv} \; (\ln\frac{\pi}{6} + 3L, \; 9\sigma^2) \tag{77}$$

6.1.2 - A property P such that $dP = dN.a.l^b$, follows a lognormal law. From (3) one can obtain:

$$\frac{1}{N} \frac{dP}{dl} = al^b \frac{d\Lambda}{dl} \; (L, \; \sigma^2)$$

$$= \frac{a \; \exp(b\ln l)}{\sqrt{2\pi}\,\sigma l} \; \exp\left|- \frac{1}{2\sigma^2} (\ln l - L)^2\right|$$

$$= \frac{a}{\sqrt{2\pi}\,\sigma l} \; \exp\left|- \frac{1}{2\sigma^2} [(\ln l^2 - 2L\ln l + L^2) - 2\sigma^2 b\ln l]\right|$$

$$= \frac{a}{\sqrt{2\pi}\,\sigma l} \; \exp\left|- \frac{1}{2\sigma^2} [\ln l - (L + b\sigma^2)]^2 + bL + \frac{1}{2}b^2\sigma^2\right|$$

$$= \frac{a\exp(bL + (1/2)b^2\sigma^2)}{\sqrt{2\pi}\,\sigma l} \; \exp\left|- \frac{1}{2\sigma^2} [\ln l - (L + b\sigma^2)]^2\right| \tag{78}$$

Then

$$\frac{1}{N}\int_0^\infty dP = a l_N^b \ \exp(\frac{1}{2}b^2\sigma^2)\int_0^\infty d\wedge_{(L+b\sigma^2, \ \sigma^2)}$$ (79)

The property P is distributed by a lognormal function multiplied by a constant term. Knowing:

$$\int_0^\infty dP = P \qquad \text{and} \quad \int_0^\infty d\wedge = 1$$ (80)

one obtain:

$$P/N = a \ \exp(bL+(1/2)b^2\sigma^2) \ = \underline{p}$$ (81)

and

$$dP/(N\underline{p}) = dP/P = d\wedge(L+b\sigma^2, \ \sigma^2)$$ (82)

By definition, $\underline{p}$ is the arithmetic mean of p. Knowing L, σ and the concentration in number N, it is possible to compute in a simple way the concentration of the property P.

6.2 Evaluation of the lognormal coefficients

6.2.1 Use of the function dψ

The variables L', σ' and V are independant and their best values are those which minimize the sum:

$$\upsilon = \Sigma \left| e_i - d\psi(L', \sigma'^2|u_i) \right|^2$$ (83)

υ is the sum of the square values of the difference between the data and the theoritical values. The computing programs which usually give the minimum of υ, such as the Fletcher-Powell method, need the knowledge of the partial derivative:

$$\frac{\partial \upsilon}{\partial L'} = -2\Sigma[e_i-f(u_i)]f(u_i)\frac{(u_i-L')}{\sigma'^2}$$ (84)

$$\frac{\partial \upsilon}{\partial \sigma'} = -2\Sigma[e_i-f(u_i)]f(u_i)\frac{(u_i-L')}{\sigma'^3}$$ (85)

$$\frac{\partial \upsilon}{\partial V} = -2\Sigma[e_i-f(u_i)]\frac{f(u_i)}{V}$$ (86)

where $f(u_i) = Vd\psi(L', \sigma'^2, |u_i)$ (87)

Another simple way to compute the lognormal coefficients is to consider the logarithm of the normal function associated with the log-normal function, which is:

$$\ln(\frac{dV}{dX}) = -\frac{1}{2\sigma'^2}X^2+\frac{L'}{\sigma'^2}X+\ln(\frac{V}{\sqrt{2\pi}\sigma'})-\frac{L'^2}{2\sigma'^2}$$ (88)

where $X = \ln(\pi/6)l^3$ (89)

Equation (88) is a polynomial function of degree 2 for variable x. The method is to fit the transformed data to the best parabola:

$$y = ax^2 + bx + c \tag{90}$$

where

$$y = \ln\left(\frac{dV}{dX}\right), \quad a = -\frac{1}{2\sigma'^2}, \quad b = \frac{L'}{\sigma'^2}, \quad c = \ln\left(\frac{V}{\sqrt{2\pi}\sigma'}\right) - \frac{L'^2}{2\sigma'^2} \tag{91}$$

the solution of this system is given by:

$$a = \frac{1}{D} \begin{vmatrix} \Sigma y_i x_i^2 & \Sigma x_i^3 & \Sigma x_i^2 \\ \Sigma y_i x_i & \Sigma x_i^2 & \Sigma x_i \\ \Sigma y_i & \Sigma x_i & K \end{vmatrix} \qquad b = \frac{1}{D} \begin{vmatrix} \Sigma x_i^4 & \Sigma y_i x_i^2 & \Sigma x_i^2 \\ \Sigma x_i^3 & \Sigma y_i x_i & \Sigma x_i \\ \Sigma x_i^2 & \Sigma y_i & K \end{vmatrix}$$

$$\tag{92}$$

$$c = \frac{1}{D} \begin{vmatrix} \Sigma x_i^4 & \Sigma x_i^3 & \Sigma y_i x_i^2 \\ \Sigma x_i^3 & \Sigma x_i^2 & \Sigma y_i x_i \\ \Sigma x_i^2 & \Sigma x_i & \Sigma y_i \end{vmatrix} \qquad D = \begin{vmatrix} \Sigma x_i^4 & \Sigma x_i^3 & \Sigma x_i^2 \\ \Sigma x_i^3 & \Sigma x_i^2 & \Sigma x_i \\ \Sigma x_i^2 & \Sigma x_i & K \end{vmatrix}$$

K is the number of data and L' , σ' and V are immediately given by a, b and c. This method is very usefull if $\sigma' > 1.2$.

6.2.2 Use of the function ψ

To use this function it is necessary to know the volume proportion of the particles having a size l_n less than that of the class n considered. At first we can consider that Pr_A and Pr_B can be neglected, then for the size l_i, the proportion p_i is:

$$p_i = \left| \sum_{j=1}^{i} \Delta V_j \right| / \left| \sum_{j=1}^{N} \Delta V_j \right| \tag{93}$$

The problem is to find the values of L' and σ' for which $Q = (X - L')/\sigma'$ is the best variable that follows the normal law $\psi(0,1)$ having a mean equal to 0 and quadratic dispersion equal to 1. This means that for each proportion p_i, there must exist a quantity q_i, called the quantile, which is the abscissa of the p_i proportion for the normal function $G(q)$ having a mean equal to 0 and a quadratic dispersion equal to 1 (Figure 8), $G(q)$ is:

$$G_{(q)} = \frac{1}{\sqrt{2\pi}} \int_{-\infty}^{q} \exp\left(-\frac{Q^2}{2}\right) dQ \tag{94}$$

By inverting the function $G(q) = p$, one can generate the pairs (x_i, q_i) equivalent to the experimental pairs (x_i, p_i). X is linearly distributed with respect to q_i:

$$Q = aX + b \quad \text{with} \quad a = 1/\sigma' \quad \text{and} \quad b = -L'/\sigma \tag{95}$$

The determination of a and b is done by linear regression from X to Q, which gives:

$$a = \frac{\Sigma x_i^2 \Sigma q_i - \Sigma x_i \Sigma x_i q_i}{N \Sigma x_i^2 - (\Sigma x_i)^2} \tag{96}$$

$$b = \frac{N\Sigma x_i q_i - \Sigma x_i \Sigma q_i}{N\Sigma x_i^2 - (\Sigma x_i)^2} \qquad\qquad (97)$$

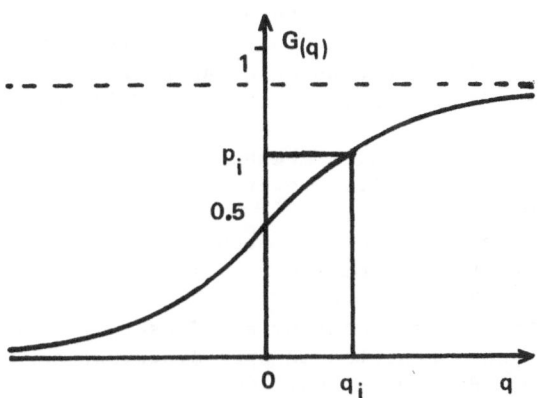

Figure 8. By inverting the function $G(q)$, one can generate q_i, knowing the proportion p_i and, then, compute a linear fit of q_i regarding x_i.

Generally with the experimental data the proportions Pr_A and Pr_B are unknown. The simplest way to solve this problem, if Pr_A and Pr_B are less than 15%, is to make a first step assuming that Pr_A and Pr_B are zero, then to determine L' and σ' and use these values to compute a first estimate of Pr_A and Pr_B. With successive iterations, two is usually sufficient, we get a good estimate of Pr_A and Pr_B, and thus we have also a good estimate of the volume of the particles having sizes less than l_A and greater than l_B.

REFERENCES

Aitchison J. and J.A.C. Brown, 1976: The Lognormal Distribution. Cambridge University Press.

Aubey O., 1976: Contribution à l'étude de la dissolution des particules de carbonate de calcium dans les eaux profondes océaniques. Thèse de 3° cycle, Université Pierre et Marie Curie, Paris.

Bacon P.M. and R.F. Anderson, 1982: Distribution of Thorium Isotopes between Dissolved an Particulate forms in the Deep Sea. Journal of Geophysical Research, vol. 87, pp 2045-2056

Bacon P.M., 1984: Radionuclide in the Ocean interior. Global Ocean Flux Study, pp 181-205

Bader H., 1970: The hyperbolic distribution of particles sizes. Journal of Geophysical Research, vol. 75, pp 2822-2830.

Berger W.H., 1967: Foraminiferal ooze: solution at depths. Sciences, vol 159, pp 383-385.

Bishop J.K.B., R.W. Collier, D.R. Ketten and J.M. Edmond, 1980: The chemistry, biology and vertical and vertical flux of particulate

matter from the upper 1500 m of Panama Basin in the Equatorial
        Ocean. Deep sea Research, vol 8A, pp 615-640.

Bishop J. K. B. and J. Marra, 1984: Variations in primary production and
        particulate carbon flux through the base of the euphotic zone at
        the site of the Sediment Trap Intercomparison Experiment (Panama
        Basin). Journal of Marine Reseach, vol 42, pp 189-206.

Brun-Cottan J. C. , 1967: Influence du marquage radioactif sur les para-
        mètres dynamiques des sédiments pélitiques. Thèse de 3° cycle en
        océanographie.

Brun-Cottan J. C. , 1971: Etude de la granulométrie des particules. Me-
        sures effectuées avec un Coulter counter. Cahiers Océanographiques,
        vol. 23, pp 193-205.

Brun-Cottan J. C. , 1976: Contribution à l'étude de la granulométrie et de
        la cinétique des particules marines. - Thèse de Doctorat d'Etat,
        Université Pierre et Marie Curie.

Brun-Cottan J. C. , 1976: Stokes settling and dissolution rate model for
        marine particles as a function of size distribution. Journal of
        Geophysical Res. vol. 81 n°9, p. 1601-1605.

Brun-Cottan J. C. , 1977: Contribution à l'étude de la granulométrie et de
        la cinétique des particules marines. Journal de Recherche Océnogra-
        phique, vol. 21, pp 41-54.

Carder K. L. ,   F. Beardsly and H. Pak, 1971: Particles size distribution
        in the Eastern Equatorial Pacific. Journal of Geophysical Research,
        vol. 76, pp 5070-5077.

Fortier A. , 1967: Mécanique des suspensions. Monographie de mécanique
        des fluides et thermiques. Masson Paris, pp 176.

Harris J. E. , 1972: Characterisation of suspended matter in the gulf of
        Mexico - 1, Spatial distribution of suspended matter. Deep Sea
        Research, vol. 19, p 719-726.

Hoffmann E. E,   J. M. Klinck, and G. A. Paffenhofer, 1981: Concentrations
        and vertical fluxes of zooplankton fecal pellets on a continental
        shelf. Marine Biology, 61, pp 327-335.

Honjo S. , 1978: Sedimentation of materials in the Sargasso Sea at 5367 m
        depth deep station. Journal of Marine Research, vol 36, pp 469-492

Hunt R. ,   1980: Prediction of Oceanic Particle Size Distribution from
        coagulation and sedimentation mechanisms. Advances in chemistry
        series, No. 189: Particulates in water. Copyright by the American
        Chemical Society.

Jerlov N. G. , 1968: Optical Oceanography. Elsevier, Amsterdam, 194 pp.

Junge C. E. , 1963: Air chemistry and radioactivity. Academic Press, New-
        York, 382 pp.

Lal D. , 1977: The Oceanic Microcosm of Particles. Sciences, vol. 198, pp
        977-1009.

Lambert C. , C. Jehanno, N. Silverberg, J. C. Brun-Cottan and R. Chesselet,
        1981: Lognormal distribution of suspended particles in the Ocean.
        Journal of Marine Research, Vol 39, 1, p77-98.

Lerman A. ,   D. Lal and M. F. Dacey, 1974: Stokes' settling and chemical
        reactivity of suspended particles, in natural waters. Edited by R.
        J. Gibbs, Plenum Press, pp 17-47.

Mac Cave I. N. , 1975: Vertical flux of particles in the Ocean. Deep-Sea
        Research, vol. 22, pp. 491-502.

Mac Cave, I. N., 1982: Size spectra and aggregation of suspended particles. Joint Oceanographic Assembly, Abstract, Dalhousie Univ.
Halifax, Canada, pp 81-82.

Martin J. M., J. M. Mouchel and A. J. Thomas, 1985: Time concepts in hydro-
dynamics systems with an application to Be7 in the Gironde estuary.
Marine Chemistry, to be published.

Morel a., 1973: Diffusion de la lumière par les eaux de mer. Résultats
expérimentaux. Interprétation théorique. Thèse de Doctorat d'Etat,
Université Pierre et Marie Curie.

Riley G. A., 1963: Organic aggregates in sea water and the dynamic of
their formation and utilisation. Limnol. Oceagr., vol 8, pp 372-
381.

Sheldon R. W., A. Prakash and S. Sutcliffe, 1972: The size distribution
of particles in the Ocean. Limnology and Oceanography, vol. 17, pp
327-340.

Silver M. W., A. L. Shanks and J. D. Trent, 1978: Marine snow: microplank-
ton and habitat source of small-scale patchiness in pelagic population. Sciences, vol 201, pp 371-373.

Slin W. G. N., 1983: Air to sea transfer of particles. In Air-Sea Exchange
of Gases and particles. NATO ASI N°108.

Turner J. T., 1977: Sinking rates of fecal pellets from the marine cope-
pod pontella meadii. Marine Biology, vol 40, pp 249-259.

Wiebe P. H., S. H. Boyd and C. Winget, 1976: Particulate matter sinking to
the deep-sea floor at 2000 m in the Tongue of the Ocean, Bahamas,
with a description of a new sediment trap. Journal of Marine Research, 34, pp 341-354.

AIR-SEA GAS EXCHANGE RATES : INTRODUCTION AND SYNTHESIS

Peter S. Liss
School of Environmental Sciences
University of East Anglia
Norwich   NR4 7TJ
United Kingdom

and

Liliane Merlivat,
Laboratoire d'Océanographie Dynamique
et de Climatologie
C.E.N. Saclay
91191 Gif-sur-Yvette
France

1. INTRODUCTION

In this chapter we attempt to present a brief introduction to the
subject of air-sea gas exchange. First the basic equations governing
such exchange are given, then a review of some models proposed to
describe the gas transfer process. Following this, experimental
approaches through both laboratory (principally using wind/water
tunnels) and field measurements are summarised. Finally, we present
what seems to us to be the best current synthesis of the wind tunnel and
field results for the prediction of gas exchange rates across the sea
surface.
     We refer rather frequently to a chapter one of us wrote for a
previous NATO ASI on 'Air-Sea Exchange of Gases and Particles' held at
the University of New Hampshire in July 1982 (Liss, 1983a). Reference
seems better than extensive repetition. However, hopefully the summary
given here stands on its own. It also serves to update the subject,
since quite a number of interesting developments have occurred in the
intervening three years. A book to which we refer several times is 'Gas
Transfer at Water Surfaces' which is the proceedings of a meeting held
at Cornell University in June 1983 (Brutsaert and Jirka, 1984). Our
hope is that the short chapter we now present will whet the reader's
appetite to delve into both of the above mentioned references as well as
other papers cited here.

P. Buat-Ménard (ed.), The Role of Air-Sea Exchange in Geochemical Cycling, 113–127.
© *1986 by D. Reidel Publishing Company.*

2. BASIC PRINCIPLES

The basic equations governing air-sea gas exchange are presented and discussed in detail in Liss (1983a). Here only the more important relationships are reproduced:

$$F = K_{(T)w}\Delta C \tag{1}$$

Where ΔC is the concentration difference driving the flux(F) and $K_{(T)w}$ is the total transfer velocity.
The concentration difference can be expressed more specifically as,

$$\Delta C = C_a H^{-1} - C_w \tag{2}$$

Where C_a and C_w are the gas concentrations in air and water, respectively, and H is the dimensionless Henry's Law Constant (expressed as the ratio of the concentration of gas in air to its concentration in unionised form dissolved in the water, at equilibrium).
     The total transfer velocity can be broken down into its component parts as follows:

$$1/K_{(T)w} = 1/\alpha k_w + 1/H k_a \tag{3}$$

Where k_a and k_w are the individual transfer velocities for chemically unreactive gases in the air and water phases, respectively; and $\alpha(=k_{reactive}/k_{inert})$ is a factor which quantifies any enhancement of gas transfer in the water due to chemical reaction.
Equation (3) is often expressed in terms of resistances as,

$$R_{(T)w} = r_w + r_a \tag{4}$$

Where $R_{(T)w}(=1/K_{(T)w})$ is the total resistance, with $r_w(=1/\alpha k_w)$ and $r_a(=1/H k_a)$ the resistances of water and air phases separately.
     By substitution of appropriate values for k_w, k_a and α in (3), it is easy to show that for many gases either r_w or r_a is dominant. Gases for which $r_a \gg r_w$ generally have low values of H (i.e. they partition dominantly into the water) and may show high values of α, and include H_2O, SO_2, NH_3 and HCl. In contrast, gases for which r_w is the dominant resistance to transfer mostly have high H (low solubility) and $\alpha \sim 1.0$, and include O_2, N_2, N_2O, the Inert Gases, CO_2, CO, CH_4, CH_3I, and $(CH_3)_2S$. It is clear from this that the majority of the gases of interest in geochemical cycling fall into the second category so that k_w is the transfer velocity controlling their air-sea exchange, i.e. in (1) $K_{(T)w} = k_w$. For this reason, the present chapter will be almost solely concerned with evaluation of k_w.
     We will also largely ignore the ΔC term in (1) since this has to be known for each gas of interest and is more appropriately dealt with in the book chapters on individual gases or groups of gases (e.g. those on S, H and C, Hg, and halogen gases by Andreae, Conrad and Seiler, Fitzgerald, and Liss, respectively).
     There are two main ways in which transfer velocities have been

investigated, laboratory, including wind tunnel, experiments and field
measurements. In parallel with these experimental approaches, several
attempts have been made to model the transfer process. We briefly
review the most widely used of these models in the next section.

3. MODELS

Modelling air-sea gas exchange serves three main purposes: i) to
increase understanding of the physics and chemistry of the transfer
process, ii) to allow measurements of k_w made using one gas to be
converted into equivalent values for another gas, and iii) to make
predictions of the value of k_w from a knowledge of other parameters,
such as wind speed, or sea state. The models constructed to date fall
into three main categories, i.e. film, surface renewal, and boundary-
layer. The first two have their roots in chemical engineering studies
of gas transfer between liquid and gas phases, whereas the boundary-
layer models come from micrometeorological research, largely into
interfacial heat and momentum transfer. Each of these classes of models
is reviewed in the next three sub-sections.

3.1 Film Model (Whitman, 1923)

Here, the main resistance to transfer is envisaged as molecular diffusion
through a stagnant layer of water adjacent to the interface. For a
given set of conditions, the thickness of the water film is invariant in
space and time. Although such spatial and temporal uniformity is
unlikely to be realised, except possibly under the calmest of
conditions, the model has been widely used. It has proved useful in
identifying whether r_a or r_w is the controlling resistance to exchange,
as discussed in Section 2, for quantifying the chemical enhancement
factor (α), and for understanding the possible role of oil and surface
active organic films in affecting gas transfer. Mathematically the
model is straightforward and it is easy to show that with the film
model k_w is proportional to the coefficient of molecular diffusivity
(D) to the power unity.

3.2 Surface Renewal Models (Higbie, 1935; Danckwerts, 1951)

In this class of models the stagnant film is retained but periodically
it is replaced by fluid from the bulk, and the film replacement rate is
then the limiting step in gas transfer. Mathematically these models
are somewhat more complex than the film model and it turns out that in
this case k_w is proportional to $D^{1/2}$. Although physically more
realistic than the film model, surface renewal models have been applied
rather little particularly in environmental investigations. This is
principally because of the difficulty of specifying the film replacement
rate, except in well defined laboratory systems.

3.3 Boundary-Layer Models (Deacon, 1977)

There is a considerable body of theoretical and experimental information
on the transfer of momentum, heat and mass at surfaces. Based on the
analogy between momentum and mass transfer, it has been shown that k_w
is proportional to the friction velocity in air (u_*). Similarly, k_w is
also proportional to the ratio of the transfer coefficients for
momentum (kinematic viscosity, ν) and mass (molecular diffusivity, D)
to the power - 2/3 (i.e. $k_w \propto Sc^{-2/3}$, where Sc is the Schmidt Number,
ν/D). This led Deacon (1977) to apply a treatment developed by Reichardt
(1951) for the velocity profile over a smooth, rigid surface to
estimation of k_w. Apart from the assumption that the surface is smooth,
it was also necessary to assume continuity of stress across the
interface in order to convert the velocity profile in the air to the
equivalent profile in the water. The relationship developed by Deacon
is as follows:

$$k_w = 0.082 \ Sc^{-2/3}(\rho_a/\rho_w)^{1/2}u_* \tag{5}$$

where ρ_a and ρ_w are the densities of air and water, respectively. It
is clear from (5) that for this model k_w is proportional to $D^{2/3}$.

The Deacon model is found to work well when the water surface is
unruffled by waves, as might be predicted from the rigid wall
assumption, but underestimates k_w when the surface ceases to be smooth.
This deficiency led Kerman (1984) to extend the micrometeorological
approach by using Yaglom and Kader's (1974) treatment of heat and mass
transfer to a rough surface. By estimating the amount of surface
roughening caused by patches of breaking waves Kerman was able to
predict k_w for a non-smooth surface.

Another effect which occurs under wave breaking conditions is the
formation of bubbles which can play a significant role in gas exchange.
Attempts to model this will be mentioned in the next section when wind
tunnel studies of gas exchange are discussed.

4. LABORATORY (WIND TUNNEL) STUDIES

Most serious attempts to examine air-water gas exchange in the
laboratory have employed wind tunnels. Results from such studies are
reviewed extensively in Liss (1983a) and only a summary and update will
be attempted here. Laboratory investigations of the chemical
enhancement term (α) and the possible role of natural and man-made
films in modifying gas exchange are also reviewed in Liss (1983a) and
are not considered further in this chapter.

In Figure 1 we have attempted to summarise, in an idealised plot
of k_w against wind speed, what the wind tunnel studies seem to tell us
about the exchange process. An idealised graph is shown rather than
actual results for two reasons: i) real data inevitably contains
experimental scatter and this tends to obscure trends, and ii) no single
wind tunnel study has yet managed to cover the full range of wind
speeds shown in Figure 1. This latter problem might have been overcome

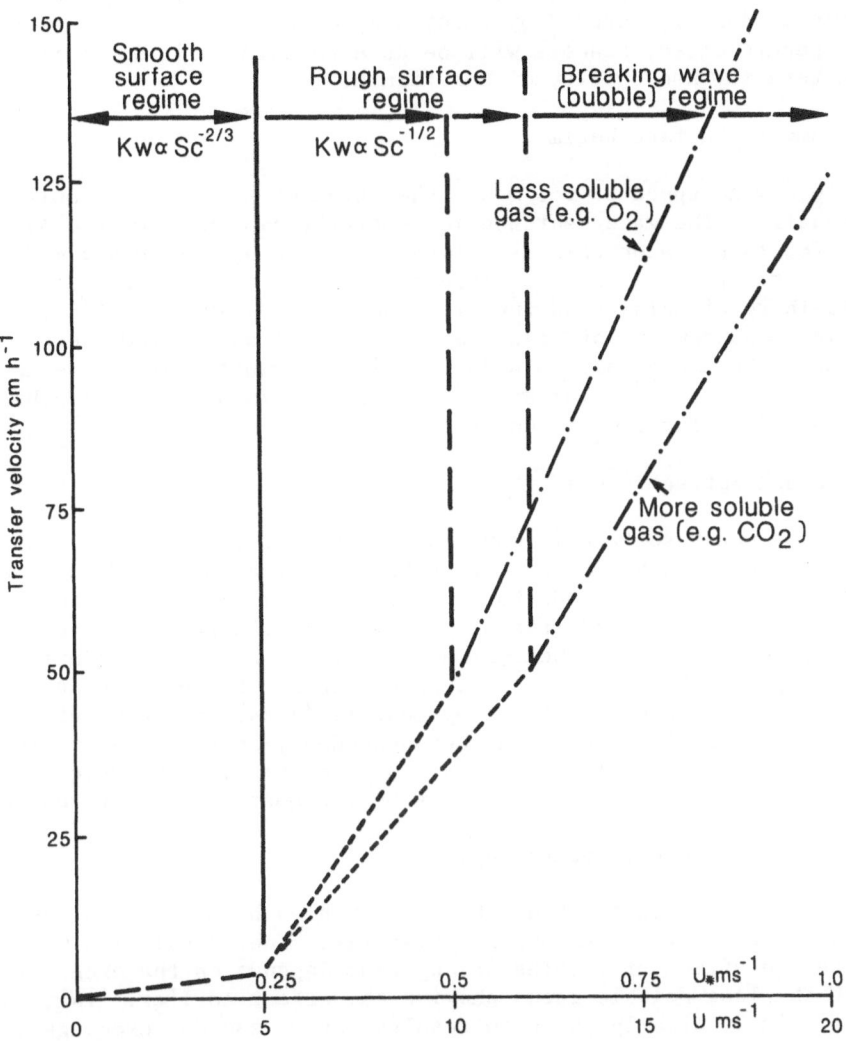

Figure 1.   Idealised plot of water phase transfer velocity
(k_W) against wind speed (u) and friction velocity (u_*).
Based on laboratory results, mainly from the Hamburg wind
tunnel (Broecker and Siems, 1984).

by amalgamating data from several studies, but this is difficult when
different wind tunnels have been used, since for a given wind speed or
friction velocity different wave fields can develop. Such bulking of
data from several tunnels is again apt to hide any trends. Of course,
the idealised plot is an oversimplification, but it does serve to
identify three regimes where different physical processes appear to be

controlling gas exchange. These three regimes will now be briefly
reviewed and illustrated by reference to real results. By consulting
the papers cited, readers will be able to form their own opinions as to
the veracity and utility of the figure.

4.1 Smooth Surface Regime

Up to a wind speed of 5 ± 3 m s^{-1} the value of k_w increases only very
gradually. The water surface is generally smooth or with only few waves
and for this reason this is called the "Smooth Surface Regime". The
best data at these low wind speeds have been obtained using the
Heidelberg circular wind tunnel (Jahne et al., 1979 and 1984). By
making measurements of heat transfer as well as gas (CO_2) exchange
these workers are able to show that in the smooth surface regime k_w is
proportional to $D^{2/3}$ (or $Sc^{-2/3}$), thus confirming the applicability of
the boundary-layer model under calm conditions.

4.2 Rough Surface Regime

Here the water surface is wave covered but it is not rough enough for
wave breaking to be common. This "Rough Surface Regime" commences at
about 5 m s^{-1} and extends to between 10 and 12 m s^{-1}, depending on the
gas being investigated (see Section 4.3). Compared with the smooth
surface case, the presence of waves produces a considerable increase in
the slope of the k_w versus wind speed graph. Recent studies of the
wave covered region are those by Ledwell (1984) and Jahne et al. (1984).
They find that the Schmidt number dependency in this region gives the
transfer velocity as proportional to D to the 0.5-0.6, which implies
that the surface renewal model may be the most appropriate one here.

4.3 Breaking Wave (Bubble) Regime

Above a wind speed of about 10 m s^{-1} bubbles become common due to wave
breaking and act to enhance gas transfer. The actual wind speed at
which the effect of bubbles is apparent depends on the exchanging gas,
with the less soluble gases showing the effect earlier. This is
particularly clearly shown in results for O_2 and CO_2 exchange measured
in the Hamburg wind tunnel by Broecker and Siems (1984). A theoretical
treatment of the enhancement of gas exchange due to bubbles has been
given by Merlivat and Memery (1983). It confirms that the enhancement
of gas transfer associated with wave breaking has to be more important
and must occur at a lower wind speed for the less soluble gas.

5. FIELD MEASUREMENTS

Methods used to estimate the water-side transfer velocity (k_w) in the
field are now reviewed briefly - much of the detail is given in Liss
(1983a) and Broecker and Peng (1984), although some newer results are
included here.

5.1 Box Method

In this approach the sea surface is covered with a floating box and the composition of the head space in the box examined as a function of time as the gas(es) of interest invade or evade into or out of the water. Early measurements are mentioned in Liss (1983a) and a more recent application is by Hartman and Hammond (1984) who looked at ^{222}Rn evasion from the waters of San Francisco Bay.

This approach suffers from two major problems: i) it is difficult to deploy the apparatus except under rather calm conditions, and ii) the normal interaction between wind and water is severely inhibited by the presence of the box. Since wind stress is, from the results of wind tunnel experiments, an important control on k_w, there are large doubts as to what the method actually measures, and therefore about its usefulness.

5.2 Dissolved Gas Balance Method

In surface seawater physical and biological processes tend to cause dissolved gases to be out of equilibrium with their atmospheric concentrations. By making time-series measurements of dissolved gas and nutrient concentrations in a discrete parcel of seawater it is possible to budget the observed concentration changes in terms of air-sea exchange and in-situ biological processes. Details of the method with the assumptions made and the uncertainties they imply, together with results obtained using it, are given in Liss (1983a). There do not appear to have been any further attempts to use the technique.

5.3 Micrometeorological Techniques

There are two micrometeorological techniques which have been adapted for gas exchange studies in the field, the profile and eddy correlation methods, more usually used for estimating momentum, heat and water vapour fluxes over surfaces.

The profile method, which involves measuring the property of interest in air as a function of height above the surface, isn't really of use for gases for which r_w is the dominant transfer resistance. This is because, by analogy with electrical resistances in series, differences in concentration are proportional to the size of the resistances, i.e. where $r_w \gg r_a$ there will be very little gradient to measure in the air. For example, as Deacon (1977) has shown, for CO_2 transfer between air and sea a typical gradient of concentration in air would be 0.05 ppm between 1 and 10 m above the sea. This is < 0.02% of the atmospheric concentration and very hard to measure accurately.

Potentially of more use in estimating k_w is the eddy correlation technique in which fluxes are obtained from the time-averaged product of measured gas concentrations in air and the vertical component of the wind. The method and some of the problems associated with it when applied to estimating gas fluxes are discussed in Liss (1983a). Suffice to say here that when applied to air-sea CO_2 exchange (Jones and Smith, 1977; Wesley et al., 1982; Smith and Jones, 1985) the results imply

fluxes which are up to 100 times higher than values obtained by other
techniques. Recently, Broecker et al. (1986) have presented a detailed
examination of the clear discrepancy between the eddy correlation
results and those obtained using isotopic techniques (natural and bomb-
produced ^{14}C and ^{222}Rn, see Sections 5.4 and 5.5). They conclude that
no reliable estimate of the air-sea flux of CO_2 has ever been produced
using the eddy correlation technique. There are clearly difficulties
with the technique which have yet to be successfully overcome. One
fundamental aspect is that, as pointed out above, the CO_2 sensor is
looking at very small differences in concentration between vertically
moving parcels of air, so that the signal to noise ratio is inevitably
unfavourable.

In contrast, the eddy correlation technique does appear to have
been successful for direct measurement of ozone fluxes from the
atmosphere to the oceans. Lenschow et al. (1982) report values for the
transfer resistance for O_3 uptake by seawater which are in good agreement
with predictions from laboratory studies and in situ experiments. As
discussed in Liss (1983b), O_3 is relatively insoluble ($H\sim3$), and although
it reacts rapidly with surface seawater (α may be as high as 20), its
exchange is still dominated by liquid phase resistance ($r_a\sim0.04R_{(T)w}$).
However, from the eddy correlation point of view, the situation is more
favourable than for CO_2 where $r_a\sim0.006R_{(T)w}$, and this must be at least
part of the reason why the technique appears to work for O_3 but not for
CO_2.

5.4 Natural and Bomb-Produced ^{14}C

Broecker and Peng (1974 and 1984) have summarised how the $^{14}CO_2$ from
both natural and bomb sources can be used to obtain global-average
values for k_w. Transfer velocities calculated using data from either
natural or bomb-produced $^{14}CO_2$ yield values which are in close
agreement, and average globally to about 20 cm h^{-1}.

It must be stressed that this is a value averaged temporally and
over all oceans, and does not tell us anything about the transfer
velocity locally or at particular times or seasons, or the relationship
between k_w and meteorological parameters such as wind speed. However,
it does impose a vital constraint on the acceptability of the results
of other techniques for estimating k_w. This is why concern was
expressed over the use of the eddy correlation technique for estimating
k_w for CO_2, since reported spot values are generally in excess of 20 cm
h^{-1}, many by more than an order of magnitude.

5.5 The Radon Deficiency Method

This is potentially the most powerful and probably the most used field
method currently available for measuring k_w. The method and assumptions
involved in its use are reviewed in Liss (1983a). The technique was
used extensively as part of the GEOSECS cruises in the Atlantic and
Pacific Oceans (Peng et al., 1979) and by Roether and Kromer (1984) in
JASIN and FGGE. Somewhat disappointingly, the approximately 100 GEOSECS
data points and the JASIN and FGGE results show no close dependence of

k_w on wind speed, when such a dependence might be expected from the wind tunnel studies reported in Section 4.

Recently, Smethie et al. (1985) have published results for k_{Rn} obtained as part of the TTO programme in the tropical Atlantic Ocean in 1982/83. These authors note that there does appear to be a general relationship in this later data set between k_{Rn} and the climatological wind regime, i.e. the highest transfer velocity values occur in areas which are climatologically the most windy and vice versa. However, a plot of the determined k_{Rn} against the wind speed measured around the time of the measurements only partially bears this out. The correlation coefficient of k_{Rn} on wind speed is about 0.5 (similar to the value obtained by Deacon (1981) in his reanalysis of the GEOSECS data).

Despite the lack of a clear wind speed dependence, the Radon Deficiency Method does give results which are on average in reasonable agreement with those from the ^{14}C techniques. The mean k_{Rn} from the GEOSECS data is 11.9 cm h^{-1} and that from the TTO tropical Atlantic results is 15.0 cm h^{-1}. Using the surface renewal model to derive the corresponding values for the transfer velocity for CO_2 gives 13.9 and 17.6 cm h^{-1} for the GEOSECS and TTO results, respectively. Although somewhat lower than the value of 20 cm h^{-1} given by ^{14}C measurements, there is no large discrepancy between the mean Rn results and the necessarily average values for ^{14}C.

5.6 Sulphur Hexafluoride

The use of artificially added tracers to obtain transfer velocities in the oceans is some way off. However, Wanninkhof et al. (1985) have recently shown the power of such techniques for studies in confined water bodies, in their case Rockland Lake in New York State. These authors added SF_6 to the lake and measured the decreasing concentrations of the gas in the water as the SF_6 was lost to the atmosphere across the lake surfaces. The results are shown in Figure 2, and demonstrate a clear relationship between transfer velocities and wind speed. However, the slope of the k_w versus wind speed plot is about a factor of two less than that found in wind tunnel studies. The apparently smaller effect of wind in increasing k_w in the field than in the laboratory is due, at least in part, to differences in wave field developed in the two situations arising from the significantly different fetch regimes.

The success of the lake experiments using SF_6 should give encouragement to those attempting the much more difficult task of using purposefully added tracers to estimate k_w values in the oceans.

5.7 Summary

A summary of much of the information on values of k_w determined at sea is given in Figure 3 which is adapted from Roether (1986). Also shown are the predictions of the Deacon smooth surface model according to (5) and a dashed line representing the wind tunnel data at intermediate wind speeds. In Section 6 we develop a predictive relationship for the dependence of k_w on wind speed and this is also shown in Figure 3. There appears to be a good relationship between our prediction and much

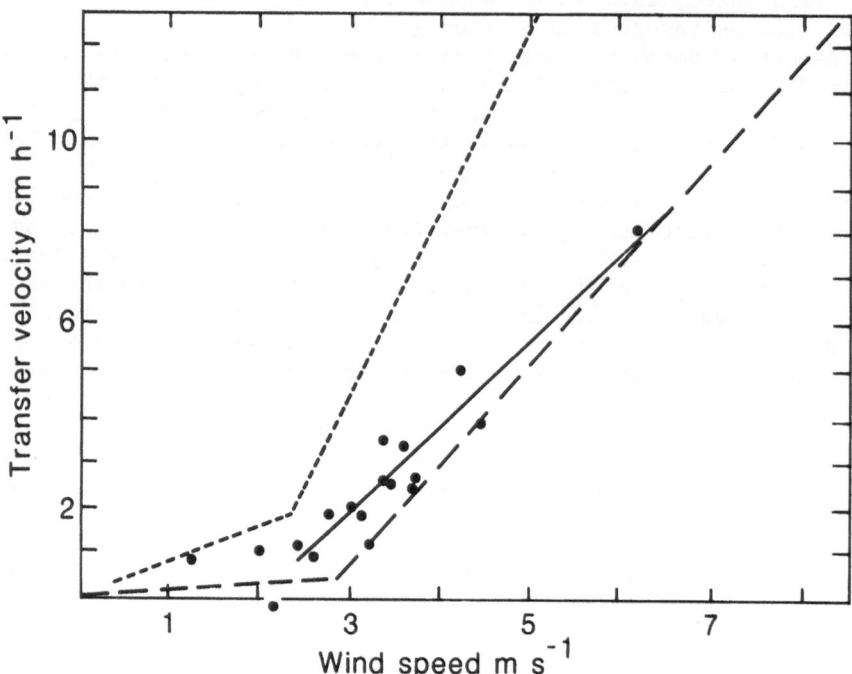

Figure 2. Transfer velocity for SF$_6$ determined on Rockland
Lake (New York State, U.S.A.) as a function of wind speed
measured at 1 m above the lake. The solid line is the least
squares fit to the data above a wind speed of 2.4 m s^{-1}. The
short-dashed curve is from the wind tunnel study of Broecker
et al. (1978). The long-dashed curve is an estimate of the
relation between the transfer velocity and wind speed if the
wind was steady. (After Wanninkhof et al., 1985.)

of the field data. However, it should be borne in mind that several
of the points are averages of larger data sets (e.g. the GEOSECS
results) and in reality show considerable dispersion in both k_w and
wind speed.

6.   SYNTHESIS

We propose to take advantage of the field data set reported by
Wanninkhof et al. (as shown in figure 2) and extrapolate it to a more
general context, i.e. for different gases and sea surface temperatures,
and for an enlarged set of values of wind speed. This extrapolation
will be based on knowledge obtained from either models or wind tunnel
experiments.
     The Wanninkhof et al. wind speeds were measured at a height of 1m

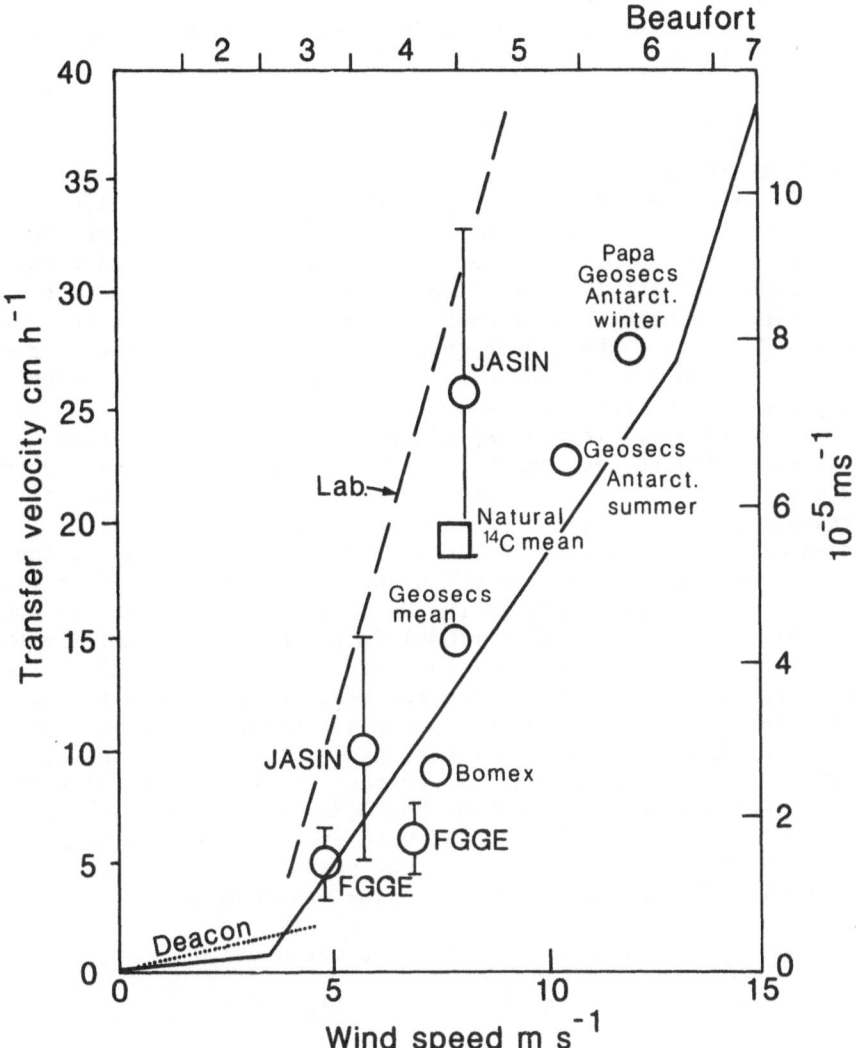

Figure 3. Oceanic measurements of the liquid phase gas
transfer velocity plotted as a function of wind speed
measured at a height of 10 m and the corresponding Beaufort
scale number. The dotted line represents predictions based
on the Deacon (1977) smooth surface model according to
equation (5). The dashed line represents the results for
intermediate wind speeds from laboratory wind tunnel studies.
All data are converted to Sc = 600, corresponding to CO_2 at
20°C, by assuming that $k_w \propto Sc^{-\frac{1}{2}}$. (After Roether, 1986.)
The predictions made in Section 6 of the present paper are
shown by the three full lines.

above the lake surface. Since most of the wind speeds measured at sea
are for a measurement height of 10 m, a value for the drag coefficient
of 1.5×10^{-3} is used to convert the Rockland Lake wind speeds to
equivalent values at 10 m height. This gives,

$$u_{z=10\ m} = 1.29\ u_{z=1\ m} \tag{6}$$

Where z is the height above the water surface.

The transfer velocities reported in Figure 2 are valid for SF_6 at
10°C, i.e. for Sc = 1600. We propose to derive expressions for k_w
adjusted to Sc = 600, the value for CO_2 at 20°C, the conditions to which
the oceanic results summarised in Figure 3 have been normalised.

Using a Schmidt Number power dependence of -2/3 for the lower wind
speed and -1/2 for the higher one, we obtain, $k_w = 0.21$ u, for u $\leqslant$ 3.6
and $k_w = 2.85$ u - 9.65, for u > 3.6; where k_w values are in cm h^{-1}, and
u is in m s^{-1} measured at z = 10 m.

In Section 4.3 it was explained that at high wind speed, when wave
breaking begins to develop, an increase in transfer velocity occurs.
For u (z = 10 m) > 13 m s^{-1} the results of wind tunnel studies will be
extrapolated following a slope similarity based on the data of Broecker
and Siems (1984). A wind speed of 13 m s^{-1} is identical to the value
of the mean wind velocity at which the change of regime (see Figure 1)
is observed for either CO_2 in the Hamburg wind tunnel or N_2O (which has
a solubility close to that for CO_2) in the Marseille facility (Merlivat
and Memery, 1983). Since no straightforward way is available to
calculate u* in these high wind regimes, we are forced to extrapolate
this value directly to the open ocean situation. This is consistent
with observations of the distribution of white caps at sea as a function
of wind speed, which indicate a coverage of 2 to 3% of the surface for
u (z = 10m) $\sim$ 13 m s^{-1} (Monahan and O'Muircheartaich, 1980).

In summary, for the range of wind speeds examined, we propose the
following three relationships for the variation of k_w (for a gas with
Sc $\sim$ 600) with wind speed in the marine environment,

$$k_W = 0.17\ u \qquad\qquad , \text{for u} \leqslant 3.6 \tag{7}$$

$$k_W = 2.85\ u - 9.65 \qquad , \text{for } 3.6 < u \leqslant 13 \tag{8}$$

$$k_W = 5.9\ u - 49.3 \qquad , \text{for u} > 13 \tag{9}$$

Where k_w is in cm h^{-1} and u is in m s^{-1} at z = 10 m.

For a given gas, the Schmidt number varies with water temperature,
decreasing as the temperature increases. Table 1 indicates some values
of Sc for different gases at various temperatures. Such values are
needed to adjust (7), (8) and (9) so that they apply where Sc is
different from 600, by assuming that k_w is proportional to $Sc^{-2/3}$ for
u (z = 10 m) $\leqslant$ 3.6 m s^{-1} and to $Sc^{-1/2}$ for higher wind speeds.

It is worthwhile to emphasize that when computing gas fluxes at the
air-sea interface at the world ocean scale, it is necessary to deal with
the full range of sea surface temperatures, extending between 0 and 30°C.
This implies, for CO_2 transfer for instance, a variation of the gas

TABLE 1 : Schmidt numbers for some gases, 0 - 40°C (from Jahne, 1980).

Temperature		Schmidt number		
(°C)	He	O_2	CO_2	Rn
0	510	1450	1860	3150
10		850	1010	1600
20	140	470	595	870
30			360	500
40	65	200	240	300

transfer velocity which is as large as 2.3 or 3, depending on the wind regime. Water temperature is thus an important parameter which has to be taken into account in the calculation of gas fluxes, not only because of its influence on solubility, but also through its effect on gas transfer velocities.

6.1 Comparison with Field Data

Our summary equations (7), (8) and (9) are plotted in Figure 3 as three solid lines. The broad agreement between the field results collected together by Roether, 1986 (some of which have been averaged from larger data sets - see Section 5.7) and the predictions of (7) - (9) seems reasonable. Since our approach is completely independent of the oceanic field results, being based on lake experiments using SF_6 and wind tunnel observations, we have some confidence that the relationships developed here are realistic and should be of utility in predicting gas transfer velocities at the sea surface.

ACKNOWLEDGEMENTS

We thank the following ASI participants for providing comments which have been helpful in preparing the chapter: Chantal Andrié, Dave Erickson, Tom Fogg, Véronique Garcon, Keith Hunter, Catherine Jeandel, John Merrill, Ed Monahan, Alex Pszenny, Howard Ross, Uli Siegenthaler, Michael Walker, Rik Wanninkhof, and Ollie Zafiriou.

REFERENCES

Broecker, H.C., J. Petermann and W. Siems, 1978: 'The influence of wind
    on CO_2-exchange in a wind-wave tunnel, including the effects of
    monolayers.' *J. Mar. Res., 36*, 595-610.
Broecker, H.C. and W. Siems, 1984: 'The role of bubbles for gas transfer
    from water to air at higher windspeeds. Experiments in the wind-
    wave facility in Hamburg.' In: *Gas Transfer at Water Surfaces*
    (W. Brutsaert and G.H. Jirka, eds.), Reidel, 229-236.
Broecker, W.S. and T.H. Peng, 1974: 'Gas exchange rates between air and
    sea.' *Tellus, 26*, 21-35.
Broecker, W.S. and T.H. Peng, 1984: 'Gas exchange measurements in
    natural systems.' In: *Gas Transfer at Water Surfaces* (W. Brutsaert
    and G.H. Jirka, eds.), Reidel, 479-493.
Broecker, W., J.R. Ledwell, T. Takahashi, R. Weiss, L. Merlivat, L.
    Memery, T.H. Peng, B. Jahne and K.O. Munnich, 1986: 'Isotopic
    versus micrometeorologic ocean CO_2 fluxes: an order of magnitude
    conflict.' Submitted.
Brutsaert, W. and G.H. Jirka (eds.), 1984: *Gas Transfer at Water
    Surfaces,* Reidel, 539 pp.
Danckwerts, P.V., 1951: 'Significance of liquid-film coefficients in
    gas absorption.' *Ind. Engng. Chem., 43*, 1460-1467.
Deacon, E.L., 1977: 'Gas transfer to and across an air-water interface.'
    Tellus, 29, 363-374.
Deacon, E.L. 1981: 'Sea-air gas transfer: the wind speed dependence.'
    Boundary-Layer Met., 21, 31-37.
Hartman, B. and D.E. Hammond, 1984: 'Gas exchange rates across the
    sediment-water and air-water interfaces in South San Francisco Bay.'
    J. Geophys. Res., 89, 3593-3603.
Higbie, R., 1935: 'The role of absorption of a pure gas into a still
    liquid during short periods of exposure.' *Trans. Am. Inst. Chem.
    Engr.. 35*, 365-373.
Jähne, B., 1980: *Zur Parameterisierung des Gasaustausches mit Hilfe von
    Laborexperimenten,* Doctoral dissertation, University of Heidelberg,
    124 pp.
Jähne, B., K.O. Munnich and U. Siegenthaler, 1979: 'Measurements of gas
    exchange and momentum transfer in a circular wind-water tunnel.'
    Tellus, 31, 321-329.
Jähne, B., W. Huber, A. Dutzi, T. Wais and J. Ilmberger, 1984: 'Wind/
    wave - tunnel experiment on the Schmidt number - and wave field
    dependence of air/water gas exchange.' In: *Gas Transfer at Water
    Surfaces* (W. Brutsaert and G.H. Jirka, eds.), Reidel, 303-309.
Jones, E.P. and S.D. Smith, 1977: 'A first measurement of sea-air CO_2
    flux by eddy correlation.' *J. Geophys. Res., 82*, 5990-5992.
Kerman, B.R., 1984: 'A model of interfacial gas transfer for a well-
    roughened sea.' *J. Geophys. Res., 89*, 1439-1446.
Ledwell, J.J., 1984: 'The variation of the gas transfer coefficient with
    molecular diffusivity.' In: *Gas Transfer at Water Surfaces* (W.
    Brutsaert and G.H. Jirka, eds.), Reidel, 293-302.

Lenschow, D.H., R. Pearson and B.B. Stankov, 1982: 'Measurements of ozone vertical flux to ocean and forest.' *J. Geophys. Res.*, 87, 8833-8837.

Liss, P.S., 1983a: 'Gas transfer: Experiments and geochemical implications.' In: *Air-Sea Exchange of Gases and Particles* (P.S. Liss and W.G.N. Slinn, eds.), Reidel, 241-298.

Liss, P.S., 1983b: 'The exchange of biogeochemically important gases across the air-sea interface.' In: *The Major Biogeochemical Cycles and Their Interactions* (B. Bolin and R.B. Cook, eds.), Wiley, 411-426.

Merlivat, L. and L. Memery, 1983: 'Gas exchange across an air-water interface: Experimental results and modelling of bubble contribution to transfer.' *J. Geophys. Res.*, 88, 707-724.

Monahan, E.C., and I. O'Muirchaertaich, 1980: 'Optimal power-law description of oceanic whitecap coverage dependence on wind speed.' *J. Phys. Oceanogr.*, 10, 2094-2099.

Peng, T.H., W.S. Broecker, G.G. Mathieu, Y.H. Li and A.E. Bainbridge, 1979: 'Radon evasion rates in the Atlantic and Pacific Oceans as determined during the GEOSECS Program.' *J. Geophys. Res.*, 84, 2471-2486.

Reichardt, H., 1951: 'Vollständige darstellung der turbulenten geschwindigkeitsverteilung in glatten leitungen.' *Z. angew. Math. Mech.*, 31, 208-219.

Roether, W., 1986: 'Field measurements of gas exchange.' In: *Dynamic Processes in the Chemistry of the Upper Ocean* (J.D. Burton et al., eds.) Plenum, in press.

Roether, W. and B. Kromer, 1984: 'Optimum application of the radon deficit method to obtain air-sea gas exchange rates.' In: *Gas Transfer at Water Surfaces* (W. Brutsaert and G.H. Jirka, eds.), Reidel, 447-457.

Smethie, W.M., T. Takahashi, D.W. Chipman and J.R. Ledwell, 1985: 'Gas exchange and CO_2 flux in the Tropical Atlantic Ocean determined from ^{222}Rn and pCO_2 measurements.' *J. Geophys. Res.*, 90, 7005-7022.

Smith, S.D. and E.P. Jones, 1985: 'Evidence for wind-pumping of air-sea gas exchange based on direct measurements of CO_2 fluxes.' *J. Geophys. Res.*, 90, 869-875.

Wanninkhof, R., J.R. Ledwell and W.S. Broecker, 1985: 'Gas exchange-wind speed relation measured with sulfur hexafluoride on a lake.' *Science*, 227, 1224-1226.

Wesley, M.L., D.R. Cook, R.L. Hart and R.M. Williams, 1982: 'Air-sea exchange of CO_2 and evidence for enhanced upward fluxes.' *J. Geophys. Res.*, 87, 8827-8832.

Whitman, W.G., 1923: 'The two-film theory of gas absorption.' *Chem. metall. Engng.*, 29, 146-148.

Yaglom, A.M. and B.A. Kader, 1974: 'Heat and mass transfer between a rough wall and turbulent fluid flow at high Reynolds and Peclet numbers.' *J. Fluid Mech.*, 62, 601-623.

THE OCEAN AS A SOURCE FOR ATMOSPHERIC PARTICLES

EDWARD C. MONAHAN
Department of Oceanography
University College
Galway
Ireland

1. INTRODUCTION

Recognizing that descriptions of the physical mechanisms whereby sea-salt particles are introduced into the atmosphere have been treated in some detail by Blanchard (1983) in an earlier publication in this series (Liss and Slinn, 1983), it seems appropriate in this contribution to critique several explicit models for the estimation of the number of sea-salt aerosols per size interval, per unit area of the sea surface, per unit time, produced by these mechanisms under specified meteorological conditions, and in so doing illustrate the state of knowledge of the various processes that link the flux of sea-salt aerosol up from the sea surface to the wind speed measured at an elevation of 10 meters.

Given the great difficulty encountered in any attempt to directly measure in situ sea surface particle production (as opposed to low-elevation aerosol concentration), it is perhaps not surprising that the two models of aerosol generation via bubble bursting to be discussed both attempt to combine laboratory measurements of the sea-salt aerosol particles produced from simulated whitecaps with field measurements of oceanic whitecap coverage to deduce indirectly the flux of sea-salt particles upwards from the surface of the open ocean.

In the first model to be considered the authors (Monahan, et al, 1979; 1982) have attempted to combine estimates of the fraction of the sea surface from which whitecap foam disappears per unit time, $\overset{\bullet}{W}$, with laboratory measurements of the number of sea-salt particles per radius increment produced throughout the lifetime of a simulated whitecap of initial unit area, $\partial E/\partial r$. In this set of laboratory experiments the simulated whitecaps are produced by causing spilling crests moving from opposite ends of a rectangular tank to interact to produce a plunging breaker.

P. Buat-Ménard (ed.), The Role of Air-Sea Exchange in Geochemical Cycling, 129–163.
© *1986 by D. Reidel Publishing Company.*

The second, alternative, model to be considered is that introduced by Cipriano (1979) and Cipriano and Blanchard (1981). In this approach what in effect is done is to combine estimates of the instantaneous fraction of the sea surface covered by whitecaps, W, with laboratory measurements of $\partial^2 E/\partial t \partial r$, the rate of production of sea-salt particles per radius increment from a unit area of a steady-state whitecap sustained by water continuously overflowing a weir.

In the following descriptions of the $\dot{W} \, \partial E/\partial r$ and $W \, \partial^2 E/\partial t \partial r$ sea-salt particle generation models the assumptions unique to each model will be discussed. Those assumptions which are common to both models will then be considered in a subsequent section.

2. THE $\dot{W} \, \partial E/\partial r$ MODEL

In all stages of the evolution of this model it is assumed that,

$$\dot{W} = W\tau^{-1} \tag{1}$$

where τ is the appropriate time constant characterising the decay of the surface area of individual whitecaps (Monahan 1971). The contention that the area of the sea surface covered by those bubbles which have their origins in a particular breaking wave event decreases exponentially with time once the wave has finished breaking is supported by the findings from several tank studies (Monahan and Zietlow, 1969; Monahan, et al, 1982). If it is acknowledged that in analysing film or video records of the sea surface (or of the water surface in a tank) only those specific areas where there exists on the surface at least a minimum areal density of bubbles with radii greater than R_m are identified as whitecap covered, and if it is recognized that bubbles of radius R_m have a terminal rise velocity v_m, then it follows for the present model that at any time t after a whitecap's formation, the horizontal cross-sectional area at depth z of the bubble cloud beneath that whitecap $A_t(z)$, is related to the surface area of the whitecap at that time $A_t(0)$, by

$$A_t(z) = A_t(0)e^{-z/D} \tag{2}$$

where D, the e-folding or scale depth of the whitecap bubble cloud is given by,

$$D = v_m \tau \tag{3}$$

It should be noted that it is not assumed that the

surface of a particular whitecap necessarily remains one
contiguous area, nor is it assumed that the cross-section
through the associated bubble cloud at any particular
depth can be represented as a single area. It is assumed
that the bubble cloud has not been subjected to any net
vertical advection associated with Langmuir cells or large
scale turbulence, possibilities discussed in detail by
Thorpe (1982, 1984a, 1984b). While it is not directly
apparent from the sonographs of bubble clouds obtained
during a series of measurements carried out in the vicinity
of the Scottish coast (Thorpe, 1982) that these clouds
attenuate exponentially with depth, the variation of
acoustic cross-section with depth determined during these
experiments would suggest that the average concentration
of bubbles with radii in the range of 50μm to 60μm does
indeed vary with depth in a manner consistent with the
exponential attenuation with depth model (Thorpe, 1985).

In developing this model, the total number of sea-salt
aerosol particles per radius increment generated during
the entire period of decay of a whitecap of initial unit
area, $\partial E/\partial r$, was determined from measurements made in the
whitecap simulation tank illustrated in Figure 1. Starting
with the sea-water level behind the two gates considerably
higher than in the central section of the tank (Fig. 1a),
two solitary waves which progressed toward the centre of
the tank (Fig. 1b) were generated by raising the gates
simultaneously. As these waves moved close to the centre,
they each spilled (Fig. 1c) and became bore-like in
character. Their collision on the surface of the central
well (Fig. 1d) resulted in a plunging breaker, and in the
entrainment of bubbles down to considerable depths below
the surface (Fig. 1e). A pair of Particle Measuring
Systems aerosol spectrometer probes, models CSASP-100 and
ASASP-300 (Schacher, et al, 1981), and on other occasions a
Royco model 225/241 particle counter, were used to
continuously measure the particle concentrations $\partial c/\partial r$ at
various heights within the hood above the tank (Fig. 1a) as
successive breaking wave events were staged at appropriate
intervals. From the resulting time series records which
clearly showed the change in particle concentration $\Delta(\partial c/\partial r)$
associated with each breaking wave event it was possible,
knowing the interior dimensions of the hood, to determine
the total number of particles per radius increment generated
during a typical event, i.e. during the decay of a standard
tank whitecap, $\Delta(\partial n/\partial r)$. The explicit expression derived
from these simulation tank experiments is given in Equation
4 (Monahan, et al, 1983), where r represents the radius of
the sea spray droplets at 80% relative humidity, expressed
in micrometers.

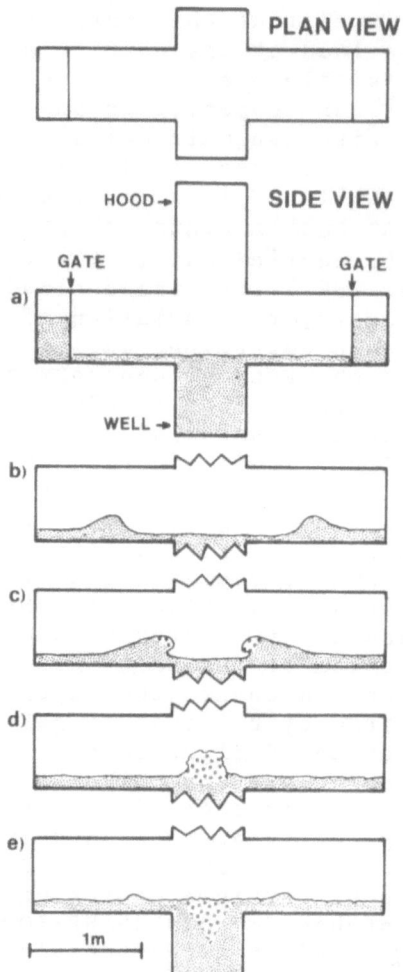

Figure 1. Whitecap simulation tank at University College, Galway. See text for description of sequence of profile views.

$$\Delta(\partial n/\partial r) = 4.40 \times 10^5 r^{-3}(1+0.057r^{1.05}) \times 10^{1.19e^{-B^2}},$$

$$B = (0.380 - \log r)/0.650 \qquad\qquad (4)$$

This expression is valid in the droplet radius range, 0.8μm to 10μm. Here again, and throughout the discussion of this model, all sea spray droplet dimensions have been adjusted to the dimensions these particles would have if they were airborne in, and in equilibrium with, an

atmosphere at 80% relative humidity. The radius of a droplet in equilibrium at 80% relative humidity is just half the
radius possessed by the droplet at the instant it was
formed at the sea surface, and just twice the effective
radius of the particle that would result if this droplet
were to be dried out.

The exponential time constant associated with the decay
of the simulation tank whitecaps, τ', and the initial
surface area of the standard tank whitecap, $A'_0(0)$, were
determined from the analysis of cine-film recordings of a
number of breaking wave events to be 3.53 s and $0.349m^2$
respectively. A more detailed description of the simulation
tank experiments can be found elsewhere (Monahan, et al,
1982).

The quantity required from the laboratory experiments,
$\partial E/\partial r$, the number of particles per increment radius
produced during the decay of a unit area of whitecap, is
then given by Equation 5.

$$\partial E/\partial r = \Delta(\partial n/\partial r)/A'_0(0) \tag{5}$$

The rate at which sea-salt particles are generated,
per radius increment per unit area of sea surface, by the
bursting of whitecap bubbles, $\partial F_0/\partial r$, is therefore, from
Equations 1 and 5,

$$\partial F_0/\partial r = \dot{W}\, \partial E/\partial r = \frac{W}{\tau}\, \frac{\Delta(\partial n/\partial r)}{A'_0(0)} \tag{6}$$

It should be recalled that $\partial F_0/\partial r$ is a statement of the
upward flux of sea-salt particles right at the sea surface,
and that the upward particle flux through any horizontal
plane at higher elevation will be less, often substantially
less, than $\partial F_0/\partial r$.

In both this model and the model of Cipriano and
Blanchard (1981) it is assumed that the respective laboratory whitecaps are valid facsimiles of open ocean whitecaps.
The specific assumption embodied in the $\dot{W}\,\partial E/\partial r$ model is
that the same number of aerosol particles, in each size
range, are produced during the decay of a unit area of the
laboratory foam patch as are generated during the decay of
a unit area of open ocean whitecap, regardless of the
relative magnitudes of the initial surface areas of the
foam patch ($A'_0(0)$) and of the oceanic whitecap ($A_0(0)$).
Before moving on to a consideration of the Cipriano and
Blanchard model it is instructive to go back and recast the
present model in slightly altered terms, i.e. by beginning
with the assumption that the number of aerosol particles
in each size range generated during the disappearance at

the water surface in the tank of a bubble cloud of initial
volume V' is proportional to the particle production
associated with the disappearance at the sea surface of a
bubble cloud of initial volume, V_0, the ratio of
proportionality being simply V_0'/V_0. For the whitecap model
characterised by an exponential attenuation of the
associated bubble cloud's cross-section with depth, it
follows from Equation 2 that,

$$V_0 = \int_0^\infty A_0(z)dz = \int_0^\infty A_0(0)e^{-z/D}dz = DA_0(0) \tag{7}$$

and similarly, that V_0' equals $D'A_0'(0)$. Now in the case of
this recast version of the model the quantity to be
evaluated from the laboratory tank results would be the
number of sea-salt particles per radius increment produced
throughout the lifetime of a simulated whitecap of initial
unit volume, $\partial G/\partial r$, which is expressed in Equation 8,

$$\partial G/\partial r = \Delta(\partial n/\partial r)/V_0' = \frac{\Delta(\partial n/\partial r)}{D'A_0'(0)} \tag{8}$$

and since the rate at which the aggregate volume of those
clouds of bubbles with radii greater than or equal to R_m,
i.e. the clouds of bubbles associated with the optically
resolvable surface whitecaps, is disappearing up through a
unit area of sea surface is simply Wv_m, the rate at which
sea-salt particles are generated, per radius increment per
unit area of sea surface, by the bursting of whitecap
bubbles, $\partial F_0/\partial r$, is in this case given by,

$$\partial F_0/\partial r = Wv_m \, \partial G/\partial r = \frac{Wv_m\Delta(\partial n/\partial r)}{D'A_0'(0)} \tag{9}$$

making use of Equation 3 this becomes,

$$\partial F_0/\partial r = W \frac{v_m}{v_m'} \frac{\Delta(\partial n/\partial r)}{\tau'A_0'(0)} \tag{10}$$

or, alternatively,

$$\partial F_0/\partial r = W \frac{D}{D'} \frac{\Delta(\partial n/\partial r)}{\tau A_0'(0)} \tag{11}$$

It should be apparent that this model does not involve
the unrealistic assumption that the size distribution and

concentration of bubbles within the cloud beneath a whitecap are uniform over time or depth. Rather, this approach
assumes that the initial volume occupied by an ocean $\langle V_0 \rangle$
or laboratory (V_0') bubble cloud, defined by the presence of
bubbles with radii of R_m or greater, is proportional to the
initial number of bubbles in each size category, not simply
those with radius R_m, present in that just formed cloud.
While it seems probable, in the absence of reliable data,
that the large-R cut-off in the bubble spectrum occurs at a
smaller radius in the case of a smaller V_0 bubble cloud than
in the case of an initially larger bubble cloud, it is to be
remembered that these very large bubbles are quite sparce in
any whitecap bubble cloud, and are not efficient generators
of jet-drops (Blanchard, 1963).

     If the scale depths of oceanic and tank bubble clouds
are the same, then Equation 11 simplifies to Equation 6.
Given the variations in the general appearance of bubble
clouds, from billow-like to column-like, which were observed
by Thorpe (1982) to be associated with changes in the
stability of the lower atmosphere as reflected in changes
in the quantity ΔT, i.e. the surface water temperature
minus the air temperature, it is probable that this
simplification of Equation 11 is not in all cases justified.

     Likewise, it is to be noted that if v_m equals v'_m, then
Equation 10, becomes,

$$\partial F_0 / \partial r = \frac{W \; \Delta(\partial n/\partial r)}{\tau' A_0'(0)} \qquad\qquad (12)$$

and we can use the tank derived exponential time constant τ'
in place of the open ocean whitecap time constant τ called
for in Equation 6. This simplification is only applicable
when the sea surface temperature is equal to the water
temperature in the simulation tank, for only in this
circumstance will the terminal rise velocities of bubbles
of radius R_m in both the ocean and the tank be the same, as
can be seen from the curves in Figure 2 (from Blanchard,
personal communication).

3. THE $W \; \partial^2 E / \partial t \partial r$ MODEL

     The measurements required to evaluate the second term
in this model, i.e. $\partial^2 E / \partial t \partial r$, were made in a circular
laboratory tank covered with a cylindrical hood. The most
distinctive feature of the laboratory set-up used by
Cipriano (1979) was a weir, which was mounted at a fixed
elevation (0.33m) over the surface of the tank. A constant
waterfall over the sill of the weir down into the center of

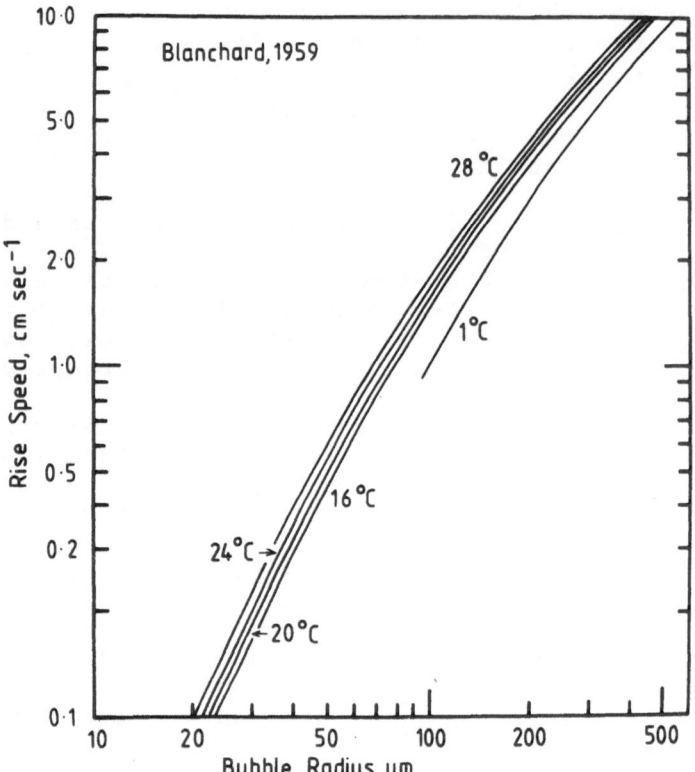

Figure 2. Influence of water temperature on the terminal
rise speed of bubbles. Personal communication from
Blanchard (1959).

the tank was achieved by continuously pumping salt water up
from the bottom of that tank into the weir. This apparatus
is illustrated in Blanchard (1983). The waterfall and
resulting bubble plume are assumed to be analogous to a
plunging breaking wave and associated whitecap.
    Filtered, aerosol-free, air was introduced into the hood
at a dilution rate, F_d or $\partial v_h/\partial t$, and air was extracted at
the same rate from the other side of the hood and passed
through a Royco model 225 particle counter to determine the
equilibrium concentration spectrum, $\partial c/\partial r$, associated with
droplets of radii greater than 0.3μm. The number of sea-
salt aerosol droplets, per radius increment, produced per
unit time by the rising bubble plume, $\partial^2 n/\partial t \partial r$, is given by
Equation 13, and the desired quantity, $\partial^2 E/\partial t \partial r$ is expressed

$$\partial^2 n/\partial t \partial r = \frac{\partial c}{\partial r} F_d = \frac{\partial c}{\partial r} \frac{\partial v_h}{\partial t} \qquad (13)$$

by Equation 14, where A_c is the constant surface area of the bubble plume. The area A_c was taken to be that of a circle

$$\frac{\partial^2 E}{\partial t \partial r} = \frac{1}{A_c} \frac{\partial^2 n}{\partial t \partial r} = \frac{F_d}{A_c} \frac{\partial c}{\partial r} = \frac{1}{A_c} \frac{\partial c}{\partial r} \frac{\partial V_h}{\partial t} \tag{14}$$

with an effective radius of 0.08 m (Cipriano, 1979). The
standard dilution rate, F_d, used in these experiments was
such that the effects of aerosol migration to the hood
surfaces and of aerosol coagulation on $\partial c/\partial r$ were negligible.
Likewise, the loss of aerosol particles by sedimentation
was negligible except in the case of the largest droplets.
By measuring $\partial c/\partial r$ for different values of F_d Cipriano
(1979) was able to infer the asymptotic value of the product
of $\partial c/\partial r$ and F_d, in the case where F_d approached infinity,
i.e. the value of $\partial^2 n/\partial t \partial r$ associated with negligible
sedimentation.
       By making use of such instruments as a Gardner counter,
an electrical aerosol analyser, and a Sinclair diffusion
battery Cipriano was able to obtain information on the sub-
micron-sized particles generated in this weir-waterfall
apparatus (Cipriano, et al, 1983).
       The rate at which sea-salt particles are produced, per
radius increment per unit area of sea surface, by the
shattering, collapsing whitecap bubbles, $\partial F_0/\partial r$, is,
according to this model,

$$\partial F_0/\partial r = W \frac{\partial^2 E}{\partial t \partial r} = \frac{W}{A_c} \frac{\partial c}{\partial r} \frac{\partial V_h}{\partial t} \tag{15}$$

       It should be noted that in this development it is
assumed that the rate of aerosol production per unit surface
area of the rising, energetic, laboratory plume is the same
as the aerosol production rate from a typical unit area of
an optically resolvable open ocean whitecap.
       An additional feature of the weir-waterfall experiments
was the introduction of bubble collectors, apparatus which
made it possible to photographically determine the bubble
concentration spectrum, $\partial C/\partial R$, at various locations relative
to the plume axis. Now the bubble arrival rate per unit
tank surface area per radius increment, $\partial^2 E'/\partial t \partial r$, is given
by Equation 16, where v_R is the terminal rise velocity of

$$\partial^2 E'/\partial t \partial r = \frac{v_R}{A_c} \int \frac{\partial C}{\partial R} dA = v_R \overline{\frac{\partial C}{\partial R}} \tag{16}$$

bubbles with radius R.
       If it is assumed in this case of jostling bubbles on

the surface above the plume produced by the weir waterfall
that the relationship between the radii of the daughter jet
droplets, r_d, and the radii of the parent bubbles, R_p, is
the same as in the case studied by Blanchard (1963) of
individual bubbles rising to a still sea water surface, then
the coefficients required to evaluate Equation 17 are
known. Relying on the isolated bubble studies of Hayami

$$r_d = aR_p + b \tag{17}$$

and Toba (1958) and Blanchard (1963) an estimate can be
made for J_R, the average number of jet droplets produced
by each bubble of radius R. Cipriano and Blanchard (1981)
assumed J_R to have an R-independent value of 5. It follows
from the above that the desired quantity, $\partial^2 E/\partial t \partial r$, can be
expressed in the manner of Equation 18.

$$\partial^2 E/\partial t \partial r \Big|_{r_d} = (J_R \partial^2 E'/\partial t \partial r)\Big|_{R_P} \frac{\partial R}{\partial r}\Big|_{r_d}$$

$$= (J_R v_R \overline{\frac{\partial C}{R \partial R}})\Big|_{R_P} \frac{\partial R}{\partial r}\Big|_{r_d} \tag{18}$$

Finally, the open ocean jet-droplet production to be
inferred from the laboratory measurements of bubble concentration spectra is given by,

$$\partial F_0/\partial r \Big|_{r_d} = W \partial^2 E/\partial t \partial r \Big|_{r_d} = W(J_R v_R \overline{\frac{\partial C}{R \partial R}})\Big|_{R_P} \frac{\partial R}{\partial r}\Big|_{r_d} \tag{19}$$

Because of the difficulty experienced in attempting to
determine the number and sizes of the film droplets arising
from the shattering of the upper surface of an isolated
bubble of given radius, the quantities comparable to J_R and
$\partial R/\partial r$ in the analogous, complementary, expression describing
the film droplet production associated with a known bubble
concentration spectrum cannot be readily evaluated. It is
to be hoped that the application of the laser holographic
techniques described by Resch (1985) will yield the required
information on the number and sizes of the film droplets
produced when an individual bubble bursts.
     In light of the considerable uncertainty associated
with any attempt to extrapolate droplet production rates
determined from experiments on isolated bursting bubbles to
mass bubbling circumstances such as those encountered in

bubble plumes, a concern first expressed by Blanchard (1963), the model represented by Equation 19 will not be considered further, and only the $W\partial^2E/\partial t\partial r$ expression given in Equation 14 will be compared in detail with the $W\ \partial E/\partial r$ expression of Equation 6, or Equation 10, if it is assumed that v_m equals v'_m.

4. COMPARISON OF $\dot{W}\ \partial E/\partial r$ AND $W\ \partial^2E/\partial t\partial r$ MODELS

It is apparent from a consideration of Equation 6 (or 12) on the one hand and of Equation 15 on the other, that the dependence on whitecap coverage is the same in both the $\dot{W}\ \partial E/\partial r$ and $W\ \partial^2E/\partial t\partial r$ aerosol generation models. If the statement of the $\dot{W}\ \partial E/\partial r$ model given in Equation 12 is adopted, it is then possible to directly compare just those portions of both models which rely upon results obtained from the respective laboratory experiments, i.e. $(\tau')^{-1}\partial E/\partial r$ can be compared with $\partial^2E/\partial t\partial r$, and for the moment the vexing problem of selection for use in these models of an appropriate expression for W can be deferred. Since the laboratory experiments of Monahan, et al (1982) were carried out at a water temperature of approximately 19^0C and the tank studies of Cipriano and Blanchard (1981) were conducted with the water at about 26^0C, in making the suggested comparison it would be simplest to assume that $(\tau')^{-1}\partial E/\partial r$ and $\partial^2E/\partial t\partial r$ are independent of water temperature (an assumption implicit in the early attempts to assess from the models global sea-salt flux, e.g. Monahan and Spillane, 1983) but more prudent to avoid this simplifying assumption and interpret the results of the comparison in light of the 7^0C temperature difference between the two tanks. The suggested comparison is shown in Figure 3. The two expressions, $(\tau')^{-1}\partial E/\partial r$ and $\partial^2E/\partial t\partial r$, predict roughly equal rates of production of sea spray droplets with radii, at 80% relative humidity, of one or two micrometers, but the curve based on the measurements taken during the water-fall-weir experiment gives a rate of production of 10µm droplets that is almost one hundred times greater than that given by the formula evaluated using the results of the two-interacting-waves experiment.

Could the relative amplitudes of the two curves, which represent for their respective models the rate of marine aerosol production per radius increment per unit area of whitecap ($W^{-1}\partial F_0/\partial r$), be explained in terms of the 7^0C difference in water temperature ? While the ratio of daughter-jet-droplet to parent-bubble radii appears to increase with increasing water temperature (Blanchard, 1963), and certainly the 10µm radius droplets observed in these experiments are such jet droplets, this slight shift could not explain the relative shapes of these production spectra

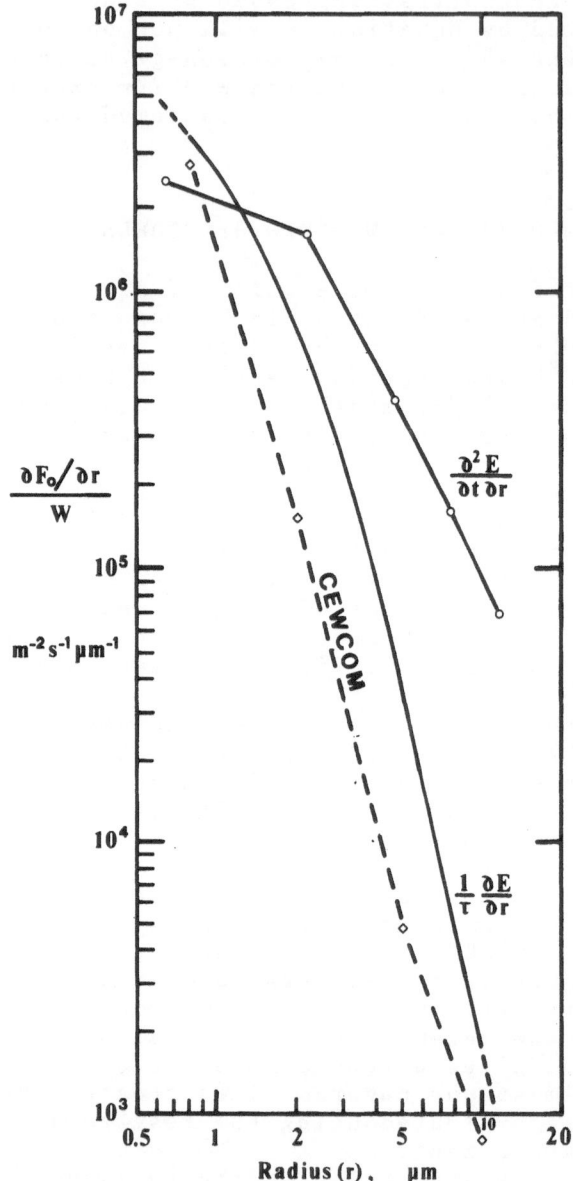

Figure 3. The incremental sea surface droplet production flux, $\partial F_0/\partial r$, for 100% whitecap coverage, as determined from the Galway ($\tau^{-1}\partial E/\partial r$) and Albany ($\partial^2 E/\partial t\partial r$) laboratory tank measurements. Curve labeled CEWCOM derived by dividing $\partial F_0/\partial r$ estimate of Fairall, et al (1983) by W-value from Equation 24 (see text for details).

unless there is a marked shift in the bubble spectrum with

temperature as well. Now several pieces of evidence have
recently emerged that suggest that the spectrum of bubbles
that result from a breaking wave does alter significantly
with water temperature. Pounder (1985), in Monahan and
Mac Niocaill (1985), has observed that as the temperature
increased from 1^0 to 28^0C the average size of the bubbles
produced in sea water decreases, and that the number of
bubbles increases, i.e. as the water temperature goes up
the air volume entrained goes into more, relatively smaller,
bubbles, than was the case for colder water. Bowyer (1983),
in Monahan et al (1983b), found that the amount of electrostatic space charge produced during the decay of a standard
whitecap formed in the wave-wave simulation tank increased
some three-fold as the sea water's temperature was raised
from near 0^0C up to 25^0C, and subsequent experiments
(Bowyer, 1984) showed a comparable increase, with increasing temperature, in the number of sea-salt aerosol
droplets with radii of 1.5μm or larger produced during the
decay of a standard simulation tank whitecap. Bowyer's
observations for droplets with radii of 2.5μm and larger are
presented in the lower portion of Figure 4. Since these
charged droplets are almost certainly jet droplets, it can
be concluded that as the temperature of the sea water in
the tank increased, the air volume entrained by a standard
breaking wave was divided up into more and more smaller and
smaller bubbles, i.e. the peak in the bubble spectrum shifted toward small-R as sea water temperature increased, while
the total volume of all the bubbles formed remained
relatively unchanged. (It would appear that the cold sea
water bubble spectrum differs from the warm sea water bubble
spectrum in the same manner as the fresh water bubble
spectrum differs from the sea water bubble spectrum at the
same temperature, as described by Monahan and Zietlow,
1969). It follows from Pounder's and Bowyer's findings that
the amplitude of the $W^{-1}\partial F_0/\partial r$ spectrum associated with
the 26^0C temperature of the waterfall-weir experiment might
well be greater than that associated with the 19^0C
temperature of the wave-wave interaction experiment, but
these same findings provide no obvious explanation for the
preferential increase in amplitude at the large-r end of the
26^0C $W^{-1}\partial F_0/\partial r$ expression when compared to that derived from
measurements made at 19^0C.
     Since the temperature difference between the waterfall-
weir and wave-wave experiments does not seem to afford an
explanation for the marked disparity between the two
laboratory-derived $W^{-1}\partial F_0/\partial r$ spectra, some other explanation
need be sought. The suggestion that the bubble spectrum in
the energetic weir waterfall plume is quite different from
the spectrum associated with spilling breaking waves on the
open ocean needs to be considered. It has been suggested
(Monahan, 1982) that the bubble spectrum resulting from

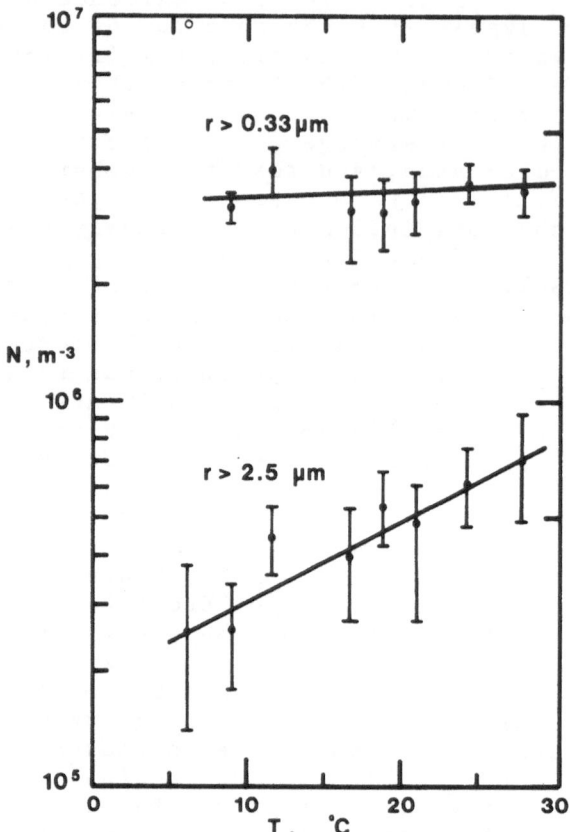

Figure 4. The number of particles per unit volume of hood,
with radius greater than r, $N = \int \Delta(\partial c/\partial r) dr$, produced by a
standard breaking wave event in the UCG whitecap simulation
tank, as a function of sea water temperature, based on
Figures 1 and 2 of Bowyer (1984), in Monahan et al (1984a).
Upper portion of figure: r equals 0.33 micrometers. Lower
portion of figure: r equals 2.5 micrometers.

wave-wave interactions in the simulation tank has more in
common with the spectra measured in the field by Blanchard
and Woodcock (1957), Kolovayev (1976), and Johnson and
Cooke (1979) than does the laboratory bubble spectrum
associated with the weir waterfall, particularly the
spectrum identified with the centre of the plume. It is
now appropriate to see if there are observational data which
can be used to determine which of these two models best
describes what is occurring in nature.
        Fairall, et al (1983) measured the aerosol spectra
within the marine atmospheric boundary layer on numerous
occasions during a 20 hour period of the Cooperative

Experiment for West Coast Oceanography and Meteorology [1] characterised by quite constant wind speeds close to 9ms^{-1}, normalised these spectra to a relative humidity of 80%, assumed a boundary layer aerosol budget expression that took into account the height of the boundary layer and the loss of particles by entrainment and fall out, and thereby calculated the average sea surface particle (droplet) flux, $\partial F_0/\partial r$, associated with this wind speed. Dividing the $\partial F_0/\partial r$ expression inferred by Fairall, et al (1983) from the CEWCOM-78 observations, by the W-value predicted for 9ms^{-1} winds by Monahan and O'Muircheartaigh (1980) (see Equation 24 in next section), yields the $W^{-1}\partial F_0/\partial r$ curve labeled CEWCOM on Figure 3. While the amplitude of the resultant expression is less than that of the $\tau^{-1}\partial E/\partial r$, i.e. Galway, expression throughout the one to ten micrometer radius range, these two curves are qualitatively similar. The CEWCOM $W^{-1}\partial F_0/\partial r$ expression differs markedly from the $\partial^2 E/\partial t\partial r$ expression determined from the Albany laboratory tank measurements.

There are certain characteristic features of the aerosol concentration spectra just above the sea surface which are reflected throughout the marine atmospheric boundary layer. These common spectral features provide an additional basis for testing the $W\tau^{-1}\partial E/\partial r$ and $W\partial^2 E/\partial t\partial r$ aerosol generation models. In figure 5 are to be found the average aerosol spectra associated with winds of 7ms^{-1} and 9ms^{-1}, as measured at 14m above the sea surface during the JASIN experiment (Monahan, et al, 1983c), and the aerosol spectra near cloud base as measured by Woodcock (1953) for comparable winds. To show the features more clearly, the quantity plotted is $\partial S/\partial r$, the fraction of the marine air volume occupied by sea-salt aerosol droplets which fall within a unit increment of radius, which is related to the numerical concentration spectrum, $\partial c/\partial r$, as shown in Equation 20.

$$\partial S/\partial r = \frac{4}{3}\pi r^3 \partial c/\partial r \tag{20}$$

All four of these spectra have peaks in the 2 to 4μm radius range, with the peaks in the cloud base spectra being somewhat broader than those in the near-sea-surface spectra.

Returning to a consideration of the quantity, whose magnitude can be predicted from the models, it is noted that $\partial F_0/\partial r$, the number of particles per unit radius increment produced per unit area of the sea surface per unit time must just be equal, under conditions of dynamic equilibrium, to the rate at which particles in each one micrometer radius band return back into the sea. Now this rate of return is equal to the product of the dry deposition

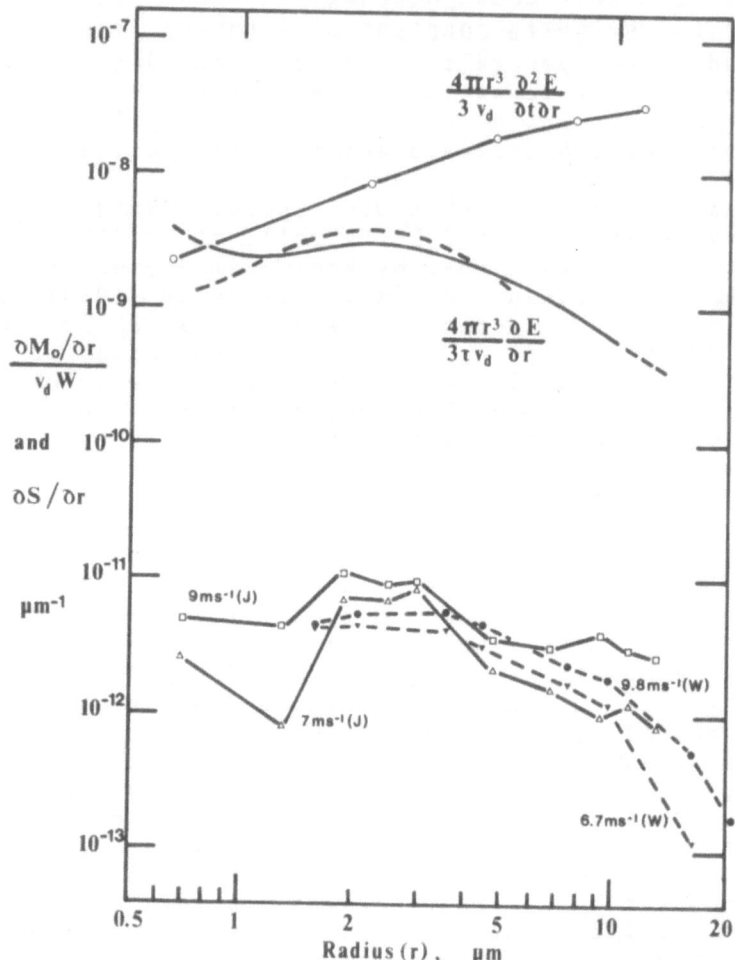

Figure 5. The volume concentration spectra at low elevation
over the sea sustainable by 100% whitecap coverage,
$(v_dw)^{-1}\partial M_0/\partial r$, as inferred from the Galway
$(4\pi r^3(3\tau v_d)^{-1}\partial E/\partial r)$, and from the Albany
$(4\pi r^3(3v_d)^{-1}\partial^2 E/\partial t\partial r)$, laboratory results, and the average
14m-elevation volume concentration spectra $(\partial S/\partial r)$ measured
during the Joint Air-Sea Interaction Experiment, (J), for
winds of $7ms^{-1}$, and for winds of $9ms^{-1}$. Near cloud-base
volume concentration spectra deduced from results in
Woodcock (1953), (W), for winds of $6.7ms^{-1}$, and for winds of
$9.8ms^{-1}$, included as dashed curves. The two solid
$(v_dW)^{-1}\partial M_0/\partial r$ curves obtained using v_d values from Slinn and
Slinn (1980) (Figure 2 of Slinn, 1983), while dashed curve
obtained using v_d values from Slinn and Slinn (1981)
(Figure 32 of Slinn, 1983).

velocity, v_d, and the numerical concentration spectrum, $\partial c/\partial r$, (Slinn, 1983), as stated in Equation 21.

$$\partial F_0/\partial r = v_d \quad \partial c/\partial r \tag{21}$$

If a quantity $\partial M_0/\partial r$ is defined as the volume of the droplets produced per unit radius increment, per unit area of the sea surface, per unit time, then it follows that,

$$\partial M_0/\partial r = \frac{4}{3}\pi r^3 \partial F_0/\partial r = \frac{4}{3}\pi r^3 v_d \partial c/\partial r = v_d \partial S/\partial r \tag{22}$$

and the fraction of the marine air volume expected to be filled with droplets in a particular one micrometer wide radius band in that unattainable circumstance where the sea is entirely covered by whitecaps, $W^{-1}\partial S/\partial r$, can be calculated as indicated in Equation 23.

$$W^{-1}\partial S/\partial r = (Wv_d)^{-1}\partial M_0/\partial r = \frac{4\pi r^3}{3Wv_d}\partial F_0/\partial r \tag{23}$$

This quantity, $(Wv_d)^{-1}\partial M_0/\partial r$, has been determined from the $\dot{W} \, \partial E/\partial r$ and $W \, \partial^2 E/\partial t\partial r$ models, making use of dry deposition velocity curves found in Slinn (1983), and the results plotted on Figure 5. The marked similarity in shape between the $(Wv_d)^{-1}\partial M_0/\partial r$ curve derived from the wave-wave interaction experiments and the $\partial S/\partial r$ curves describing the field observations is immediately apparent. Particular-ly of note is the peak at two to three micrometers radius. The $(Wv_d)^{-1}\partial M_0/\partial r$ curve based on the weir waterfall experiments does not conform in shape to the $\partial S/\partial r$ curves, and the $W \, \partial^2 E/\partial t\partial r$ model will therefore at this point be set aside in favour of the $\dot{W} \, \partial E/\partial r$ model. (In reviewing this chapter Cipriano has written that it is possible that the discrepancy between the Galway and Albany results is in part or in whole an artifact associated with the particular instruments used to measure the droplets in the Albany experiment).

Recognizing that the bigger bubbles, with their large film areas, are effective sources of film droplets, and that the smaller bubbles are the efficient generators of jet drops, it is not surprising that Cipriano and Blanchard (1981) inferred from the results of the large-bubble-rich weir waterfall experiment that in all droplet radius (at 80% relative humidity) categories up to 3.75µm (which corresponds to a sea water droplet diameter of 15µm) the droplets produced are primarily film droplets, and only in size ranges above 3.75µm radius do the jet drops dominate,

while from a consideration of the variation with
radius of the temperature dependence of $\partial E/\partial r$ found during
the wave-wave whitecap simulation tank experiments (Figure
4) Bowyer (1984) concluded that the "cross-over" from
predominately film droplets to mainly jet drops occurs at a
80% R.H. radius of only 1.5μm. When it is noted that
Woodcock (1972) suggested, based on his studies of Hawaiian
marine aerosols, that the transition from bubble-film
produced particles to bubble-jet produced particles falls
in the 10^{-17} to 10^{-16} kg (0.2 to 0.5μm 80% R.H. radius)
range, it again appears that the $\dot{W}\,\partial E/\partial r$ laboratory approach
comes closer to duplicating what happens on the sea surface
than does the $W\partial^2 E/\partial t\partial r$ approach.

The amplitude at 2μm radius of the JASIN $\partial S/\partial r$ curve
for 7ms^{-1} winds shown in Figure 5 is 2.35 X 10^{-3} times the
amplitude of the $\dot{W}\,\partial E/\partial r$ model's $(Wv_d)^{-1}\partial M_0/\partial r$ curve at 2μm
radius, i.e. the whitecap coverage for winds of 7ms^{-1} must
be 2.35 X 10^{-3}, or 0.24%, if the aerosol production as given
by Equation 6 (or 10) is to be that required to sustain the
average low elevation aerosol concentration observed at
7ms^{-1} winds during the JASIN experiment. In the following
sections the appropriate expressions describing whitecap
coverage, W, as a function of such variables as 10m-eleva-
tion wind, U, will be considered in detail. It will then
be feasible to test the conclusion reached above, as to the
whitecap coverage required when the winds are 7ms^{-1} to
produce the observed aerosol concentration, against the
independently arrived at expressions for W(U) based on
extensive shipboard observations of oceanic whitecaps.

5. OCEANIC WHITECAP COVERAGE

In order to estimate the rate of production of marine
aerosol droplets, per radius increment, per unit area of
sea surface, arising from the bursting of whitecap bubbles,
$\partial F_0/\partial r$, all that remains is to identify the W-value
appropriate for the time and place in question.

The first quantitative description of the wind
dependence of oceanic whitecap coverage appeared in
Blanchard (1963). From the analysis of photographs from a
Naval Weather Squadron manual (anon., 1952), Blanchard
concluded that whitecap coverage was proportional to the
square of the near-surface wind speed, and suggested that
therefore W was proportional to sea surface wind stress.

That whitecap coverage depends on the stability of the
lower atmosphere, as reflected by the previously introduced
quantity ΔT, as well as on the deck-height wind speed, was
inferred by Monahan (1969) from his quantitative observations of whitecapping on the Laurentian Great Lakes of North
America. This result confirmed numerous earlier qualitative

observations made at sea (Roll, 1965), and has subsequently
been supported by the results of a statistical analysis of
numerous whitecap coverage measurements made in diverse
regions of the world ocean (Monahan, et al, 1981). The
dependence of oceanic whitecap coverage on the atmospheric
stability as represented by ΔT, as well as on the 10m-
elevation wind speed, is graphically portrayed in Figure 6
(Monahan and O'Muircheartaigh, 1985). It is clear from this

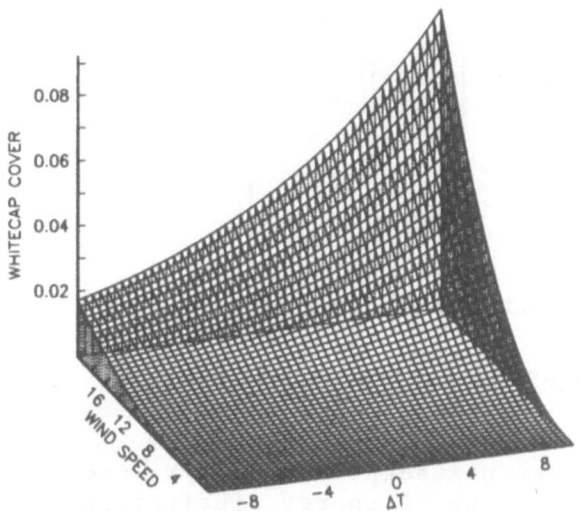

Figure 6. Oceanic whitecap coverage as a function of 10m-
elevation wind speed, in ms^{-1}, and of the temperature
difference between water and air, in C^0. Note that for a
given wind speed, a much smaller fraction of the sea
surface is covered by whitecaps when the lower atmosphere
is stable (ΔT negative) than when it is unstable (ΔT
positive).

figure that, for a given wind speed, there is much greater
whitecapping when the atmosphere is unstable (i.e. ΔT is
positive) than when the lower atmosphere is stably
stratified (ΔT negative).
        Cardone (1969) suggested that the instantaneous
whitecap coverage is proportional to the rate at which
energy is being dissipated out of the fully developed, high
frequency, end of the wave spectrum, and tested this hypothesis using the Great Lakes data set (Monahan, 1969), which
included estimates of fetch, to calculate for each whitecap
observation interval the rate of energy transfer from the
wind via a combined Miles-Phillips instability mechanism to
the fully developed spectral components of the wave field.
The results of this exercise, reproduced in Figure 7, fully

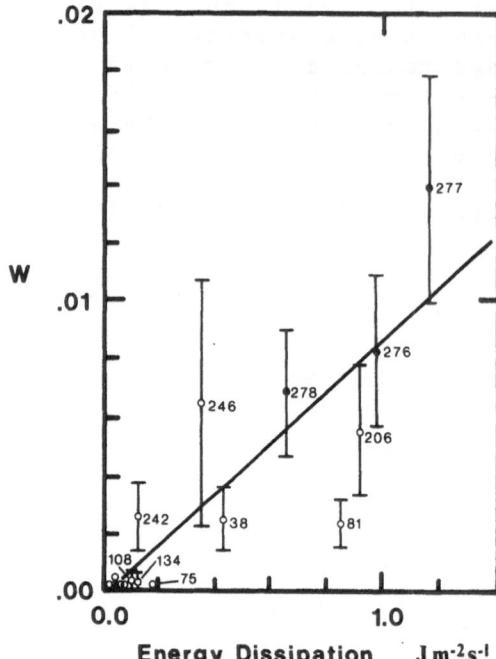

Figure 7. Fresh water whitecap coverage, W, from Monahan (1969), versus ε, the rate of energy dissipation out of the fully developed, high frequency, end of the wave spectrum. Redrawn from Cardone (1969). Note the simple proportionality between W and ε, as reflected by Cardone's straight line.

support Cardone's hypothesis. This is a most encouraging result. Since the wave generation theory utilised by Cardone is cast in terms of deck-height wind speed, atmospheric stability, fetch, and wind duration, it is now possible to estimate W in any circumstance where these four parameters are known. It is reasonable to suggest that it is not in fact the instantaneous whitecap coverage, but rather the rate of whitecap formation (which is just equal under steady state conditions to the rate of whitecap decay, $W\tau^{-1}$), that is proportional to the rate of energy dissipation from the fully developed portion of the wave spectrum (Monahan, 1971).

The influence of fetch on whitecap coverage is reflected in the results obtained from the analysis of some 1500 visual observations of whitecapping logged at the Alte Weser lightstation offshore from Bremerhaven (Monahan and Monahan 1985). For the equilibrium case of a fully developed sea, where fetch and duration are effectively infinite, Wu (1979) suggested that whitecap coverage would be proportional to the product of wind stress and surface drift velocity, i.e. to the rate at which work is done by the wind on a unit area of the sea surface. This work done by the wind is therefore proportional to the cube of the friction velocity, u_*, and Wu concluded, following on his earlier observation that the drag coefficient is proportional to the square root of the 10m-elevation wind speed, U (Wu, 1969), that this work would also be proportional to $U^{3.75}$. A review of all the earlier attempts to describe the wind dependence of whitecap coverage, W(U), by such a power-law, and a statistically optimal power-law expression for W(U), are to be found in Monahan and O'Muircheartaigh (1980). Applying the technique of robust biweight fitting to the whitecap data sets of Monahan (1971) and Toba and Chaen (1973), they obtained the optimal expression given in Equation 24, where

$$W = 3.84 \times 10^{-6} U^{3.41} \tag{24}$$

W is as before the fraction of the sea surface covered by whitecaps and U is the 10m-elevation wind speed expressed in ms^{-1}. Further discussion of these results is to be found in Wu (1982) and Monahan and O'Muircheartaigh (1982). A concomitant of the whitecap bubble cloud model presented previously, in which each individual bubble cloud diminishes exponentially with depth, is that the aggregate volume of all the bubble clouds beneath a unit area of ocean surface is simply proportional to W, the fraction of the sea surface covered by whitecaps, and hence the number of bubbles in each size category under a unit area of sea surface, $\partial Q/\partial R$, is likewise simply proportional to W. It is thus appropriate to consider the evidence for the U-dependence of $\partial Q/\partial R$, to see if it accords with the U-dependence of W as described above. By combining ambient noise measurements made in Queen Charlotte Sound, British Columbia, with a model in which the sound, which is assumed to be uniformly generated over the sea surface, is scattered and absorbed by a thin layer of bubbles, Farmer and Lemon (1984) concluded that the $\partial Q/\partial R$ associated with bubbles that resonate at 25.0 kHz, i.e. bubbles of 132 micrometers radius, and the $\partial Q/\partial R$ for bubbles that resonate at 14.5 kHz, i.e. have a radius of about 229 micrometers, both varied as u_*^3 over the wind speed range of 10 to 16 ms^{-1}. It should be noted that for wind speeds greater than about 5ms^{-1}, Kerman, et al (1983) reported that the intensity of ambient under-water sound in the kilohertz range, as measured at a depth of 4 km in the equatorial Pacific Ocean, was proportional to U^2, or $u_*^{1.5}$, while Shooter and Gentry (1981) found the ambient noise in the 0.15 to 0.5 kHz range to vary at a deep oceanic site as $U^{2.8}$ for winds greater than 9ms^{-1}.

It should be noted that the use of Equation 24 yields a

whitecap coverage of 0.00292, or 0.29%, for winds of $7 ms^{-1}$.
This is in good agreement with the value of W (0.00235) that
had to be introduced in Equation 6 to give the low elevation
aerosol concentration observed during the JASIN experiment
for such winds (see Figure 5). It is therefore appropriate
to suggest that Equation 24 be combined with Equation 10
(or Equation 6) and Equation 4, to provide an explicit
expression (Equation 25) for the rate of sea salt aerosol
production, $\partial F_0/\partial r$, for a given value of U.

$$\partial F_0/\partial r = 1.37\ U^{3.41} \frac{v_m}{v_m'}\ r^{-3} (1 + 0.057\ r^{1.05})\ \times$$

$$\times\ 10^{1.19} e^{-B^2} \tag{25}$$

$$B = (0.380 - \log r)\ /\ 0.650$$

In the instance where v_m can be taken equal to v_m'
Equation 25 is identical to Equation 12 of Monahan, et al
(1983a).

In a recent paper Burk (1984) has described the results
obtained from a set of experiments with a numerical model
of the marine atmospheric boundary layer that incorporated
as the sea surface aerosol production term the $\partial F_0/\partial r$
expression that appeared as Equation 12 in Monahan, et al
(1983a), an equation that differs from Equation 25 only by
the omission of the multiplier v_m/v_m'. Starting with an
aerosol free boundary layer, the model was integrated until
a quasi-equilibrium aerosol particle population, $\partial c/\partial r$, was
obtained. The aerosol spectra at 25m-elevation predicted by
these experiments were described by Burk as bearing a
"striking resemblence" in "magnitude and shape" to the
spectra obtained from the near-cloud-base measurements of
Woodcock (1953).

It has often been suggested that there is a threshold
wind speed below which no whitecaps are formed (e.g. Munk,
1947; Blanchard, 1963; Monahan, 1971). Recent results
(Monahan and O'Muircheartaigh, 1985) indicate that there is
no single threshold velocity, but rather that this Beaufort
velocity, U_B, is a function of atmospheric stability and sea
surface temperature.

The T_w-dependence of whitecap coverage at wind speeds
above U_B has been well documented by Bortkovsky (1983), some
of whose findings are reproduced in Figure 8. Qualitatively
similar results have recently arisen out of the Marginal
Ice Zone Experiment (Monahan, et al, 1984). The observed
increase in W with increasing T_w and constant U can be
explained in terms of the shift in bubble spectrum with
changing T_w mentioned previously. One working hypothesis

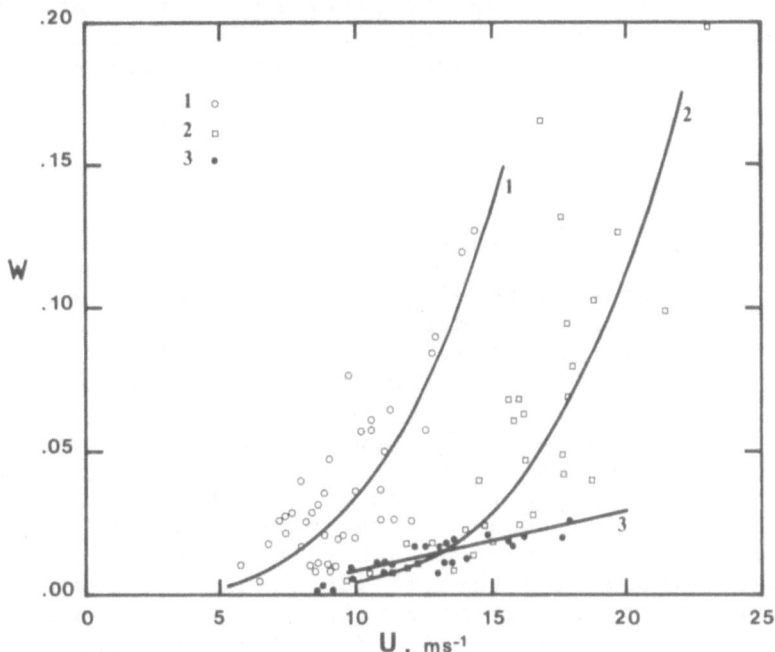

Figure 8. Proportion of the sea surface covered by foam and
whitecaps, as a function of wind speed. Based on Figure
2.3 of Bortkovsky (1983). Open circles and curve 1, for
water temperature, T_w, of 27°C. Squares and curve 2, for
T_w between 3° and 15°C. Filled circles and curve 3, T_w less
than or equal to 3°C.

is that the effectively smaller bubbles of the warm water
whitecap bubble cloud take longer to reach the surface, in
spite of the relatively lower viscosity at high T_w, than
do the larger bubbles that characterise the cold water
bubble cloud. It follows that for the same rate of whitecap formation, $W\tau^{-1}$, the effectively larger τ in warm water
leads to a larger instantaneous W than in the cold water
case. This explanation, based on the dependence of τ on
the bubble spectrum, is similar to the one put forth to
explain why W is less on the Great Lakes than on the ocean
for the same U, and presumably the same $W\tau^{-1}$ (Monahan,
1971).
    One additional factor influencing W is the organic
content of the surface sea water. Organic material can
alter the terminal rise velocity of bubbles (Detwiler and
Blanchard, 1978; Thorpe, 1982), and their persistence at the
sea surface (Blanchard and Syzdek, 1978), thus influencing
τ and hence W. Organic material may alter $\partial E/\partial r$, by
changing the efficiency with which bubbles produce droplets

and the size of the droplets produced, as well as altering
the height to which jet drops are ejected (Blanchard and
Hoffman, 1978). Sufficient information is not yet available
to quantify the dependence of W and $\partial E/\partial r$ on the surface sea
water concentration of typically encountered organic
substances.

6. GLOBAL SEA-TO-AIR SALT FLUX

With a marine aerosol generation formula such as that
given by Equation 25 to hand, it becomes possible, given an
adequate meteorological data set, to calculate the annual
sea-to-air flux of salt associated with jet drops for the
entire world ocean. It is of course not sufficient to have
a knowledge of the distribution of monthly, or annual, mean
winds over the earth's surface. The pitfalls associated
with trying to deduce the mean of U^2, which is markedly
different from the square of the mean value of U, from
climatological charts of mean U values were pointed out by
Blanchard (1963), and the difficulties encountered in
attempting to arrive at mean values of U^3 from such charts
were discussed by Monahan (1983) and by Blanchard (1985).
Fortunately Hellerman has summarised, by 2^0x 2^0 geographical "squares", the mean monthly values of such
quantities as U^3, U^4, and $U^3\Delta T$, determined from the TDF-11
file of some 30 million shipboard observations (Hellerman
and Rosenstein, 1983). Hellerman has also summarised
(personal communication) for each such geographical region
and for each month the rate at which the wind does work on
a unit area of the sea surface, P, a quantity he determined
by the use of Equation 26.

$$P = \rho \quad C_D(U,\Delta T) \ U^3 \qquad\qquad\qquad (26)$$

(It is to be noted that this equation differs from the
previously described expression of Wu (1979) for the same
physical quantity, in that Hellerman states that P is simply
proportional to the drag coefficient C_D, while Wu contended
that P was proportional to $C_D^{1.5}$). The wind-and stability-
dependent drag coefficient, C_D (U,ΔT), used by Hellerman
was that given in Bunker (1976). While O'Muircheartaigh
(personal communication) found that the warm water whitecap
observations of Monahan (1971) and Toba and Chaen (1973)
could be described using a U^3 expression without undue
reduction in the goodness-of-fit, Spillane et al (1985)
chose to use the conceptually more satisfying expression
given in Equation 27, where the coefficient of P was
determined using these same two warm water W, U data sets.

$$W = 6.667 \times 10^{-6}P \qquad\qquad (27)$$

In figure 9 are reproduced the mean monthly global whitecap charts for January and July extracted from the

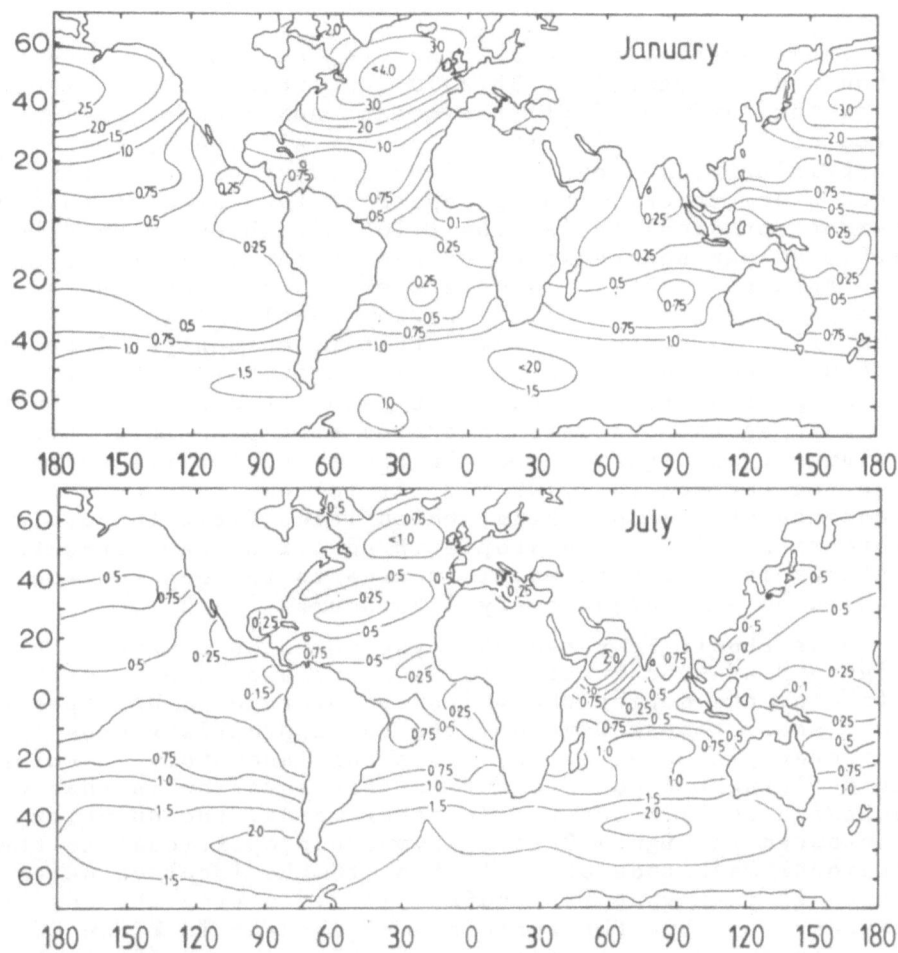

Figure 9. Charts of world ocean showing monthly average whitecap coverage, in percent, for January, and for July, from Spillane et al (1985).

comprehensive whitecap atlas found in Spillane, et al (1985). Using Equation 28, which differs but slightly from Equation 25, Monahan and Spillane (1983) determined that the annual salt flux from the world ocean between $70°N$ and $70°S$, associated with droplets whose radii fall between 0.8 and 10 micrometers, is 3.50×10^{12} kg, i.e. 3.5×10^{9} metric tons. The total mass of sea water contained at the instant

$$\partial F_0 / \partial r = 3.585 \times 10^5 \ W \ r^{-3} \ (1+0.0033r^{1.179})x$$

$$x \ 10^{1.19e^{-B^2}} \tag{28}$$

$$B = (0.380 - \log r) / 0.650$$

of their generation in all the droplets in this size range produced per year is 9.98×10^{13} kg, which if distributed over the surface of all the oceans and seas of the earth, an area of 3.61×10^{14} m^2 (Sverdrup, et al, 1942), would, quite surprisingly, be represented by a layer of sea water less than 0.3mm thick. The aggregate surface area of this annual harvest of droplets is 4.64×10^{16} m^2, which amounts to 128.5 m^2 per square meter of sea surface. If these droplets are assumed to be jet-droplets whose radii are approximately one-tenth the radii of the parent bubbles (Blanchard, 1963), and if it is further assumed that five such jet-droplets are produced per bubble (Cipriano and Blanchard, 1981), then the total area of the new surfaces associated with the bubbles as well as the droplets generated each year in a vertical column of one square meter in cross-section rising through the sea surface is 2,700 m^2. It follows that as a consequence of wave breaking, which leads first to bubble formation and then to droplet production, new air-water interfacial surfaces equal in area to the sea surface are generated on average every 3.25 hours.

It should be noted that the whitecap coverage charts of Figure 9 are uncorrected for sea surface temperature (T_w) effects, and depict the whitecap coverage to be expected on warm seas. These W-values are the appropriate ones to incorporate in Equation 28, since the laboratory τ'-value used in defining the coefficient in this equation is that which pertains to 19°C water. In other words, the quantity contoured in Figure 9 is everywhere proportional to the regional magnitude of $W\tau^{-1}$, i.e. to the local value of $\dot{W}$, which is just what is required in evaluating the $\partial F_0 / \partial r$ expression. But the influence of changes in T_w on $\Delta(\partial n/\partial r)$, i.e. on $\partial E/\partial r$, associated presumably with shifts in the spectrum of the bubbles in the whitecap cloud, is not taken into account in Equation 28, and hence is not reflected in the global results quoted from Monahan and Spillane (1983). Incorporating the T_w-dependence of $\Delta(\partial n/\partial r)$ found by Bowyer (1984) for droplet sizes corresponding to the peak in the $\partial M_0/\partial r$ production spectrum, Monahan, et al (1984b) found that the annual sea-to-air salt flux for the North Atlantic Ocean would be reduced to 2.04×10^{11} kg, i.e. would be reduced 17.8%, from the previous estimate of 2.47×10^{11} kg. If this adjustment is taken as typical for the world ocean,

then the annual salt flux from the sea surface between 70^{0}N
and 70^{0}S associated with droplets in the radius range of
0.8 to 10 micrometers would be reduced to 2.89 x 10^{12}kg,
and the mass of sea water injected into the atmosphere each
year would then be 8.24 x 10^{13} kg.

Having calculated the annual global salt flux by
integrating the sea surface aerosol generation expression
given in Equation 28 (or Equation 25) over droplet radii
from 0.8 to 10 micrometers, over the surface of the world
ocean from 70^{0}N to 70^{0}S, and over the period of a calendar
year, it is now possible to subject the $\partial F_0/\partial r$, or $\dot{W} \partial E/\partial r$,
model to one further test. Several workers have inferred
the annual sea-to-air salt flux by calculating the annual
removal of sea salt from the atmosphere via such mechanisms
as dry deposition (fallout) and rain scrubbing (washout),
and by making the assumption that the tropospheric sea salt
load does not change significantly from year to year. In
this manner Eriksson (1959) concluded that 10^{9} tons, i.e.
10^{12} kg, of salt are injected into the air from the sea each
year, while Blanchard (1963) inferred that this figure
should be 10^{10} tons, i.e. 10^{13}kg. Noting that the result
obtained using the $\partial F_0/\partial r$ generation model, approximately
3 x 10^{12}kg, falls near the geometric mean of these two
atmospheric salt budget estimates, and is equally close to
the several other recent estimates of annual salt production
summarised by Blanchard (1985), it could be argued that the
validity of $\partial F_0/\partial r$ generation model has been at least
provisionally demonstrated.

But if it is acknowledged that many of the sea spray
droplets generated at the sea surface never rise to
significant heights above the waves, contributing therefore
only to the low-elevation high salt concentration zone
described in Blanchard and Woodcock (1980) and not to the
general sea salt aerosol population of the marine atmospheric boundary layer whose maintenance was the subject of
Eriksson's and Blanchard's budget calculations, then it may
be inferred that the $\partial F_0/\partial r$ model is lacking in some manner,
even if it appears to quite adequately explain the sea-salt
aerosol population under moderate wind conditions, e.g
7 ms^{-1}.

7. TOWARD A COMPREHENSIVE MARINE AEROSOL GENERATION MODEL

The JASIN aerosol spectra (Monahan, et al, 1983c)
associated with relatively high winds, i.e. U greater than
about 11 ms^{-1}, show an increase in $\partial S/\partial r$ as the droplet
radius increases beyond 10 micrometers. This increase in
$\partial S/\partial r$, as seen clearly in the aerosol spectra identified
with winds of 13 ms^{-1} and 15 ms^{-1}, continues to the upper
limit of the range of measurement, i.e. to r equals 15

micrometers. It is apparent that for these winds a second,
as yet unresolved, peak in the $\partial S/\partial r$ spectrum exists at a
droplet radius larger than 15 micrometers. While the
ubiquitous $\partial S/\partial r$ peak at 2 to 3 micrometers is identified
with the jet droplets that arise from the collapse of the
whitecap bubbles, this second $\partial S/\partial r$ peak at large radius is
the signature of the spume drops, those drops produced by
the direct mechanical disruption of the wave crests by the
wind (as depicted on the far right of Figure 10), a

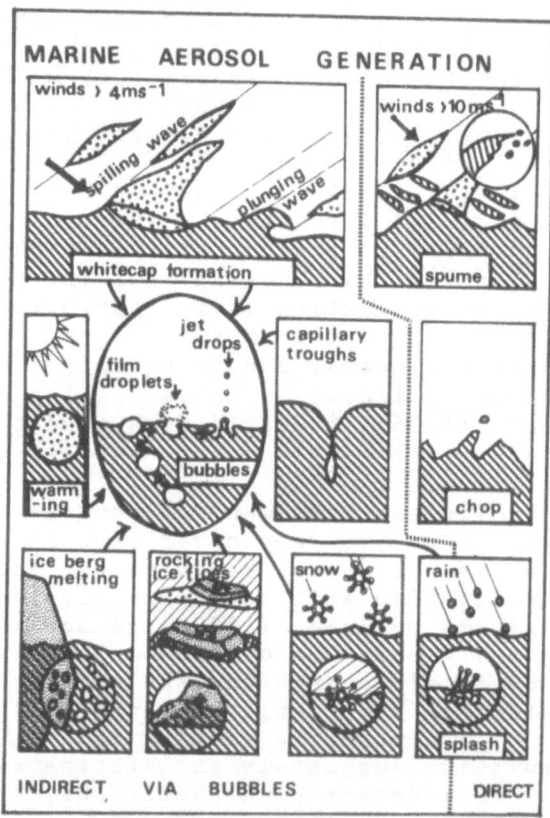

Figure 10. Direct and indirect mechanisms that result in
the production of sea-salt aerosol particles at the ocean
surface.

mechanism that only becomes a significant contributor of
aerosol particles at high winds (Wang and Street, 1978).
Synthesising certain of the wind flume results of Wu (1973)
and Lai and Shemdin (1974), Monahan, et al (1983a) attempted
to give crude by explicit form to this direct aerosol
production component, $\partial F_1/\partial r$, of the comprehensive aerosol
generation model outlined in Equation 29. Burk (1984), in

describing the results of his recent numerical experiments, reported that the inclusion of elements of this initial $\partial F_1/\partial r$ expression in the sea surface aerosol production function led to a far too high steady state concentration of particles larger than 10 micrometers for constant winds of 15 ms^{-1}, and concluded that the functional dependence on wind speed incorporated in this $\partial F_1/\partial r$ term is much too strong. A revised, but

$$\partial F/\partial r = \partial F_0/\partial r + \partial F_1/\partial r \qquad\qquad (29)$$

still conjectural, expression for $\partial F_1/\partial r$ is to be found in Monahan, et al (1985). This still unproven expression is included here as Equation 30 solely to provide a point of

$$\partial F_1/\partial r = 6.45 \times 10^{-4}\ e^{2.08}\ U\ r^{-3} e^{-D^2} \qquad (30)$$

$$D = 2.18(1.88 - \log r)$$

departure for discussion. Some such Junge-type production spectrum modified by a Gaussian envelope should prove to be an appropriate description of spume drop production. While it is hoped that further insights into spume drop production will come from on-going wind flume studies, low-elevation measurements at sea of the concentration of large drops during high winds, involving the use of such devices as de Leeuw's wave riding Rotorod sampler (de Leeuw, et al, 1984), will perhaps shed the clearest light on the direct spray production mechanism.

To complete the description of the indirect aerosol production mechanism, i.e. to refine the current $\partial F_0/\partial r$ expression as given in Equation 25 (or Equation 28), it will be necessary to extend the whitecap simulation tank measurements to include those jet droplets which are too large to be counted by the aerosol spectrometer probes now in use. Likewise, the laboratory tank observations need to be extended to include the sub-micrometer droplets too small to be detected with these instruments. In this fashion the $\partial F_0/\partial r$ production term can be expanded to include the film-droplet component.

Every effort should be made to carry out both laboratory and field measurements over the greatest possible range of water temperatures, so that the T_w-dependence of the $\partial F_0/\partial r$ and $\partial F_1/\partial r$ terms can be accurately assessed.

Mechanisms beyond those associated with winds and waves can give rise to bubbles and hence spray droplets. Blanchard and Woodcock (1957) have identified the super-saturation of the surface waters during spring warming, and

the impingement of raindrops and snow flakes on the sea
surface, as additional mechanisms that will produce bubbles.
When the glacial ice of ice-bergs melts numerous bubbles are
released (Joseberger, 1980), which will, when they reach the
sea surface, contribute to the production of marine aerosols.
Likewise, when ice floes rock back and forth on the sea
surface in response to waves passing beneath them they
agitate the surface and produce many bubbles (Stabeno,
1984), which will also produce sea-salt aerosol particles
when they burst. All these mechanisms need be included in
a truly comprehensive model describing the generation of
aerosol particles at the sea surface.

This discussion has been restricted to a consideration
of sea-salt particle production models appropriate to the
open ocean. A supplementary model accounting for the
generation of spray droplets in the surf zone will also be
required to predict the aerosol population of the
atmospheric boundary layer in coastal regions.

ACKNOWLEDGMENTS

We are most grateful for the sustained collaboration
of Prof. K.L. Davidson and his colleagues at the Naval
Postgraduate School, Monterey, California, in our
observational and modeling programmes, and for the aid of
Prof. I.G. O'Muircheartaigh and Dr. M.C. Spillane of
University College, Galway, in the interpretation of the
field results. The co-operation of the members of the UCG
Physical Oceanography Unit, particularly the work of
Ms. M.R. Higgins, Mrs. R.T. Brock, Mr. T. Luibheid, and the
others who analyse the numerous whitecap video tapes and
films, of Mrs. F.A. Yates who typed the drafts of this
chapter, and of Mr. D.M. Doyle and Mrs. M.M. Watson who
helped in the preparation of the illustrations, is
acknowledged with thanks, as is the assistance with the
simulation tank and field experiments, received from the
technical staffs of the U.C.G. departments of Oceanography
and Experimental Physics.
We thank Mr. S. Hellerman of GFDL, Princeton, for a
most useful summary of the TDF-11 file, and Dr. H. Gienapp
of DHI, Hamburg, for the extensive Alte Weser Whitecap log.
We are grateful to the various participants at the Bombannes
Advanced Study Institute for their comments on this chapter,
and to D. Ericksson and R.J. Cipriano in particular for
their reviews.
Our research on whitecapping has been supported
throughout by the U.S. Office of Naval Research. The on-
going whitecap programme at U.C.G. is supported via ONR
Grant N00014-78-G-0052. Our recent working visit to the
Woods Hole Oceanographic Institution was made possible

by ONR Contract N00014-85-M-0065 with W.H.O.I.

REFERENCES

Anonymous, 1952: Wind estimations from aerial observations
of sea conditions. (unpublished report) Weather
Squadron Two (VJ-2), Jacksonville, Florida.
Blanchard, D.C., 1963: The electrification of the atmosphere
by particles from bubbles in the sea. Prog. Oceanog.,
1, 71-202.
Blanchard, D.C., 1983: The production, distribution, and
bacterial enrichment of the sea-salt aerosol. In Liss
and Slinn, 1983 (op.cit.), 407-454.

Blanchard, D.C., 1985: The oceanic production of atmospheric
sea salt. J. Geophys. Res., 90, 961-963.
Blanchard, D.C., and E.J. Hoffman, 1978: Control of Jet Drop
Dynamics by Organic Material in Sea Water. J. Geophys.
Res., 83, 6187-6191.
Blanchard, D.C., and L.D. Syzdek, 1978: Seven problems in
bubble and jet drop researches. Limnol. Oceanogr.,
23, 389-400.
Blanchard, D.C., and A.H. Woodcock, 1957: Bubble formation
and modification in the sea and its meteorological
significance. Tellus, 9, 145-158.
Blanchard, D.C., and A.H. Woodcock, 1980: The production,
concentration and vertical distribution of the sea-
salt aerosol. Annals N.Y. Acad. Sci., 338, 330-347.
Bortkovsky, R.S., 1983: Heat and Moisture Exchange between
Atmosphere and Ocean under Storm Conditions, Hydro-
meteorological Pub. House, Leningrad, 160pp.
Bowyer, P.A., 1983: New estimate of the space charge
produced by a whitecap in the laboratory. In Monahan,
et al, 1983b (op.cit.), 60-67.
Bowyer, P.A., 1984: Aerosol production in the whitecap
simulation tank as a function of water temperature, and
space charge measurements during MIZEX 84. In Monahan,
et al, 1984a (op.cit.), 95-103.
Bunker, A., 1976: Computations of surface energy flux and
annual air-sea interaction cycles of the North Atlantic
Ocean. Mon.Wea. Rev., 104, 1122-1140.
Burk, S.D., 1984: The generation, turbulent transfer and
deposition of the sea-salt aerosol. J. Atmos. Sci.,
41, 3040-3051.
Cardone, V.J., 1969: Specification of the wind distribution
in the marine boundary layer for wave forecasting.
Tech. Rep. GSL-69-1, New York University, 1-131.
Cipriano, R., 1979: Bubble and Aerosol Spectra Produced by a
Laboratory Simulation of a Breaking Wave. Ph.D. Thesis,
State University of New York at Albany.
Cipriano, R.J., and D.C. Blanchard, 1981: Bubble and aerosol

spectra produced by a laboratory 'breaking wave'. J.
    Geophys. Res., 86, 8085-8092.
Cipriano, R.J., D.C. Blanchard, A.W. Hogan, and G.G. Lala,
    1983: On the production of Aitken nuclei from breaking
    waves and their role in the atmosphere. J. Atmos. Sci.
    40, 469-479.
de Leeuw, G., L.H. Cohen, and M.M. Moerman, 1984: Survey of
    aerosol and lidar measurements at the North Atlantic
    (25 May-28 June, 1983). Physics Laboratory, TNO, the
    Hague, the Netherlands, 98 pp.
Detwiler, A., and D.C. Blanchard, 1978: Aging and bursting
    bubbles in trace-contaminated water. Chem. Engin. Sci.,
    33, 9-13.
Eriksson, E., 1959: The yearly circulation of chloride and
    sulfur in nature; meteorological, geochemical and
    pedological implications. Part 1. Tellus, 11, 375-403.
Fairall, C.W., K.L. Davidson, and G.E. Schacher, 1983: An
    analysis of the surface production of sea-salt aerosols.
    Tellus, 35B, 31-39.
Farmer, D.M., and D.D. Lemon, 1984: The influence of bubbles
    on ambient noise in the ocean at high wind speeds. J.
    Phys. Oceanogr., 14, 1762-1778.
Hayami, S., and Y. Toba, 1958: Drop production by bursting
    of air bubbles on the sea surface (1) experiments at
    still sea water surface. J. Ocean. Soc. Japan, 14,
    145-150.
Hellerman, S., and M. Rosenstein, 1983: Normal monthly wind
    stress over the world ocean with error estimates. J.
    Phys. Oceanogr. 13, 1093-1104.
Johnson, B., and R.C. Cooke, 1979: Bubble populations and
    spectra in coastal waters; a photographic approach. J.
    Geophys. Res., 84, 3761-3766.
Joseberger, E.G., 1980: The effect of bubbles released from
    a melting ice wall on the melt-driven convection in
    salt water. J. Phys. Oceanogr., 10, 474-477.
Kerman, B.R., D.L. Evans, D.R. Watts, and D. Halpern, 1983:
    Wind dependence of underwater ambient noise. Bound.
    Layer Meteor., 26, 105-113.
Kolovayev, P.A., 1976: Investigation of the concentration
    and statistical size distribution of wind-produced
    bubbles in the near-surface ocean layer. Oceanology,
    15, 659-661.
Lai, R.J., and O.H. Shemdin, 1974: Laboratory study of the
    generation of spray over water. J. Geophys. Res., 79,
    3055-3063.
Liss, P.S., and W.G.N. Slinn, eds., 1983: The Air-Sea
    Exchange of Gases and Particles. D. Reidel, Dordrecht,
    Holland, 561 pp.
Monahan, E.C., 1969: Fresh water whitecaps. J. Atmos. Sci.,
    26, 1026-1029.
Monahan, E.C., 1971: Oceanic whitecaps. J. Phys. Oceanog.,1,

139-144.

Monahan, E.C., 1979: The influence of whitecaps on the marine atmosphere, Annual report for the year ending 30 June 1979. University College, Galway, Ireland, 171pp.

Monahan, E.C., 1982: Comments on "Bubble and aerosol spectra produced by a laboratory 'breaking wave'" by R.J. Cipriano and D.C. Blanchard. J. Geophys. Res., 87, 5865-5867.

Monahan, E.C., 1983: Positive charge flux from the world ocean resulting from the bursting of whitecap bubbles. in L.H. Ruhnke and J. Latham, Eds. Proceedings in Atmospheric Electricity, A. Deepak Publishing, Hampton, Virginia, 85-87.

Monahan, E.C., P.A. Bowyer, and M.C. Spillane, 1984b: The Temperature Dependence of Whitecap Aerosol Productivity and the Implications for Regional Sea-Air Salt Fluxes, S10.10. Terra Cognita, 4, 347-348.

Monahan, E.C., K.L. Davidson, and D.E. Spiel, 1982: Whitecap aerosol productivity deduced from simulation tank measurements. J. Geophys. Res., 87, 8898-8904.

Monahan, E.C., C.W. Fairall, K.L. Davidson, and P. Jones Boyle, 1983c: Observed inter-relations between 10m winds, ocean whitecaps and marine aerosols. Quart. J. Roy. Met. Soc., 109, 379-392.

Monahan, E.C., and G. Mac Niocaill, 1986: Oceanic Whitecaps and their role in Air-Sea Exchange Processes. D. Reidel Publishing Co., Dordrecht, Holland.

Monahan, E.C., and C.F. Monahan, 1986: The influence of fetch on whitecap coverage as deduced from the Alte Weser lightstation observer's log. In Monahan and Mac Niocaill, 1986 (op.cit.).

Monahan, E.C., I. O'Muircheartaigh, 1980: Optimal power-law description of oceanic whitecap coverage dependence on wind speed. J. Phys. Oceanogr., 10, 2094-2099.

Monahan, E.C., and I.G. O'Muircheartaigh, 1982: Reply (to comments of J. Wu on "Optimal power-law description of oceanic whitecap coverage dependence on wind speed"), J. Phys. Oceanogr., 12, 751-752.

Monahan, E.C., and I.G. O'Muircheartaigh, 1985: The significance of whitecaps to marine remote sensing. Int. J. of Remote Sens., (submitted).

Monahan, E.C., I.G. O'Muircheartaigh, and M.P. FitzGerald, 1981: Determination of surface wind speed from remotely measured whitecap coverage, a feasibility assessment. Proceedings of an EARSeL-ESA Symposium, Application of Remote Sensing Data on the Continental Shelf, Voss, Norway, 19-20 May, 1981, European Space Agency SP-167, 103-109.

Monahan, E.C., B.D. O'Regan, and K.L. Davidson, 1979: Marine aerosol production from whitecaps. Inter-Disciplinary

Symposia, Abstracts and Timetable, International Union
of Geodesy and Geophysics, XVII General Assembly,
Canberra, Australia, 3-15 December, 1979, 423.

Monahan, E.C., D.E. Spiel, and K.L. Davidson, 1983a: Model
of marine aerosol generation via whitecaps and wave
disruption. Ninth Conference on Aerospace and Aero-
nautical Meteorology, 6-9 June, 1983, Omaha, Nebraska,
American Meteorological Society, Preprint Volume,
147-158.

Monahan, E.C., D.E. Spiel, and K.L. Davidson, 1986: A model
of marine aerosol generation via whitecaps and wave
disruption. In Monahan and Mac Niocaill, 1986 (op.
cit.).

Monahan, E.C. and M.C. Spillane, 1983: The role of oceanic
whitecaps in the exchange of mass across the air-sea
interface, 20/23. IUGG Inter-Disciplinary Symposia,
Programme and Abstracts, Vol. 2, International Union
of Geodesy and Geophysics, XVIII General Assembly,
Hamburg, F.R.G., 15-27 August, 1983, 884.

Monahan, E.C., M.C. Spillane, P.A. Bowyer, D.M. Doyle and
P.J. Stabeno, 1983b: Whitecaps and the Marine Atmos-
sphere, Report No. 5, to the Office of Naval Research,
from University College, Galway, 93pp.

Monahan, E.C., M.C. Spillane, P.A. Bowyer, M.R. Higgins, and
P.J. Stabeno, 1984a: Whitecaps and the Marine
Atmosphere, Report No. 7, to the Office of Naval
Research from University College, Galway, 103pp.

Monahan, E.C., and C.R. Zietlow, 1969: Laboratory
comparisons of fresh-water and salt-water whitecaps.
J. Geophys. Res., 74, 6961-6966.

Munk, W.H., 1947: A critical wind speed for air-sea
boundary processes. J. Marine Res., 6, 203-218.

Pounder, C., 1986: Sodium chloride and water temperature
effects on bubbles. In Monahan and Mac Niocaill, 1986
(op.cit.).

Resch, F., 1986: Oceanic air bubbles as generators of
marine aerosols. In Monahan and Mac Niocaill, 1986
(op.cit.).

Roll, H.U., 1965: Physics of the Marine Atmosphere.
Academic Press, New York. 426 pp.

Schacher, G.E., K.L. Davidson, C.W. Fairall, and D.E. Spiel,
1981: Calculation of optical extinction from aerosol
spectral data. Appl. Optics, 20, 3951-3957.

Shooter, J.A., and M.L. Gentry, 1981: Wind generated noise
in the Parece Vela Basin. J. Acoust. Soc. Am., 70,
1757-1761.

Slinn, S.A., and W.G.N. Slinn, 1980: Predictions for
particle deposition on natural waters. Atmos. Env.,
14, 1013-1016.

Slinn, S.A., and W.G.N. Slinn, 1981: Modeling of atmospheric
particulate deposition to natural waters. In

Atmospheric Pollutants in Natural Waters, S.J. Eisenreich, ed., 1981, Ann Arbor Sci. Pub., Ann Arbor, Michigan, 23-53.

Slinn, W.G.N., 1983: Air-to-sea transfer of particles. In Liss and Slinn, 1983, (op.cit.), 299-405.

Spillane, M.C., E.C. Monahan, P.A. Bowyer, D.M. Doyle and P.J. Stabeno, 1986: Whitecaps and global fluxes. In Monahan and Mac Niocaill, 1986 (op.cit.).

Stabeno, P.J., 1984: Formation of foam about ice floes. In Monahan, et al, 1984a (op.cit.), 44-59.

Sverdrup, H.U., M.W. Johnson, and R.H. Fleming, 1942: The Oceans, their Physics, Chemistry, and General Biology. Prentice-Hall, Inc., Englewood Cliffs, N.J., 1087 pp.

Thorpe, S.A., 1982: On the clouds of bubbles formed by breaking wind-waves in deep water, and their role in air-sea gas transfer. Phil. Trans. R. Soc. London, A304, 155-210.

Thorpe, S.A., 1984a: A model of the turbulent diffusion of bubbles below the sea surface. J. Phys. Oceanogr., 14, 841-854.

Thorpe, S.A., 1984b: On the determination of K_v in the near-surface ocean from acoustic measurements of bubbles. J. Phys. Oceanogr., 14, 855-863.

Thorpe, S.A., 1986: Bubble clouds: a review of their detection by sonar, of related models, and of how K_v may be determined. In Monahan and Mac Niocaill, 1986 (op.cit.).

Toba, Y., and M. Chaen, 1973: Quantitative expression of the breaking of wind waves on the sea surface. Records of Oceanographic Works in Japan, 12, 1-11.

Wang, C.S., and R.L. Street, 1978: Measurements of spray at an air-water interface. Dyn. Atmos. Oceans, 2, 141-152.

Woodcock, A.H., 1953: Salt nuclei in marine air as a function of altitude and wind force. J. Met., 10, 362-371.

Woodcock, A.H., 1972: Smaller salt particles in oceanic air and bubble behavior in the sea. J. Geophys. Res., 77, 5316-5321.

Wu, J., 1969: Wind stress and surface roughness at air-sea interface. J. Geophys. Res., 74, 444-455.

Wu, J., 1973: Spray in the atmospheric surface layer: Laboratory study. J. Geophys. Res., 78, 511-519.

Wu, J., 1979: Oceanic whitecaps and sea state. J. Phys. Oceanogr., 9, 1064-1068.

Wu, J., 1982: Comments on "Optimal power-law description of oceanic whitecap coverage dependence on wind speed". J. Phys. Oceanogr., 12, 750-751.

THE OCEAN AS A SINK FOR ATMOSPHERIC PARTICLES

Patrick Buat-Ménard
Centre des Faibles Radioactivités,
Laboratoire mixte CNRS-CEA,
Domaine du CNRS, BP N°1
91190, Gif sur Yvette,
France.

1. OVERVIEW

The understanding of the geochemical transport of materials from land to sea and from sea to land via the atmosphere depends to a great degree on a quantitative assessment of the processes by which aerosol particles are removed from the marine atmosphere. Such an understanding requires simultaneously to differentiate amongst the various sources which can contribute to the atmospheric burden of materials attached to such particles. Indeed, areosol particles over the ocean, which we will refer to as marine aerosols, are not only derived from the ocean. They can be produced by a variety of sources on land, natural and/or anthropogenic (i.e., soil erosion, volcanic activity, emissions from the terrestrial biomass and human activity through industrial and agricultural practices). Such a production can be either direct, by mechanical processes, or indirect, due to gas to particle conversion (Prospero et al., 1983). In the size-range of 0.1 to 20 μm where most of the mass of the atmospheric aerosol is found long-range transport (>1000 km) of materials over ocean waters can be expected before their deposition onto the ocean surface by dry and wet removal processes. Dry deposition, which refers here to all deposition processes which occur in the absence of precipitation, may be more or less continuous whereas the removal of aerosol particles by precipitation is essentially discontinuous. The theoretical aspects of dry and wet deposition of aerosol particles to the ocean surface will not be reviewed here since this topic has already been treated extensively (see e.g. Slinn, 1983).

In principle, knowledge of aerosol particle deposition processes combined with the use of atmospheric transport models should allow an evaluation to be made of the net air to sea transfer rate of particulate materials, provided accurate source-emission inventory data are available. At the present time such an approach is not possible and we must rely on experimental field measurements for such an assessment. One major conclusion which has emerged in the recent years is that the deposition rate of particulate material from the atmosphere to the ocean exhibits a strong spatial and temporal variability. This

165

P. Buat-Ménard (ed.), The Role of Air-Sea Exchange in Geochemical Cycling, 165–183.
© *1986 by D. Reidel Publishing Company.*

is primarily because of a) meteorological factors (air-mass circulation patterns; see e.g. Merrill this volume), b) the geographical variability of continental source strengths and c) the fact that aerosol particles have very short residence times in the troposphere (1 day to 1 month) and cannot therefore be homogeneously distributed (Prospero et al., 1983).

It should also be pointed out that geographical variability is also a characteristic of the oceanic system especially for trace elements (Buat-Ménard, this volume). Thus any reliable assessement of the role of air to sea transfer of particles on ocean chemistry cannot be obtained simply by a consideration of global mass balance. This applies also to the problem of the assessment of the oceanic source strength for some atmospheric compounds (e.g. biogenic sulfur) when it is inferred from the estimates of the deposition to the ocean surface.

Since the 1960's extensive data have been obtained for wet and dry fallout of aerosol particles from measurements and analyses of artificial and natural radionuclides (NAS, 1978). However such materials generally have source functions in the upper troposphere or in the stratosphere (gas to particle conversion) whereas most of the production of continental aerosol particles occurs at or close to the land-air interface. It would therefore seem premature to use such data for assessing the relative importance of wet and dry deposition processes over the ocean for all materials attached to aerosol particles.

A major factor to consider is that both wet and dry deposition of particles are dependent on particle size (Slinn, 1983). Therefore total deposition assessment requires knowledge of the mass particle-size distribution of each element or compound. Moreover, these data are required to aid in source identification of these materials because various sources generate aerosol particles with different mass-size functions (Buat-Ménard, 1983).

Another major concern is the "recycling" of materials between the ocean and the atmosphere, i.e. material associated with sea-salt particles (Duce, 1983; Buat-Ménard, 1983; Buat-Ménard and Duce, 1985). In most cases, this recycled component is associated with particles much larger than those transported over thousands of kilometers from the continents. As a consequence, recycled material may be the major contributor to the total wet deposition, and particularly dry deposition for many trace elements (Buat-Ménard, 1983; Arimoto et al., 1985). Direct measurements of gross wet and dry deposition may thus give inacurate estimates of the net input to the ocean. Techniques to evaluate this recycled fraction must therefore be developed in parallel.

Finally, because the concentration levels of atmospheric particulate materials in air and rain are often extremely low over remote oceanic areas, state of the art sampling and analytical techniques must be used together with stringent precautions against sample contamination problems.

For all these reasons, few reliable data on wet and dry deposition rates for atmospheric particulate materials are available at present. I will try to show, based on our present, albeit imperfect, knowledge, what kinds of feasible experimental field measurements need to be undertaken to assess quantitatively air to sea particulate fluxes and how such an assessment is heavily dependent on accurate aerosol concentration data, especially as a function of particle size.

2. ASSESSMENT OF WET DEPOSITION

Obviously, the first step towards evaluation of the wet deposition of atmospheric particulate material to the surface of the ocean is collection of precipitation samples. This would appear to be a trivial task, but it is perhaps here that research studies in the marine atmosphere run aground. We will consider here only true wet deposition, as determined by sampling specific precipitation events, as opposed to total deposition measured by continuously open collectors. This type of sampling allows us to investigate how the chemical composition of rain varies with the kind of precipitation system (i.e. trade wind showers, mesoscale storm system in mid-latitudes) and with parameters such as duration, intensity and droplet size of the precipitation. Great care has to be taken during collection to sample the entire rain duration period. Indeed, sequential analysis of rains has shown that the concentration of trace materials in rain is substantially greater during the early period of the rainfall, possibly caused by partial evaporation of drops en route to the earth's surface, through air not yet saturated, and/or by the clouds first use of the largest condensation nuclei (Slinn, 1983).

Contamination of rain samples collected in remote open ocean is a fundamental problem that is often given relatively little attention. Contamination can occur as a result of (1) unsatisfactory collection material, (2) inadequate cleaning of the collection system, (3) improper storage of the system between rain events, and (4) collection of locally produced material in the sampling system which is not representative of the marine rain being investigated. It must be emphasized that collections materials and sampling protocols depend on the substance being investigated. In any case collection and analysis of rain sample is not a routine exercice and requires the use of ultra-clean procedures (Peltzer et al., 1984; Arimoto et al., 1985; Buat-Ménard and Duce, 1985).

Samples representative of rain falling to the ocean can be obtained along windward coastlines and/or from ships. In both cases however, the sampling is fraught with pitfalls ! Along a coastline, local contamination from sea spray, sand and soil, can occur so that it is preferable to collect the rain from a tower several meters above ground level. Sampling must be performed only when the wind is off the ocean and has been for some time. This is difficult in convective rains, when the wind circulation in and around the shower itself is quite complex and impredictable. Table 1 illustrates how marine rain can be contaminated with locally derived Pb and Cd at American Samoa during the passage of trade-wind showers. Obviously improper sampling may lead to wet deposition fluxes overestimated by as much as a factor of 10.

Collection of precipitation samples from ships also provides ample opportunity for contamination as a result of eddying of the wind around ship structure. However, the maneuverability of a ship often provides the opportunity to avoid these problems provided collection is performed from the bow with the ship's exhaust plume being carried downwind.

TABLE I. Atmospheric Pb and Cd in Samoa, 1981[1]
(pg m^{-3})

SAMPLE N°	Pb	Cd
1-0[2]	25 $+$ 5	2 $+1$
1-L[3]	120 $\overline{+}20$	10 $\overline{+}3$
2-0	22 $+$ 5	4 $+2$
2-L	130 $\overline{+}20$	14 $\overline{+}4$
3-0	1.5 $+$ 0.8	8 $+2$
3-L	210 $\overline{+}40$	8 $\overline{+}2$
4-0	5 $+$ 2	<1
4-L	290 $\overline{+}50$	33 $\underline{+8}$

[1] Duce, unpublished data.
[2] Collected only when the air flow was off the ocean and condensation nuclei counts were <600 cm^{-3}.
[3] Collected only when the air flow was off the island or the CN counts were >600 cm^{-3}.

At a given site, large variations of the concentrations of trace elements or substances are generally observed from one rain sample to another (Jickells et al., 1984; see table 2). To obtain accurate wet deposition estimates on a year basis, integration over long time sampling periods is therefore required. For example, over the North Pacific ocean where the intensity of the transport of mineral aerosol particles from Asian deserts varies seasonnally each year, Arimoto et al. (1985) have found higher concentrations of trace elements associated with such particles in rain samples during the "high-dust" season (spring and early summer).

Since it is almost impossible to collect all precipitation at a given site, an approximation of the total wet flux can be obtained using mean values of the volume weighted data and the amount of rainfall during a sampling period characterized by a given meteorological regime. We have to be aware that representative wet flux estimates are extremely difficult to obtain however especially in some oceanic regions such as high latitudes, where high winds often make sampling impossible, and in arid regions where the occurence of rain is very sporadic. In the latter case such events ideally should be sampled since they may constitute a major fraction of the total flux on a yearly basis. This is for example most likely the case over the Mediterranean sea when "red rains" events occur during episodes of transport of Saharan dust (Ganor and Mamane, 1982; Chester et al., 1984; Arnold, 1985).

TABLE 2. Range of concentrations measured in Bermuda
Rainwater, November 1981 to October 1982
(from Jickells et al., 1982)

	minimum concent- ration	maximum concent- ration	volume weighted average concentration
H^+ $\mu E\ 1^{-1}$	0.5	50.1	14.6
SO_4 $\mu E\ 1^{-1}$	4.0	48.1	12.7
Na mg 1^{-1}	0.67	17.9	2.9
Mg mg 1^{-1}	0.08	2.2	0.37
Cd $\mu g\ 1^{-1}$	0.02	0.11	0.06
Cu $\mu g\ 1^{-1}$	0.09	1.6	0.66
Fe $\mu g\ 1^{-1}$	2.2	13.0	4.8
Mn $\mu g\ 1^{-1}$	0.1	0.83	0.27
Ni $\mu g\ 1^{-1}$	0.09	0.8	0.21
Pb $\mu g\ 1^{-1}$	0.19	2.4	0.71
Zn $\mu g\ 1^{-1}$	0.3	2.75	1.15

Average of 18 samples. Rainfall range per sample:
0.5 to 7 cm.

Another method that is often used to calculate the total wet deposition is based on the washout factors or scavenging ratios, which involves the ratio of the rain concentration to the atmosphere concentration of the substance of interest near the surfaces. We will define the washout factor, W, for any material M, as $(W)_M$ given by:

$$(W)_M = [(C)_{M,rain} \times \rho]\ /(C)_{M,air} \qquad (1)$$

where $(C)_{M\ rain}$ is the concentration of material M in rain in g kg^{-1}
$(C)_{M\ air}$ is the concentration of metal M in the air in g m^{-3} and
ρ is the density of air (1.20 kg m^{-3}) at standard conditions.

In this form, washout factors are dimensionless, but the density term is not used by all authors. If a relationship between the concentration of a given material in the air and in the rain has been established, one can utilize atmospheric concentrations to predict the concentrations in air and the total wet deposition flux F_W as follows:

$$(F_W)_M = (W)_{M,T} \times (C)_{M, \text{ air, } t} \times P_t/\rho ; \qquad (2)$$

where $(W)_{M,T}$ is the mean washout factor for material M during an interval of time T, $(C)_{M,\text{air},t}$ is the mean concentration of M in atmospheric particles during time t within the interval T, P_t is the climatological mean rainfall for the subinterval t, and ρ is the density of air. Note that the term $(W)_{M,T} \times P_t$ has the dimension of a velocity. This product is often called wet deposition velocity, v_w and can be compared easily to dry deposition velocity v_d.

Washout factors are generally found to be in the range of 100 to 2000. They have been extensively used over land because they tend to be less variable than the concentrations in rains themselves. This empirical approach has been often criticized because it implicitly assumes that the concentrations of the elements in rain and air are linearly related. The existence of such a relationship is not obvious at all. Problems arise because of possible differences in the removal efficiency of particles as a function of particle size, chemical composition and rain droplets size (Slinn, 1983). Rain concentrations represent an integrated removal throughout the water column through which the rain is falling, while the air concentration is determined only at the surface. Thus particulate elements or substances with different vertical concentration profiles but similar surface air concentrations might have very different washout factors. An additional problem faced in truly remote marine regions is that it takes many hours, even days, to collect an aerosol particle sample large enough to analyze, whereas the rain sample may be collected over a period of minutes or hours. Scott (1981) also points out that smaller washout ratios would be expected in "cold" clouds, i.e. those where a much greater fraction of the precipitation occurs as a result of vapor growth rather than droplet coalescence processes. The latter apparently dominates in "warm" rains, i.e. those not containing ice particles or supercooled droplets. Washout factors might be as low as 100 in cold rains and as high as 1000 in marine rains.

An interesting finding in urban or near continental marine areas (Cawse, 1974; Gatz, 1975; Duce et al., 1979, Scott, 1981) was that of an apparent relationship between the washout factor and the mass median diameter of the particles containing the elements (see Figure 1). In general, washout factors increased with particle size. That was explained by the lower collection efficiency of the rain drops for the smaller particles, with the chemical composition, particularly solubility, of the different size particles also playing a role. In this context, the few results obtained up to now in the remote marine atmosphere of the Tropical Pacific by Arimoto et al. (1985) and Buat-Ménard et al. (1983) are quite puzzling since such a relationship is not observed, i.e. there was no significant variation of W with particle size (Figure 2). The causes of these differences between continental and remote marine regions are uncertain. Slinn (1983) has suggested that, in relatively clean remote areas clouds use essentially

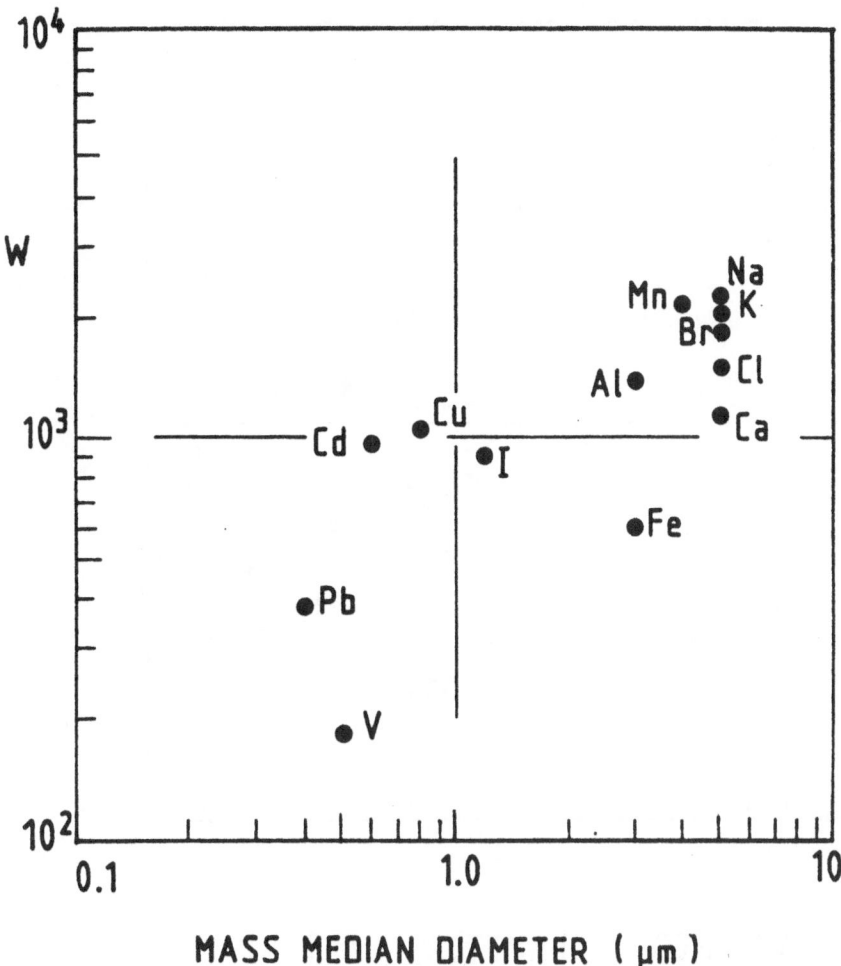

Figure 1. Variation of washout factor or scavenging ratio, W, as a function of particle size for various trace elements in the Florida Keys (Adapted from Duce et al., 1979).

all available particles as cloud condensation nuclei; thereby no particle-size dependance of W would be expected. This suggestion deserves further investigation but there are other possible causes for such differences such as a) differences in vertical distributions of the materials over land and over sea, b) differences in precipitation patterns, in the vertical and horizontal extent of cloud systems and in duration, intensity and droplet size of the precipitation. For the same reasons a geographical variability of W may be expected over marine areas. In any case, it is clear that any relationship developped between washout factors and particle size distribution in urban or continental regions cannot be simply applied over tropical open ocean regions. Many additional studies are required among which priority should be given to measurements of the vertical distribution of atmospheric concentrations and careful studies of the micrometeorology

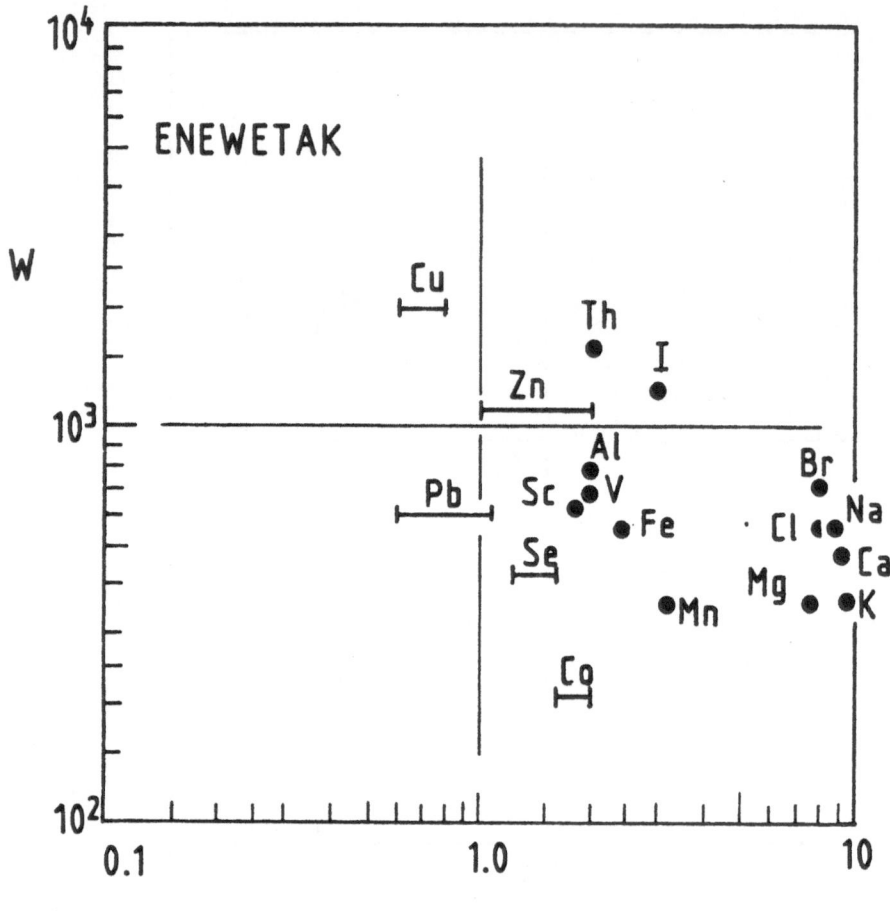

Figure 2. Variation of washout factor or scavenging ratio, W, as a function of particle size for various trace elements at Enewetak Atoll (Adapted from Buat-Ménard and Duce, 1985).

during rainstorms. Until such studies are performed, large uncertainties up to a factor of ten will remain for washout factor values from one oceanic region to another. Great care should also be taken when extrapolating data obtained for a given tracer element or substance to the element or substance being investigated, without consideration of possible differences in particle size distributions, sources and source regions, vertical concentration profiles etc.

3. FIELD APPROACH TO DRY DEPOSITION

Dry deposition of aerosol particles includes essentially three processes: the settling of particles to the surface by gravitation, the impaction and the diffusion of particles to surfaces. The theory is very complex because each of these processes acts simultaneously, and because each of them is dependent on a number of variables (i.e. wind speed, particle-size, relative humidity, air viscosity, sea-surface roughness etc...). Even the most recent models (Slinn and Slinn, 1980; Williams, 1982) do not include all the relevant physics. In such models the atmosphere is separated in two layers (Figure 3). In the "constant flux layer", atmospheric turbulence and gravitational settling govern particle transport, and in the second layer, the "deposition layer" which is just above the air-sea interface, particles grow in response to the higher relative humidity there. The overall resistance to air to sea particulate transfer is treated as several resistances in series.

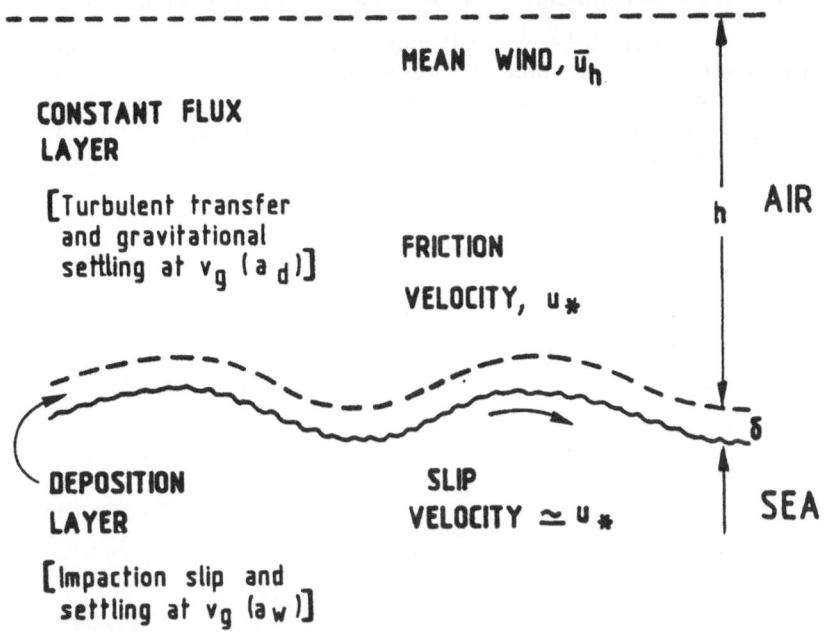

Figure 3. A simple two layers model of the marine atmosphere close to the air-sea interface. (Adapted from Slinn and Slinn, 1980).
h, the reference height, is equal to 10 m. According to Slinn (1983), δ is approximately given by the ratio of the kinematic viscosity to the friction velocity and is of the order of 100 µm.

For example in the Slinn and Slinn model, the deposition velocity (v_d) is:

$$1/v_d = 1/K_C + 1/K_D - v_g(a_d)/K_C K_D \qquad (3)$$

where $1/K_C$ is the resistance in the constant flux layer:

$$K_C = 1/(1-\kappa) \ C_D u + v_g \ (a_d) \qquad (4)$$

and $1/K_D$ is the resistance in the deposition layer:

$$K_D = cm'' + (1/\kappa)C_D \ u \ Sc^{(-1/2)} + 10^{(-3/St)} + v_g(a_w) \qquad (5)$$

where: a_d = dry particle radius

a_w = wet particle radius

$v_g(a_d)$ = gravitational settling velocity of the dry particle

$v_g(a_w)$ = gravitational settling velocity of the wet particle

u = wind speed at 10m

C_D = the drag coefficient (1.3×10^{-3})

κ = Von Karman's constant (0.4)

c = a constant   m" = rate of water evaporation

Sc = particle Schmidt number; St = particle Stokes number.

The overall resistance is reduced by gravitational settling. The predictions of such a model can be seen in Figure 4 and clearly illustrate the influence of wind speed and relative humidity for particle deposition.

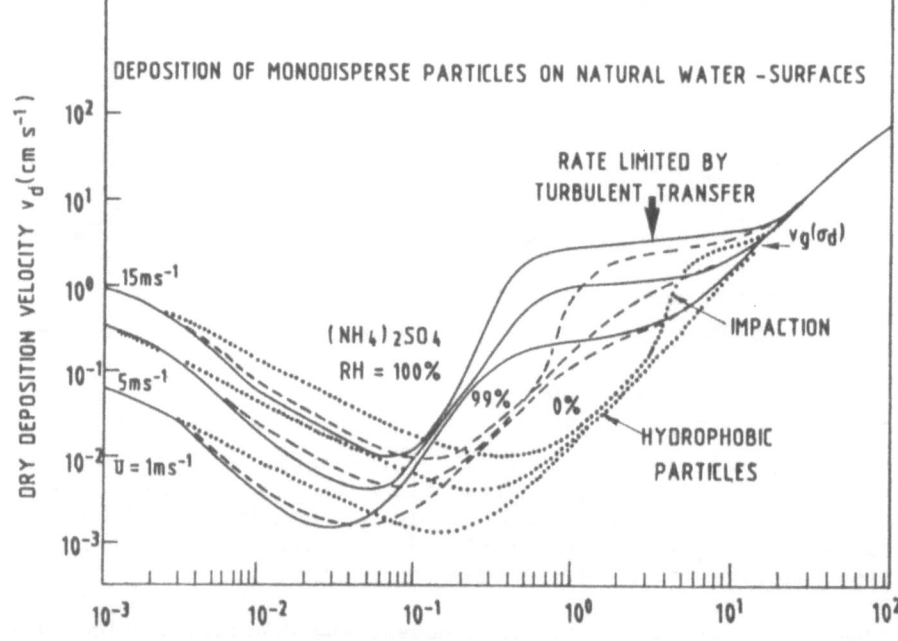

Figure 4. Plots of Equation (3), using Equations (4) and (5), for three wind speeds and three types of particles: dotted curves –hydrophobic (dry) particles; dashed curves –for particles that grow to equilibrium size for $(NH_4)_2SO_4$ particles exposed to a relative humidity of 99%; solid curves –expected behavior of $(NH_4)_2SO_4$ particles for the case of deposition to lakes with relative humidity near 100% in the deposition layer. Note that for a range of particle radii near 1 μm, it is predicted that the dry transfer of hygroscopic particles can be rate limited by turbulent transfer through the constant flux layer; i.e., for these particles $v_d \simeq C_D u_h$ (Adapted from Slinn and Slinn, 1980).

In the last years, several studies (Moller and Schumman, 1970; Sehmel and Sutter, 1974; Wesely and Williams, 1981; Sievering et al., 1982) have attempted direct measurements of deposition velocities to water surfaces from water-wind tunnel measurements and using the eddy flux technique. Under some conditions, wind-tunnel data are valuable, especially if the controlling resistance is in the deposition layer next to the surface; however, extrapolations from wind-tunnel data to estimates of the fluxes to the oceans must be suspect if the state of the sea surface (including waves and spray) is not adequately duplicated. Eddy flux measurements have been rather inconclusive as a result of particle growth and decay initiated by the steep relative humidity gradients present over water. Almost no attempts using the gradient technique have been reported, perhaps because of the difficulty in resolving concentration differences that are on the order of 1% if v_d (<0.5µm radius) is as small as 0.1 cm.s^{-1}. It seems therefore that the best way to estimate dry deposition is the use of available dry deposition models to a water surface as a function of particle size. The use of such models requires an accurate measurement of the mass-size distribution of the trace metals. In addition, other parameters, particularly relative humidity and wind speed, have to be measured in the field since they influence markedly deposition velocities (Figure 4).

Model calculations of dry deposition have been rather successful for sea-salt aerosol particles, provided extreme care was taken to obtain air samples that represent the true sea-salt particle mass and size distributions (Mc Donald et al., 1982). Acceptably accurate size distributions can be deduced from the use of cascade impactors, when proper corrections are made for particle loss in the largest size range. It has to be emphasized however that, even at low wind speeds (<5 m s^{-1}), the large particles (>20µm radius) dominate the deposition though they are present in relatively low concentrations. The use of such sampling devices (especially high-volume cascade impactors) may, however, cause serious problems for other types of aerosol particles to which trace elements or substances may be attached. For example, it appears that particle "bounce" effects, especially for dry particles such as alumino-silicate minerals, can result in an apparent shift of the size distribution to smaller sizes (Buat Ménard et al., 1983). The situation is even worse for other materials, trace elements such as Pb, As, Cd, excess sulfate, organic compounds, because a major fraction of their mass occurs on particles less than 0.25 µm radius, and cascade impactors currently in use do not separate particles below that size range. For such materials a further complication arises from the large differences in the deposition velocities than can be derived from the models for particles with dry radii of about 0.1µm. As an exemple, for particles of this approximate size, the velocities calculated from a model by Williams (1982), which allows for differences in transfer velocities to smooth and broken water surfaces, are higher than those calculated from the Slinn and Slinn model, which only considers a smooth water surface, with a maximum difference of more than 30 fold, at 15m s^{-1} wind speed, 20% relative humidity and assuming a broken surface transfer coefficient of 10 cm s^{-1}. This suggests that the dry deposition of small particles can be enhanced by breaking waves and scavenging by sea-spray. There is unfortunatly very few experimental data which would allow to quantify such processes. Air entrained by waves may be effectively stripped of all particles. Based on the laboratory experiments of Cipriano and Blanchard (1981), it does not seem however that this process would solely explain the differences in dry deposition velocities to smooth and broken water surfaces. With respect to scavenging by sea-spray, Slinn (1983) has argued against the assumption made by Williams that the broken surface transfer coefficient is particle-size independant. Slinn (1983) suggests that this coefficient should be much lower for particles with dry radii of 0.1 to 1 µm than those for other sizes of particles as a result of a decrease of the collision efficiency with decreasing particle-size.

Arimoto and Duce (1985) have compared total dry deposition rates for Na (sea-salt particles), Al (mineral aerosol particles) and ^{210}Pb (small particle element) from measured mass-particle-size distributions and these two deposition models. Calculations using William's model give always dry deposition rates higher than or equal to those using Slinn and Slinn model. The maximum difference for Na is only 5%, which indicates that the dry deposition of sea-salt particles is mainly controlled by large particles whose deposition velocities are limited by the resistances in the constant flux layer. For Al which has a smaller mass mean radius than Na, the maximum difference was found to be a factor of two. For these two types of particles it can therefore be concluded that neither the broken surface effects nor the relative high humidity in the deposition layer greatly affect the mass transfer of these substances by dry deposition. On the other hand these effects are significant for small particle elements since a maximum difference of a factor 5 was found for ^{210}Pb total dry deposition rate when such effects were taken into account in the model calculations. As a consequence, large uncertainties still exist on the dry deposition rates of materials attached to small particles. This is precisely the size range where most of the mass of materials derived from anthropogenic emissions (heavy metals) or from naturally occuring processes through gas to particle conversion is found. Together with model's imperfections, other questions still need to be answered. Improvements of the quality of data on the mass-particle-size distributions are needed especially for giant particles (>10 µm radius). Also, shifts in the mass-particle size distributions may occur within the constant flux layer as a function of height and wind speed and need therefore to be documented. Finally, temporal changes in the mass size distributions may occur during periods of high inputs of continental aerosols, especially in pericontinental areas and regional seas which implies that non steady state transfer needs to be investigated.

Another method to estimate dry deposition rates, which has been extensively used in the past, consists of measurements of the amount of materials deposited to a surrogate surface such as a rimless plastic plate (Settle and Patterson, 1982; Arimoto et al., 1985). The use of such surfaces has been seriously criticized because they do not accurately mimic the characteristics (micro-structure, roughness, etc...) of natural water surfaces (Hicks et al., 1980). But, if the gravitational settling of large particles is responsible for a major fraction of the dry removal, then the deposition rates to the surrogate and water surfaces may not differ greatly over a broad range of wind speeds and relative humidities. This provides a clue to the relatively good agreement (within a factor of 3) which has been found between results from model calculations and direct measurements of the dry deposition rate using surrogate surfaces (Arimoto et al., 1985). Such field measurements can therefore be undertaken. However, they should systematically be performed in parallel with determinations of the mass-size distributions of the elements or substances of interest. Indeed for many "small particle" elements or substances, their large particle fraction, which may dominate their total dry deposition, has a different origin than the submicrometer fraction. For example, anthropogenic lead is primarily associated with the submicrometer fraction, whereas lead from crustal origin or associated with sea-salt is present on much larger particles.

As an example, if the measured Pb concentrations as a function of particle size is applied to the theoretical deposition velocity as a function of particle size relationship (Slinn and Slinn, 1980), the small percentage of Pb on the largest particles (which may simply represent recycled Pb from the sea surface) controls the total Pb deposition. Therefore, a knowledge of the mass size function of Pb is necessary to assess the relative contributions of the various Pb sources to the total dry deposition of this element (Duce, 1983).

Moreover, the atmospheric concentrations of particles derived from different sources vary temporally, so the relative contributions of the different sources to the total dry deposition will also vary temporally. Sampling must be undertaken at various times of the year to obtain representative annual dry deposition data at a given location. However, individual dry deposition samples should be collected over short enough time intervals to indicate the natural temporal variability in the deposition. Again care should also be taken in extrapolating data from open ocean regions to coastal environments or regional seas or vice versa. The mass-size functions of continental aerosol particles may vary markedly with increasing distance from the source. Studies of mineral aerosols transported from the Sahara desert over the Atlantic Ocean have clearly shown that up to a distance of 1000 km from the African coast, the mass-size function exhibits a gradual shift to smaller sizes (Schutz, 1980).

4. ACCURATE DEPOSITION MEASUREMENTS DO NOT GUARANTEE ACCURATE NET AIR TO SEA TRANSFER RATES.

While an accurate measurement of the concentration of materials in rain and dry deposition is a necessary first step in evaluating fluxes of these materials to the ocean, it is not sufficient in itself. The total amount of rainfall in the period of interest, per year for example, must be known. Potential problems arise concerning variability of rainfall amount and intensity with season and how this will affect the material concentrations, as well as how seasonal changes in wind flow patterns or particle production processes will affect atmospheric concentrations at the marine location. Even taking these factors into consideration, significant problems remain. In the marine environment the gross deposition of some materials (e.g. trace metals) to the ocean is composed of a net input as well as a component associated with recycled sea spray. The importance of the atmosphere as a transport path for material from the continents to the ocean can only be assessed accurately if the relative contributions of the net and recycled components can be distinguished (Settle and Patterson, 1982; Buat Ménard, 1983; Jickells et al., 1984; Arimoto et al., 1985). For example, there is strong evidence that atmospheric sea salt particles produced by bubbles bursting at the sea surface contain many metals in concentrations considerably higher than would be expected on the basis of the metal-to-sodium ratio of the near-surface water. It is apparent that some fraction of these metals is associated with surface active organic material and is scavenged by the rising bubbles and concentrated on the sea salt particles produced both from the air-sea and bubble interface when the bubbles burst (Weisel et al., 1984). If this fractionation is not taken into account, the calculated net deposition to the ocean will be anomalously high.

As one approach to this problem, Arimoto et al. (1985) determined the metal/Na ratio on the largest atmospheric particles at Enewetak Atoll in the Marshall Islands, using only particles collected on the first stage of a cascade impactor. This represented, in general, particles with radii greater than about 3.5 μm. Most particles in this size range are sea salt, so this size range gave a good estimate of the metal/Na ratio in particles produced by bursting bubbles. These ratios also agreed well with direct measurements of the metal/Na ratios on sea salt particles produced and collected artificially using the Bubble Interfacial Microlayer Sampler in the North Atlantic by Weisel et al. (1984) (see also Buat-Ménard, 1983 for an extensive discussion of this problem). The recycled component for any metal in the rain was calculated by:

$$(M)_{recycled} = (M/Na)_{Stage\ 1} \times (Na)_{rain} \qquad (6)$$

where $(M)_{recycled}$ is the concentration in rain of metal M recycled from the sea, in g kg^{-1}.

$(M/Na)_{stage\ 1}$ is the metal/Na ratio on stage 1 of the impactor, and $(Na)_{rain}$ is the sodium concentration in rain in g kg^{-1}.

Recycled components in wet deposition that were calculated in this way for several metals ranged from 15% to about 50% and are comparable to the recycled fraction calculated using a similar approach by Jickells et al. (1984) at Bermuda. For Pb, about 30% was recycled in Enewetak rain according to Arimoto et al. (1985), compared with 10% at Enewetak calculated by Settle and Patterson (1982) and 17% at Bermuda determined by Jickells et al. (1984).

Recycled components calculated in the same way for dry deposition measurements to a surrogate surface were shown to range from 12 to

100%. For Pb, the percentage recycled varied considerably between samples. In two of the samples, approximatly 50 to 60% of the gross Pb deposition was due to recycled material, but in a third sample, essentially all of the Pb could be attributed to sea spray. Recent data based of stable lead isotope measurements have provided unambiguous and direct evidence of the importance of this recycling process for pollution lead (Patterson, pers. comm.).

These examples give an indication of the state-of-the-art in this area. Clearly, more sophisticated techniques for accurately evaluating this recycled fraction must be developed. Future work should focus in two areas a) the use of adequate tracers (stable or radioactive) during field measurements, b) an improvement of our knowledge of metal/Na ratios as a function of sea-salt particle size through carefully designed in situ or laboratory experiments.

5. RELATIVE IMPORTANCE OF WET AND DRY REMOVAL RATES.

Because of insufficient knowledge and data, it is very difficult if not impossible to make a general statement about the relative importance of wet and dry removal for aerosol particles over the ocean. For example Uematsu et al. (1985) found that over the North Pacific Ocean dry deposition contributes on the average to only about 20% of the total flux of mineral aerosol. This contribution was found to vary temporally at each sampling site which clearly illustrates the need for measurements over long periods of time. Also a spatial variability of the total flux was found from one site to another (Figure 5). This can be explained by the fact that differences in total precipitation amount, frequency of precipitation, frequency of non steady state inputs of materials to the atmosphere may generate a significant variability of the ratio between wet and dry deposition rates. However, the most important finding of this work is that the periods of high dust flux were generally observed during periods of high rainfall and atmospheric dust loadings, corresponding to dust storm outbreaks in Asia during the spring. This suggests that, over the ocean, wet removal should dominate dry removal for that class of aerosol particles except where little or no precipitation occurs.

For sea-salt particles dry and wet deposition over tropical and mid-latitude appear to be of nearby equal importance. It must be stressed however that data from high latitudes, where precipitation patterns and average wind speeds are different might give a different picture. Also we know almost nothing about removal rates during hurricanes or severe storms because of the impossibility of sampling.

For small particle elements, it seems clear that (in areas of intense precipitation) net dry deposition fluxes are much less than wet deposition fluxes. This is probably why the residence time of such elements or substances in the troposphere is very close to that of water vapor (about 1 week) as has been shown from studies based on measurements of artificial or natural radionuclides (NAS, 1978; Lambert et al., 1983). Again, variations however must be expected from one

oceanic region to another, depending on the amount and frequency of
rainfall.

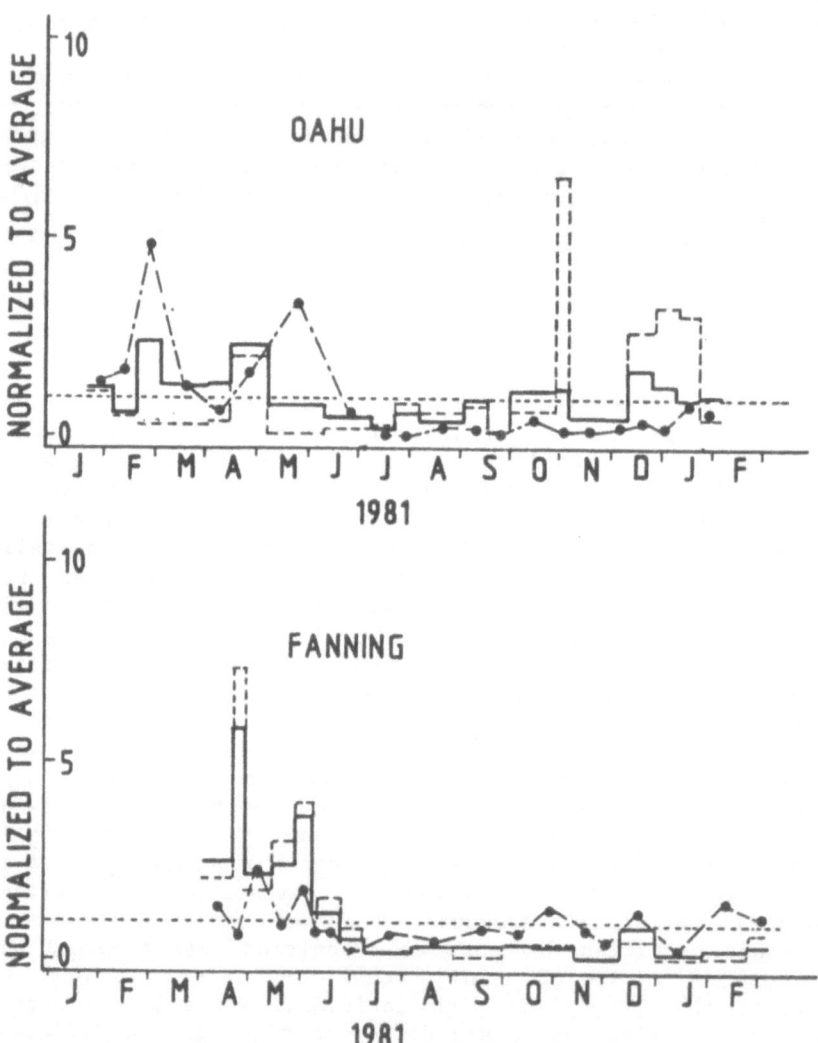

Figure 5. Normalized temporal variations of dust flux for particles
smaller than 20 μm diameter (solid line), atmospheric dust
concentration (circles), and precipitation amount (dashed line) in 1981
and part of 1982, at Oahu (21°20'N and 157°42'W) and Fanning (3°55'N
and 159°20'W). At Fanning the correlation between the dust flux and the
precipitation amount is much stronger than that found at Oahu. Redrawn
from Uematsu et al. (1985).

6. CONCLUSION

Together with the obvious need for more data of high quality on air and rain concentrations, future studies should also tend to make more and more use of atmospheric transport and removal models in order to evaluate net air to sea particulate fluxes at a given oceanic region. Especially for some anthropogenic materials source-emission inventory data are becoming progressively available. Also, an improvement in the modeling of atmospheric transport and removal of continental aerosols to the ocean can be expected through the use of specific tracers and source markers (i.e. ^{210}Pb, stable isotopes of C, N, S and Pb compounds, biogeochemical source markers) for model validation. Finally, for the next decades, it can be expected that significant advances in flux prediction will come from the use of remote sensing techniques conducted from aircrafts or satellites. Indeed such techniques already give us and will give us some of the important parameters needed for such predictions, i.e. precipitation patterns, whitecap coverage, sea-surface roughness, extent of aerosol-plumes following desert-dust outbreaks events, volcanic eruptions, intense biomass burning activities. Obtaining the latter information may be one of the highest priorities for future studies because of their possible major impact on atmospheric and ocean chemistry.

ACKNOWLEDGEMENTS

I wish to thank the people at the ASI who contributed comments and suggestions to this manuscript, especially R. Arimoto and D. Blanchard. Thanks are also due to R. Duce for exchange of thoughts and ideas over these last years. This is C.F.R. contribution n° 728.

REFERENCES

Arimoto, R., R.A. Duce, B.J. Ray and C.K. Unni, 1985: 'Atmospheric trace elements at Enewetak Atoll: 2. Transport to the ocean by wet and dry deposition'. J. Geophys. Res., 90, 2391-2408.

Arimoto, R. and R.A. Duce: 'Dry deposition models and the air/sea exchange of trace elements'. J. Geophys. Res. (in press).

Arnold, M., 1985: Géochimie et transport des aérosols métalliques au-dessus de la Méditerranée occidentale. Ph. D. Thesis, University of Paris VII, 226 pp.

Buat-Ménard, P., 1983: 'Particle geochemistry in the atmosphere and oceans'. In: Air-Sea Exchange of Gases and Particles, (P.S. Liss and W.G.N. Slinn, eds), D. Reidel Publishing Company, 455-532.

Buat-Ménard, P., U. Ezat and A. Gaudichet, 1983: 'Size distribution and mineralogy of alumino-silicate dust particles in tropical Pacific air and rain'. In: Precipitation Scavenging, Dry Deposition and Resuspension, Vol. 2, (H.R. Pruppacher, R.G. Semonin and W.G.N. Slinn, eds.) Elsevier, New York, 1259-1270.

Buat-Ménard, P. and R.A. Duce, 1985: 'Metal transfer across the air-sea

interface: myths and mysteries'. In: Metal cycling in the
Environment (T.C. Hutchinson, ed.,), Wiley, New York (in press).
Cawse, P.A., 1974: A survey of atmospheric trace elements in the United
Kingdom. AERE Harwell Report R7669, UKAEA, HMSO, London.
Chester, R., E.J. Sharples, G.S. Sanders and A.C. Saydam, 1984:
'Saharan dust incursion over the Thyrrenian Sea'. Atmos.
Environ., 18, 929-935.
Cipriano, R.J. and D.C. Blanchard, 1981: 'Bubble and aerosol spectra
produced by a laboratory "breaking wave"'. J. Geophys. Res., 86,
8085-8092.
Duce, R.A., C.K. Uni, P.J. Harder, B.J. Ray, C.C. Patterson, D.M.
Settle and W.F. Fitzgerald, 1979: Wet and dry deposition of trace
elements and halogens in the marine environment. Presented at the
IAMAP/CACGP Symposium on the Budget and Cycles of Trace Gases and
Aerosols in the Atmosphere. University of Colorado, Boulder, 12-18
August.
Duce, R.A., 1983: 'Biogeochemical cycles and the air-sea exchange of
aerosols'. In: The major biogeochemical cycles and their
interactions (B. Bolin and R.B. Cook, eds.) SCOPE 21, Wiley,
427-456.
Ganor, E. and Y. Mamane, 1982: 'Transport of Saharian dust across the
Eastern Mediterranean'. Atmos. Environ., 16, 581-587.
Gatz, D.F., 1975: 'Scavenging ratio measurements in Metromex'. In:
Precipitation scavenging - 1974 (R.W. Beadle and R.G. Semonin
eds.) ERDA Symposium series, CONF-741014, 71-87.
Hicks, B.B., M.L. Wesely and J.L. Durham, 1980: Critique of methods to
measure dry deposition. Workshop Summary.EPA 1600.9-80-050, 70pp.
Jickells, T.D., A.H. Knap and T.M. Church, 1984: 'Trace metals in
Bermuda Rainwater'. J. Geophys. Res., 89, 1423-1428.
Lambert, G., J. Sanak and G. Polian, 1983: 'Mean residence time of the
submicrometer aerosols in the global troposphere'. In:
Precipitation Scavenging, Dry deposition and Resuspension, (H.R.
Pruppacher, R.G. Semonin and W.G.N. Slinn, eds.) Vol. 2,
Elsevier, New York, 1353-1358.
Mc Donald, R.L., C.K. Unni and R.A. Duce, 1982: 'Estimation of
atmospheric sea salt dry deposition: wind speed and particle size
dependance'. J. Geophys. Res., 87, 1246-1250.
Moller, V. and G. Schumman, 1970: 'Mechanisms of Transport from the
atmosphere to the earth's surface'. J. Geophys. Res., 75,
3013-3019.
N.A.S., 1978: The Tropospheric Transport of Pollutants and other
Substances to the Ocean (J.M. Prospero ed.), National Academy of
Science, Washington DC, 243 pp.
Peltzer, E.T., J.B. Alford and R.B. Gagosian, 1984: Methodology for
sampling and analysis of lipids in aerosols from the remote marine
atmosphere. Woods Hole Oceanographic Institution, Technical Report
WHOI-84-9, 104 pp.
Prospero, J.M., R.J. Charlson, V. Mohnen, R. Jaenicke, A.C. Delany, J.
Moyers, W. Zoller and K. Rahn, 1983: 'The Atmospheric Aerosol
system: An overview'. Rev. Geophys. Space Phys., 21, 1607-1629.
Scott, B.C., 1981: 'Modeling of atmospheric wet deposition'. In:

Atmospheric Pollutants in Natural Waters (S.J. Eisenreich ed.) Ann
Arbor Science, Ann Arbor, MI, 3-21.

Schutz, L., 1980: 'Long range transport of desert dust with special
emphasis on the Sahara', Ann. N.Y. Acad. Sci., 338, 515-532.

Sehmel, G.A., and S.L. Sutter, 1974: 'Particle deposition rates on a
water surface as a function of particle diameters and air
velocity'. J. Rech. Atmos., 8, 911-920.

Settle, D.M. and C.C. Patterson, 1982: 'Magnitudes and sources of
precipitation and dry deposition fluxes of industrial and natural
leads to the North Pacific at Enewetak'. J. Geophys. Res., 87,
8857-8869.

Sievering, M., J. Eastman and J.A. Schmidt, 1982: 'Air-sea particle
exchange at a near-shore oceanic site'. J. Geophys. Res., 87,
11027-11037.

Slinn, W.G.N., 1983: 'Air to Sea transfer of particles'. In: Air-Sea
Exchange of Gases and Particles (P.S. Liss and W.G.N. Slinn eds.),
D. Riedel Publishing Company, 299-405.

Slinn, S.A. and W.G.N. Slinn, 1980: 'Predictions for particle
deposition on natural waters'. Atmos. Environ., 14, 1013-1016.

Uematsu, M., R.A. Duce and J.M. Prospero, 1985: 'Deposition of
Atmospheric Mineral Particles to the North Pacific Ocean'. J.
Atmos. Chem. 3, 123-138.

Weisel, C.P., R.A. Duce, J.L. Fasching and R.W. Heaton, 1984:
'Estimates of the transport of trace metals from the ocean to the
atmosphere'. J. Geophys. Res., 89, 11607-11618.

Wesely, M.L. and R.H. William, 1981: Eddy correlation measurements of
particle fluxes over Lake Michigan. Annual Rept. of Argonne Nat'l
Lab. Radiolog. and Environ. Div., Argonne, IL, 36-38.

Williams, R.M., 1982: 'A model for the dry deposition of particles to
natural water surfaces'. Atmos. Environ., 16, 1933-1938.

ATMOSPHERIC, OCEANIC, AND INTERFACIAL PHOTOCHEMISTRY AS FACTORS
INFLUENCING AIR-SEA EXCHANGE FLUXES AND PROCESSES

Oliver C. Zafiriou
Department of Chemistry
Woods Hole Oceanographic Institution
Woods Hole, MA 02543
U. S. A.

1. INTRODUCTION

The conceptual framework used to approach complex problems influences
the outcome. Analyzing the role of photochemical processes in modifying air-sea exchange processes certainly presents difficulties in this
regard, because no single point of view seems best for accomplishing
the objectives of this chapter: first to review the fundamentals of
environmental photochemistry, and second to consider in more detail
our knowledge of the impact of photochemstry on air-sea exchange.
    For presenting an initial overview, the usual geochemical approach
of subdividing the environment into macroscopic reservoirs is both
natural and fruitful. Thus, in considering air-sea exchange we require
both ocean and atmosphere reservoirs, as well as a continental source/
sink. The ocean and the atmosphere need to be partitioned into more
homogenous sub-units: the photochemically active stratosphere, troposphere, and surface oceans.
    However, this framework is counterproductive for discussing how
photoreactions influence the air-sea exchange processes at the mechanistic level because of the extreme heterogeneity of environments critically affecting the exchange mechanisms. Gas exchange, wet deposition, dry deposition, and bubble-bursting all involve microenvironments
in which crucial mechanistic events influence fluxes of materials.
Discussing all these processes in a reservoir-based model requires sub-
compartments for raining and nonraining clouds, below-cloud air, the
interfacial region between sea and air, etc. - a profusion of subdivi-
sions that impairs rather than facilitates understanding.
    Thus Section 2 presents a condensed introduction to environmental
photochemistry relevant to air-sea exchange processes, organized by
major environmental 'boxes.' Those familiar with these areas, or able
to read the key references cited below, may dispense with this section
and proceed directly to Section 3. Section 3 shifts to a more mechanistic organization in order to consider the effects of photochemistry
on exchange processes in terms of exchange mechanisms, not by geochemical compartments.
    Although it is difficult to cite a few 'key' references giving

P. Buat-Ménard (ed.), The Role of Air-Sea Exchange in Geochemical Cycling, 185–207.

thorough coverage of the subject addressed here, the following are
among the more inclusive references available: the Dahlem Conference
Volume **Atmospheric Chemistry** (Goldberg, 1982), and Logan et al. (1981)
for summaries of much basic atmospheric chemistry, Zafiriou et al.
(1984) for an overview of natural water photochemistry, and other chapters in the book edited by Liss and Slinn (1982), and in this volume
for an understanding of the various air-sea exchange processes. For
secific examples of photochemical effects, the papers by Andreae (this
volume, COS section), Calvert et al. (1985), Chameides and Davis
(1982), Conrad et al. (1982); and Garland and Curtis (1981) are recommended to those unable to read more widely.

2. ENVIRONMENTAL PHOTOCHEMISTRY

2.1. Stratospheric Photochemistry

Although stratospheric photochemistry affects air-sea exchange, the
connections are spatially separate and therefore indirect. The most
relevant feature of the stratosphere is its ozone layer, a dynamic feature formed there in consequence of the availability of light energy
of short enough wavelength to photolyze atmospheric O_2. Aside from its
temperature effects, there are two major consequences of the ozone
layer: it screens out almost all solar radiation below roughly 300 nm,
preventing short-wavelength photochemistry from occuring in the the
troposphere, and it is a source of ozone to the troposphere. This
stratospheric ozone plays an important role in initiating tropospheric
photochemistry.
      The stratosphere is also the ultimate sink for those surface-
derived materials that enter the troposphere and are long-lived due to
their biogeochemical and photochemical refractoriness in the lower
atmosphere. Since particles are not thought to have long tropospheric
residence times, this sink primarily affects gases such as the natural
compounds nitrous oxide and methyl chloride, and the anthropogenic
chlorofluoromethanes (Cicerone, 1981). Although methane and hydrogen
react primarily in the troposphere, these gases are sufficiently abundant that the fraction reaching the stratosphere is important; for
methane, the current estimate is that roughly 10% is oxidized there.
      Trace gases diffusing to the stratosphere and supplying upon des-
truction their halogen atoms, nitrogen oxides, and water vapor also
have important influences on stratospheric chemical dynamics, so that
the indirect connection between oceanic sources and sinks of materials
is a two-way street. In this context, methyl chloride is the major
source of natural Cl to the stratosphere. As Cl atoms have significant
effects there, the oceanic methyl chloride source is perhaps the single
strongest connection in the direction ocean-stratosphere; as we shall
see, the ozone source from the stratosphere is among the most important
indirect flux in the opposite sense.

2.2. Homogenous Tropospheric Photochemistry

This area has been the subject of a great deal of research in the last decade, some of which is probably familiar to most readers. Basically, the simple pictures of the workings of tropospheric photochemistry out-lined by Levy (1971, 1972) and McConnell et al. (1971) have proved qualitatively and even quantitatively reasonable; however, enormous extensions, refinements, and observational validations of these early ideas have occurred (see Logan et al., 1981; Cicerone et al., 1982). The following account is an oversimplified primer.

The general character of tropospheric photochemical reactions – which are actually a complex and coupled set of photochemical and thermal chemical processes – can be summarized as follows: photochem-ically initiated, short-chain radical oxidation reactions set the con-centrations and speciations of many trace components. However, the reaction pathways are complex, involving numerous cycling intermediates with lifetimes ranging from about one second (OH and other radicals) to several years (CH_4). The involvement of coupled cycles in these pro-cesses, which do nevertheless result in net reactions, makes it diffi-cult to represent them accurately with a small number of simple stoi-chiometric equations.

The hydroxyl radical, OH, is a species of central importance (though by no means the only important one in these processes). It provides the principal sink for many trace reduced species because of its generally high reaction rate constants for reaction with all organic compounds that are unsaturated or have C-H bonds, as well as with a number of inorganic species.

Some of the major pathways involved in initiating tropospheric photochemistry are shown in Figure 1A. Although a number of processes form and consume OH in the troposphere, the simplest one conceptually starts at dawn with the onset of photolysis of ozone, which has a life-time of several days in the troposphere and is present in traces almost everywhere. Half or more of this ozone is thought to originate from the stratosphere via stratosphere-troposphere exchange; the rest is synthesized by tropospheric chain reactions. The inital reaction is:

$$O_3 + h\nu \ (290 - 310 \ nm) = O_2 + O \ (^1D)$$

$$O_3 + h\nu \ (>310 \ nm) = O_2 + O \ (^3P)$$

The O (^1D) atoms (but not the O (^3P)) go on in part to react with water vapor to form OH, in competition with reactons with air to form O (^3P):

$$O \ (^1D) + HOH = 2 \ OH$$

$$O \ (^1D) + N_2 \ or \ O_2 = O \ (^3P) + N_2 \ or \ O_2 + heat$$

The most important fate of the O (^3P) atoms is to re-form ozone:

$$O \ (^3P) + O_2 = O_3$$

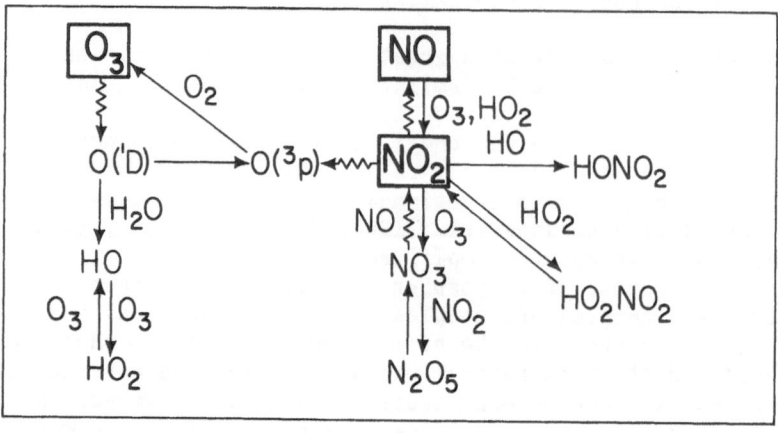

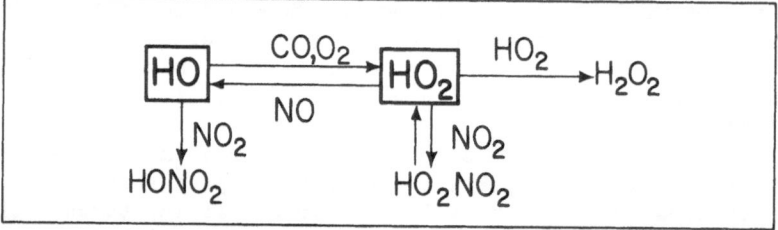

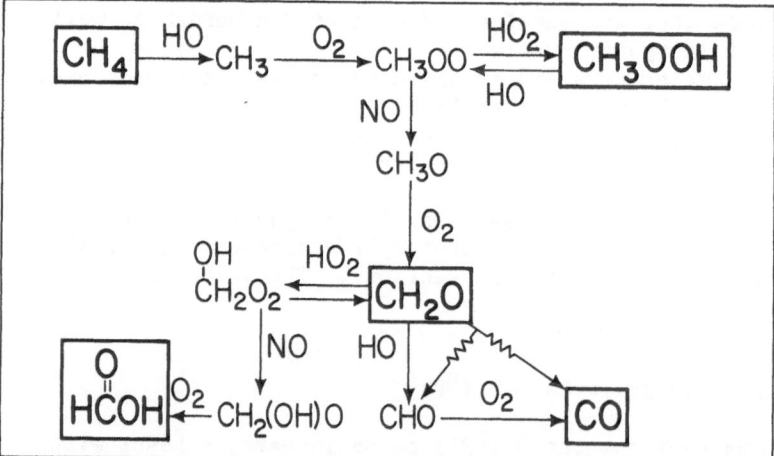

Figure 1.   Some Tropospheric Photochemical Interconversions
   A. Formation of O_3 and OH radicals.
   B. Interconversion of NO and CO by OH and HO_2 radicals.
   C. Steps in the tropospheric oxidation of methane.

Since, in addition to ozone, there are usually other photochemical sources of O (^{3}P), such as photolabile trace nitrogen compounds (e.g. NO, NO_2, NO_3, and N_2O_5), more ozone can form by these paths as well.

Almost all OH sources, other than the O (^{1}D) reaction, involve molecules which themselves are formed by processes involving OH (as is ozone, but only partly); additionally, these other molecules are highly water-soluble (HOOH, CH_3OOH, HNO_2, HNO_3) or reactive species (NO, HO_2); therefore, their atmospheric distributions and relative importance are extremely heterogeneous compared with that of O_3 (which itself shows order-of-magnitude variability in relatively unpolluted air).

The subsequent reactions of OH are closely coupled to the rest of the tropospheric chain reactions, and provide the principal sink term for numerous trace species as well. Figure 1B symbolizes the interconversion of OH radical with the HO_2 radical in reactions involving CO and NO; this subset drives the chain oxidation of CO to CO_2, while some of the other reactions consuming these important radicals are also shown.

Finally, Figure 1C shows how these and other subcycles combine to drive the oxidation of atmospheric methane, producing and consuming a number of important stable molecules in the process. Some of the more important radical reaction types are also given in chemical equation form in Table 1.

TABLE 1: SOME IMPORTANT REACTIONS OF OH IN TROPOSPHERIC CHEMISTRY

- with other radicals:

    NO + OH + M = HONO
    NO_2 + OH + M = HNO_3

- with ozone:

    OH + O_3 = O_2 + HOO·

- with organic and other reduced compounds:

    OH + H_2 = H_2O + H·
    OH + CO = CO_2 + H·
    OH + CH_4 = CH_3 + HOH
    OH + H_2CO = HOH + HCO·

- with oxygen:

    CH_3 + O_2 = CH_3O_2·
    HCO = CO + HO_2·

Figure 2 shows the results of one model calculation of the steady-state concentrations (lifetimes in the vicinity of 1 second) of the radicals OH and HO_2. Some of the major parameters setting these concentration levels are: the gaseous concentrations of O_3, HOH, CH_4, CO, NO_x, and the light intensity/spectrum. More recently, an extensive review has considered the available OH concentration measurements and suggested that the 24-hour average value globally is in the range 0.3 - 3 x 10 + 6/cm^{-3} (Hewitt and Harrison, 1985).

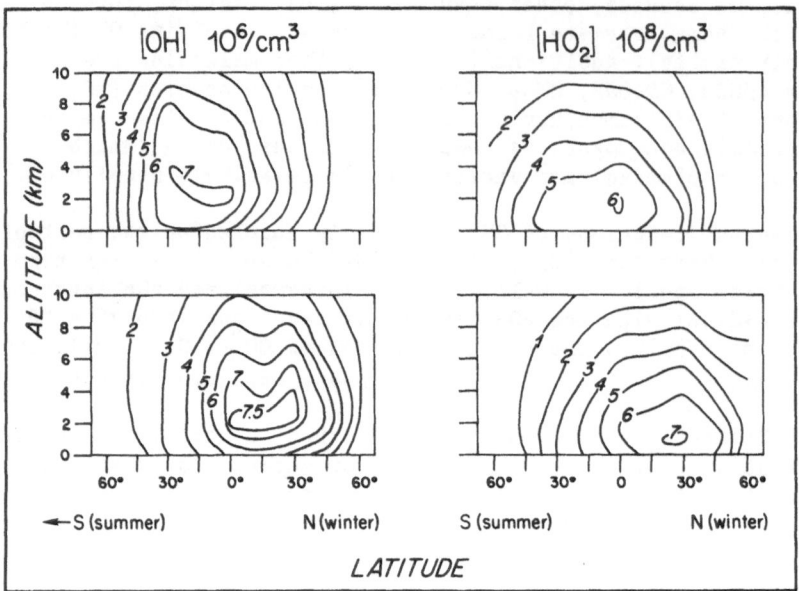

Figure 2.   Modelled noontime concentrations of OH and HO$_2$ radicals in
            the troposphere. Modified from Logan et al. (1981).

Although in this simplified summary we have stressed OH as the
key species, the reaction set involves numerous other highly reactive
short-lived radicals, including:

H, HOO, CH$_3$OO, CH$_3$O, HCO, NO, NO$_2$, NO$_3$

Of these, NO$_3$ is of particular interest because its importance as a
reactant emerges <u>at night</u>: it is formed both day and night by reaction
of NO$_2$ with ozone. However, during the day it is rapidly photolyzed,
while at night much of the (NO + NO$_2$) pool is converted to NO$_3$, which
then participates in important dark reactions; for example, the oxidation of dimethylsulfide (Andreae, this volume).
      Several of the more abundant trace species that are chemically
stable enough to be measured and are controlled in the troposphere
largely by photochemical chain reactions are summarized in Table 2.

TABLE 2:   SOME TRACE GASES CONTROLLED IN THE TROPOSPHERE BY CHAIN
           OXIDATION REACTIONS.

Produced	Consumed	Photochemical Steady-State
CO$_2$, O$_3$, HNO$_2$, HNO$_3$, CO	CH$_4$, CO, most trace organic gases, NO$_x$	H$_2$, HCHO, CH$_3$OOH, HOOH

Here the categories are approximate, stressing that the process can be a net source, sink, or both for various trace materials. For example, CO has major photochemical sources and sinks, but there are important nonphotochemical sources as well. Table 3 summarizes the expected fluxes due to atmospheric chemistry for a number of the important trace gases as given by Logan et al. (1981).

TABLE 3: TRACE GAS PHOTOCHEMICAL SINKS[*]

Gas	Mixing Ratio (PPBV)	Loss Rate (MT/Y)	Turnover Time (Y)
H_2	550	21	9
CO**	100	3,170	0.25
CH_4	1,650	600	8
CH_3Cl	0.6	5	1
CH_3Br	0.02	0.3	1
C_2H_6	1	31	0.2
Isoprene	?	?	<0 001

* = Simplified from Logan et al. (1981), Tables 8 and 18.
** = CO shows major interhemispheric differences.

To summarize, the methane oxidation chain has major implications for the trace gases involved in numerous elemental cycles, particularly C, H, O (documented above) as well as N (production and deposition of HNO_3/HNO_2), S, (see chapter by M. O. Andreae) and other elements which may have significant organic-bound forms, such as the methyl derivatives of Cl, I, and some other metals and metalloids (Jorgensen, 1982; Brinckmann et al., 1982). Additionally, we have paid very little attention to polluted air, where oxidation processes are intensified and altered significantly in form. Here again, OH is a key intermediate, though O_3 increases in relative importance and aerosol formation/ evolution processes are critical.

2.3. HETEROGENEOUS TROPOSPHERIC PHOTOCHEMISTRY

The extent, nature, and significance of heterogeneous photoprocesses in the atmosphere are much less well-known than that of the homogenous processes. This is currently a very active area of investigation, with papers appearing almost monthly. These processes certainly contribute substantially to the process of chemical modification of some trace gases and particles in the atmosphere. Although these complex hetero- geneous processes may actually comprise a continuum, at this stage of our knowledge the generation of particles by nucleation phenomena, the growth and chemical modification of aerosol under noncondensing condi- tions, and the 'wet' chemistry of cloud/smog/fog droplets are treated somewhat separately in the literature.

The transitional state of the art in this area is well illustrated by excerpts from the 1982 Dahlem Conference volume Atmospheric Chemistry. The abstract of the Working Group report on 'Aqueous Chemistry in the Atmosphere' stated that '... Among the areas judged likely to be of importance but where the understanding is presently meager are: knowledge of the details of the atmospheric water cycle, the convective transport of trace components through clouds, the incorporation of trace gases into atmospheric water, the role of catalysis in atmospheric water chemistry, the role of photochemical processes, and the efficiency of gas and aerosol particle scavenging by the Earth' (Graedel et al., 1982). Concurrently, the Working Group on 'Tropospheric Gases, Aerosols, and Photochemical Reactions,' included photochemical aspects of aerosol research in 6 out of 17 goals: (4) ... explore mechanisms for in situ aerosol formation..., (6) ...develop a new generation of models to treat cloud chemistry, (7) Assess importance of reactions between radicals and aerosol and droplet surfaces both theoretically and experimentally ..., (12) Determine the chemical nature of tropospheric ions ..., (15) Determine atmospheric fate of organic emissions from plants..., (16) Determine relative yields of gaseous CO and of organic aerosols..., (17) Determine mechanisms of chlorine loss and iodine enrichment in marine aerosols' (Cicerone et al., 1982).

Concerning de novo formation of atmospheric particulate matter, Duce (1978) suggested that there exists an uncharacterized source of secondary aerosol that forms ca 70-130 x 10^{12} g organic C/y and speculated that efficient gas-to-particle conversion processes might be responsible. Since that time, there has been little progress in quantifying or identifying this presumed source. Much of this material could be fine-mode soot formed by combustion processes; the other plausible source would be photochemical formation of aerosol material, which is well-known to occur under polluted 'smog' conditions and is strongly suspected for some vegetation emissions (Altschuller, 1983; Schuetzle and Rasmussen, 1978; Hooker et al., 1985). However, the nature and importance of this process on regional and global scales and in clean air is still an open question.

Likewise, the photochemical modification of 'dry' aerosol that has formed by other mechanisms almost surely occurs, but is relatively poorly understood. Since these materials have an atmospheric residence time against deposition processes of about a week, adequate opportunity exists over most of the globe for major modifications of materials.

The heterogeneous photochemistry of clouds, fogs, and precipitation is somewhat better studied, though still essentially a mystery. Some basic facts about clouds, primitively stated, will clarify how crucial cloud-driven chemical transformations probably are. Clouds are obviously the mothers of wet deposition (neglecting 'washout'), but that is not their usual fate: the average cloud does not disappear by raining itself out of existence; it evaporates (Pruppacher and Klett, 1980). An often-cited guesstimate of Junge's (Slinn, 1982) is that on the average tropospheric water vapor molecule undergoes ten condensation-evaporation cycles for every removal as rain! Thus, the potential of clouds to function as enormously active chemical factories influencing hydrophilic gases and particles in the atmosphere is enormous. Photoreactions are just one of the array of transformations that operate in this dynamic regime; the entire process is an area of immense scientific complexity and opportunity.

Several modelling studies have suggested, however, that photochemistry is important within clouds. In a pioneering study, Graedel and Weschler (1981) first modelled coupled photochemistry in both aqueous and gas phases. This study revealed the possibility of major effects due to the uptake of photochemically generated gas-phase species into droplets as well as to photolysis within droplets. Chameides and Davis (1982) addressed similar questions using a more elaborate treatment (especially of the crucial transport/uptake phenomena involved), and also found potentially important effects in both categories.

More recently, Chameides (1984) modelled the coupled gas-liquid photochemistry of an insolated stratiform cloud in the remote marine atmosphere in greater depth, using a time-dependent box model and emphasizing possible pH-controlling processes. In the model, air was suddenly cooled and water condensed to form initially 120 10-micron droplets per cubic centimeter, which then evolved chemically in time in response to the gaseous and aqueous phase chemistry and the coupling between the two. The model calculations indicating the droplet pH as a function of droplet age showed an increase in acidity even without any aqueous phase photochemistry included; with photochemistry, the acidity at droplet ages longer than about 100 s increased, reaching twice the $[H^+]$ of the 'without' model at about 3000s.

The aqueous phase photochemistry contributed significantly to several processes, primarily the oxidation of sulfur dioxide to sulfate and of formaldehyde to formic acid. Hydrogen peroxide was important in both these oxidations; its concentration (compared to that modelled without in-droplet photochemistry) also increased dramatically beyond droplet lifetimes of a few hundred seconds. The only aqueous primary photoreaction included was hydrogen peroxide photolysis. This primitive assumption may substantially underestimate in-droplet photochemistry.

The principal conclusions of this study were that even under remote conditions, aqueous-phase chemistry and photochemistry in cloud droplets could be significant. Chameides emphasizes that as the sophistication of the models increases, it seems untenable to neglect the aqueous phase of clouds and that as a geochemical corollary, rainwater analyses for redox-active substances cannot be interpreted as reflecting the unaltered chemistry of the ambient atmosphere.

Chatfield and Crutzen (1984) stressed the interaction of cloud transport and photochemistry in a related paper; they show that the gaseous photochemistry and the vertical transport in tropical cloud systems may strongly interact. The process has major implications for promoting long-range transport (and hence implicitly for greatly increasing exposure of compounds to photolysis via increased residence time in clear air). Schwartz (1984) provides a careful examination of the physical chemistry of just one radical, HO_2, in model cloudwater-tropospheric air systems. He concludes that major effects are feasible but that the current knowledge of boundary conditions does not permit accurate modelling of this aspect of the problem.

Calvert et al. (1985) review the chemical mechanisms of acid generation in the troposphere and find that both gaseous and liquid-phase reactions are implicated, with suspended aerosol alteration a less certain mechanism. The data for the eastern U.S. strongly support the notion that in-cloudwater oxidation of SO_2 by hydrogen peroxide scavenged from the gas phase is an important mechanism of acid and excess sulfate generation. Recently, Graedel at al. (1985) have suggested that although SO_2-HOOH interactions are a major pathway, at pH 4 some 30-35% of the SO_2 oxidation may be mediated by in-droplet processes involving complexed transition metals.

In the larger perspective, the photochemistry of aerosols, fogs and rain cannot be separated from the total ongoing chemistry, of which it seems an important part for some species. The entire area is in productive ferment.

2.4. SEAWATER PHOTOCHEMISTRY

The photochemistry of fresh waters and the oceans has been studied experimentally much more intensively than has cloudwater chemistry. An introductory review (Zafiriou et al., 1984) and bibliography (Zafiriou, 1984) are available. See these sources for a more complete discussion and references. The following material is highly condensed.

Photochemistry is only one of an array of biological and chemical processes influencing natural water bodies. Furthermore, there is no strong evidence that photochemistry alone can <u>control</u> the geochemical cycle of any element (defining photosynthesis to be a biological process). Nevertheless, photochemical reactions can be important sources and sinks for a variety of materials.

Before entering detailed consideration of aqueous-phase photochemistry, a few basic points may serve as useful orientation:

1. Aqueous photochemistry cannot usually be predicted, though results can often be rationalized qualitatively or quantitatively.
2. Thus, aqueous photochemistry is an experimental/observational science; the primary measurement is of some differential effect between light and dark conditions.
3. Photochemistry begins with absorption of a photon by a chromophore. The chemical structures of natural water chromophores are largely unknown. Some hypothesized structures are shown in Figure 3, more to highlight the variety of possibilities and lack of knowledge than to suggest that these are representative structures.
4. Photoproducts that are no longer electronically or thermally excited - secondary products - are often nevertheless still highly reactive intermediates that form the observed products.
5. The presence of dissolved oxygen opens several pathways that generate reactive species. Thus, while there is enormous diversity in natural water photochemistry, perhaps the simplest generalization is that net oxidative redox processes are often important.

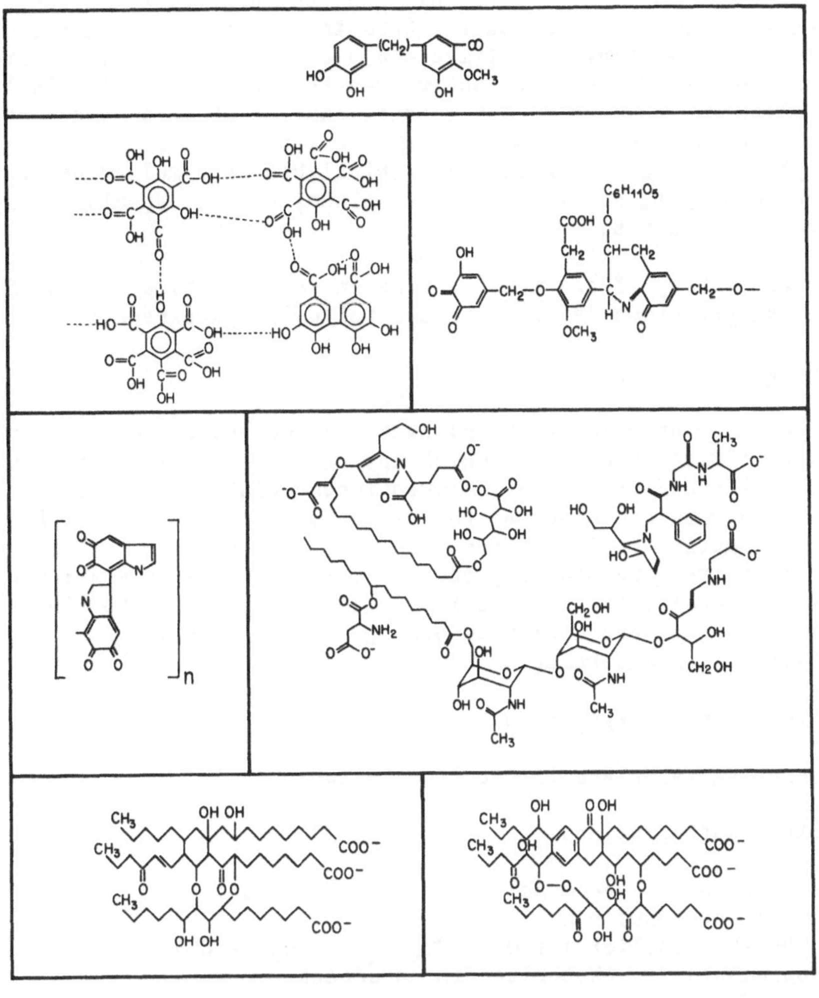

Figure 3. Some proposed structures for colored organic materials in natural waters. Modified from Zafiriou et al. (1984). In this review we also adopt the nomenclature proposed in that article: 'UC' = unknown chromophore; 'UPC' = unknown photoreactive chromophore.

6. Organic and inorganic particulate materials participate in photochemical processes. The relative importance of particulate **vs** homogeneous chemistry and the question of which of these effects are primary **vs** secondary photochemistry are unresolved issues.
7. Photosynthesis gives numerous light/dark effects not easily differentiated from photochemistry in systems containing particles.
8. Natural water photochemical systems can be highly sensitive to trace organic, inorganic, and particulate speciation.

9. Although the underwater light field and its involvement in photochemistry are extremely complex, there is a strong general trend
   for blue-violet-ultraviolet light (ca 300-500 nm) to be most effective and for the depth of effective penetration (90% extinction) to
   be of the order of 1-20 m near coasts, and 15-100 m in 'blue water.'
10. Many synthetic organic chemicals absorb sunlight well and are much
    more photochemically active than typical biogenic molecules of low
    molecular weight, so direct photochemistry can be especially impor-
    tant for xenobiotics.

A brief summary of the major actors in various key elements of
photochemical processes is given in the following paragraphs:

Chromophores

Important chromophores in marine systems include unknown chromophores
('UPC's' and 'UC's'; Figure 3), which may be closely related to humic
and fulvic acids and/or marine 'heteropolycondensate,' as well as known
organic and inorganic compounds; for example, nitrite, nitrate, hydrogen peroxide, methyl iodide, trace vitamins (biotin, B_{12}, thiamine),
flavinoid pigments, and probably organometallic complexes are important. Clay minerals, Fe and Mn colloids and coatings, sorbed 'UC's,'
algal cell surfaces, and nonliving biopigments are likely to be important particulate chromophores.

Primary and Secondary Processes

Several reaction types are illustrated in Table 4.

Particulate transformations

Multiple pathways - no model equation. Examples:
$$Particulate\ MnO_2 \rightarrow Mn^{2+}_{aq}$$
$$Fe\text{-}Org\text{-}PO_4\ (colloid) + O_2 \rightarrow PO_{4(aq)} + Fe^{2+}_{aq} + CO_2 + Org'$$
$$Biopigments \rightarrow \rightarrow photobleaching\ products$$

Short-lived reactive species

In complex oxygenated systems many important transformations result
from secondary reactions of short-lived (on the timescale of the
observations) intermediates. Some of the more important ones in marine
systems (see also Table 4 equations) are:

Excited states: UPC^* (triplet state), O_2^* (singlet state)
Free radicals: OH, O_2^- , Br_2^-, $ROO\cdot$ (peroxy radical)
Reactive molecules: $HOOH$, R_2SOO (see energy transfer example)

TABLE 4: SOME NATURAL WATER PHOTOREACTION TYPES

Photoionization: $X \rightarrow X^+ + e^-_{aq}$

   UPC (phenols?) → radical cations + solvated electrons

Bond homolysis to free radicals: $X - Y \rightarrow X\cdot + Y\cdot$

   $CH_3I \rightarrow CH_3\cdot + I\cdot$

   NO_2^- (aq) → NO + OH

Photonucleophilic substitution: $Ar-X + N^- \rightarrow Ar-N + X^-$

   Pentachlorophenol + H_2O → tetrachlorohydroquinone + Cl^- + H^+

Molecular rearrangements: $X \rightarrow X'$ (isomer)

   R══R (E or Z configuration) →

   R══R (Z or E double bond isomer)

Energy transfer: $D + h\upsilon \rightarrow D^*$ (singlet) → D^* (triplet) $D^* + A \rightarrow A^* + D$

   UPC + $h\upsilon$ → UPC* (triplet)

   UPC* + O_2 → O_2^* (singlet oxygen) + UPC

Singlet Oxygen Reactions: $O_2^* + X \rightarrow$ products

   $O_2^* + 2 R_2S \rightarrow R_2SOO + R_2S \rightarrow 2 R_2SO$

Secondary radical reactions: Radical-Radical: $X\cdot + Y\cdot \rightarrow$ Products:

   $O_2^- + O_2^- + H_3O^+ \rightarrow HOOH + O_2 + H^+$

   Primary → Secondary Radical (S = solute): $R\cdot + S \rightarrow R'\cdot$

   $\cdot OH + 2 Br^- \rightarrow Br_2^- + OH^-$

   $CH_3\cdot + O_2 \rightarrow CH_3OO\cdot$

   $e_{(aq)}^- + O_2 \rightarrow O_2^-$

Rates and Concentrations

It is difficult to generalize about these topics. Singlet oxygen formation is probably the largest photochemical flux in seawater; it
varies roughly in proportion to the optical absorption coefficient of
the water at 350 nm in freshwater. However, most of this very large
flux does not react chemically, but decays to ground-state oxygen.
   Photochemical hydrogen peroxide formation also involves UPC and
thus may roughly parallel singlet oxygen generation rates. Both generation rates would then be expected to be higher in coastal than in blue
water; however, the variation in HOOH loss rates may be equally important in regulating the observed surface water concentrations. Thus,
low formation rates need not correspond to low concentrations.
   Nitrite photolysis rates are proportional to nitrite concentrations and water transparency, and hence are primarily important in
coastal waters, estuaries, and temperate ocean waters before summer

stratification.

A reasonable speculation appears to be that in coastal and estuarine waters the unknown dissolved organic chromophores (UPC's, Fig.
3), and perhaps suspended particulates, dominate the photochemistry,
while in the already dominant open ocean such processes as biopigment
and nitrite photolysis combine with an unknown - possibly minor - level
of 'UPC'-driven photochemistry to give an overall effect which may be
rather different.

2.5 SOIL PHOTOCHEMISTRY

Almost nothing is known about soil photolysis. By analogy with freshwater photochemistry (Zafiriou et al., 1984), it seems that soil humic
and fulvic materials and some minerals react in soil, perhaps resem-
bling on a microscale aquatic colloidal/particulate material. These
processes might be quite intense at the soil surface because of the
relatively high concentration of colored organic matter there. If so,
they may result in the continental, easily deflatable material already
being photochemically modified before becoming airborne under conditions suited to long-range transport. Then aerosol content of photolysis products would not be a simple indicator of atmospheric residence
time. However, the 'freshness' of soil-derived primary aerosols with
respect to photochemical alteration is uncertain; they could range from
highly reactive to 'burnt out.'

3. INTERACTION OF PHOTOCHEMISTRY WITH AIR-SEA EXCHANGE PROCESSES

The foregoing introductory survey shows both the complexity and the
relative scarcity of quantitative information concerning the rates and
detailed chemistry of environmental photoprocesses, excepting gas-phase
tropospheric chemistry. Other chapters in Liss and Slinn (1982) reveal
that the exchange mechanisms themselves are also mechanistically complex and only partly understood. Therefore, it is not surprising that
the effects of photochemical processes on exchange fluxes and processes
are poorly understood. The following sections discuss the possibilities and the evidence for them, arranged according to the author's subjective estimate of decreasing certainty concerning our knowledge and
probable importance of the impact on the mechanism under discussion.

3.1. AIR-SEA GAS EXCHANGE

Air-sea gas exchange as modelled by the two-film stagnant boundary
layer model is certainly the best understood process, and also the one
most certainly and pervasively affected photochemically. There seem
to be four mechanisms of interaction; three are geochemically significant, while the fourth is speculative. These are:

1. Modifications of gas-exchange fluxes by alteration of bulk tropos-

pheric concentrations of gases.

2. Modifications of fluxes by alteration of bulk seawater concentrations of volatile compounds.

3. Modifications of fluxes by processes altering the local concentrations and speciations of exchanging materials <u>within</u> the boundary layers.

4. Modification of the physical properties of the air-water interface, leading to altered exchange rates.

3.1.1. GAS EXCHANGE - MODIFIED BULK AIR CONCENTRATIONS

As developed earlier, stratospheric and especially tropospheric photoprocesses dramatically modify the average tropospheric concentrations, hence fluxes, of a variety of trace gases, most especially: ozone (+), methane (-), carbon monoxide (+) hydrogen (+), nitrous oxide (-) formaldehyde (+), hydrogen peroxide (+) , and all trace organic compounds bearing C-H bonds (-). Here the signs following the gases indicate whether flux <u>into</u> the sea tends to be increased (+) or decreased (-). In general, these effects are so large that the photochemistry is a major factor setting the atmospheric level and hence the air-sea exchange flux sign/intensity.

For example, in the absence of tropospheric methane oxidation, tropospheric levels would be set by some combination of transport into the stratosphere and biological oxidation in soils and the hydrosphere; the steady-state tropospheric level could easily be at least an order of magnitude higher. If the stratosphere were then the major sink, the water generated there would make the world quite a noticeably different place!

This interaction clearly has important effects on trace species in the C, H, and O cycles, and probably significantly perturbs the N (NO, NO_2, HNO_3), S (CH_3SCH_3), and gaseous halogen cycles. It plays a crucial role in the transformation of carbon compounds from anthropogenic gaseous emissions (freons excepted).

3.1.2. GAS EXCHANGE - MODIFIED SURFACE WATER CONCENTRATIONS

Seawater photochemistry also modifies the bulk surface-water concentrations of some substances that undergo air-sea gas exchange, thus modifying the sign/intensity of the exchange flux in analogy to the tropospheric effects discussed above. These effects are far less general than those in the air. Carbon monoxide is produced photochemically in seawater (Conrad et al., 1982; see also the chapter by Seiler in this volume) by photolysis of UPC; some concurrent biological production occurs as well. The seawater photolysis renders the sea a CO source in tropical regions (where the atmospheric level is also strongly modified by photochemistry). However, the sea-air flux is not a major term in the global CO budget.

Likewise, in restricted regions at least there is evidence for formation of COS in seawater (Chesapeake Bay) (see discussion by Andreae, this volume) and for sufficient production of NO in some areas (equatorial Pacific) to reverse the sign of the expected air-to-sea flux (Zafiriou et al., 1980) for this compound from a marine sink to a source. However, as discussed later, this estimate presumes that NO can behave conservatively in the boundary film, which may be incorrect due to still other photochemically driven fluxes.

The major photoproducts singlet oxygen and hydrogen peroxide are respectively too short-lived and too water soluble to have a major direct influence on the gaseous air-sea exchange fluxes, but their subsequent reactions may be significant sources of exchangeable volatiles.

3.1.3. GAS EXCHANGE - INTERACTIONS IN BOUNDARY LAYERS

It is important to remember that the stagnant-film view of air-sea exchange is merely a construct that probably has limited basis in physical reality (Fig. 4A). It is still a useful framework for thinking about the interactions that occur near the interface.

The only case for which there is strong (though indirect) evidence for such specifically interfacial interactions is the enhanced reaction rate of ozone diffusing to the sea surface (Aldaz, 1969) where it reacts according to model studies with dissolved iodide at a sufficient rate to produce molecular iodine, which evaporates in part - at least in the laboratory (Garland et al., 1980; Garland and Curtis, 1981).

Thompson and Zafiriou (1983) carried out a more general survey estimating fluxes of reactive atmospheric species to the air-sea interface. They found that in addition to ozone, there should be rather large fluxes of HOOH and the radicals CH_3O_2 and HO_2, which might also generate volatile iodine species, and further suggested that I_2 might not be the relevant volatile species. An heuristic 'cartoon' of some of the interactions near the air-sea interface by these authors is reproduced here in Figures 4B and C. If the stagnant film layer were to exist physically, numerous chemical reactions of these and other atmospherically derived species could take place on physical and temporal scales small compared to diffusive transport through the layer. However, the physics and chemistry of the situation may be so complex that actual impacts are not known, except for the laboratory studies on ozone and iodine.

Another likely effect is annihilation reactions of atmospherically and oceanically derived species as they attempt to diffuse past one another in the near-surface film; for example, Thompson and Zafiriou suggested that scavenging of NO by atmospherically derived radicals could minimize or even nullify the sea-air NO flux expected on the basis of bulk-phase concentration estimates (see also Figure 4C).

Furthermore, enhanced concentrations of surface-active organic matter and of chromophores have been reported in sea-surface films (Gershey, 1983), suggesting that the interface is the site of intense direct photochemistry altering concentrations and fluxes.

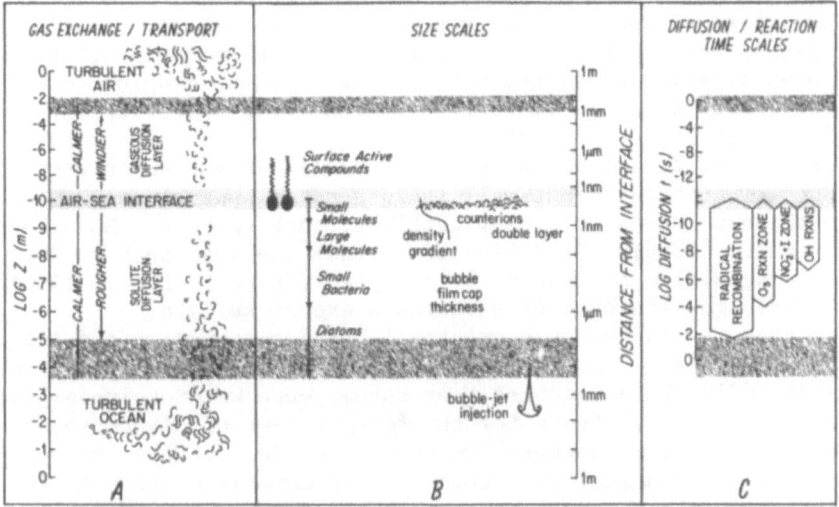

Figure 4.   A 'cartoon' of processes near the air-sea interface, modi-
             fied from Thompson and Zafiriou (1983). Note logarithmic
             distance scale and discontinuity at interface. A. The 'two-
             layer model,' with 'real' turbulence disturbing the diffu-
             sion layers. B. Some relevant size scales. C. Distance
             scale for some chemical processes, if defined by molecular
             diffusive transport.

        To summarize, the notion of significant effects on air-sea
exchange occurring within the boundary layer, while not proven, seems
plausible and is backed by laboratory studies in the ozone-iodine case.
However, the physical model of transfer processes, the chemical dynamics, and their coupling are not yet sufficiently understood to permit
much further analysis. Experimental techniques to facilitate observations in this regime are likewise lacking.

3.2. RAINOUT-WASHOUT DEPOSITION PROCESSES

The only certain effect of photochemistry on rainout-washout processes
at this time appears to be the formation of highly water-soluble
substances by tropospheric photochemical reactions; these are then
incorporated into precipitation and hence enter the sea. Some of the
more geochemically interesting species affected by this pathway are:
formaldehyde, hydrogen peroxide, nitric acid, and atmospheric iodine
species (Chameides and Davis, 1980). These fluxes may be quite large;
for example, the formaldehyde flux may approach 1% of primary production in oligotrophic areas (Zafiriou et al., 1980) and hydrogen peroxide may approach 100 μM (!) concentration in rain (Zika et al., 1982).
Nitrate removal is the dominant depletion term for atmospheric NO_x and
hence strongly affects the tropospheric reaction cycle (Logan et al.,

1981; Thompson and Cicerone, 1982; Heikes and Thompson, 1983), and sup-
plies some fixed nitrogen to the sea (Duce, this volume).

Of less certain significance is the coupling between photochemistry and precipitation chemistry due to aqueous-phase droplet reactions
discussed in Section 2.3. It has the potential for altering the chemistry of relatively water-soluble materials, as well as the involatile
materials forming the cloud condensation nuclei around which droplets
form. Such alterations might well include redox speciation changes of
trace metals not locked into clay lattices, such as Fe and Mn oxides
(Graedel et al., 1985). Oxidation of aqueous species, such as SO_2
and formaldehyde, are now generally accepted (Calvert et al., 1985) to
influence hydrogen ion concentrations and excess sulfate.

This area of cloudwater chemistry is both speculative and likely
to be quite important. Since in general as reduced trace species begin
to be oxidized in the atmosphere they become more water-soluble, there
might well be a general trend for the final stages of alteration to
occur in aqueous phases wherever these are abundant in the atmosphere,
thus altering the chemical speciation of the materials input to the
oceans (presumably tending to increase their degree of oxidation) and
the form in which they are delivered (rainout vs dry depostion of soluble materials).

3.3. DRY DEPOSITION

There is no firm evidence that photochemical processes directly affect
dry deposition. Although the physical mechanisms of dry deposition
are complex and only partially understood (Liss and Slinn, 1982), it
seems unlikely the actual mechanisms can be much influenced by photochemistry. Two plausible exceptions are (1) the reworking of materials
by a variety of processes in evaporating clouds, as discussed in the
previous section, and (2) the photochemical alteration of surface-
active organic matter at the air-sea interface, which may alter
capillary-wave properties and hence the microphysics of the deposition
process.

More likely effects of photochemistry involve altering the chemical speciation of the aerosol available for deposition. The clear-air
aerosol is undoubtedly influenced by photolysis processes that resemble
to some extent those occurring in cloud droplets, which are similarly
ill-explored. Since material undergoing long-range mid-tropospheric
transport spends much more time in the clear-air noncondensing regime
than in cloud droplets, this process could conceivably be as important
as in-cloud transformations.

The major aerosol processes that require study to define their
importance are: (1) formation of secondary aerosols from gaseous materials, and (2) alteration of pre-existing aerosols both by photolysis
(such as of accreted HOOH, of primary soil minerals, and humic materials), and by secondary reactions due to dissolution/reaction of ozone,
radicals, etc.

3.4. MARINE AEROSOL GENERATION

The marine aerosol is generated by bubbles bursting at the air-sea interface, largely as a consequence of the injection of air into the surface ocean by breaking waves (MacIntyre, 1974; Lion and Leckie, 1981). The bubbles are enriched in surface-active materials which may be adsorbed during bubble rising, or be derived from the sea-surface by the 'microtome effect.' The transport of trace elements and organic compounds to the interface by rising bubbles has been demonstrated (e.g. Wallace and Duce, 1978).

Thus, photochemistry of the surface-active organic matter at the interface or in the upper few meters of the sea may modify the amounts and chemical nature of materials subject to aerosol formation by break-ing bubbles. There is indirect evidence for such photoreactions, though there is no information on their effect on surface activity or on colloid-bound trace element levels and speciation. The 'UC' chromo-phores, such as structures shown in Figure 3, are both surface-active and photolabile. Also, well-known surface-active compounds such as long-chain fatty acids are photolabile in the marine environment (Wheeler, 1972). Finally, there are enhanced concentrations of chromo-phores in bubble-ejected aerosols (Garvey, 1983).

However, all these indications are indirect, and there is no unequivocal environmental evidence (possibly excepting the iodine cycle, (see papers by Seto and Duce, 1972; Miyake and Tsunogai, 1963; and Korzh, 1984) for the importance of these processes.

4. SUMMARY

The case for photochemical effects on exchange mechanisms other than gas exchanges influenced by changes in their bulk-reservoir concentra-tions ranges from indirect evidence to outright speculation. A major factor in this situation is that these other mechanisms, involving het-erogeneous chemistry at the molecular level, are very intractable prob-lem areas - both experimentally and for modelling. The difficulty in evaluating some of the effects cited is in part due to the complexities of environmental photochemistry, but is more fundamentally a problem of our inability to deal cleanly with the other physical and chemical aspects of heterogeneous geochemical processes. Hopefully, this chap-ter can serve a purpose by alerting workers who may have relevent ob-servational evidence or new and more powerful experimental tools for probing effects of some of the photoprocesses we have speculated about, since these issues will finally be resolved by measurements rather than verbal or mathematical speculations.

5. ACKNOWLEDGEMENTS

The preparation of this chapter was funded by NSF grant OCE-85-17770. D. Blanchard, A. Carlier, Q. I. T. Espey, M. Ewald, P. S. Liss, and J. Tokos contributed useful information and criticisms.

6. REFERENCES

Aldaz, L., 1969: 'Flux measurements of atmospheric ozone over land and water,' J. Geophys. Res., 74, 6943-6946.

Andreae, M.O., F. Beuner, N.R. Anderson, O.H. Ehlalt, W. Balzer, R.O. Hallberg, H.G. Bingemer, B.B. Jorgensen, P.S. Liss, J.E. Lovelock and U. Schmidt, 1982: 'Biogenic contributions to atmospheric chemistry,' in: Atmospheric Chemistry (E.D. Goldberg, ed.), Springer-Verlag, Berlin, pp. 251-272.

Altschuller, A.P., 1983: 'Natural volatile organic substances and their effect on air quality in the United States,' Atmos. Environ. 17, 2131-2165.

Brinckman, F.E., G.J. Olson and W.P. Iverson, 1982: 'The production and fate of volatile molecular species in the environment: metals and metalloids,' in: Atmospheric Chemistry (E.D. Goldberg, ed.), Springer-Verlag, Berlin, pp. 231-249.

Calvert, J.G., A. Lazrus, G.L. Kok, B.G. Heikes, J.G. Walega, J. Lind and C.A. Cantrell, 1985: 'Chemical mechanisms of acid generation in the troposphere,' Nature 317, 27-35.

Chatfield, Robert B. and Paul J. Crutzen, 1984: 'Sulfur dioxide in remote oceanic air:cloud transport of reactive precursors,' J. Geophys. Res. 89, 7111-7132.

Chameides, W.L. and D.D. Davis, 1980: 'Iodine: its possible role in tropospheric photochemistry,' J.Geophys. Res. 85, 7383-7398.

Chameides, W.L. and D. Davis, 1982: 'The free-radical chemistry of cloud droplets and its impact upon the composition of rain,' J. Geophys. Res., 87, 4863-4877.

Chameides, W.L., 1984: 'The photochemistry of a remote marine stratiform cloud,' J.Geophys. Res., 89, 4739-4755.

Cicerone, R.J., 1981: 'Halogens in the atmosphere,' Rev. Geophysics and Space Physics 19, 123-139.

Cicerone, R.J, F. Arnold, P.J. Crutzen, D.F. Hornig, R. Jaenicke, F.X. Meixner, M.J. Molina, H. Niki, S.A. Penkett, F.S. Rowland, R.E. Zellner and P.R. Zimmerman, 1982: 'Topospheric gases, aerosols and photochemical reactions,' in: Atmospheric Chemistry (E.D. Goldberg, ed.), , Springer-Verlag, Berlin, pp.357-372

Conrad, R., W. Seiler, G. Bunse and H.Giehl, 1982: 'Carbon monoxide in seawater (Atlantic Ocean),' J.Geophys.Res. 87, 8839-8852.

Duce, R.A., 1978: 'Speculations on the budget of particulate and vapor phase non-methane organic carbon in the global troposphere,' Pageoph. 116, 244-273.

Garland, J.A. and H. Curtis, 1981: 'Emission of iodine from the sea surface in the presence of ozone,' J. Geophys. Res. 86, 3183-3186.

Garland, J.A., A. Elzerman and S.A.Penkett, 1980: 'The mechanism for dry deposition of ozone to seawater surfaces,' J. Geophys.Res. 85, 7488-7492.

Gershey, R.M., 1983: 'Characterization of seawater organic matter carried by bubble-generated aerosols,' Limnol. Oceanogr. 28, 309-319.

Goldberg, E.D. (ed.), 1982: Atmospheric Chemistry, Springer-Verlag, Berlin, 385 pp.

Graedel, T.E. and C.J. Weschler, 1981: 'Chemistry within aqueous atmospheric aerosols and raindrops,' Rev. Geophys. Space Phys. 19, 505-539.

Graedel, T.E., G.P. Ayers, R.A. Duce, H.W. Georgii, D.G.A. Klockow, J.J. Morgan, H. Rodhe, B. Schneider, W.G.N. Slinn and O.C. Zafiriou, 1982: 'Aqueous chemistry in the atmosphere,' in: Atmospheric Chemistry (E.D. Goldberg, ed.), Springer-Verlag, Berlin, pp. 93-118.

Graedel, T.E., C.J. Weschler and M.L. Mandich, 1985: 'The influence of transition metal complexes on atmospheric droplet acidity,' Nature (in press).

Heikes, B.G. and A.M. Thompson, 1983: 'Effects of heterogeneous processes on NO_3, HONO, and HNO_3 chemistry in the troposphere,' J. Geophys.Res. 8, 10883-10895.

Hewitt, C.N. and R.M. Harrison, 1985: 'Tropospheric concentrations of the hydroxyl radical - a review,' Atmospheric Environment 19, 545-554.

Hooker, C.L., H.H. Westberg and J.C. Sheppard, 1985: 'Determination of carbon balances for smog chamber terpene oxidation experiments using a ^{14}C tracer technique,' J. Atmospheric Chem. 2, 307-320.

Jorgensen, B.B., 1982: 'The production and fate of reduced C, N, and S gases from oxygen-deficient envioronments,' in: Atmospheric Chemistry (E.D.Goldberg, ed.), Springer-Verlag, Berlin, pp.215-229.

Korzh, V.D., 1984: 'Ocean as a source of atmospheric iodine,' Atmospheric Environment 18, 2707-2710.

Levy H., II, 1971: 'Normal atmosphere: large radical and formaldehyde concentrations predicted,' Science 173, 141-143.

Levy H., II, 1972: 'Photochemistry of the lower troposphere,' Planet.
        Space Sci. 20, 919-935.

Lion, L.W. and J.O. Leckie, 1981: 'The biogeochemistry of the air-sea
        interface,' Ann. Rev. Earth Planet. Sci. 9, 449-486.

Liss, P.S., 1982: Air-Sea Exchange of Gases and Particles (P.S. Liss
        and W.G.N. Slinn, eds.), Dordrecht:D. Reidel Publ. Co., 561 pp.

Logan, J.A., Michael J. Prather, Steven C. Wofsy and Michael B.
        McElroy, 1981: 'Tropospheric chemistry: a global perspective,' J.
        Geophys. Res. 8, 7210-7254.

MacIntyre, F., 1974: 'Chemical tractionation and sea-surface microlayer
        processes,' in The Sea 5 (E.D. Goldberg, ed.), pp. 245-299.

McConnell, J.C., M.B. McElroy and S.C. Wofsy, 1971: 'Natural sources
        of atmospheric CO,' Nature 233, 187-188.

Miyake, Y. and S. Tsunogai, 1963: 'Evaporation of iodine from the
        ocean,' J.Geophys. Res. 68, 3989-3993.

Niki, H., 1982: 'Homogeneous gas phase oxidation processes in the
        troposphere,' in: Atmospheric Chemistry (E.D. Goldberg, ed.),
        Springer-Verlag, Berlin, pp. 301-312.

Pruppacher, H.R. and J.D. Klett, 1980: Microphysics of clouds and
        precipitation, D. Reidel, Hingham, MA., 714 pp.

Schuetzle, D. and R.A. Rasmussen, 1978: 'The molecular composition of
        secondary aerosol particles formed from terpenes,' J. Air Pollut.
        Control Assoc. 28, 236-240.

Schwartz, S.E., 1984: 'Gas- and aqueous- phase chemistry of HO_2 in
        liquid water clouds,' J. Geophys Res. 89, 11589-11598.

Seto, F.Y.B. and R.A. Duce, 1972: 'A laboratory study of iodine enrich-
        ment on atmospheric sea-salt particles produced by bubbles,' J.
        Geophys. Res. 77, 5339-5345.

Slinn, W.G.N., 1982: 'Some influences of the atmospheric water cycle
        on the removal of atmospheric trace constituents,' in: Atmospheric
        Chemistry (E.D.Goldberg, ed), Springer-Verlag, Berlin, pp. 57-90.

Thompson, A.M. and R.J. Cicerone, 1982: 'Clouds and wet removal as
        cause of variability in the trace-gas composition of the marine
        troposphere,' J. Geophys. Res. 87, 8811-8826.

Thompson, A.M., and O.C. Zafiriou, 1983: 'Air-sea fluxes of transient
        atmospheric species,' J.Geophy. Res. 88, 6696-6708

Wallace, G.T., Jr. and R.A. Duce, 1978: 'Open-ocean transport of particulate trace metal by bubbles,' Deep-Sea Res. 25, 827-835.

Wheeler, J., 1972: 'Some effects of solar levels of ultraviolet radiation on lipids in artificial sea water,' J. Geophys. Res. 77, 5302-5306.

Zafiriou, O.C., J. Alford, M. Herrera, E.T. Peltzer and R.B. Gagosian, 1980: 'Formaldehyde in remote marine air and rain: Flux measurements and estimates,' Geophys. Res. Lett. 7, 341-344.

Zafiriou, O.C., 1984: A Bibliography of Refrences in Natural Water Photochemistry, Technical Memorandum, WHOI-2-84 (Woods Hole Oceanographic Institution), 51 pp.

Zafiriou, O.C., J. Joussot-Dubien, R.G. Zepp , and R. G. Zika, 1984: 'Photochemistry of natural water,' Environ. Sci. Tech. 18, 358A-371A.

Zika, R., E. Saltzman, W.L. Chameides and D.D. Davis, 1982: J. Geophys. Res. 87, 5015-5017.

CARBON DIOXIDE: ITS NATURAL CYCLE AND ANTHROPOGENIC PERTURBATION

Ulrich Siegenthaler
Physics Institute, University of Bern
Sidlerstrasse 5
3012 Bern, Switzerland

1. INTRODUCTION

Carbon dioxide is, besides water, the main nutrient for plants and therefore for life on earth. In consequence of fossil fuel burning and human impact on the land biota, the atmospheric concentration of carbon dioxide is steadily increasing, which may lead to long-lasting changes of the global climate. These two facts explain the strong interest of scientists from many disciplines in this gas and its natural cycle.

This chapter deals with the cycling of CO_2 between the atmosphere and other global reservoirs. Its atmospheric concentration is largely determined by the ocean, so that emphasis is naturally given to air-sea exchange and the oceanic carbon cycle. However, the discussion would not be complete without also shortly considering the exchange with the terrestrial biosphere. Other carbon-containing compounds, such as CH_4 or CO, are not discussed here, since CO_2 behaves essentially as an inert gas in the atmosphere, decoupled from the cycles of other trace gases.

After a general description of natural reservoirs and fluxes, the anthropogenic CO_2 increase and its modelling are discussed, as well as CO_2-induced global warming. Finally, natural variations of atmospheric CO_2 are considered. Obviously, this chapter cannot be a comprehensive treatment of all important aspects of carbon dioxide in nature. Rather, the aim is to provide some insight into the processes that determine the natural level of atmospheric CO_2 and its variability as well as the fate of man-made CO_2, and to report on some recent developments in the field. The interested reader can find more detailed information in a number of reviews. "Tracers in the Sea" [Broecker and Peng, 1982] is an excellent textbook and review of the oceanic cycles of carbon and other elements. SCOPE 16 [Bolin, 1981] deals with modelling aspects of the anthropogenic perturbations and contains compilations of oceanic and atmospheric data. Sundquist [1985] critically reviewed carbon cycle data. "Carbon Dioxide Review 1982" [Clark, 1982], "Changing Climate" [National Research Council, 1983], and the book by Liss and Crane [1983] are all excellent reviews, the former two with emphasis on climatic effects of CO_2 and trace gases.

P. Buat-Ménard (ed.), The Role of Air-Sea Exchange in Geochemical Cycling, 209–247.

2. THE NATURAL CYCLE OF CARBON DIOXIDE

2.1. Reservoirs, Fluxes, Residence Times

Data on the distribution and fluxes of carbon are given in Table 1. The numbers are affected with uncertainty, some more than others; more information on this can be found in the review-papers cited above.

   The amounts of carbon in the various reservoirs and the fluxes between these must have fluctuated somewhat already before the perturbation by human activities, but here are reasons (for instance sedimentary records) for assuming that they were, if averaged over appropriate time periods, more or less constant. Therefore, steady state is often assumed in geochemical studies. This permits to apply the useful concept of mean residence time. The mean residence time of molecules entering a reservoir is defined in the usual statistical way, as the average time-span between entrance and exit. It can be shown [Nir and Lewis, 1975] that the mean residence time is given by

$$\tau = M/F$$

where M is the mass in the reservoir and F the (total) influx (= out-flux) into the reservoir. Sometimes, residence times are calculated with respect to only one specific influx or outflux; in this way the importance of different fluxes can easily be compared. In Table 1, some residence times are also indicated.

   More than 99 percent of the carbon present in the atmosphere is in the form of CO_2. The pre-industrial concentration must have been near 280 ppm by volume (see paragraph 3.1). The present atmospheric concentration is about 20 percent above that value. Of the rapidly exchanging reservoirs, the ocean is the largest one, containing about 60 times as much carbon as the atmosphere, most of which is dissolved inorganic carbon or "total CO_2" (ΣCO_2), including HCO_3^-, $CO_3^=$ and CO_2. The sediments are by far the largest carbon pool, but the exchange with the other reservoirs is so slow that it can be neglected on time scales of 10^3 years or less.

Table 1. Major global carbon reservoirs and fluxes. Main sources:
Bolin et al. [1979]; Bolin [1981]; Clark [1982].

Reservoirs			10^{15} g C
Atmosphere: CO_2:	Before 1850 ca. 280 ppm		594
	1980	338 ppm	717
Other gases: CH_4:		1.5 ppm	4
CO:		0.1 ppm	
(Troposphere: 80 %, stratosphere: 20 % of atmospheric mass)			
Oceans:	Inorganic C (ΣCO_2)		37,400
	Dissolved organic matter	ca.	1,000
	Biomass		3
Land biosphere:	Living		560
	Soil, humus (pre-historic: 200-500·10^{15} g C more)		1,500
Groundwater:			450
Sediments:	Inorganic C	ca.	60,000,000
	Organic C	ca.	12,000,000
Fossil fuels:		ca.	5,500

Fluxes (gross)		10^{15} g C/yr
Atmosphere-ocean, CO_2 exchange		78
Atmosphere-land biota, photosynthesis/respiration (NPP)		65
Marine photosynthesis		45
Sedimentation in oceans		0.2
Volcanism	ca.	0.9
Fossil fuel combustion, 1980		5.3

Residence times: τ = mass/flux

Atmosphere (pre-industrial):	total exchange	4 yr
	exchange with ocean only	8 yr
	exchange with biosphere only	9 yr
Living land/biosphere: photosynthesis/respiration		11 yr
Marine biosphere: photosynthesis/respiration		0.07 yr
Oceans: exchange with atmosphere, total flux		490 yr
sedimentation only		180,000 yr
Atmosphere + biosphere + oceans: sedimentation		210,000 yr

The residence time of CO_2 in the atmosphere with respect to exchange with the oceans is 8 years and that with respect to exchange with the biosphere is similar, so that the overall atmospheric residence time is about 4 years. The figures indicated for photosynthesis on land correspond to net primary production (NPP), excluding the short-time (hours to days) fixation and respiration: these are included in the gross primary production (GPP) rate which is about twice as large as NPP

that only NPP is efficient for biological uptake of a perturbation like bomb-produced ^{14}C, but not fixed carbon that is returned to the atmosphere by plant respiration after a very short period. Compared to the land biosphere, the mass of the marine biosphere is negligible. Marine organisms have a very short lifetime but a large productivity, so that the rate of photosynthesis in the sea is comparable to that on land. The material from dead marine organisms is recycled with high efficiency, and only 0.1 to 0.2 percent of the organic carbon produced totally settles down and is buried in the sediments.

The isotopes of carbon provide valuable information on processes and time rates. Isotope fractionation connected to physico-chemical and biological processes is responsible for differences in natural $^{13}C/^{12}C$ ratios. These are usually given in delta-notation

$$\delta^{13}C = \frac{^{13}R(sample) - ^{13}R(standard)}{^{13}R(standard)} \cdot 1000 \text{ \%o} \qquad (1)$$

with $^{13}R = [^{13}C]/[^{12}C]$. The standard used internationally is PDB (originally for "Pee-Dee belemnite" [Craig, 1957]).

The radioactive carbon isotope ^{14}C has a half-life of 5730 yr (mean life 8267 yr, decay constant $\lambda = 1/8267$ yr). ^{14}C is, as ^{13}C, subject to isotope fractionation processes which affect the isotope ratio $^{14}C/^{12}C$ ($= ^{14}R$) twice as much as the ratio $^{13}C/^{12}C$. A ^{13}C corrected $^{14}C/C$ ratio is often considered

$$^{14}R_{corr} = ^{14}R (1 - 2 [\delta^{13}C + 25 \text{ \%o}])$$

i.e. the actual $^{14}C/C$ ratio is corrected to a $\delta^{13}C$ value of -25 %o, as typical for wood. In geophysical work, ^{14}C concentrations are often given as

$$\Delta^{14}C = (^{14}R_{corr}/^{14}R(standard) - 1) \cdot 1000 \text{ \%o} \qquad (2)$$

where the standard activity approximately corresponds to the natural atmospheric ^{14}C level (i.e. $\Delta^{14}C \approx 0$).

Some typical isotopic values are given in Table 2. The ^{14}C age of mean deep sea water with respect to the warm surface is about 1000 yr.

Table 2. ^{13}C and ^{14}C in natural reservoirs (pre-industrial situation).

	$\delta^{13}C$ (%o)	$\Delta^{14}C$ (%o)
Atmospheric CO_2	-6.4	0
Land vegetation	-25	~ 0
Ocean, ΣCO_2: warm surface water	2	- 50
deep sea (average)	0.5	-160

2.2. Air-sea Exchange of CO_2

There is a continuous exchange of CO_2 between atmosphere and ocean, and under natural conditions, the global invasion and evasion fluxes must very nearly balance each other. The magnitude of the exchange flux can be estimated in three ways: using natural or bomb-produced ^{14}C, and using radon-derived air-sea transfer velocities.

    1. In steady state, the decay of ^{14}C in the ocean must be balanced by the net inflow from the atmosphere. The latter is equal to the difference between gross inflow, $R_a F_o A_s$, and gross outflow, $R_s F_o A_s$, where R_a, R_s are the mean (^{13}C-normalized) ^{14}C concentrations in atmosphere and in surface ocean water, F_o is the mean (gross) CO_2 exchange flux per m^2 of ocean surface, and A_s the global ocean surface. The balance can be written as:

$$\lambda R_{oc} N_{oc} = F_o (R_a - R_s) A_s$$

whence:

$$F_o = \frac{\lambda R_{oc} N_{oc}}{(R_a - R_s) A_s} \qquad (3)$$

λ = is the ^{14}C decay constant, R_{oc} and N_{oc} are the average ^{14}C concentration and the total mass of carbon dissolved in the ocean. The atmospheric ^{14}C concentration, R_a, is set 100 %; then $R_s \approx 95$ % and $R_{oc} \approx 84$ %. With $N_{oc} = 3.84 \cdot 10^{19}$ g C and $A_s = 3.62 \cdot 10^{14}$ m^2, we obtain $F_o = 216$ g C m^{-2} $yr^{-1} = 17.9$ mol m^{-2} yr^{-1}, or a global air-sea exchange rate of $78 \cdot 10^{15}$ g C yr^{-1}. The error of this number is about ±25 % and mainly stems from the uncertainty of about 1 % in R_s and therefore in R_a-R_s (= 5 ± 1 %).

    2. In an analogous way, the mean air-sea exchange flux can be estimated from the inventory of bomb-produced ^{14}C in the ocean, as obtained based on the GEOSECS tracer data. Again the driving force is proportional to the difference between the ^{14}C concentrations of atmosphere and surface waters, which now are functions of time and are reasonably well known from observation (Figure 1).

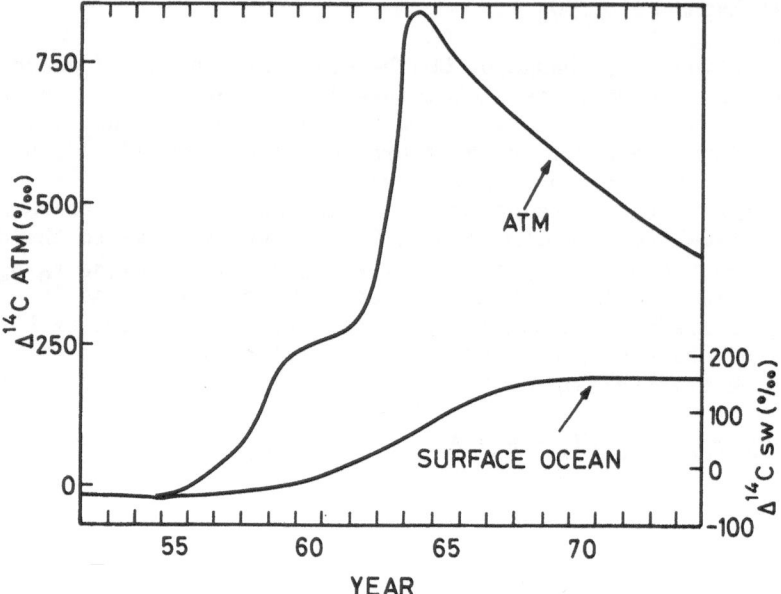

Figure 1. $\Delta^{14}C$ of atmospheric CO_2 (left-hand scale) and in ocean
surface water (right-hand scale) in response to the input of bomb-
produced ^{14}C. After Broecker and Peng [1982].

Different estimates of the mean air-sea CO_2 exchange flux, based on
bomb-^{14}C, converge in the range 20-23 mol m^{-2} yr^{-1} [e.g. Stuiver et al.,
1981]. These values are valid for the period 1962-1972, when atmospheric
bomb-^{14}C was high above oceanic values; then the CO_2 concentration was
about 320 ppm. The corresponding pre-industrial value is obtained by
noting that the flux is proportional to the partial pressure of CO_2, the
pre-industrial value of which was about 280 ppm. Correspondingly the
above range corresponds to a pre-industrial exchange flux of 17.5 to 20
mol m^{-2} yr^{-1}. This is in good agreement with the result from the
distribution of natural ^{14}C. Note that oceanic bomb-^{14}C represents an
integral over about 10 years, and natural ^{14}C over about 500 years
(oceanic residence time of CO_2). Thus we conclude that the mean pre-
industrial CO_2 exchange flux was 18 mol m^{-2} yr $\pm$ 25 %. For a pre-
industrial atmospheric concentration of 280 ppm (in dry air) and
calculated with a global mean sea surface temperature of 20°C, eq.(4)
(below) yields an approximate mean transfer velocity of $4.9 \cdot 10^{-5}$ m s^{-1}
(= 17.5 cm h^{-1}).

    3. The radon-deficit method yields local and instantaneous gas
transfer velocities (see Liss and Merlivat, this volume). Based on the
results that Peng et al. [1979] obtained during GEOSECS, a mean pre-
industrial CO_2 exchange flux of about 12 mol m^{-2} yr^{-1} can be derived
(see next paragraph). This is somewhat lower than the ^{14}C-derived
values; a probable reason is that the radon-derived transfer velocities

are not representative for mean global conditions, especially not for
high-wind conditions occurring in winter in temperate and high latitudes.

2.3. Regional Variability of Air-sea Flux

In a natural steady-state situation the global net CO_2 flux from air to
sea must have been near zero, except for a very small unbalance connected to river input minus sedimentation of carbon. Regionally, however,
surface water can be supersaturated or undersaturated in CO_2 with
respect to the atmospheric concentration. Then there is a non-zero net
flux between air (index a) and sea (index s) which can be represented as
difference between invasion flux $F_{a,s}$ and evasion flux $F_{s,a}$:

$$F_{net} = F_{a,s} - F_{s,a} = k(s \, C_a - C_w) = k \, s(C_a - pCO_2) \tag{4}$$

where k is the gas transfer velocity related to liquid-phase concentrations, s the solubility of CO_2, C_a and C_w are CO_2 concentrations in air
and in bulk surface water. [See Liss, 1983 and Liss and Merlivat, this
volume, for a review of the physics of gas exchange.]
Instead of C_w, usually the equilibrium partial pressure $pCO_2 = s^{-1} C_w$ is
considered; correspondingly the gas-phase related transfer velocity,
k s, instead of the liquid-phase related transfer velocity, k, should be
considered.
    C_a and pCO_2, as used in (4), are partial pressures in moist air of
100 % relative humidity at the temperature of surface water, since the
air is saturated with water vapour directly at the interface. Therefore,
atmospheric concentrations referring to dry air must be converted to C_a
according to

$$C_a = C_{a,dry} (1 - e_s/p_a) \tag{5}$$

where e_s is the saturation water vapour pressure and p_a the barometric
pressure at sea-level. The correction may be important for the net
fluxes calculated using eq.(4).

2.3.1 <u>Dependence of transfer velocity on temperature and wind</u>. The evasion flux (for given C_w) is proportional to k (transfer velocity with
respect to liquid-phase concentrations), the invasion flux (for given
C_a) proportional to k s (transfer velocity w.r.t. gas-phase concentrations). The transfer velocity k can approximately be represented as
$k = (D/\nu)^n$ f(turbulence), where D = diffusion coefficient for CO_2 in
water, ν = kinematic viscosity of water (ν/D = Sc, Schmidt number),
$n \approx 0.5$ in presence of capillary waves [Jähne et al., 1984] and probably
for average oceanic conditions, and f(turbulence) summarizes the dependence of k on the near-interfacial turbulence caused for instance by
wind. $(D/\nu)^{0.5}$ increases, solubility decreases with rising temperature.
Between 0°C and 30°C, s varies by a factor 0.40, $(D/\nu)^{0.5}$ by a factor
2.2, and their product by a factor 0.87. Thus, the direct effect of
temperature is small for k s and for the CO_2 flux for given pCO_2 and
atmospheric concentration. Laboratory experiments [see Liss and

Merlivat, this volume] and field work [Wanninkhof et al., 1985] show
that transfer velocity increases with wind more strongly than proportional to wind speed. Annual mean wind speed above the ocean is nearly a
factor 2 larger in high latitudes (>50°) than in tropical regions. Thus,
we may expect that k s is higher in high than in equatorial latitudes by
roughly a factor of two due to the effect of wind, while the temperature
difference should cause a slight reduction. The expected overall effect
is an increase towards high latitudes.

2.3.2. <u>Estimate of regional variability</u>. Some of the quantities determining the CO_2 flux exhibit considerable regional variability. In order
to see the influence of that variability, I estimated mean values for
the Atlantic and Pacific Oceans and for the latitude zones >50°N,
50°N - 10°N, 10°N - 10°S, 10°S - 50°S and >50°S (excluding the Pacific
region >50°N for which there are essentially no data on pCO_2 and k); the
subdivision was taken over from SCOPE 16 [Bolin, 1981]. The available
field data are not complete enough to allow the calculation of reliable
regional mean values, and the exercise should therefore only be considered as a sort of sensitivity test.
    Basic data and results are given in Figure 3. CO_2 transfer veloci-
ties were obtained by averaging the radon-based results of Peng et al.
[1979] (values for actual water temperatures, not normalized to 20°C)
and multiplying by $(D(CO_2)/D(Rn))^{0.5} = 1.36^{0.5}$. Sources for pCO_2 data
are a map of GEOSECS data of Takahashi et al. [1983] (a meridional profile for the Pacific Ocean is shown in Figure 2), results obtained by

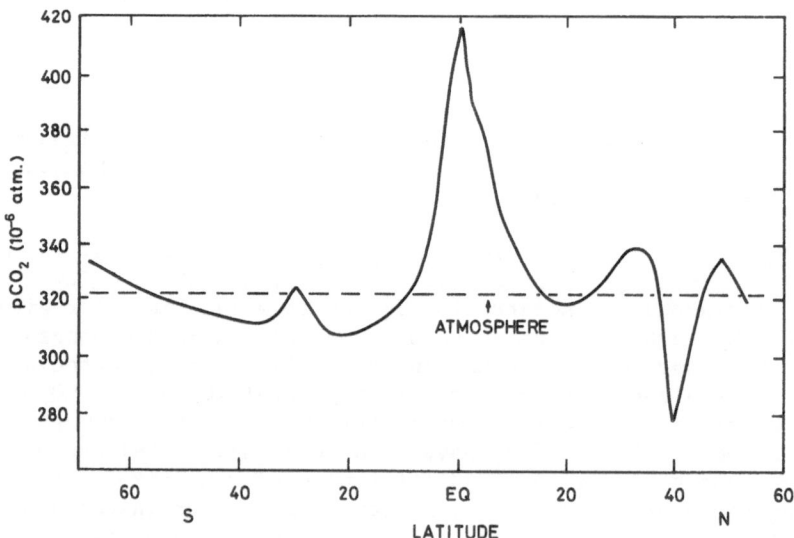

Figure 2.   North-south section of pCO_2 in surface water of the Pacific
Ocean. The equatorial maximum is due to upwelling of CO_2-rich subsurface
water. After Broecker et al. [1980].

R. Weiss during the FGGE experiment in 1979 [Keeling and Heimann, 1985] and the papers of Roos and Gravenhorst [1984] and Smethie et al. [1985]. pCO_2 data were converted to ppm in moist air where necessary and corrected to 1973. Areas and temperatures were calculated from the atlas of Levitus [1982].

Exchange fluxes and transfer velocities. The average k values derived from radon data vary between about $2.5 \cdot 10^{-5}$ and $5 \cdot 10^{-5}$ m s^{-1} (9 and 18 cm h^{-1}), the invasion fluxes between 9 and 29 mol m^{-2} yr^{-1}. Mean values (areally weighted) are $3.9 \cdot 10^{-5}$ m s^{-1} (14 cm h^{-1}) and 13.9 mol m^{-2} yr^{-1}. For a pre-industrial concentration of 280 ppm (instead of 329 ppm in 1973), the invasion flux would be 11.8 mol m^{-2} yr^{-1}, considerably less than the value obtained from ^{14}C, as mentioned above. The present discussion is based on the (unproved) assumption that the radon results still approximately reproduce the geographical variation of mean annual transfer velocities.

The transfer velocity k does not show a pronounced latitude dependence; high values are observed in the southern temperate and the Antarctic part of the Pacific Ocean, but not in the Atlantic Ocean (Figure 3). Thus, the direct temperature dependence of k may be partly compensated by higher wind speeds in cooler regions. In contrast, the gas-phase related transfer coefficient, k s, and the gross CO_2 fluxes are significantly larger in high southern latitudes than elsewhere. The calculated mean invasion flux at high northern and southern latitudes (>50°) is 22.1 mol m^{-2} yr^{-1}, a factor of 1.77 higher than for the rest of the ocean. The different behaviour of k and k s (or gross flux) is a consequence of the temperature dependence of the solubility.

Net air-sea fluxes. Large positive fluxes, i.e. net evasion from the ocean, occur in tropical latitudes because of high pCO_2 in surface water (Figure 3). In temperate latitudes the ocean is generally a natural CO_2 sink. The high latitude regions are, according to Figure 3, strong sinks. However, pCO_2 data are rather sparse there, so that the calculated fluxes are affected with large uncertainty. The calculated global net flux is very near to zero; it corresponds to a net oceanic source of $0.3 \cdot 10^{15}$ g C yr^{-1}, while a sink of about $2 \cdot 10^{15}$ g C yr would be expected from carbon cycle models (see below). The reason for the discrepancy is probably that the assumed regional values of pCO_2 and transfer velocity are affected with error; furthermore, the Indian Ocean is not included in this analysis.

It should be emphasized that the calculated net fluxes do not indicate in any way whether the considered regions are efficient or not as sinks of anthropogenic CO_2. It can be expected that the presently observed map of pCO_2 differences to the atmospheric concentration roughly corresponds to the natural pCO_2 distribution.

Takahashi [1985] also estimated net air-to-sea fluxes and ended up with a mean global value of 0.49 ± 0.49 mol m^{-2} yr^{-1}. His calculations are based on slightly different pCO_2 values and on a constant gas-phase related transfer velocity k s.

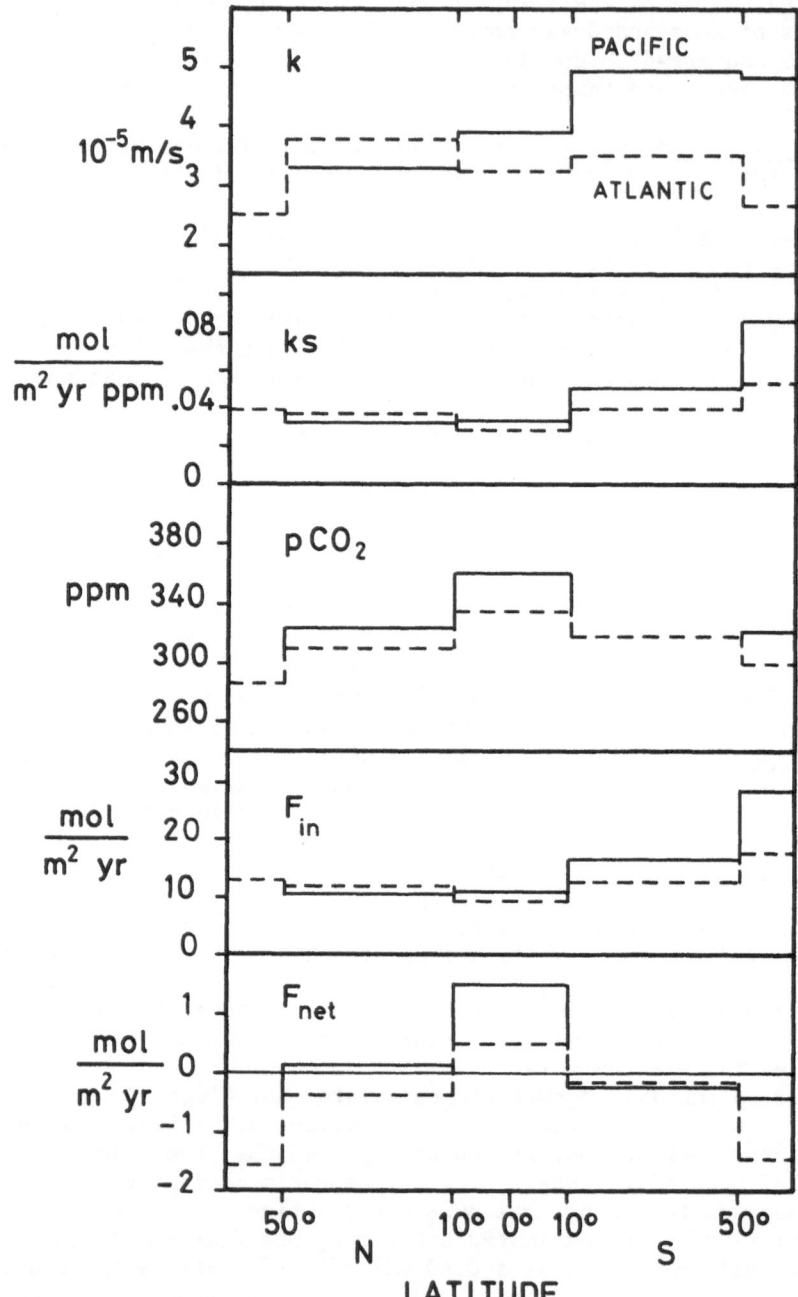

Figure 3. Estimated mean values for different latitude bands of CO_2 transfer velocities k and k s (related to liquid-phase and gas-phase concentrations, respectively), pCO_2, invasion flux of CO_2 and net flux from ocean to atmosphere. Solid line: Pacific Ocean, dashed line: Atlantic Ocean.

2.4. Marine Carbonate Chemistry

The chemical equilibria between the dissolved carbonate species
can be written as follows (for a detailed discussion see e.g. Broecker
and Peng [1982]):

$$CO_{2,aq} + H_2O \overset{K_1}{=} H^+ + HCO_3^- \tag{6}$$

$$HCO_3^- \overset{K_2}{=} H^+ + CO_3^= \tag{7}$$

where $CO_{2,aq}$ stands for dissolved CO_2 including H_2CO_3. K_1 and K_2 are
equilibrium constants. In addition, the solubility s links pCO_2 to
concentration of $CO_{2,aq}$. Instead of the concentrations of $CO_{2,aq}$, HCO_3^-,
$CO_3^=$ and H^+, a set of four other parameters can conveniently be observed: pCO_2, ΣCO_2, alkalinity A and pH = $-10 \log [H^+]$. ΣCO_2 and alkalinity (negative charge contributed by weakly dissociated acids) are
defined by

$$\Sigma CO_2 = [CO_{2,aq}] + [HCO_3^-] + [CO_3^=] \tag{8}$$

$$A = [HCO_3^-] + 2[CO_3^=] + B(OH)_4^- + \dots \approx [HCO_3^-] + 2[CO_3^=] \tag{9}$$

There exist two equilibrium constraints, equations (6) and (7). Therefore, the whole carbonate system is defined if two of the four observable quantities, e.g. ΣCO_2 and alkalinity, are measured.
    ΣCO_2 is changed by CO_2 exchange with air and by formation or des-
truction of organic matter of solid carbonate particles. Alkalinity is
not affected by gas exchange, but by dissolution of $CaCO_3$ particle and,
to a minor degree (via NO_3^- and PO_4^{3-}), of organic particles. Of special
interest for air-sea exchange are processes affecting pCO_2. pCO_2 can be
considered a function of ΣCO_2, alkalinity and temperature; other factors
like salinity are of minor influence. In connection with anthropogenic
CO_2, the relation between pCO_2 and ΣCO_2 (for constant A and T) is important. If the atmospheric CO_2 level varies, the equilibrium concentration of dissolved CO_2 in surface sea water varies proportionally, but
not the concentrations of bicarbonate and carbonate ions because of
chemical constraints. The resulting relative change of total dissolved
inorganic carbon is smaller than that of CO_2 gas alone. This is taken
into account by introducing a buffer factor ξ: if the CO_2 partial
pressure is increased by p percent, then ΣCO_2 increases by p/ξ percent
only. This can be expressed as

$$\frac{\Delta CO_2}{\Sigma CO_2} = \frac{1}{\xi} \frac{\Delta pCO_2}{pCO_2} \tag{10}$$

The buffer factor (also called Revelle factor) is a function of ΣCO_2 and
alkalinity, to a lesser degree also of temperature [Bacastow in Bolin,
1981, pp. 95-101; Takahashi et al., 1980]. On the average it is about 9
for temperate and equatorial, and about 14 for high-latitude surface
water.

For constant ΣCO_2 and temperature, pCO_2 decreases as alkalinity increases. The analogue to the buffer factor for alkalinity, i.e. the ratio of relative pCO_2 change to relative alkalinity change, is about -5. This becomes of interest when e.g. considering carbonate sediment dissolution due to the acidification of sea-water by fossil CO_2.

For given ΣCO_2 and alkalinity, pCO_2 increases with temperature by 4 to 5 percent per degree Celsius.

2.5. The Oceanic Carbon Cycle

Cycling of carbon in the oceans is governed by transport by water and by formation and redissolution of organic and carbonate particles. Information on the time scales of water circulation is obtained from the distribution of natural ^{14}C in the ocean. The natural $^{14}C/C$ ratio of warm surface water was near 95 % (atmosphere: 100 %), and near 90 % or 93 % (Antarctic Ocean and northern North Atlantic, respectively) in surface water in regions of deep water formation. Values for deep water below 1500 m range from about 90 % in the North Atlantic to 76 % in the North Pacific, corresponding to mean ages (with respect to the corresponding cold surface waters) of 200 - 300 up to 1400 years. The distribution of natural and bomb-distributed ^{14}C has been described in detail by Broecker and Peng [1982] and by Siegenthaler [1986].

Total CO_2 and alkalinity, but also the concentrations of elements like P and N, are depleted in surface water (Figure 4) because organic and carbonate particles carry them to depth, where organic particles are oxidized by microbial action and carbonate shells are redissolved.

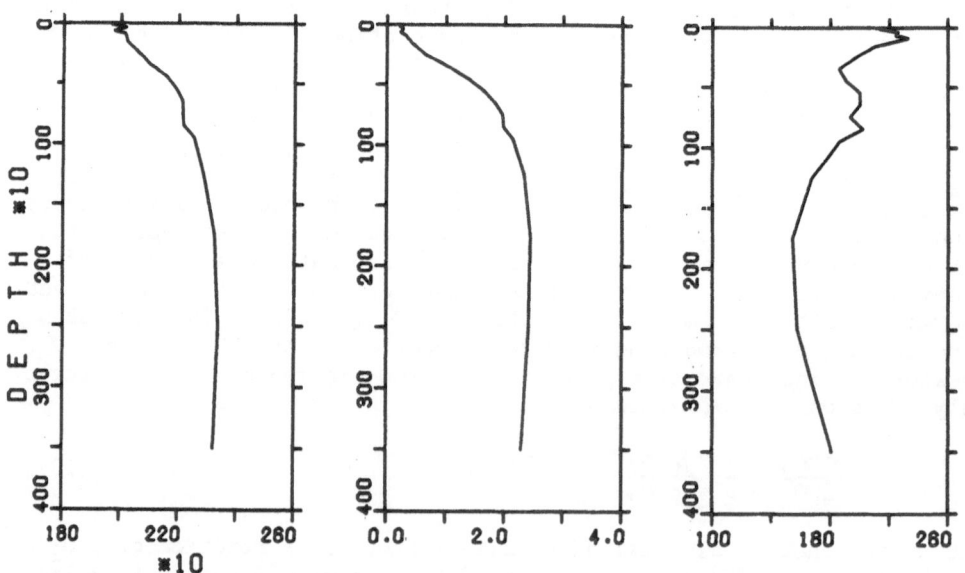

Figure 4. Vertical profiles in the Pacific Ocean, 10° - 50°S, from 0 - 4000 m depth, of (from left to right) ΣCO_2, phosphate and dissolved oxygen. Concentration units: μmol/kg (for ΣCO_2: unit = 10 μmol/kg); depth unit: 10 m.

The surface depletion of ΣCO_2 is about 10 to 20 percent of the deep water value, which can be interpreted in a first order estimate such that particle transport of carbon is about 10 to 20 percent of the transport by water circulation. To a minor degree, the surface - deep water difference is also due to the fact that the solubility of CO_2 is higher in the cold water of the deep-sea outcrop areas than in warm surface water.

The marine biological productivity depends on the availability of the elements N and P, for diatoms also of Si. These act as limiting nutrients in many oceanic regions. The mole ratios (Redfield ratios) of P, N and C are found to be relatively constant in biological material. A recent careful analysis of the variations of the composition of sea-water along isopycnal (equal density) surfaces by Takahashi et al. [1985] yielded best values for the ratios of concentration changes due to oxidation of organic matter of

$$P : N : C : -O_2 = 1 : 16 : (103-140) : 172 \qquad (11)$$

while the classical Redfield ratios are 1 : 16 : 106 : 138. The value for O_2 indicates how much oxygen is used for the oxidation of a given amount of organic matter; this includes the oxidation of phosphorus and nitrogen to phosphate and nitrate. The P : C ratio, as observed directly, was found to be 1 : 103, but this must be too large because of the influence of anthropogenic CO_2. The ratio 1 : 140 was obtained indirectly by assuming that the observed oxygen decrease was due to oxidation of organic carbon and NH_3; it would be larger (i.e. 1 : <140) if actually oxygen would be used also for oxidizing hydrogenated organic molecules. Thus the P : C ratio is not precisely known, but probably significantly lower than the classical value of 1 : 106.

Carbonate dissolution also adds total CO_2 to deep water; this contribution can be estimated from the concomitant alkalinity increase. On a global average, the ratio of carbon fluxes by organic and carbonate particles has been estimated to about 4 : 1 by Broecker and Peng [1982], while Takahashi et al. [1985] found about 10 : 1 at the density horizons they considered. However, this ratio varies regionally; in areas of high productivity (equatorial upwelling zones, Antarctic ocean) diatoms, which produce opal, are predominant and the carbonate/organic particle ratio is correspondingly small.

2.6. The Cycle of Oxygen

The example of oxygen illustrates how cycles of different elements are coupled. Dry air contains 21 percent oxygen, and there are $1.2 \cdot 10^{21}$ g or $3.7 \cdot 10^{19}$ mol O_2 in the atmosphere. The photosynthetic reaction, which creates oxygen, can be written in a simplified way as

$$CO_2 + H_2O \rightarrow CH_2O + O_2 \qquad (12)$$

where CH_2O is a sum formula for organic material (neglecting minor elements like nitrogen and phosphorus). Thus one molecule of oxygen gas is created for every carbon dioxide molecule; because of this one-to-one

relationship it is convenient to indicate masses in units of moles
rather than grams.

Plant respiration (decomposition of organic matter) is just the
reverse of photosynthesis:

$$CH_2O + O_2 \rightarrow CO_2 + H_2O \tag{13}$$

Equations (12) and (13) show that the cycles of carbon and oxygen are
directly coupled. Atmospheric oxygen must, therefore, exhibit variations
of about the same absolute size but opposite sign as CO_2. They have not
been directly observed so far because of their relative smallness - only
a few ppm out of 210 000 ppm.

For calculating the mean atmospheric residence time of oxygen,
gross, rather than net, primary productivity of terrestrial vegetation
must be considered, since an oxygen molecule that has left a plant leaf
is mixed into the bulk atmosphere within minutes. Assuming that global
GPP on land is twice NPP yields a total terrestrial plus marine flux of
about $175 \cdot 10^{15}$ g C yr^{-1} or $1.5 \cdot 10^{16}$ mol yr^{-1} (of C or O_2), which gives a
mean atmospheric residence time for oxygen, with respect to cycling
through the biosphere, of 2500 years. This is approximately the time-lag
with which the isotopic composition of atmospheric oxygen follows global
changes in $\delta^{18}O$ of the oceans: $\delta^{18}O$ of O_2 is linked via photosynthesis
and respiration to the global water cycle and thus to the glacial-
interglacial isotopic variations of sea-water [Bender et al., 1985].

The global rates of photosynthesis and plant respiration are nearly
equal, so that the net oxygen flux is very small. According to equation
(12), net production of oxygen occurs if organic matter is removed from
the atmosphere-biosphere system, which happens during sedimentation. The
corresponding sedimentation rate is about $6 \cdot 10^{12}$ mol/yr [Sundquist,
1985, Table 19]. This yields a residence time of $6 \cdot 10^6$ yr for oxygen in
the atmosphere-biosphere system. The total atmospheric mass of O_2
corresponds to only about 4 % of the estimated mass of sedimentary
organic carbon. This indicates that most photosynthetic oxygen produced
during geological time has been removed again, mostly as oxides of
sulphur and iron [Garrels and Perry, 1974].

According to Table 1, the total amount of recoverable fossil fuels
plus biomass make up about $6.2 \cdot 10^{17}$ mol C. If all this carbon was burnt
completely, this would use up about 2 percent of the atmospheric oxygen.
For comparison, the barometric pressure - and with it the partial
pressure of oxygen - decreases by 10 percent for an altitude increase of
1000 m. Thus, human activity does not create any problem of atmospheric
oxygen. The impacts of energy use are different: problems on a local or
regional scale due to production of pollutants like NO_x or SO_2, and the
threat to global climate by the world-wide increase of CO_2 and infrared-
active trace gases.

3. ANTHROPOGENIC INCREASE OF ATMOSPHERIC CO_2

3.1. Observations and Airborne Fraction

In 1958, C.D. Keeling started continuous measurements of the atmospheric carbon dioxide concentration by means of non-dispersive infrared spectroscopy at the observatory on Mauna Loa, Hawaii, and - using flask air samples - at the South Pole [Keeling et al., 1976, 1982]. The results, and later observations at many other stations [Komhyr et al., 1985], have indicated a steady increase, presently at a rate of ca. 1.5 ppm/yr (Fig. 5). The increase is due to the production of CO_2 by the combustion of fossil fuels and, to a lesser degree, also to deforestation and land use (plowing leads to faster oxidation of the organic carbon stored in soils).

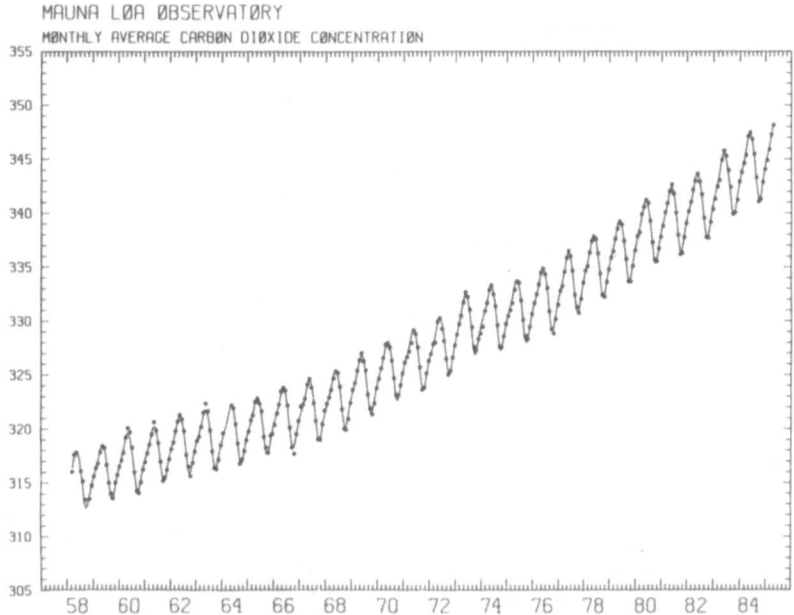

Figure 5. Atmospheric CO_2 concentration at Mauna Loa Observatory, Hawaii. Dots: monthly average values; smooth curve is a fit. Courtesy of C.D. Keeling.

Measurements of atmospheric CO_2 performed before 1958, generally by chemical techniques, were not reliable enough to accurately document the CO_2 increase. Different evaluations of 19th century measurements were undertaken [e.g. Callendar, 1958]. Based on a well-documented French series of data, an estimate for the 1880's is 285 to 290 ppm [Siegenthaler, 1984]. However, there are uncertainties with all the old direct measurements, and it is questionable whether reliable absolute values can be reconstructed based on these data sets. The best method available at present for reconstructing the atmospheric concentration of CO_2 (and other trace gases) in the past is the analysis of air bubbles trapped in

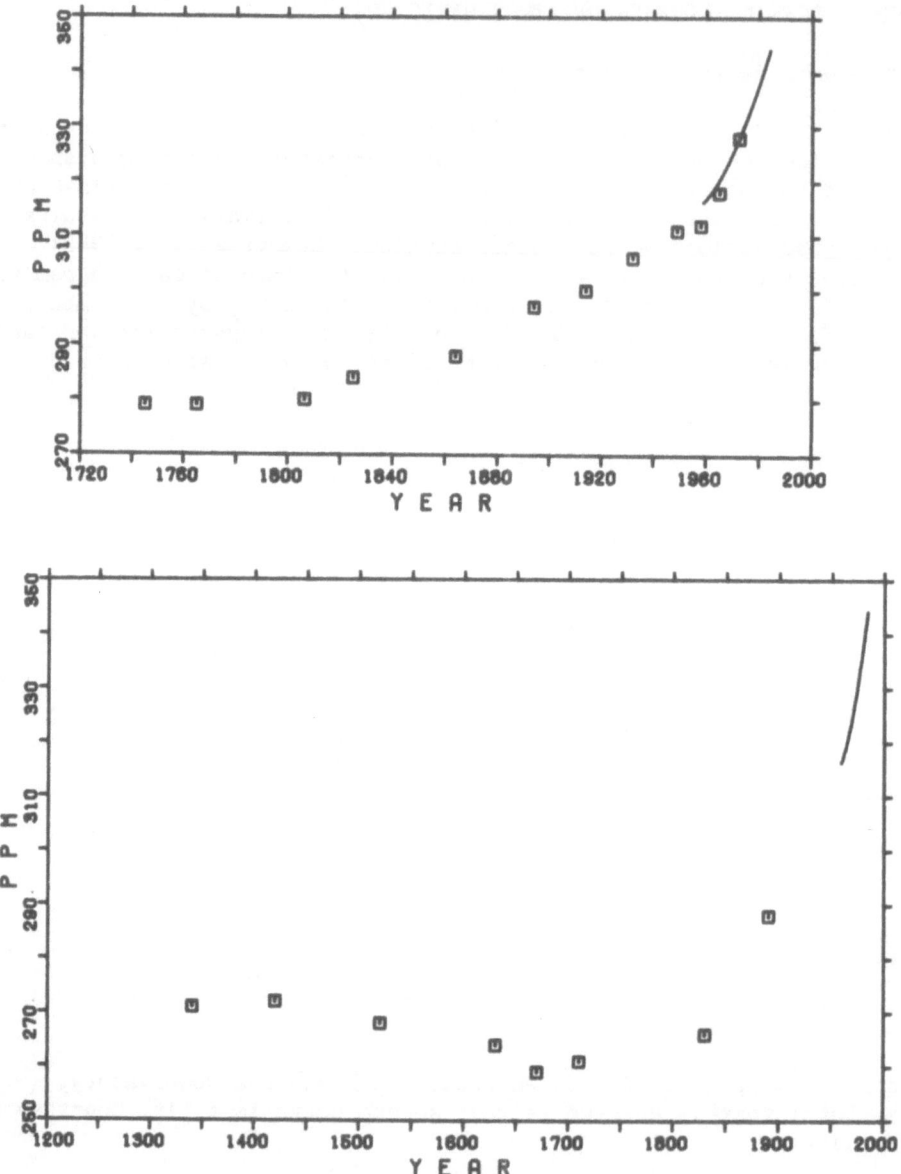

Figure 6. CO_2 concentration measured in air bubbles trapped in old polar ice from Antarctica (a) (top): Results from an ice core from Siple Station [Neftel et al., 1985]; (b) (bottom): Results from Station D57 [Raynaud and Barnola, 1985]. Solid lines: Mauna Loa results.

old polar ice [Neftel et al., 1985; Raynaud and Barnola, 1985]. For this
purpose, only ice from cold regions (mean annual temperature below ca.
-25°C), where no snow melting in summer occurs, is suited, because high
CO_2 concentrations in melt layers, due to the high solubility of CO_2 in
water, may lead to erroneous results. The results of the group at Bern
(Fig. 6a) indicate a value of about 280 ppm before about 1800. The
results of the Grenoble laboratory (Fig. 6b) are by about 10 to 20 ppm
lower for the same time. The difference seems to be due, at least
partly, to an experimental problem - selective gas-phase transport of
CO_2 in the presence of water vapour; and based on a thorough recent
intercomparison between the two laboratories the higher values of about
280 ppm appear to be nearer to the truth.

The age of the trapped air is smaller than that of the surrounding
ice, because the pores in the firn are closed only at a depth of
typically between 50 and 100 m. The age difference between ice and
occluded air was for instance found to be 82 ± 10 yr for the Siple ice
core (Fig. 6a); the range of ±10 yr corresponds to the actual time
interval of enclosure, different bubbles in the same ice layer having
been closed at different depths.

Based on the ice core results, it can be concluded that the 18th
century CO_2 concentration was near 280 ppm, possibly somewhat lower.

During the period 1860 to 1984, $183 \pm 15 \cdot 10^{15}$ g C were released
into the atmosphere due to fossil fuel combustion; the current rate
of release is $5.2 \cdot 10^{15}$ g C. The amount of CO_2 produced by deforestation
and land use in the same period is, according to the best current
estimates, $150 \pm 50 \cdot 10^{15}$ g C, and the corresponding current rate is
$1.6 \pm 0.8 \cdot 10^{15}$ g C [Bolin, 1985]. The atmospheric increase from 280 ppm
to 344 ppm in 1984 corresponds to an airborne fraction of 41 percent of
the total emission. The increase since the start of the Mauna Loa measurements corresponds to about 55 % of the _fossil_ CO_2 released during
that period; this value is often quoted because the actual net biospheric release is not well known.

3.2. Modelling the Oceanic Response to Carbon Cycle Perturbations

The major sink for excess CO_2 is the ocean. It contains about 63 times
as much inorganic carbon as the preindustrial atmosphere. Taking into
account the buffer factor $\xi \approx 9$, about $63/\xi = 7$ parts of all excess CO_2
will reside in the sea and 1 part in the air once the whole ocean has
equilibrated with the atmosphere. Thus, after a long time the final
airborne fraction will be roughly 12 percent. This value will, however,
be reached only after about 10^3 years, when anthropogenic CO_2 has pene-
trated to the large volumes of the deep ocean. The time-dependent response to the anthropogenic input must be calculated using an atmosphere-
ocean model, simulating the finite air-sea-gas exchange rate and the
mixing of CO_2-laden surface water to depth. Since satisfactory ocean
circulation models are not yet available for this purpose [see
Sarmiento, 1985], simple perturbation models have been used in which the
atmosphere is usually represented by one well-mixed box, the ocean by
two or more boxes or by a diffusive reservoir.

The thermocline, extending from the bottom of the surface mixed layer to about 1 km depth, is the zone of the ocean most important in connection with the CO_2 increase. It has a characteristic mixing time-scale of a few decades. Few-box models have too coarse a resolution to simulate well the transport in the thermocline. A better approach is eddy diffusion. Rapid variations cause steeper concentration gradients and therefore larger diffusive fluxes, which cannot be simulated by first-order exchange between few boxes. This was the basis for the box-diffusion model of Oeschger et al. [1975], in which the ocean consists of a mixed layer, assumed 75 m deep, and a deep sea in which transport occurs by vertical diffusion with constant eddy diffusion coefficient. The box-diffusion model has two adjustable transport parameters that have to be determined by means of a suitable tracer, usually natural or bomb-produced ^{14}C. These are (1) an air-sea gas exchange coefficient (see above), (2) a coefficient describing interior mixing, i.e. the vertical diffusion coefficient K of the deep sea.

The box-diffusion model is a perturbation model, not designed for simulating the whole steady state carbon cycle, but only the transport of excess CO_2 and of isotopic perturbations. Features supposed not to change with the anthropogenic CO_2 invasion are not considered. This can be illustrated by the carbon balance for some box i containing a carbon mass N_i at steady state and $[N_i + n_i(t)]$ in a perturbed state; the flux from box i to box j is represented by $k_{ij} N_i$, and there is a carbon source S_i into box i, e.g. due to particle dissolution:

Steady state:
$$\frac{dN_i}{dt} = 0 = \Sigma \, k_{ji} \, N_j - \Sigma \, k_{ij} \, N_i + S_i$$

Perturbated state:
$$\frac{d(N_i + n_i)}{dt} = \Sigma \, k_{ji}(N_j + n_j) - \Sigma \, k_{ij}(N_i + n_i) + (S + s_i)$$

Perturbation only:
$$\frac{dn_i}{dt} = \Sigma \, k_{ji} \, n_j - \Sigma \, k_{ij} \, n_i + s_i$$

Thus, if the source term is constant with time ($s_i = 0$), it does not appear in the perturbation equation (= difference equation). This is specifically the case for the oceanic particle fluxes, since the biological productivity is governed by limiting nutrients like phosphate or by light, but not by the abundant carbon, and therefore does not respond to the CO_2 increase.

The concept of eddy diffusion is based on the assumption that the flux of matter can be described by $F = - K \, \partial C / \partial z$ (C = concentration). It is a simple parameterized way to describe the continuous character of mixing. The eddy diffusion coefficient K can be estimated based on the distribution of natural ^{14}C or of bomb-produced ^{14}C or tritium. Calibration with natural ^{14}C, by demanding that the surface value and the mean oceanic ^{14}C concentration be correctly simulated, yields $K = 1.3 \ cm^2 s^{-1}$ [Oeschger et al., 1975]. However, calibration with bomb-produced isotopes is preferred, since they have time-scale of change comparable to that of anthropogenic CO_2. A useful concept in connection with bomb-produced isotopes is the mean penetration depth, defined as the ratio of water-column inventory to surface concentration. Broecker et al. [1980] estimated a global mean penetration depth of 375 m for tritium in about

1972, corresponding to K = 1.7 cm^2 s^{-1}. The same K value was obtained by Li et al. [1984] from an evaluation of all GEOSECS tritium data. A calibration with bomb-^{14}C yielded 2.4 cm^2 s^{-1} or 7700 m^2 yr^{-1} [Siegenthaler, 1983].

The step that mainly limits the oceanic uptake of anthropogenic CO_2 is vertical mixing. Thus, according to the box-diffusion model results, the present-day excess CO_2 concentration in surface water is about 80 % of its equilibrium value, i.e. gas exchange is rapid enough for chemical equilibrium to be nearly established. The fossil CO_2 production has increased in an approximately exponential way, with an e-folding time $\tau \approx 25$ yr, and the mean penetration depth of fossil CO_2 into the deep sea is $\sqrt{K\tau} = 438$ m (for K = 2.4 cm^2 s^{-1}), to which the depth of the mixed layer of 75 m has to be added. The pre-industrial atmosphere contained as much CO_2 as an ocean layer 67 m thick. Taking into account a degree of equilibration of surface water of 0.80 and a buffer factor of 9, the airborne fraction turns out to be

$$\frac{67 \text{ m}}{0.80 \cdot 513 \text{ m}/9 + 67 \text{ m}}$$

The exact numerical calculation using the actual CO_2 production history yields a value of 0.61.

The ratio of observed CO_2 increase to fossil fuel input is about 0.55, and a lower airborne fraction is obtained if biospheric CO_2 emission is also considered. Thus, the box-diffusion model yields a higher atmospheric increase rate than actually observed. One possible reason for this discrepancy is that the horizontally averaged models do not appropriately include the rapid vertical exchange occurring at high latitude which may represent an additional channel for oceanic CO_2

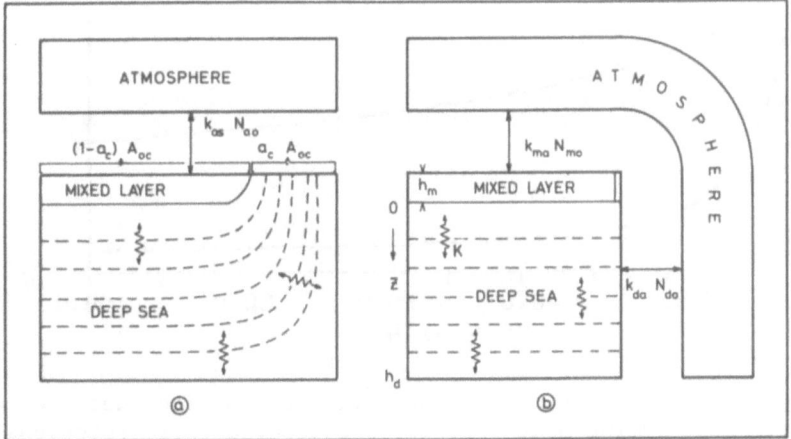

Figure 7.   Schematic representation of outcrop-diffusion model of the ocean-atmosphere system [Siegenthaler, 1983]. Left: physical idea; right: mathematical representation.

uptake. Siegenthaler [1983] extended the box-diffusion model to include a high-latitude outcrop, so that deep water is partly directly venti-lated from the atmosphere (outcrop-diffusion model, Fig. 7). With an outcrop covering 10 % of the ocean and a calibration using bomb-produced ^{14}C, the outcrop-diffusion model yields an airborne fraction of 0.52. In the outcrop region, the degree of equilibration with the atmospheric CO_2 excess is much lower than in warm surface water, i.e. the pCO_2 differ-ence air-sea that drives net CO_2 invasion is larger. Therefore the outcrop is an efficient sink for excess CO_2.

For comparing different model versions, it is essential that they are always calibrated in a consistent way. Bomb-produced ^{14}C is often used for calibration, because its invasion into the ocean is partly ana-logous to that of excess CO_2. The analogue is, however, not perfect because the two perturbations behave differently with respect to air-sea exchange, since the buffer factor affects the uptake of CO_2, but not of ^{14}C [Broecker and Peng, 1974]. The equilibration time between a 100 m thick oceanic mixed layer for an isotopic perturbation is about 10 yr. For a CO_2 perturbation the equilibration time is only about 1 yr, because not all CO_2 must be exchanged between the two reservoirs as for ^{13}C or ^{14}C, but establishment of the new chemical equilibrium requires the transfer of only a fraction of the CO_2. The difference is illustra-ted by Figure 8, showing the atmospheric responses to pulse inputs of CO_2 and of ^{13}C (^{14}C would behave very similarly) according to the box-

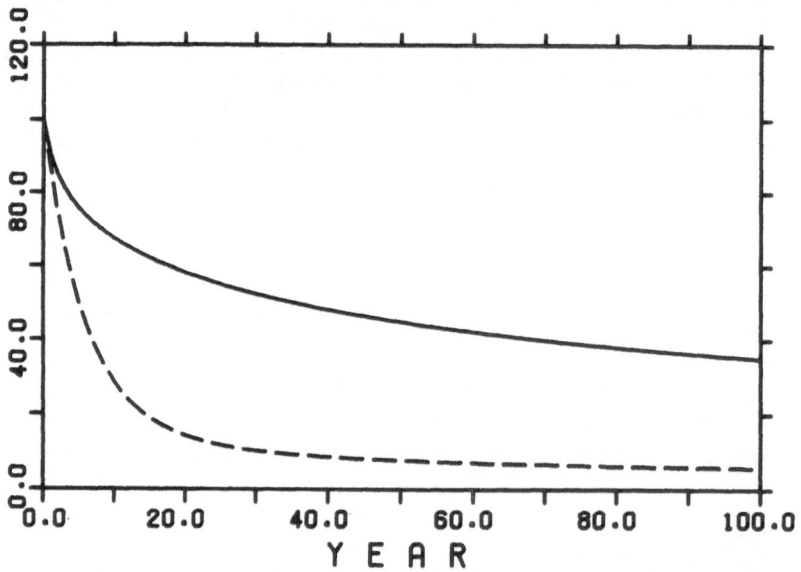

Figure 8. Model responses of the box-diffusion model of Oeschger et al. [1975], calibrated using bomb-^{14}C, for a pulse input of isotopically labelled CO_2 into the atmosphere. Solid line: CO_2, dashed: $\delta^{13}C$. Initial values assumed as +100 ppm and +100 permil.

diffusion model. Note also that the decrease is first rapid and then slower and slower, i.e. it is not governed by one time constant, since after entering the upper ocean layers the perturbation penetrates more and more slowly into the thermocline and the deep sea.

3.3. CO_2 Release from the Terrestrial Biosphere and the "Missing CO_2 Sink"

The outcrop-diffusion model with an outcrop area of 5 to 10 percent can reproduce the atmospheric CO_2 increase assuming that fossil fuel burning is the only cause. However, ecologists' estimates of present CO_2 emission due to deforestation and land use are in the range $1.6 \pm 0.8 \cdot 10^{15}$ g C yr^{-1} [Bolin, 1985] which is not negligible compared to fossil fuel consumption ($5 \cdot 10^{15}$ g C yr^{-1}). Thus, in the budget of excess CO_2, the total emissions seem not to be matched by the increase in atmosphere and ocean. Part of this problem may be non-existent, because the values of the airborne fraction mentioned above refer to simulation runs with fossil, i.e. quasi-exponentially increasing input. The results can be

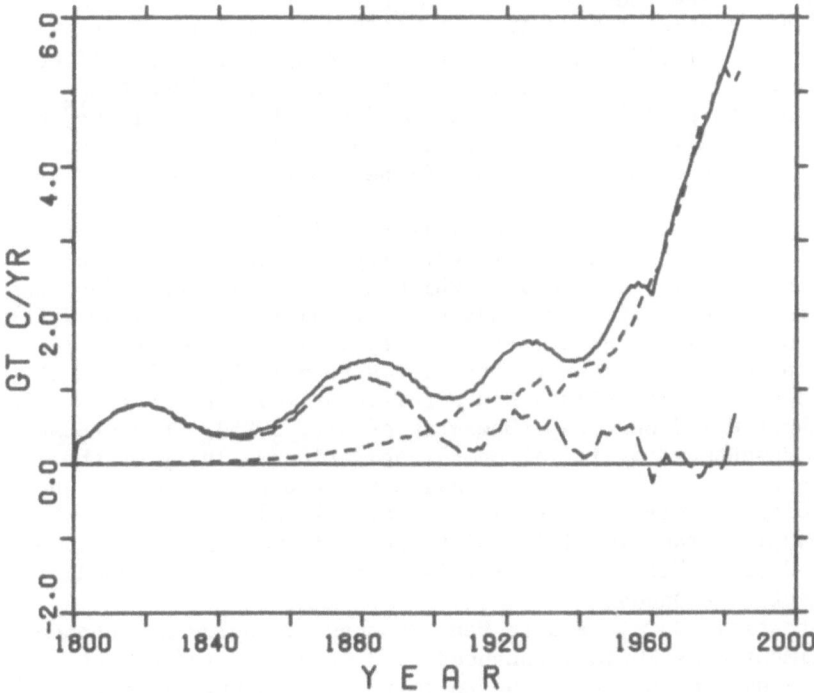

Figure 9. Total CO_2 production rate (solid line) calculated by deconvolving the CO_2 concentration history reconstructed from the Siple ice core results of Figure 6a. Short-dashed: fossil input rate [Rotty and Marland, 1984]; long-dashed: difference = non-fossil input rate.

different if the input history deviated significantly from an exponential increase. If a large input of biospheric CO_2 occurred in the past,
it is still influencing the present CO_2 concentration by providing a
decreasing atmospheric baseline, since excess CO_2 is still being taken
up by the ocean. This can be seen when considering the model response to
a pulse input of atmospheric CO_2 (Figure 8). In this case, the ratio of
atmospheric increase to present input rate becomes smaller.

Instead of determining the atmospheric concentration for a prescribed production history, we can calculate, by means of a carbon cycle
model, the production rate from a prescribed concentration history. This
has been done by Siegenthaler et al. [1986] for the CO_2 concentrations
as measured in the Siple ice core (cf. Figure 6a). Figure 9 shows the
results. The calculated total production rate (solid line) minus the
known fossil input (short-dashed) yields the non-fossil, i.e. probably
biospheric CO_2 release rate (long-dashed). According to the box-diffusion model simulation, a major non-fossil input occurred in the last
century, $66 \cdot 10^{15}$ g C until 1900 compared to $12 \cdot 10^{15}$ g C from fossil
fuels; for the period 1900 to 1980, the total non-fossil input was
$20 \cdot 10^{15}$ g C (fossil: $148 \cdot 10^{15}$ g C). Carrying out the deconvolution with
the outcrop-diffusion model ($a_c = 0.1$) yields a higher non-fossil input
($156 \cdot 10^{15}$ g C till 1980) because of the higher flux into the ocean. The
calculated cumulative non-fossil input by 1980 is compatible with the
ecological estimate of $150 \pm 50 \cdot 10^{15}$ g C. According to the box-diffusion
and outcrop-diffusion model deconvolution, the net input from the
biosphere during the past 20 - 30 years would have been 0 to $1 \cdot 10^{15}$ g C
yr, compared to the direct estimate of $1.6 \pm 0.8 \cdot 10^{15}$ g C yr^{-1}. Thus,
the "missing sink" is rather small and of the order of 0 to $2 \cdot 10^{15}$ g C
yr^{-1}.

An additional effect that is possibly significant as a sink of
excess CO_2 is the fact that plant productivity is stimulated by increased CO_2 concentration in the air. While this fertilizing effect has
been clearly found in laboratory studies, it is difficult to assess its
importance in nature where plant productivity is limited by other
factors like space, light or nutrients. The seasonal CO_2 amplitude at
Mauna Loa (and at other stations) has significantly increased from 1958
to 1982, by about 0.7 % per year [Bacastow et al., 1985]. This probably
indicates a strenghtened assimilation - respiration activity of the
northern hemisphere land biota, which might be accompanied by a higher
biospheric storage of carbon. Kohlmaier et al. [1986] suggest that the
additional annual storage in living biota and soil may amount to about
1 to $2 \cdot 10^{15}$ g C yr^{-1}. With a sink of this size, the excess CO_2 budget
would be perfectly balanced.

Other potential CO_2 sinks have been discussed, including enhanced
carbonate sediment dissolution, enhanced production and sedimentation of
particulate organic carbon (POC) due to fertilizer input, or carbon
transport by rivers. The invasion of excess CO_2 leads to a decrease of
the concentration of $CO_3^=$ ions in sea-water which consequently becomes
more aggressive for $CaCO_3$ dissolution. Enhanced $CaCO_3$ dissolution leads
to an increase of alkalinity and therefore to a higher CO_2 uptake capacity (lower buffer factor; see paragraph 2.4). Waters above 2 - 4 km
depth are supersaturated with respect to calcite (the predominant form

of $CaCO_3$ in sediments). Enhanced calcite dissolution will only be
effective for CO_2 uptake when water containing excess CO_2 has been
transported to depth, taken up additional alkalinity (i.e. $CO_3^=$ ions)
and again returned to the surface. Thus, this process can only become
active after a period corresponding to several ocean turnover times,
i.e. about 10^3 years [Broecker and Peng, 1982, pp. 552 ff].

Recent sediment trap observations in the North Pacific have indicated that the carbonate flux by aragonitic pteropods is considerably
larger than previously assumed [Betzer et al., 1984]. Aragonite is more
soluble than calcite, and the particles dissolve rapidly at relatively
shallow depth. Acidification of the water by anthropogenic CO_2 may lead
to dissolution at even shallower depths and thus to an increase of
alkalinity. It is not easy to assess the possible importance of such an
effect without a model. Intuitively, one would judge it as not very
significant, since no additional alkalinity is added to the ocean (as is
the case for sediment dissolution), but the alkalinity is just injected
somewhat nearer to the surface.

Due to man's activity, the amounts of phosphate and nitrate brought
into the ocean by rivers has increased, which may act as fertilizers and
lead to enhanced biological production in the sea. The additional phosphate input has been estimated as $3.2 \cdot 10^{10}$ mol yr^{-1}, the nitrate input
as roughly $5 \cdot 10^{11}$ mol yr^{-1} [Meybeck, 1982]. With the Redfield ratios
P:N:C = 1:16:140, the phosphate and nitrate inputs would correspond to a
maximum additional production of $5.4 \cdot 10^{13}$ g C or $1.0 \cdot 10^{14}$ g C. Only
organic matter that escapes oxidation and is buried in the sediment
corresponds to a carbon sink; this is the case for about 10 % of the
total particulate flux [Emerson, 1985]. Thus, this mechanism provides
for a sink of something like $0.5 - 1 \cdot 10^{13}$ g C yr^{-1}, which is negligible
compared to the fossil CO_2 production rate of about $5 \cdot 10^{15}$ g C yr^{-1}. The
hypothesis of Walsh et al. [1981] that burial of organic carbon on the
continental shelf is an effective carbon sink seems to be due to a
confusion between anthropogenic fluxes and natural fluxes that have
nothing to do with man's activity.

Kempe [1984] has suggested that increased transport of POC to the
ocean by rivers and increased deposition in lakes could provide a sink
of up to $1 \cdot 10^{15}$ g C yr^{-1}. It is, however, uncertain what fraction of
these fluxes actually corresponds to a sequestration of carbon, and how
much is subject to recycling by oxidation. In addition, Kempe's estimate
of up to $1 \cdot 10^{15}$ g C yr^{-1} appears to be biased towards high values.

3.4. Scenarios for Future CO_2 Concentrations

Future atmospheric CO_2 concentrations primarily depend on release rates
by fossil energy consumption, but also on the uptake by the oceans.
Figure 10 illustrates the influence of direct ventilation of the deep
sea through an outcrop, covering 10 percent of the world ocean, for a
high CO_2 production scenario [Siegenthaler, 1983]. The assumed final
cumulative production corresponds to about 8 times the pre-industrial
atmospheric CO_2 mass and is probably unrealistically high; it is shown
as a sensitivity study. The example demonstrates how slow the decrease

will be even after production has stopped completely. There is a large
difference between the cases with and without outcrop; this clearly
shows that it is necessary to develop reliable ocean models for long-
term estimates of future CO_2 concentrations that take into account in a
realistic way the formation of thermocline and deep waters.

For the next several decades, however, existing models are sufficient for reasonably accurate predictions; uncertainties mainly stem
from the scenarios of future CO_2 emissions. These have changed considerably in the past few years [see National Research Council, 1983].
According to the study of the National Research Council [1983], the CO_2
concentration is expected to have doubled (reached 600 ppm) with 50 percent probability between 2050 and 2100. Bolin [1985] concludes that a
doubling will probably not occur before 2050, possibly even after 2100.
The corresponding warming will be about 1° to 2°C due to CO_2 alone;
anthropogenic trace gases will cause an additional warming of approximately the same size (see below).

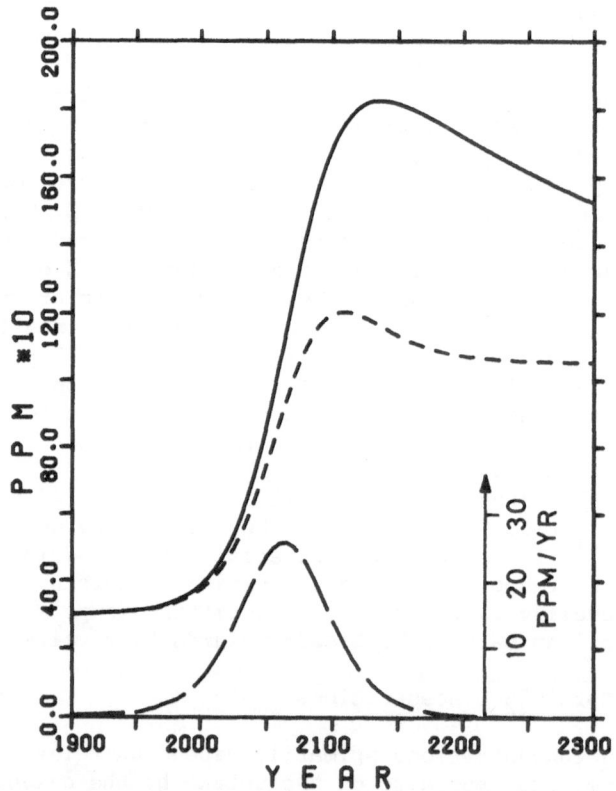

Figure 10. Model-calculated atmospheric CO_2 levels (upper curves; scale
from 0 to 2000 ppm) for an upper-limit production scenario (lower, long-
dashed curve) according to which $4924 \cdot 10^{15}$ g C (~ 8 times pre-industrial
atmosphere) are finally burnt. Simulations by means of box-diffusion
model (solid line) and outcrop-diffusion model with 10 percent deep-sea
outcrop (dashed line), both bomb-^{14}C calibrated. Note the large
influence of deep-sea ventilation by outcrop. From Siegenthaler [1983].

3.5. Carbon Isotope Perturbations

Fossil fuel does not contain any ^{14}C, since it has been isolated from the atmosphere long enough that all ^{14}C has decayed. Thus, the emission of anthropogenic CO_2 caused a decrease of the $^{14}C/C$ and of the $^{13}C/^{12}C$ ratio in atmospheric CO_2 ("Suess effect").

By 1950 the cumulative fossil production corresponded to 10 % of the atmospheric CO_2 mass. If all this CO_2 had remained airborne, $\Delta^{14}C$ of atmospheric CO_2 would have decreased from 0 to -91 %o. As a result of exchange with ocean and biosphere, the Suess effect was, however, considerably less. The decrease found by tree-ring studies is 20 %o, and 17 %o if corrected for presumed natural fluctuations [Stuiver and Quay, 1981] (Figure 11). Calculations using the box-diffusion and the outcrop diffusion model yield a change of 18 to 20 %o, in excellent agreement with observation.

The ^{14}C Suess effect reflects the CO_2 production from fossil fuel consumption, but not that from biomass changes. Both emissions, however, cause a $\delta^{13}C$ decrease in atmospheric CO_2; air-CO_2 has $\delta^{13}C \approx -7$ %o, fossil and biospheric carbon values near -25 %o. A model simulation based on assuming only fossil carbon release predicts a decrease of 1.1 %o until 1980. Additional biospheric CO_2 inputs must have led to a stronger $\delta^{13}C$ decrease. Several attempts have been made to reconstruct the atmospheric $\delta^{13}C$ history by means of tree-ring analyses and then to estimate past biospheric CO_2 releases [e.g. Peng et al., 1983]. However, the $\delta^{13}C$ records of tree-rings exhibit much scatter; obviously, they do not reflect only atmospheric isotope variations, but are influenced by other environmental and plant-physiological factors. Stuiver et al.

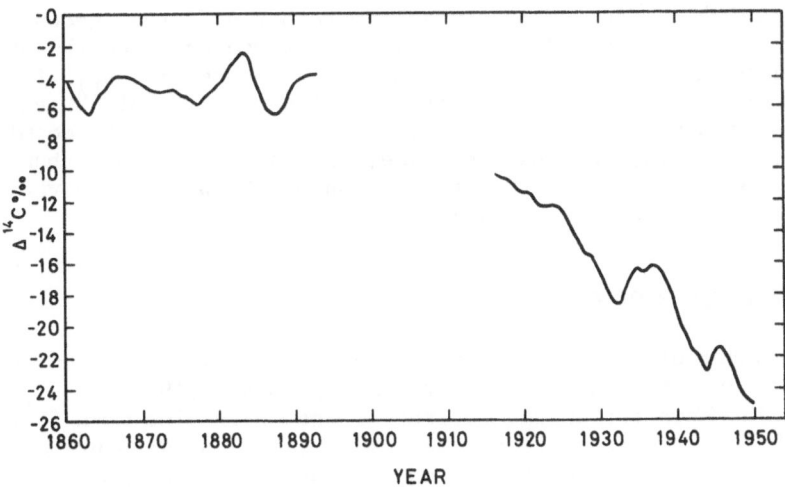

Figure 11.   Decrease of ^{14}C in the atmosphere due to fossil fuel burning (Suess effect), reconstructed by high-precision measurements on tree-rings of known age. After Stuiver and Quay [1981].

[1984] attempted to correct the $\delta^{13}C$ values for such influences by using a correlation with tree-ring width; they obtained an average decrease of ca. 1.0 %o until 1980.

A reliable $\delta^{13}C$ reconstruction is possible again by analyzing air trapped in old polar ice [Friedli et al., 1984]. First measurements performed at the University of Bern indicate a decrease of slightly more than 1 %o from 1800 to 1980, which is consistent with the CO_2 increase documented in Fig. 6a.

Tree-ring measurements have shown that the natural ^{14}C concentration was not quite constant in time [e.g. Damon et al., 1978]. It was about 8 % higher 6000 to 8000 yr B.P. (before present) than in the 19th century. As a possible reason variations of the earth's magnetic field intensity have been discussed: radioisotope production in the atmosphere is due to cosmic rays, and the geomagnetic field partly shields the earth from cosmic radiation. Palaeomagnetic data seem to support this hypothesis, but the representativeness of these data for the actual geomagnetic field intensities has been challenged.

Short-period variations, "wiggles", of ca. 10 %o amplitude and ca. 200 yr quasi-period are a recurring feature of the tree-ring record. They are probably due to variations of the sun's magnetic field, as indicated by a good correlation with sunspot number [Stuiver and Quay, 1980]. Model calculations show that these variations are attenuated, compared to the production changes, by a factor of about 20, because the perturbations are diluted by the carbon in the exchanging reservoirs.

The concentration of the cosmic-ray produced radioisotope ^{10}Be in ice cores also exhibits short-term fluctuations (Figure 12, top) that most probably are due to production rate changes. Taking these ^{10}Be concentrations as representative for radioisotope production rates and using the box-diffusion model of CO_2 exchange yields simulated ^{14}C variations well comparable to the observed ones (Figure 12, bottom) [Beer et al., 1984]. The two isotopes have rather different geochemical behaviour - ^{10}Be is attached to aerosols and washed out from the atmosphere within weeks to a few years -, therefore the consistency of the two records is a strong indication that the fluctuations of ^{14}C and of ^{10}Be are indeed due to a common cause, i.e. modulation of the cosmic ray flux by solar activity. At the same time, the fact that the carbon cycle model seems to simulate well the attenuation of ^{14}C production variation provides a positive test of its validity.

4. CLIMATIC EFFECTS OF CO_2 INCREASE

There is a number of reviews of the carbon dioxide/climate issue, e.g. Clark [1982], Hansen et al. [1984] or Liss and Crane [1983].

Figure 13 shows the energy balance of the earth. The main factor determining global temperature is the solar radiation. Without re-emission of energy, the temperature of the surface would increase continuously, but the earth emits thermal radiation at a rate proportional to the fourth power of its absolute temperature.

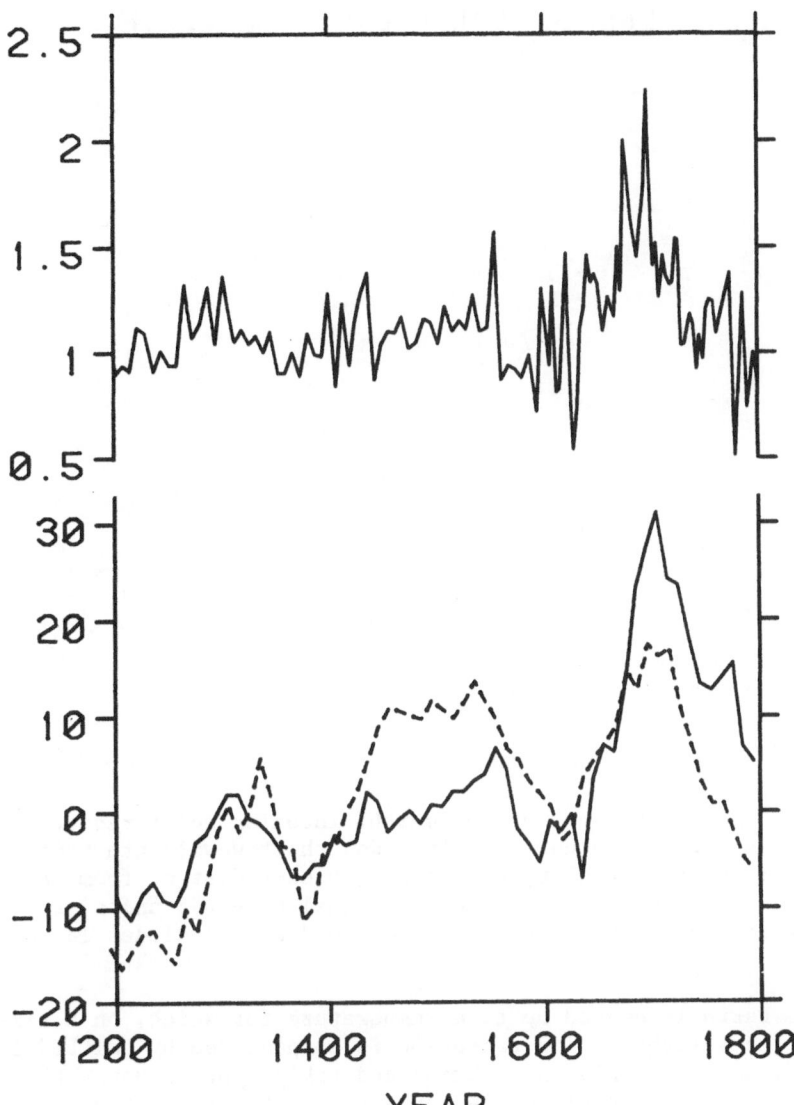

Figure 12. Top: ^{10}Be concentration (unit: 10^4 atoms/g ice) versus age in ice from Milcent, Greenland. Bottom: Δ^{14}C in the atmosphere, tree-ring measurements (dashed curve) [Stuiver and Quay, 1980] and model-calculated (solid curve) based on the assumption that ^{10}Be variations are directly proportional to production rate. From Beer et al. [1984].

ENERGY BALANCE OF THE EARTH

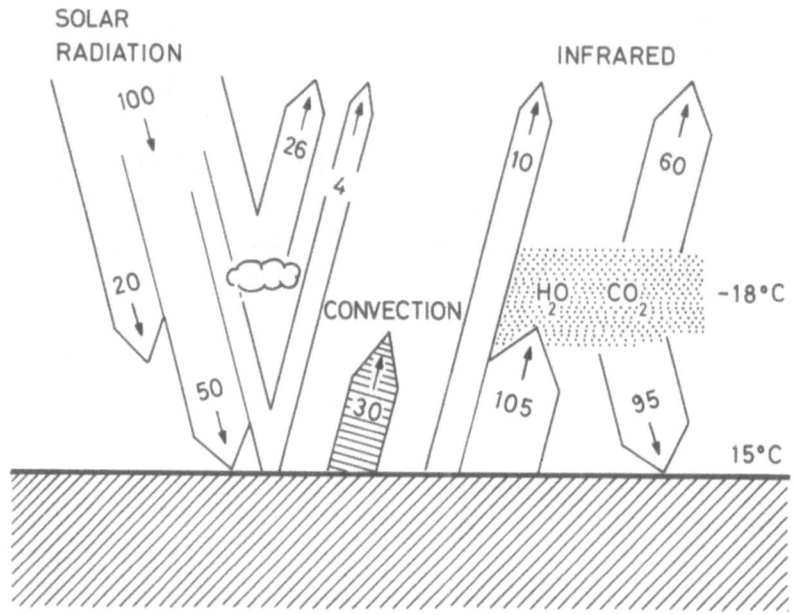

Figure 13. Energy balance of the earth. Incoming solar radiation intensity is set to 100 units. Left side: short-wavelength radiation; right side: infrared radiation. Infrared back-radiation from the atmosphere to the surface is of nearly equal size (95 units) as solar radiation at top of atmosphere. "Convection" also includes latent heat exchange.

Thus, the earth is heated up to a temperature for which, on a time-space average, the absorbed solar radiation is compensated by emitted infrared radiation. Based on this equilibrium and taking into account that about 30 % of the incident sunlight is not absorbed but reflected by clouds and the surface, an effective temperature (planetary temperature) of the earth can be calculated. This mean planetary temperature is 255 K or -18°C.

    The actual mean surface temperature is, however, about 15°C and thus considerably higher. The difference is due to the fact that most of the infrared radiation to space is not emitted by the surface but by higher (and colder) atmospheric layers, because - in contrast to the visible sunlight which penetrates the cloud-free atmosphere without much loss - infrared radiation emitted by the surface is absorbed, mainly by water vapour, but also by carbon dioxide, ozone and other trace gases (right-hand side of Figure 13). The absorbed energy heats the lower atmosphere which re-emits infrared radiation partly back to the surface.

In this way the surface receives, in addition to the solar energy, a considerable amount of thermal radiation and is heated up to an average temperature higher than the effective planetary temperature. This phenomenon, that the atmosphere is transparent to visible light but absorbs in the infrared and thus acts as a spectrally selective thermal isolation, is called greenhouse effect.

The infrared absorption by water vapour leaves a spectral window at wavelengths between 9 and 15 µm. In this range, absorption occurs by CO_2 and other trace gases but not by water vapour, and it is incomplete so that some radiation from the surface penetrates the atmosphere unaffected. An atmospheric CO_2 increase reduces the transmissivity in this spectral window and enhances the greenhouse effect. But also in other spectral regions, the greenhouse effect is amplified by a higher CO_2 concentration, because the radiation emitted by the surface is absorbed more strongly and the back-radiation originates from lower, and therefore warmer, atmospheric layers. In this way, an increase of CO_2 (and other infrared-active trace gases) leads to higher surface temperatures.

Climate models indicate that for a doubled CO_2 concentration, on a global average the downward infrared flux increases by 4 Wm^{-2} [Augustsson and Ramanathan, 1977]. For the whole earth this corresponds to an additional energy flux at the surface of $2000 \cdot 10^{12}$ W. This indirect heating due to the greenhouse effect can be compared with the direct anthropogenic energy input of $8 \cdot 10^{12}$ W at present, showing that on a global scale the direct warming by energy use is much weaker than the CO_2-induced warming.

To estimate temperature changes caused by increased atmospheric CO_2 concentrations, one has to consider, beside the changed radiation balance due to CO_2 alone, a number of feedback mechanisms. A strong positive feedback is the increase in absolute humidity going along with the warming of the earth's surface and the lower atmospheric layers. Another positive feedback effect results from changes in the earth's reflectivity for solar radiation (albedo) owing to a decrease of the snow and ice cover in polar areas caused by the warming. The combined temperature effect, including these feedback processes, is two to three times that caused by CO_2 alone.

The influence of clouds is difficult to model. Increasing cloud cover on the one hand leads to a higher albedo but on the other hand also to an increase of downward thermal radiation. The two effects partly compensate each other. At present it is not known if a higher surface temperature will lead to a change of cloud cover.

Calculations by means of a variety of climate models of different complexity indicate for a CO_2 doubling an average global temperature rise by 1.5° to 4°C, assuming that the energy fluxes at the earth's surface balance each other, i.e. that there is thermal equilibrium on a global scale.

While simple climate models provide estimates of CO_2-induced changes and permit to study the influence of specific processes and assumptions, reliable forecasts can finally only be expected from general circulation models [e.g. Manabe and Stouffer, 1980]. These models predict that in high latitudes the warming is significantly larger than on average. This amplification in polar regions is partly

due to the snow-albedo feedback mentioned above. At high latitudes there
is also a seasonal asymmetry, with large temperature changes mainly in
winter.

Temperature is not the only climate parameter of interest; changes
in precipitation and evaporation are of equal significance and for some
regions even more important. The results of Manabe and Stouffer [1980]
indicate more precipitation in high latitudes. There are, however, still
large problems with the simulation of the atmospheric water cycle which
lead to uncertainties in the prediction.

The climate model studies discussed so far assume that the earth
has adapted to the new radiative conditions and on average has reached
thermal equilibrium. In reality, however, the world ocean has a considerable heat capacity, and while the temperature is rising, there is a
relatively large flux of heat into the ocean needed to warm it up, so
that the radiation budget is not balanced. Only recently has this
aspect, which causes a delay in the global temperature increase, been
included in the model discussion in a quantitative way [Hansen et al.,
1981; Hoffert et al., 1980]. Since heat is transported in the ocean by
the same processes as CO_2 - currents and turbulent water motions -, the
box-diffusion ocean model can be also used for simulating the penetration of the temperature increase into the ocean [Siegenthaler and
Oeschger, 1984]. Figure 14 shows a model simulation of the CO_2-induced
mean atmospheric warming for the CO_2 increase of Figure 6a, reconstructed from the Siple ice core (solid line in Figure 14) and for a model-
predicted CO_2 increase due to fossil fuel emissions only (pre-industrial
concentration 297 ppm; dashed line). In addition, observed mean northern
hemisphere temperature anomalies are shown. The model simulations are
for constant cloud-top temperature and a ratio of land area to ocean of
1:1; see Siegenthaler and Oeschger [1984] for technical details. For a
reasonable range of climate model parameters, the calculated warming

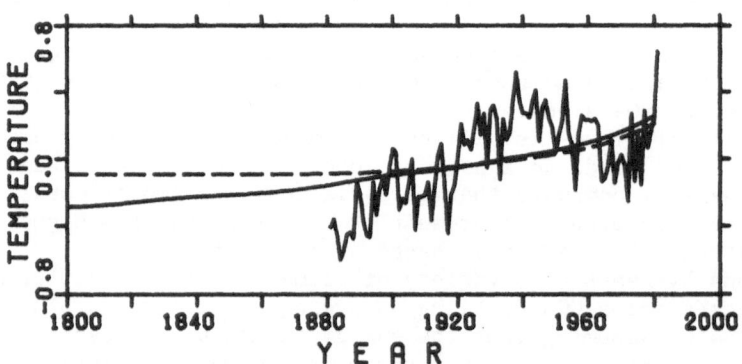

Figure 14. CO_2-induced global warming, calculated using the box-
diffusion model of vertical oceanic heat transport of Siegenthaler and
Oeschger [1984]. Dashed line: for CO_2 increase due to fossil fuels only;
smooth solid line: for CO_2 history according to the Siple ice core
results of Figure 6a. Irregular curve: Observed mean Northern Hemisphere
temperature anomalies [Jones et al., 1982].

until 1980 is 0.39° to 0.54°C for the Siple CO_2 history, and 0.23° to 0.31°C for the case with fossil input only. These results were calculated taking the ocean's thermal inertia into account; they are about 30 to 50 percent lower than for radiative equilibrium. The observations do indicate a long-term temperature in- crease, but it is superimposed by relatively large noise, and it is not possible to attribute it with certainty to CO_2.

In addition to CO_2, also other anthropogenic trace gases (methane, CFMs, tropospheric ozone etc.) contribute to the greenhouse effect. Ramanathan et al. [1985] have estimated that for the period 1980 to 2030, the temperature change due to these other trace gases may be about equal to that due to CO_2. During the last 10^5 years, the mean global temperature was at most 1 to 2°C higher than at present. The probability is large that in 100 years this natural range will be exceeded due to anthropogenic emissions of CO_2 and trace gases. The problem is especially serious when considering that this heating cannot simply be turned off, since even after a complete production stop, atmospheric concentrations of CO_2 and trace gases decrease only very slowly. Thus, these anthropogenic climate changes appear as partly irreversible during human time-scales.

5. NATURAL CO_2 VARIATIONS

5.1. Seasonal Variations

Besides the continuous increase, atmospheric CO_2 records at many sta- tions exhibit regular seasonal variations (Figures 5, 15). They are especially prominent in the northern hemisphere and reflect the bio- spheric cycle of growth and decay. In the growing season, the vegetation withdraws CO_2 from the atmosphere for the photosynthetic production of organic matter, and the atmospheric CO_2 level decreases. In the cold season, an equivalent amount of CO_2 is produced by soil respiration and the decay of plant material. On an annual basis, there is a nearly perfect balance between photosynthesis and decay, and the vegetation is neither a source nor a sink for atmospheric CO_2; this was at least the case before the biosphere was altered by human activities.

The biospheric origin of the seasonal variations is confirmed by a very good correlation of CO_2 with $\delta^{13}C$, which is due to the fact that the added or withdrawn biospheric CO_2 has a $\delta^{13}C$ value near -25 ‰, i.e. considerably lower than that of air-CO_2 [Mook et al., 1983].

In equatorial regions and in the southern hemisphere, the seasonal amplitude becomes rather small. In the southern hemisphere, the conti- nental area and correspondingly the vegetation biomass is relatively small. The seasonal amplitudes due to southern hemisphere vegetation, to interhemispheric air exchange and to air-sea exchange are comparable there and partly cancel each other. By means of two- and three-dimen- sional circulation models, the seasonal CO_2 variations can be used as tracers of atmospheric mixing [Pearman et al., 1983; Keeling and Heimann, 1985; Heimann et al., 1986].

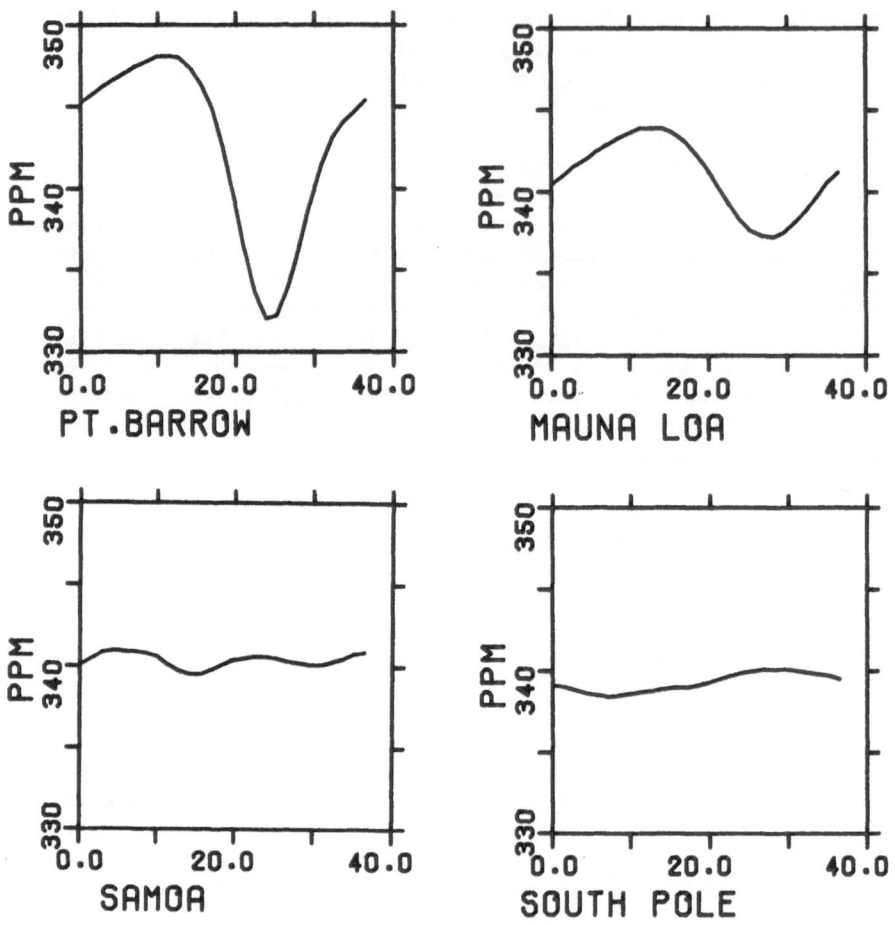

Figure 15.   Seasonal CO_2 variations, 1982, for Pt. Barrow, Alaska (71.3°N), Mauna Loa, Hawaii (19.5°N), American Samoa (14.2°S), and South Pole (90°S). Data from Komhyr et al. [1985]. Horizontal axis: day of the year (unit: 10 days).

5.2. Correlation with El Niño

The atmospheric CO_2 increase (Figure 5) is not quite regular from year to year. A correlation exists between growth rate and Southern Oscillation Index (SOI). As SOI, the difference of barometric pressure across the Pacific Ocean, between Darwin, Australia, and Easter Island, is taken. El Niño years coincide with periods of unusually low SOI. Bacastow [1976] and Bacastow and Keeling [1981] found that minima of SOI often coincide with a time of particularly high atmospheric CO_2 growth rate (Figure 16).

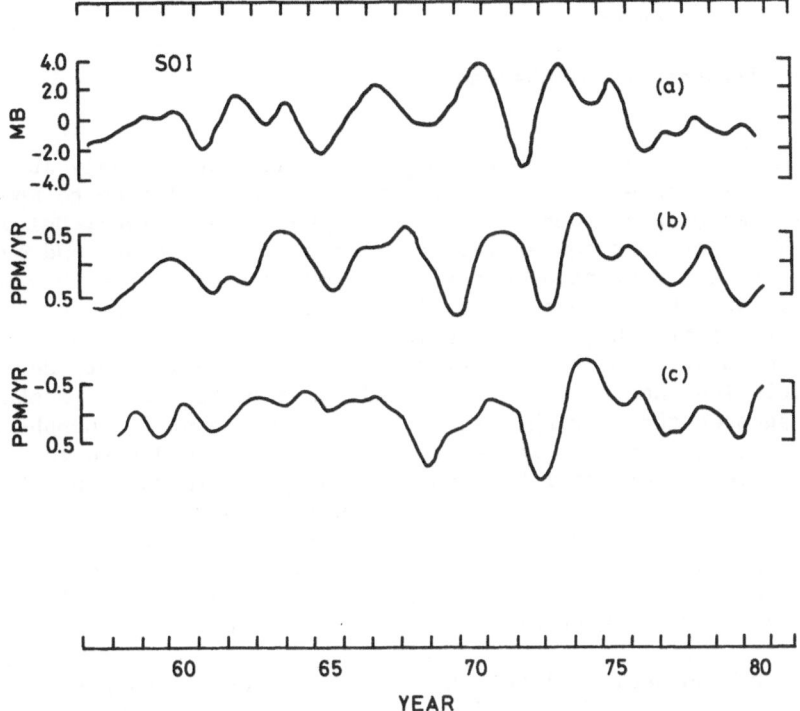

Figure 16. (a) Southern Oscillation Index, (b) and (c) time derivatives of seasonally adjusted and long-term detrended CO_2 concentrations at South Pole (b) and Mauna Loa (c). After Bacastow and Keeling [1981].

Komhyr et al. [1985] found, based on data from the NOAA network, that during the abnormally strong El Niño/Southern Oscillation event 1982/83 and also during the 1972 event, the CO_2 growth rate first decreased before increasing to higher than average values.

The phenomenon is not yet fully understood. There are various effects, partly acting in different directions. During El Niño years, equatorial upwelling of CO_2-rich water is weakened in the Pacific Ocean in response to changed atmospheric forcing. Presumably then the equatorial pCO_2 maximum (Figure 2) decays, and correspondingly the CO_2 source to the atmosphere is temporarily weakened [Keeling and Revelle, 1985]. A simple box-model, that also includes the warming of the equatorial surface water during El Niño, predicts a rapid atmospheric concentration drop, followed by a slow increase [Wenk et al., 1986], which, however, does not agree well with observation. On the other hand, $\delta^{13}C$ measurements seem to point to changes in the land biosphere affecting also the atmosphere [Keeling and Revelle, 1985].

The El Niño-related CO_2 oscillations are small, but of interest for the understanding of the factors that govern the atmospheric CO_2 concentration. In order to really understand them, it will be necessary to

obtain detailed chemical and nutrient data from the Equatorial Pacific
during an El Niño event.

5.3. Glacial/Interglacial Changes

Analyses of air bubbles trapped in old polar ice have shown that during
the last ice age, the atmospheric CO_2 level was probably only about 200
ppm and that it increased at the transition to the Holocene to approximately its recent pre-industrial value [Neftel et al., 1982; Delmas et
al., 1980]. Hypotheses concerning the causes of this change and evidence
from deep-sea sediment studies are discussed by Duplessy in this volume;
for this reason the subject is considered only shortly here.

The atmospheric CO_2 concentration is strongly influenced by the
chemical composition of ocean surface water, that in its turn depends on
marine biological activity, as described in paragraph 2.5. Broecker
[1982] suggested that during glacial times, the amount of phosphate in
the ocean was considerably larger than now, which would have led to a
stronger depletion of total CO_2, and thus of pCO_2, in surface water. He
suggested that the origin of this additional phosphate was erosion of
nutrient-rich organic sediments on the continental shelves, deposited
there during an earlier interglacial period of high sea level. This
hypothesis can explain many of the observations, but the CO_2 level would
change rather slowly which seems to be in contrast to the ice core
results.

Another hypothesis relates the CO_2 variations to changes in the
system of physical, chemical and biological processes in the ocean. In
large oceanic areas, the biological productivity is limited by the
availability of phosphate and nitrate. In these regions, a stronger vertical circulation would in a first approximation not change the surface
concentrations of total CO_2 and of alkalinity. Increased upwelling of
nutrients would lead to an enhanced flux of biogenic particles, but
because of the fixed Redfield ratios between P, N and C, the additional
particle flux of carbon would just be balanced by the increased upwelling rate of total CO_2. This is, however, not the case in the
Southern Ocean around Antarctica where phosphate and nitrate are abundant in surface waters; biological activity is obviously governed by
other factors there, like lack of light or rapid vertical mixing. This
observation led to the hypothesis that CO_2 variations were caused by
changes in ocean circulation that affected the distribution of chemical
properties, including pCO_2, in the ocean surface [Siegenthaler and Wenk,
1984; Sarmiento and Toggweiler, 1984; Knox and McElroy, 1984]. A decrease in vertical mixing in the Southern Ocean, with constant biological productivity, would lead to lower surface concentrations of
nutrients and of total CO_2 and consequently to a lower atmospheric CO_2
level. In spite of its relatively small area, the Southern Ocean turns
out to be important for atmospheric CO_2, because it is in rapid exchange
with the large volumes of the deep sea. This mechanism could lead to
rather rapid CO_2 changes (within a few 100 years, Wenk and Siegenthaler
[1985]), in contrast to Broecker's shelf sediment hypothesis.

It has not yet been possible to clearly decide which of the various
hypotheses is the true explanation of the glacial/interglacial CO_2

variations. Besides this question, the studies have clearly demonstrated that physical, chemical and biological processes in the ocean do not act independently on the global carbon cycle, but that we have to consider them as a highly interactive system. On the experimental side, numerous studies on the history of the carbon cycle, e.g. on deep-sea sediments, have been induced by the detection of the natural CO_2 variations, as e.g. documented by the volume edited by Sundquist and Broecker [1985]. Thus, the results of the ice core studies have led to large research activities that now go far beyond the issue of glacial/postglacial changes of atmospheric carbon dioxide.

ACKNOWLEDGMENTS

I thank S. Bohler for her skill and patience when typing the manuscript, and K. Hänni for drafting several figures.

REFERENCES

Augustsson, T. and V. Ramanathan, 1977: A radiative convective model study on the CO_2 climate problem. J. Atmos. Sci., 34, 448-451.

Bacastow, R. B., 1978: Modulation of atmospheric carbon dioxide by the Southern Oscillation. Nature, 261, 116-118.

Bacastow, R. B. and C. D. Keeling, 1981: Atmospheric CO_2 and the Southern Oscillation: effects associated with recent El Niño events. In: Proc. WMO/ICSU/UNEP Conference on Analysis and Interpretation of Atmospheric CO_2 Data, 109-112, World Meteorological Organization, Geneva.

Bacastow, R. B., C. D. Keeling, and T. P. Whorf, 1985: Seasonal amplitude increase in atmospheric CO_2 concentration at Mauna Loa, Hawaii, 1959-1982. J. Geophys. Res., 90, 10,529-10,540.

Beer, J., U. Siegenthaler, H. Oeschger, M. Andrée, G. Bonani, M. Suter, W. Wölfli, R. C. Finkel, and C. C. Langway, 1984: Temporal ^{10}Be variations. In: Proc. 18th Internat. Cosmic Ray Conf., Bangalore, 9, 317-320.

Bender, M., L. D. Labeyrie, D. Raynaud, and C. Lorius, 1985: Isotopic composition of atmospheric O_2 in ice linked with deglaciation and global primary productivity. Nature, 318, 349-352.

Betzer, P. R., R. H. Byrne, J. G. Acker, C. S. Lewis, R. R. Jolley, and R. A. Feely, 1984: The oceanic carbonate system: a reassessment of biogenic controls. Science, 226, 1074-1077.

Bolin, B. (ed.), 1981: Carbon Cycle Modelling. SCOPE 16, Wiley.

Bolin, B., 1985: How much CO_2 will remain in the atmosphere? In: International Assessment of the Role of Carbon Dioxide and Other Greenhouse Gases in Climate Variations and Associated Impacts. World Meteorological Organization, Geneva, in press.

Bolin, B., E. T. Degens, S. Kempe, and P. Ketner (eds.), 1979: The Global Carbon Cycle. SCOPE 13, Wiley, New York.

Broecker, W. S., 1982: Ocean chemistry during glacial times. Geochim. Cosmochim. Acta, 46, 1689-1705.

Broecker, W. S. and T.-H. Peng, 1974: Gas exchange rates between air and
    sea. Tellus, 26, 21-35.

Broecker, W. S. and T.-H. Peng, 1982: Tracers in the Sea.
    Eldigio Press, Lamont-Doherty Geological Observatory, Palisades,
    NY 10964.

Broecker, W. S., T.-H. Peng, and R. Engh, 1980: Modelling the carbon
    system. Radiocarbon, 22, 565-598.

Callendar, G. S., 1958: On the amount of carbon dioxide in the atmos-
    phere. Tellus, 10, 243-248.

Clark, W. C. (ed.), 1982: Carbon Dioxide Review 1982. Clarendon
    Press, Oxford.

Craig, H., 1957: Isotopic standards for carbon and correction factors
    for mass-spectrometric analysis of carbon dioxide. Geochim.
    Cosmochim. Acta, 12, 133-149.

Damon, P. E., J. C. Lerman, and A. Long, 1978: Temporal fluctuations of
    atmospheric ^{14}C: causal factors and implications. Ann. Rev. Earth
    Planet. Sci., 6, 457-494.

Delmas, R. J., J.-M. Ascensio, and M. Legrand, 1980: Polar ice evidence
    that atmospheric CO_2 20,000 years BP was 50 % of present. Nature,
    284, 155-157.

Emerson, S., 1985: Oceanic carbon preservation in marine sediments. In:
    Sundquist and Broecker, 1985, pp. 78-87.

Friedli, H., E. Moor, H. Oeschger, U. Siegenthaler, and B. Stauffer,
    1984: $^{13}C/^{12}C$ ratios in CO_2 extracted from Antarctic ice. Geophys.
    Res. Lett., 11, 1145-1148.

Garrels, R. M. and E. A. Perry, 1974: Cycling of carbon, sulphur, and
    oxygen through geologic time. In: The Sea, vol. 5 (E. G. Goldberg,
    ed.), pp. 303-336. Wiley.

Hansen, J., A. Lacis, D. Rind, G. Russell, P. Stone, I. Fung, R. Ruedy,
    and J. Lerner, 1984: Climate sensitivity: analysis of feedback
    mechanisms. In: Climate Processes and Climate Sensitivity
    (J. E. Hansen and T. Takahashi, eds.), Geophysical Monograph 29,
    pp. 130-163, American Geophysical Union.

Heimann, M., C. D. Keeling, and C. J. Tucker, 1986b. A three dimensional
    model of atmospheric CO_2 transport based on observed winds. 2.
    Analyses of the seasonal cycle of CO_2. Tellus, in press.

Hoffert, M. I., A. J. Callegari, and C. T. Hsieh, 1980: The role of deep
    sea heat storage in the secular response to climatic forcing.
    J. Geophys. Res., 85, 6667-6679.

Jähne, B., W. Huber, A. Dutzi, T. Wais, and J. Ilmberger, 1984: Wind
    wave-tunnel experiments on the Schmidt number and wave field
    dependence of air/water exchange. In: Gas Exchange at Water Surface
    (W. Brutsaert and G. H. Jirka, eds.), pp. 303-309. Reidel.

Jones, P. D., T. M. L. Wigley, and P. M. Kelly, 1982: Variations in
    surface air temperatures. Part 1. Northern hemisphere, 1881-1980.
    Monthly Weather Rev., 110, 59-69.

Keeling, C. D., J. A. Adams, C. A. Ekdahl, and P. R. Guenther, 1976:
    Atmospheric carbon dioxide variations at the South Pole. Tellus,
    28, 552-563.

Keeling, C. D., R. B. Bacastow, and T. P. Whorf, 1982: Measurements on
    the concentration of carbon dioxide at Mauna Loa Observatory,
    Hawaii. In: Carbon Dioxide Review 1982, W.C. Clark (ed.),
    pp. 377-384.
Keeling, C. D. and M. Heimann, 1985: Meridional eddy diffusion model
    of the transport of atmospheric carbon dioxide. 2. The mean annual
    carbon cycle. J. Geophys. Res., in press.
Keeling, C. D. and R. Revelle, 1985: Effects of El Niño/Southern
    Oscillation on the atmospheric contetn of carbon dioxide.
    Meteoritics, 20, 437-450.
Kempe, S., 1984: Sinks of the anthropogenically enhanced carbon cycle
    in surface fresh waters. J. Geophys. Res., 89, 4657-4676.
Knox, F. and M. McElroy, 1984: Changes in atmospheric CO_2: influence
    of marine biota in high latitudes. J. Geophys. Res., 89, 4629-4637.
Kohlmaier, G. H., A. Janecek, C. D. Keeling, and R. Revelle, 1986:
    Analysis of the CO_2 stimulation effect of vegetation in connection
    with the atmospheric CO_2 amplitude increase. Tellus, in press.
Komhyr, W. D., R. H. Gammon, T. B. Harris, L. S. Waterman, T. J. Conway,
    W. R. Taylor, and K. W. Thoning, 1985: Global atmospheric CO_2
    distribution and variations from 1968-1982 NOAA/GMCC CO_2 flask
    sample data. J. Geophys. Res., 90, 5567-5596.
Levitus, S., 1982: Climatological atlas of the World Ocean. NOAA
    Professional Paper 13.
Li, Y.-H., T.-H. Peng, W. S. Broecker, and H. G. Oestlund, 1984: The
    average vertical mixing coefficient for the oceanic thermocline.
    Tellus, 36B, 212-217.
Liss, P. S., 1983: Gas transfer: experiments and geochemical impli-
    cations. In: Air-Sea Exchange of Gases and Particles (P. S. Liss
    and W. G. N. Slinn, eds.), pp. 241-298. Reidel.
Liss, P. S. and A. J. Crane, 1983: Man-made carbon dioxide and climatic
    change. Geo Books, Norwich.
Manabe, S. and R. J. Stouffer, 1980: Sensitivity of a global climate
    model to an increase of CO_2 concentration in the atmosphere.
    J. Geophys. Res., 85, 5529-5554.
Meybeck, M., 1982: Carbon, nitrogen and phosphorus transport by world
    rivers. Am. J. Sci., 282, 401-450.
Mook, W. G., M. Koopmans, A. F. Carter, and C. D. Keeling, 1983: Sea-
    sonal, latitudinal and secular variations in the abundance and
    isotopic ratios of atmospheric carbon dioxide. J. Geophys. Res.,
    88, 10,915-10,933.
National Research Council, 1983: Changing Climate. National Academy
    Press, Washington.
Neftel, A., H. Oeschger, J. Schwander, B. Stauffer, and R. Zumbrunn,
    1982: Ice core measurements give atmospheric CO_2 content during
    the past 40,000 years. Nature, 295, 222-223.
Neftel, A., E. Moor, H. Oeschger, and B. Stauffer, 1985: Evidence from
    polar ice cores for the increase in atmospheric CO_2 in the past
    two centuries. Nature, 315, 45-47.
Nir, A. and S. Lewis, 1975: On tracer theory in geophysical systems in
    the steady and non-steady state. Part I. Tellus, 27, 372-383.

Oeschger, H., U. Siegenthaler, U. Schotterer, and A. Gugelmann, 1975:
    A box diffusion model to study the carbon dioxide exchange in
    nature. Tellus, 27, 168-192.
Pearman, G. I., P. Hyson, and P. J. Fraser, 1983: The global distri-
    bution of atmospheric carbon dioxide. Aspects of observations and
    modeling. J. Geophys. Res., 88, 3581-3590.
Peng, T.-H., W. S. Broecker, G. G. Mathieu, and Y.-H. Li, 1979: Radon
    evasion rates in the Atlantic and Pacific Oceans as determined
    during the GEOSECS program. J. Geophys. Res., 84, 2471-2486.
Peng, T.-H., W. S. Broecker, H. D. Freyer, and S. Trumbore, 1983:
    A deconvolution of the tree-ring based $\delta^{13}C$ record. J. Geophys.
    Res., 88, 3609-3620.
Ramanathan, V., R. J. Cicerone, H. B. Singh, and J. T. Kiehl, 1985:
    Trace gas trends and their potential role in climatic change.
    J. Geophys. Res., 90, 5547-5566.
Raynaud, D. and J. M. Barnola, 1985: An Antarctic ice core reveals
    atmospheric CO_2 variations over the past few centuries. Nature,
    315, 309-311.
Roos, M. and G. Gravenhorst, 1984: The increase in oceanic carbon
    dioxide and the net CO_2 flux into the North Atlantic. J. Geophys.
    Res., 89, 8181-8193.
Rotty, R. M. and G. Marland, 1984: Production of CO_2 from fossil fuel
    burning by fuel type, 1860-1982. Report NDP-006, Carbon Dioxide
    Information Center, Oak Ridge National Laboratory, Oak Ridge.
Sarmiento, J. L., 1985: Three-dimensional ocean models for predicting
    the distribution of CO_2 between the ocean and atmosphere. In: The
    Global Carbon Cycle: Analysis of the Natural Cycle and Implica-
    tions of Anthropogenic Alterations for the Next Century. Proc.
    6th ORNL Life Sciences Symp. Springer (in press).
Sarmiento, J. L. and J. R. Toggweiler, 1984: A new model for the role
    of the oceans in determining atmospheric pCO_2. Nature, 308, 621-
    624.
Siegenthaler, U., 1983: Uptake of excess CO_2 by an outcrop-diffusion
    model of the ocean. J. Geophys. Res., 88, 3599-3608.
Siegenthaler, U., 1984: 19th century measurements of atmospheric CO_2 -
    a comment. Climatic Change, 6, 409-411.
Siegenthaler, U., 1986: ^{14}C in the oceans. In: Handbook of Environ-
    mental Isotope Geochemistry, vol. 3 (J.-Ch. Fontes and P. Fritz,
    eds.). Elsevier, in press.
Siegenthaler, U. and H. Oeschger, 1984. Transient temperature changes
    due to increasing CO_2 using simple models. Annals of Glaciology,
    5, 153-159.
Siegenthaler, U. and T. Wenk, 1984: Rapid atmospheric CO_2 variations
    and ocean circulation. Nature, 308, 624-626.
Siegenthaler, U., H. Oeschger, and M. Heimann, 1986: Biospheric CO_2
    sources since 1800 AD reconstructed by deconvolution of ice core
    CO_2 data. Tellus, in press.
Smethie, W. M., T. Takahashi, and D. W. Chipman, 1985: Gas exchange
    and CO_2 flux in the tropical Atlantic Ocean determined from ^{228}Rn
    and pCO_2 measurements. J. Geophys. Res., 90, 7005-7022.

Stuiver, M. and P. D. Quay, 1980: changes in atmospheric carbon-14
    attributed to a variable sun. Science, 207, 11-19.
Stuiver, M. and P. D. Quay, 1981: Atmospheric ^{14}C changes resulting
    from fossil fuel CO_2 release and cosmic ray flux variability.
    Earth Planet. Sci. Lett., 53, 349-362.
Stuiver, M., H. G. Oestlund, and T. A. McConnaughey, 1981: GEOSECS
    Atlantic and Pacific ^{14}C distribution. In: B. Bolin (Editor),
    Carbon Cycle Modelling. Wiley, SCOPE 16, 201-221.
Stuiver, M., R. L. Burk, and P. D. Quay, 1984: ^{13}C/^{12}C ratios in tree-
    rings and the transfer of biospheric carbon to the atmosphere.
    J. Geophys. Res., 89, 11,731-11,748.
Sundquist, E. T., 1985: Geological perspectives on carbon dioxide and
    the carbon cycle. In: Sundquist and Broecker [1985], pp. 5-60.
Sundquist, T. T. and W. S. Broecker (eds.), 1985: The Carbon Cycle
    and Atmospheric CO_2: Natural Variations Archean to Present. Geo-
    phys. Monogr. 32, American Geophysical Union.
Takahashi, T., 1985: Geographical and time variability of partial
    pressure of CO_2 in surface waters of the Atlantic Ocean. In: The
    Global Carbon Cycle: Analysis of the Natural Cycle and Implica-
    tions of Anthropogenic Alterations for the Next Century. Proc.
    6th ORNL Life Sciences Symposium. Springer, in press.
Takahashi, T., W. S. Broecker, S. R. Werner, and A. E. Bainbridge,
    1980: Carbonate chemistry of the surface waters of the world
    ocean. In: Isotope Marine Chemistry. Uchida Rokakuho, Tokyo,
    pp. 291-326.
Takahashi, T., W. S. Broecker, and A. E. Bainbridge, 1981: Supplement
    to alkalinity and total carbon dioxide concentration in the world
    ocean. In: B. Bolin (ed.), Carbon Cycle Modelling. SCOPE 16,
    159-199, Wiley.
Takahashi, T., D. Chipman, and T. Volk, 1983: Geographical, seasonal
    and secular variations of pCO_2 in surface waters of the North
    Atlantic Ocean. In: Carbon Dioxide Science and Consensus,
    pp. II.123 - II.145, CONF-820970, US Dept. of Commerce, Spring-
    field, VA 22161.
Takahashi, T., W. S. Broecker, and S. Langer, 1985: Redfield ratio
    based on chemical data from isopycnal surfaces. J. Geophys. Res.,
    90, 6907-6924.
Walsh, J. J., R. L. Rowe, R. L. Iverson, and C. P. McRoy, 1981: Bio-
    logical export of shelf carbon is a sink of the global CO_2 cycle.
    Nature, 291, 196-201.
Wanninkhof, R., J. R. Ledwell, and W. S. Broecker, 1985: Gas exchange
    - wind speed relation measured with sulphur hexafluoride on a
    lake. Science, 227, 1224-1226.
Wenk, T. and U. Siegenthaler, 1985: The high-latitude ocean as a
    control of atmospheric CO_2. In: Sundquist and Broecker [1985],
    pp. 185-194.
Wenk T., H. Oeschger, and U. Siegenthaler, 1986: Simulation of atmos-
    pheric CO_2 response to El Niño events by means of an ocean-
    atmosphere box model. Tellus, in press.

CO$_2$ AIR-SEA EXCHANGE DURING GLACIAL TIMES:
IMPORTANCE OF DEEP SEA CIRCULATION CHANGES.

Jean-Claude Duplessy

Centre des Faibles Radioactivités
Laboratoire mixte CNRS-CEA, BP N°1
91190, Gif sur Yvette, France.

1. INTRODUCTION

Over the last 730,000 years (the Brunhes geomagnetic epoch), the earth's climate has alternated between ice ages and interglacial conditions. 9 interglacial stages have been recognized in the $\delta^{18}O$ record, but the total duration of warm conditions, roughly similar to those of today, did not represent more than 10% of the time. Glacial conditions are characterized by temperature colder than today and extensive dryness (CLIMAP, 1976). More recently, studies from ice cores and from deep sea sediments have indicated that the global carbon cycle was also very sensitive to climatic changes. In this paper we shall review the evidence of the variability of the carbon cycle and discuss the various theories proposed to explain this variability.

2. EVIDENCE FROM POLAR ICE CORES

2.1 Data

Ice formed by sintering of dry cold snow contains air with atmospheric composition in its bubbles. This discovery was first used to estimate the atmospheric pressure at the surface level of the ice-caps (and therefore their past altitude) by analyzing the amount of air included in a measured volume of ice (Raynaud and Lebel, 1979). Then attempts were made to reconstruct the variations of the chemical composition of the atmosphere by analyzing the gas in ice samples of known age. CO$_2$ measurements have been performed on ice samples from Greenland (Camp Century and Dye-3) and Antarctica (Byrd, D-10, Dome C) (Berner et al., 1980; Delmas et al., 1980; Neftel et al., 1982). These measurements suggest that the atmospheric CO$_2$ concentration was approximately 200 µAtm during the last glacial maximum, about 80 µAtm lower than the XIXth century value (assumed pre-industrial value).

These data, which suggest that the atmospheric CO$_2$ concentration is highly variable, suffer from two major problems: First, not enough data are available for the last glacial to interglacial transition; the Dome C

249

P. Buat-Ménard (ed.), The Role of Air-Sea Exchange in Geochemical Cycling, 249-267.
© *1986 by D. Reidel Publishing Company.*

record shows sharp variations in a few thousand years, which have to be documented in other cores. Second, the Dye-3 Greenland record exhibits rapid CO_2 variations of large amplitude (50 µAtm), which are associated with short $\delta^{18}O$ events (Stauffer et al., 1984). Since the CO_2 record should be global, these rapid variations should be present in other cores. Despite efforts to find them in Antarctic ice cores, they have not been found.

2.2 Discussion: Is the ocean able to absorb the missing CO_2?

The solubility of CO_2 increases when the water temperature decreases. The CLIMAP (1976) reconstruction indicates that the mean temperature of the ice-age ocean was 2°C lower than the present one. This would produce a pCO_2 decrease of only 16 µAtm. The solubility of CO_2 also depends on the salinity. The glacial ocean was more saline (about 35.9 per mil) than the present one (about 34.7 per mil), because the sea level dropped by about 120 m as fresh water was blocked over the continents as ice-caps. The pCO_2 thus tended to increase both because the solubility dropped by 1% as a result of the salinity increase and because all the concentrations of the dissolved species increased in the ratio 35.9/34.7=1.035. Taking into account all these effects, the net pCO_2 decrease for glacial conditions is 7 µAtm, a value far lower than that suggested by ice core studies. As the ocean contains 61 times more CO_2 than the atmosphere, the atmospheric pCO_2 is slave to the ocean chemistry. We are thus obliged to explain its variations by important changes in ocean chemistry.

3. EVIDENCE FROM DEEP SEA SEDIMENTS

3.1 Data

Deep sea sediments contain shells which are the remains of animals and plants, which have lived in the past ocean. The $^{18}O/^{16}O$ ratio ($\delta^{18}O$) of the carbonate shells of foraminifera is classically used as a proxy indicator for salinity and its variations primarily reflect those of the volume of ice stored over the continents. The $\delta^{13}C$ of these shells reflects more or less that of the total CO_2 dissolved in sea water.

Two kinds of foraminifera are present in the ocean. The planktonic forms are free floatting animals that live in the upper water masses and are transported by currents. The isotopic composition of their shells is thus assumed to reflect upper water conditions. The second kind of foraminifera constitute benthic forms, which live at the surface of the sediment at all depths. Their isotopic composition therefore should reflect bottom and deep water conditions (Duplessy et al., 1984).

Despite possible complications due to the fact that some species of benthic foraminifera might live inside the sediment and deposit their shell from carbonate ions dissolved in sediment interstitial waters, $\delta^{13}C$ analyses of benthic foraminifera in deep sea cores from the Pacific and eastern Atlantic ocean have been used to estimate that the $\delta^{13}C$ of glacial specimens was about 0.7 per mil lower than that of Holocene samples (Shackleton, 1977). This author suggested that the forest-soil carbon reservoir (which is characterized by δ^{13}C values about 25 per mil lower than that of ocean bicarbonate) was smaller during glacial than during interglacial times. This model is qualitatively supported by paleoclimatic reconstructions: during glacial conditions, temperate and boreal forests of the northern hemisphere disappeared, deserts expanded in the tropical belt and the equatorial rainforest was less dense than today as a consequence of the dryness.

3.2 Various hypotheses explaining the sedimentary record

The present size of the carbon reservoirs is as follows:

Reservoir	size (10^{16} Moles C)
Atmosphere	5
Forest	6
Soils	20
Ocean	305
Shelf deposits	?

Forests-Soils hypothesis: a simple mass balance calculation shows that a δ^{13}C decrease of 0.7 per mil represents a transfer of 8.4 10^{16} Moles of carbon from the continental biosphere to the ocean. A large part of the carbon should come from the soil reservoir. Duplessy and Shackleton (1984) developed a more complete sediment data base and calculated that the δ^{13}C decrease was smaller than previously estimated and may be estimated between 0.3 and 0.45 per mil. These revised estimates would correspond to carbon transfer of 3.5 and 5.3 10^{16} Moles from the continental biosphere to the ocean. This hypothesis implies a noticeable decrease of the volume of the continental biosphere. It also fits poorly with polar ice evidence since a net transfer of CO$_2$ from the biosphere to the continents would increase the atmospheric CO$_2$ content rather than decrease it.

Shelf sediment hypothesis: Broecker (1982) suggested that the deposition of organic rich sediments on the continental shelf during the marine transgression which accompanied the deglaciation would contribute to an important storage of organic carbon (with a low δ^{13}C) and give the same signal in the benthic δ^{13}C record as the forests-soils hypothesis. Assuming that shelf deposits have a δ^{13}C close to -20 per mil, the lowest and highest δ^{13}C estimates of Duplessy and Shackleton would correspond to net carbon storage of 4.6 and 7 10^{16} Moles.

The distinction between a terrestrial and a shelf origin for the δ^{13}C shift is important because the removal of nutrient Phosphate (P) and Nitrate (N) would accompany shelf deposition, but not forest and soil formation because P and N cannot be transferred through the atmosphere from the sea to the continent. Therefore, in Shackleton's hypothesis (1977) the amount of nutrient available in the ocean is not dependent on climate, whereas in Broecker's hypothesis glacial ocean deep waters contained 2.65 to 2.83 µMoles P/kg instead of 2.25 µMoles P/kg under present conditions (assuming that the C/P ratio remained constant in the organic matter),

Changing Redfield ratio hypothesis: under present conditions, the chemical composition of the organic soft tissue formed by plants is relatively constant. On the mean, for every atom of P in this tissue, there are on average 15 atoms of N and 105 atoms of C. The approximate ratio of the same elements dissolved in sea water is 15 atoms of N and 1000 atoms of C for every atom of P. Therefore, when deep water upwells to the surface, by the time all of its dissolved P has been consummed, so that its dissolved N but only 105 out of the available 1000 atoms of C have been consummed (Broecker and Peng, 1982).

However, large variations of the C/N/P ratio are observed on individual samples from the modern ocean and are poorly understood. If during glacial time on the average more than 105 atoms of C are fixed for 1 atom of P and 15 atoms of N, more organic carbon depleted in ^{13}C will be oxidized in the deep sea and the $\delta^{13}C$ signal observed in benthic foraminifera may be easily generated. The P distribution will remain basically the same, regardless of the climatic conditions.

3.3 Cadmium as a proxy-indicator for past Phosphate

Cadmium and Phosphate have similar distributions in today's ocean. Because $CdCO_3$ and $CaCO_3$ form a continuous solid solution series, certain benthic foraminifera show a consistent relationship between the Cd/Ca ratio of the bottom water and that of their calcite shells (Hester and Boyle, 1982). As Ca is uniformly distributed through the ocean, temporal changes in the Cd/Ca ratio of benthic foraminifera should reflect changes of the Cd content of bottom water at the core site and therefore those of P dissolved in bottom water.

Only 2 cores were analyzed for Cd/Ca ratio in benthic foraminifera (Boyle and Keigwin, 1985). Since one core was raised in the equatorial Pacific ocean and the other one at the Mid Atlantic Ridge, it was possible to calculate a rough mean value for glacial Cd/Ca ratio of benthic foraminifera for the world ocean. It appears that this ratio was 22 ± 5% higher than today, suggesting that the P content of bottom water was also 22 ± 5% higher than today. The present day mean value of bottom water P concentration is 2.25 µMole/kg. The glacial P concentration can be estimated as 2.75 ± 0.11 µMole/kg, a value in rather good agreement with the independent estimate derived from $\delta^{13}C$ changes in benthic foraminifera in the shelf deposit hypothesis.

In summary, the Cd evidence suggests that the P concentration of the ocean has changed by 22 ± 5% between glacial and interglacial conditions. It thus favours the shelf sediment hypothesis. All the presented hypotheses that offer an explanation for the drop in $\delta^{13}C$ imply oxidation of organic matter. Since one should from that expect a rise in pCO_2, it contradicts the evidence from the ice cores. This apparent contradiction needs to be resolved by detailed modelling of the ocean's non-linear chemical response to changes in the inventories of oxidized and reduced carbon.

4. BROECKER'S TWO BOX MODEL FOR THE CO_2 CYCLE

Broecker (1982) treats the ocean as a pair of well mixed reservoirs, an upper box of warm water where photosynthesis occurs and a lower box of cold water where chemistry is dominated by the effects of respiration and decay of biogenic products. Nutrient N and P enter the warm water box from below and are assummed to be used to depletion forming organic material with the Redfield C/N/P ratios. The organic material falls back to deep water accompanied by a constant fraction of carbonate (foraminifera and coccolith shells). This material decomposes at depth returning C,N,P to the lower box. The flux of upwelling water is compensated by deep water formation which occurs in polar regions where the water is isochemically cooled and sinks to depth. The input of the model is the chemical composition of the lower box and the output is the equilibrium pCO_2 of the upper box (Fig. 1).

This model describes pretty well the modern conditions and simulates a pCO_2 value close to the pre-industrial one. Glacial conditions may be estimated when the following parameters are known: temperature, salinity, volume of oceanic carbon during glacial conditions, phosphate and carbonate ion content of deep waters. Two cases are discussed: the first one assumes that the only changes in the oceanic CO_2 content are linked to the deposition of organic matter at the end of the glacial period. The second one takes into account the compensation of the organic C loss by carbonate deposition. As a matter of fact, the removal of C at the end of glacial times would have produced a strong increase in the carbonate ion concentration of deep waters (represented in the sediment as a carbonate preservation spike). Such a change would throw the carbonate budget out of balance, leading to an excess of $CaCO_3$ accumulation, which would gradually bring down the carbonate ion concentration and restore the balance between river input and sediment loss. For each mole of C removed as organic matter, about one mole would be removed as $CaCO_3$. In both cases, it may be observed that the pCO_2 drop simulated by the model is far too low to explain polar ice results.

Since it is not possible to model the changes in ocean chemistry with a two-box model of the ocean, Siegenthaler and Wenk (1984) establish a simple four-box model to test the hypothesis that CO_2 variations might arise from changes in ocean circulation that affected the distribution in chemical properties in the various water masses (Siegenthaler this issue).

5. EVIDENCE FOR DEEP WATER CIRCULATION DURING THE LAST CLIMATIC CYCLE

5.1 Geochemical basis

The modern distribution of the $\delta^{13}C$ of the total CO_2 dissolved in deep and bottom waters of the ocean shows that TCO_2 $\delta^{13}C$ can be used as a tracer of the deep water circulation on the following conditions:a)

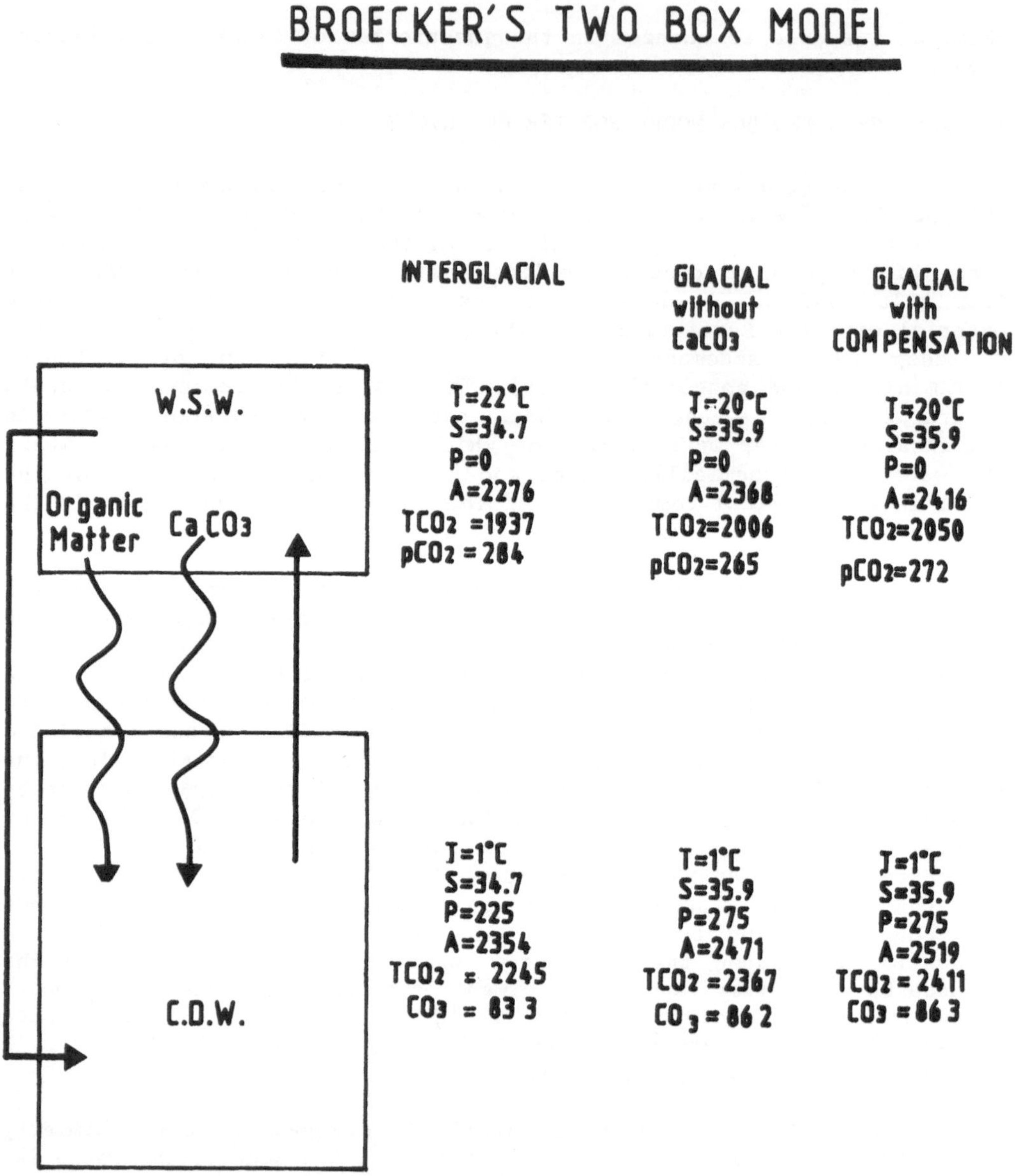
BROECKER'S TWO BOX MODEL
INTERGLACIAL
GLACIAL without CaCO3
GLACIAL with COMPENSATION
W.S.W.
Organic Matter
CaCO3
T=22°C
S=34.7
P=0
A=2276
TCO2 =1937
pCO2 = 284
T=20°C
S=35.9
P=0
A=2368
TCO2=2006
pCO2=265
T=20°C
S=35.9
P=0
A=2416
TCO2=2050
pCO2=272
C.D.W.
T=1°C
S=34.7
P=225
A=2354
TCO2 = 2245
CO3 = 83 3
T=1°C
S=35.9
P=275
A=2471
TCO2 =2367
CO3 = 86 2
T=1°C
S=35.9
P=275
A=2519
TCO2 = 2411
CO3 = 86 3

Figure 1: Estimates of pCO_2 in the warm surface water reservoir as a function of ocean chemistry (see text). Units: pCO_2 in μAtm, carbonate ion content CO_3 μeq/kg, Alkalinity A in μeq/kg, Phosphate content in μMole/kg.

within a deep water mass, $\delta^{13}C$ decreases with increasing oxidation of organic matter, such that the $\delta^{13}C$ decrease is related both to the surface productivity and to the time elapsed since the deep water was isolated from the atmosphere; b) when two water masses are to be compared, the $\delta^{13}C$ value in each water mass depends not only on the residence time at depth, but also on the ratio between the CO_2 produced from organic matter oxidation and that derived from carbonate dissolution; c) mixing between two water masses results in $\delta^{13}C$ values intermediate between the two original components, regardless of chemical reactions occurring at depth.

Changing oceanographic conditions that result in changes in both dissolved O_2 content and $\delta^{13}C$ of total CO_2, may be reconstructed from the down core ^{13}C record in a deep sea core.

Figure 2 compares four carbon and oxygen isotope records of benthic foraminifera in deep sea cores raised from various oceanic basins for the period 135000–107000 year B.P.. Core CH73–139C comes from Rockall Plateau, and its hydrological setting on the path of the Norwegian Sea Overflow Water offers a favourable location for the study of deep water production in the Northeastern Atlantic Ocean. Core M12–392 was raised from the continental margin of Africa in the Tropical Eastern Atlantic Ocean. Cores MD77–202 and V19–30 come respectively from the Northern Indian Ocean and from the Panama basin in the Eastern Equatorial Pacific, where deep waters are much older and much more depleted in dissolved oxygen and carbon-13 than those of the Atlantic.

The four oxygen isotope records show a strong similarity, in agreement with the hypothesis that the major signal in these records originates from the waxing and waning of large volume of isotopically light water in the northern hemisphere ice sheets during the major part of Pleistocene times. By contrast, the carbon isotope records exhibit noticeable differences: the amplitude of $\delta^{13}C$ variations is much larger in the Atlantic cores than in the Indian and Pacific cores, as discussed by Shackleton et al (1983). Also during the transition between glacial stage 6 and interglacial stage 5, about 128000 years ago, $\delta^{13}C$ values become isotopically heavier in the four cores, but the $\delta^{13}C$ record lags the $\delta^{18}O$ record in the Atlantic while the records are in phase in the Pacific and Indian oceans.

Since the $\delta^{13}C$ value of benthic foraminifera reflects that of the total CO_2 dissolved in the oceanic deep water, the observed differences between the various $\delta^{13}C$ records indicate that the $\delta^{13}C$ of TCO_2 in the past ocean has changed not only because of changes in the amount of carbon stored in the exchangeable reservoirs (atmosphere, biosphere and ocean), but also because of changes in the deep water circulation and in the oceanic carbon chemistry. Consequently, the major trends of the deep water circulation and chemistry in the world ocean during climatic conditions different from those of today can be reconstructed by comparing maps of the $\delta^{13}C$ of benthic foraminifera that lived at specified times in the past, since $\delta^{13}C$ values can only decrease along the flow lines of a water mass.

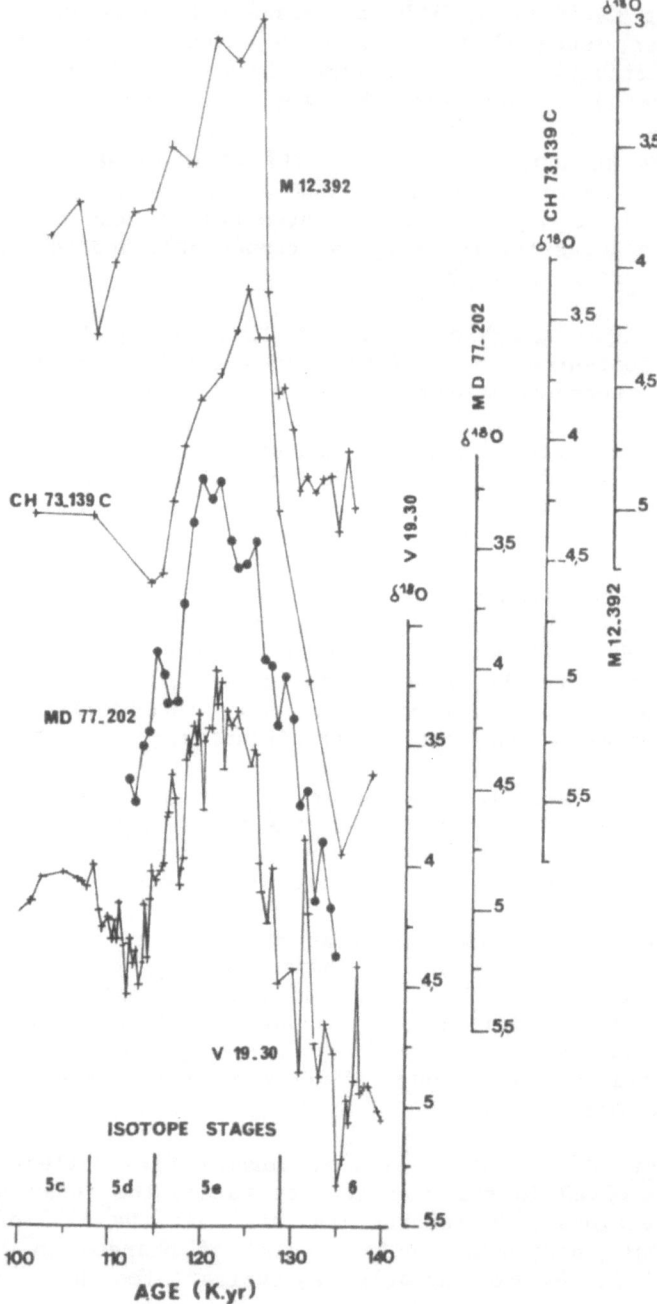

Figure 2A: Down core ¹⁸O record from 135 to 107 Kyr in four cores
originating from different oceanic basins. Note the
similarity of the temporal changes.

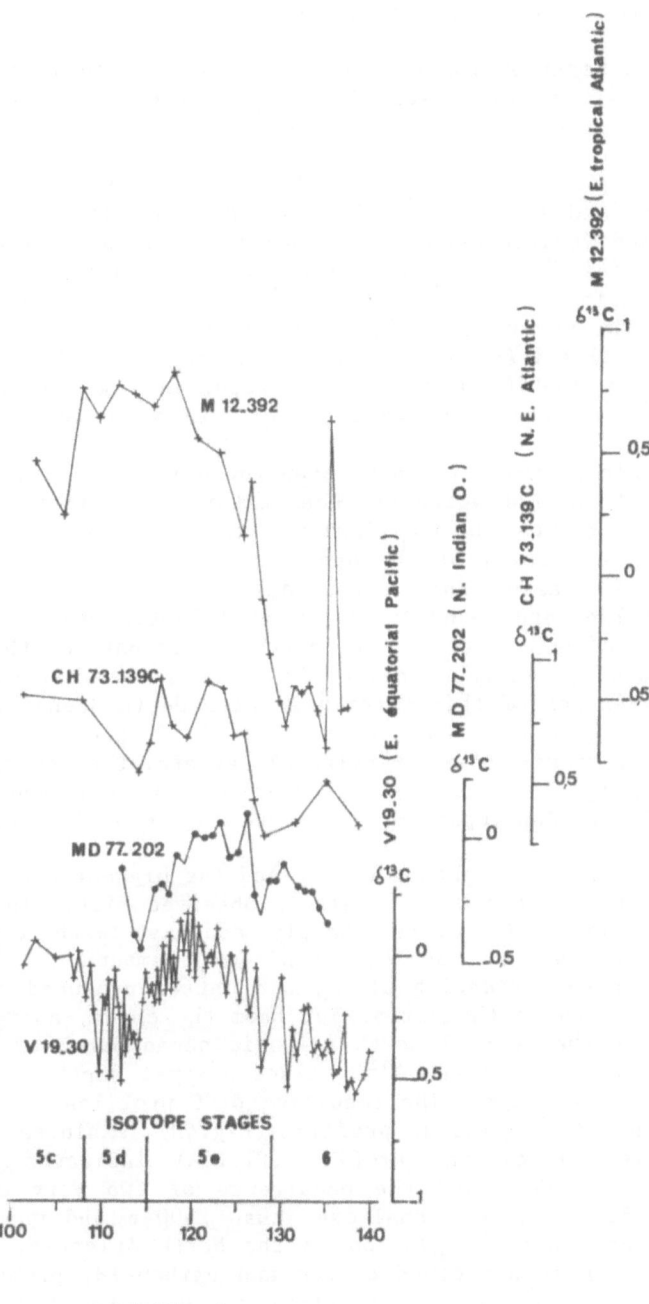

Figure 2B: Down core ¹³C record from 135 to 107 kyr in the same four cores as in fig 2A. Note the differences in the ¹³C temporal changes as opposed to the similarity of the ¹⁸O records.

5.2 Glacial to interglacial contrasts

The $\delta^{13}C$ distribution of the total CO_2 dissolved in the deep ocean during the last interglaciation (isotopic stage 5e, about 120000 years ago) has been reconstructed (Fig. 3a). This reconstruction shows that deep water sources were active both in the northern and southern hemispheres, since the heaviest isotopic values are found in the Norwegian sea and since the $\delta^{13}C$ values decrease from south to north in the Indian and Pacific deep waters. However, the production of Antarctic Bottom Water (AABW) was stronger than that of North Atlantic Deep Water (NADW), resulting in an east-west $\delta^{13}C$ gradient in the Atlantic Ocean higher than the modern one. The $\delta^{13}C$ value of the benthic foraminifera for each core location during the preceeding glacial maximum (isotopic stage 6, about 145000 years ago) is reported Fig. 3b and the glacial anomaly ($\delta^{13}C$ difference between the glacial value and the interglacial value) in Fig. 3c.

No benthic foraminifera have been found in Norwegian Sea cores. This indicates that the Norwegian Sea Deep Water was sufficiently depleted in oxygen and nutrients to drastically reduce the benthic productivity. This implies no renewal of the deep water and hence that no deep water formed in this basin. The isotopically heaviest $\delta^{13}C$ values and the smaller anomalies are found in the North Atlantic Ocean, which indicates the formation of deep water in the northernmost part of this basin, close to the limit of permanent sea-ice. However in the Western Atlantic, the southward extension of this water mass at a depth deeper than 3000 m was limited to the high latitudes, because low $\delta^{13}C$ values are found south of 50°N. By contrast with the interglacial pattern, the Western Atlantic was therefore mostly filled by water low in dissolved oxygen and carbon-13, most probably originating from the Antarctic Ocean, like AABW today.

The isotopically lightest values and the highest anomalies are found in the Eastern Atlantic Ocean. This is observed first along the coast of Africa. The low $\delta^{13}C$ values sharply contrast with the interglacial pattern and indicate the production of large amount of ^{13}C-depleted CO_2 resulting from the oxidation of organic matter produced by the enhanced upwelling activity in this area. Far from the coast, another area of low $\delta^{13}C$ values is the central north Atlantic ocean north of 40°N and below 3000 meters. We plotted the $\delta^{13}C$ values against depth for the 12 cores north of 20°N (Fig 4A,B). The resulting $\delta^{13}C$ profiles show very distinct features: the interglacial profile (Fig.4B) exhibits constant $\delta^{13}C$ values, while the glacial profile (Fig.4A) indicates a strong $\delta^{13}C$ minimum around 3200 m and the occurrence of two very different water masses: the first one was shallower than 2500 m and was oxygen rich - high $\delta^{13}C$ water which originated in the North Atlantic. A deeper water mass much poorer in dissolved oxygen and carbon-13, probably originated from the Western Atlantic Deep Water, crossing the Mid Atlantic Ridge through the equatorial fracture zones like today. Within this water mass, strong in situ consumption of oxygen resulted in a well-developped $\delta^{13}C$ minimum. In the Indian and Pacific oceans, the magnitude of the $\delta^{13}C$ anomalies is smaller than that of the Atlantic Ocean, suggesting that the stage 6 pattern of deep water circulation was not very different from the interglacial one.

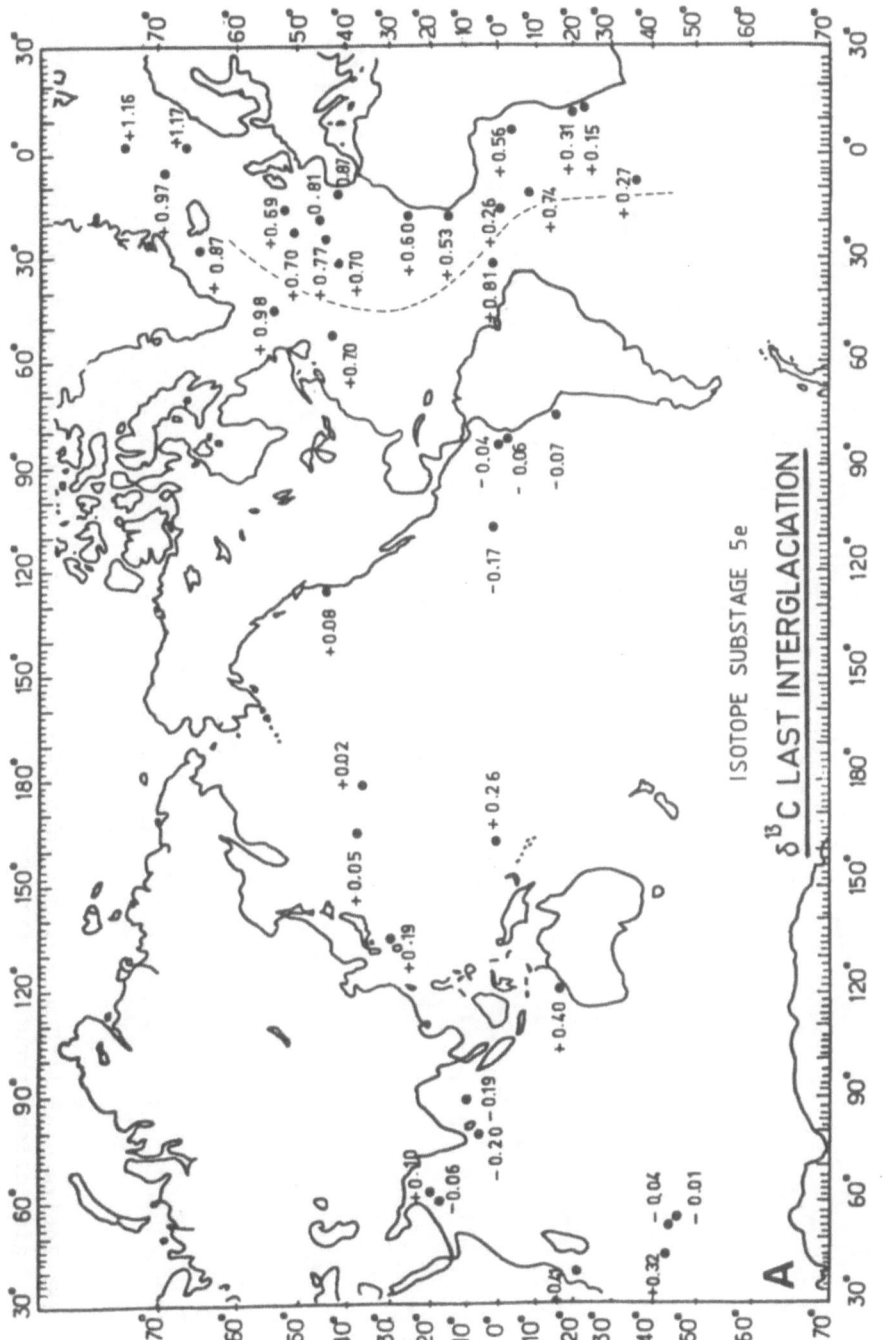

FIGURE 3A: Estimates of $\delta^{13}C$ values of total CO_2 in the world ocean deep water for the last interglaciation, about 120 Kyr ago.

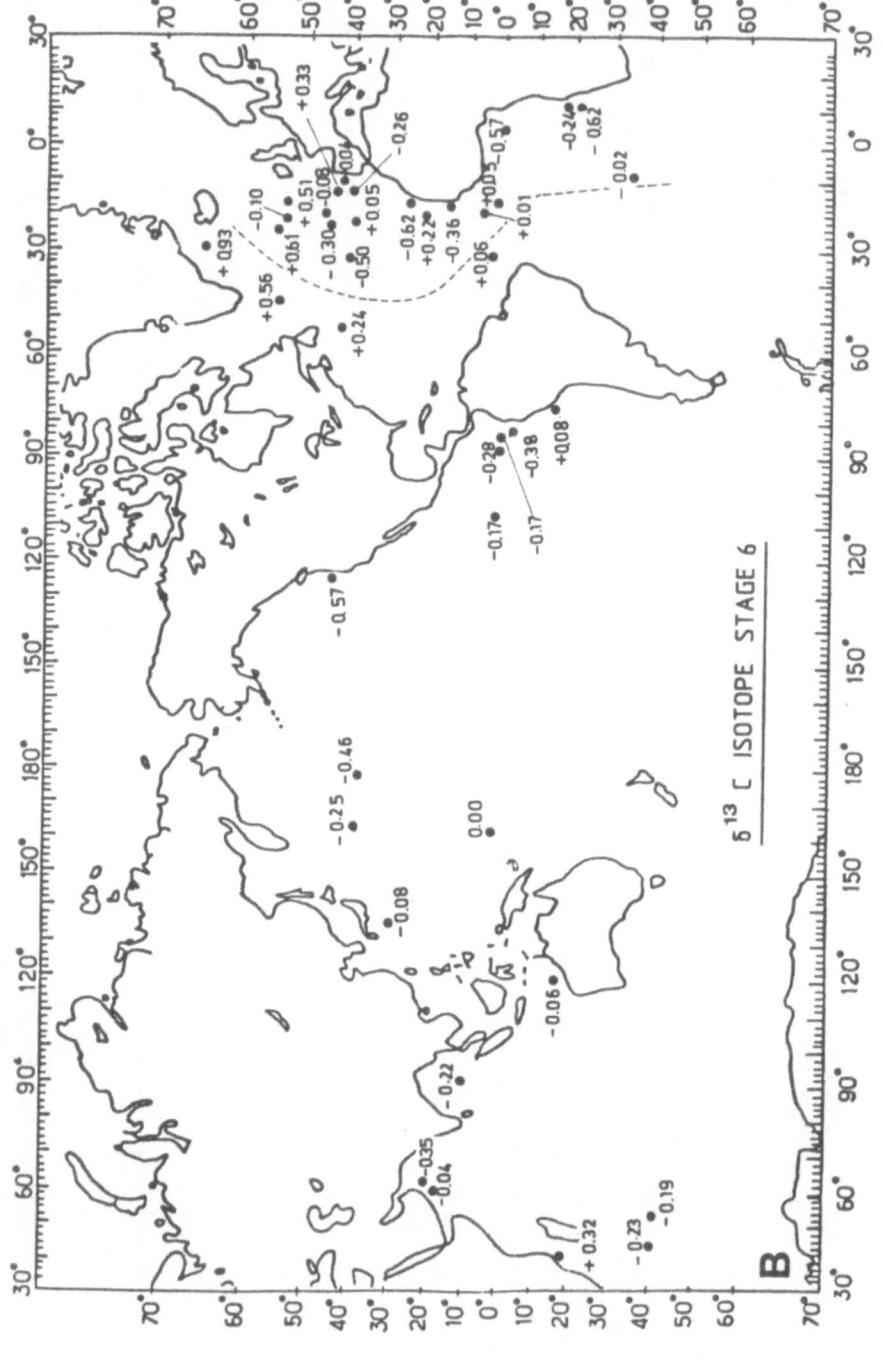

Figure 3B: Estimate of $\delta^{13}C$ values for total CO_2 in the world ocean deep water for the penultimate glaciation, about 135 Kyr ago.

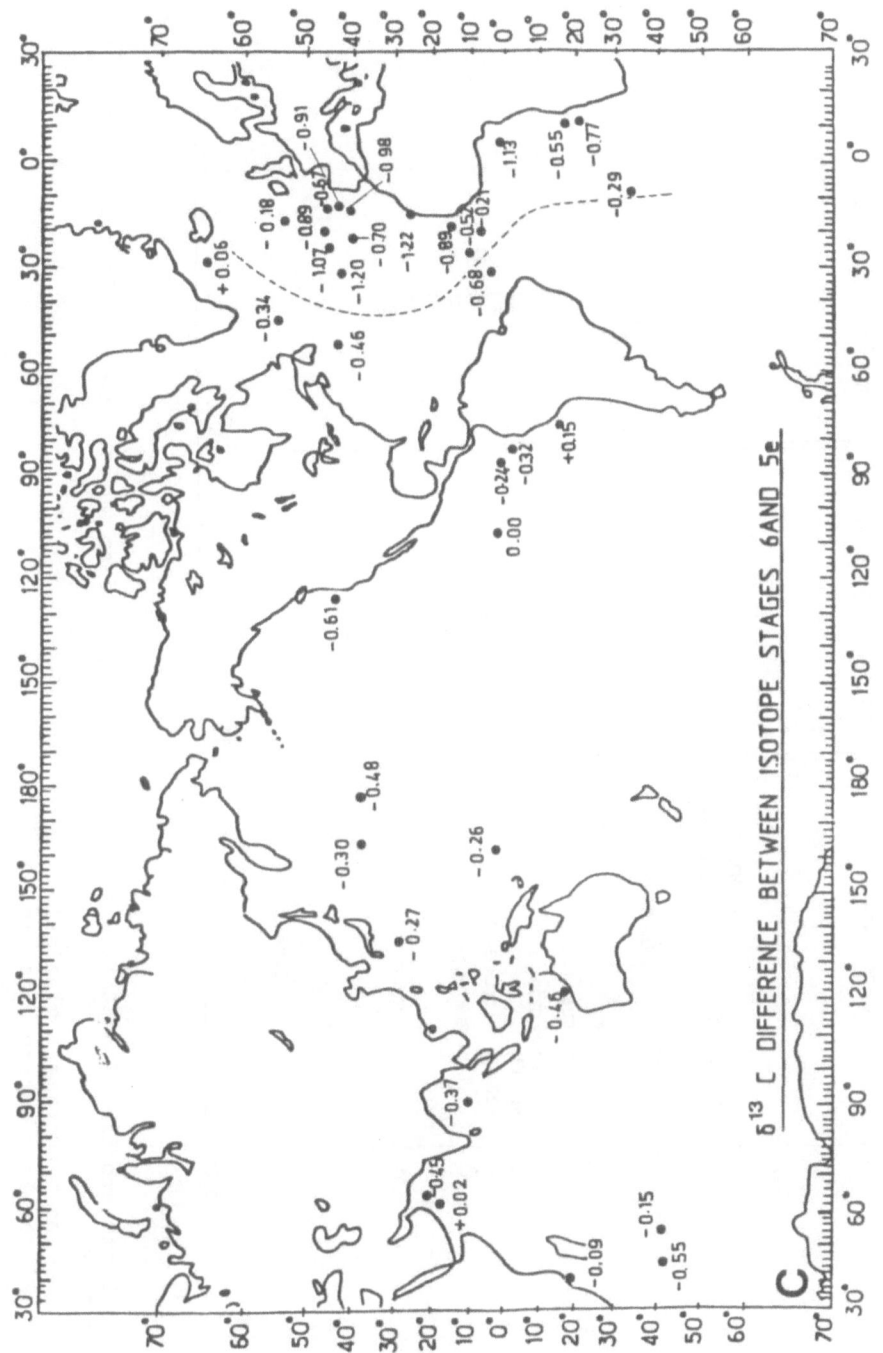

Figure 3C: δ¹³C difference between glacial and interglacial values at the same core site. Note the large anomaly in the eastern Atlantic.

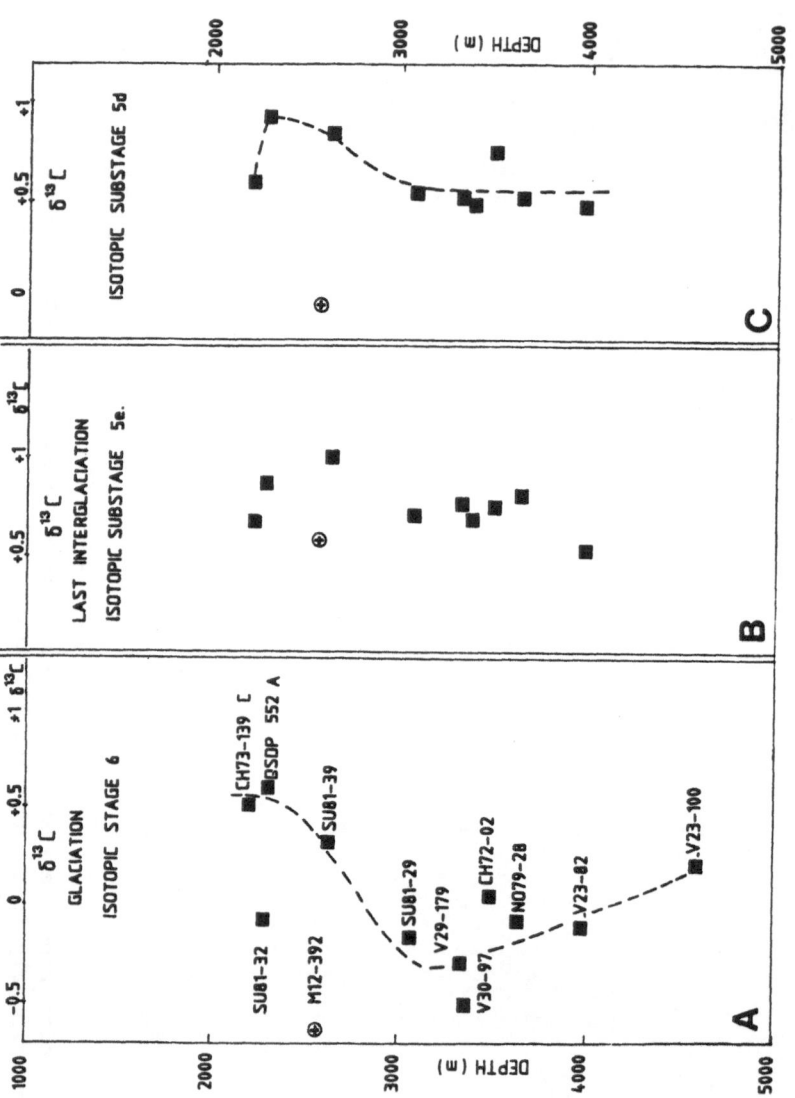

Figure 4: Depth profile of δ¹³C estimates for total CO_2 in deep waters from the Eastern Atlantic Ocean north of 20°N for:
A) the penultimate glaciation
B) the last interglaciation
C) the beginning of the glaciation (isotope substage 5d, about 110 Kyr ago).

In summary, the pattern of deep water circulation for glacial isotopic stage 6 is characterized by weaker fluxes than those of interglacial time and an important reduction of the amount of deep water formed in the northern hemisphere so that most oceanic basins were filled with water originating from the Southern Ocean.

5.3 Disappearance of NADW during the Glacial to Interglacial transition

To understand the impact of the deglaciation on the deep water circulation, Duplessy and Shackleton (1985) selected in the cores the level characterized by a $\delta^{18}O$ value, which would correspond to about 75% of total deglaciation. In many cores, no benthic foraminifera from that time have been found. This time was unfavourable for benthic productivity, because of either poor oxygenation of the bottom or reduced input of nutrients if the surface productivity was as low as that suggested by Ruddiman et al., (1980).

The $\delta^{13}C$ value of benthic calcite is reported in figure 5a. During the deglaciation, the northernmost core exhibits a very low $\delta^{13}C$ value, indicating a very low renewal of the deep water, as would be expected in response to the stratification due to the southward displacement of the limit of sea ice (Ruddiman et al., 1980). Further south, the $\delta^{13}C$ contrast between the northern and southern Atlantic Ocean has disappeared. The eastern Atlantic Ocean no longer shows isotopically light $\delta^{13}C$ values, indicating a sharp drop in the productivity of this basin. The comparison with the pattern for isotope stage 6 shows that the most dramatic $\delta^{13}C$ change during the deglaciation occurred in those areas where well oxygenated water was forming under glacial conditions. I interpret this as an indication that no deep water was forming in the Northern Atlantic Ocean 126000 yr ago and probably over the whole period of continental ice-melting. In the Indian and Pacific oceans, benthic foraminifera that lived during the deglaciation are also scarce. However, the measured $\delta^{13}C$ values for the benthic calcite are generally isotopically heavier than those of stage 6. As a $\delta^{13}C$ increase in the Pacific and Indian deep waters cannot occur without deep water circulation, these results demonstrate that some deep water formed in the Southern Ocean during the glacial to interglacial transition, when the North Atlantic source had disappeared.

5.4 Enhanced NADW formation during the inception of the glaciation

During the phase of rapid ice-growth which followed the last interglaciation, the subpolar North Atlantic Ocean maintained warm sea surface temperatures comparable to those of today's ocean (Ruddiman and Mc Intyre, 1979). This pattern constitutes an optimal configuration for forming deep water in the Northern Atlantic Ocean and in the Norwegian Sea. The ocean surface finally cooled late in isotopic substage 5d, several thousand years after the mid-point of the ice-growth phase. The $\delta^{13}C$ pattern of TCO₂ during the maximum of isotopic substage 5d, about 107000 years ago, has been reconstructed, because this level was more accurately recognized in the deep sea cores. At that time, a volume of ice that reached about 50% of the stage 2 full glacial maximum was

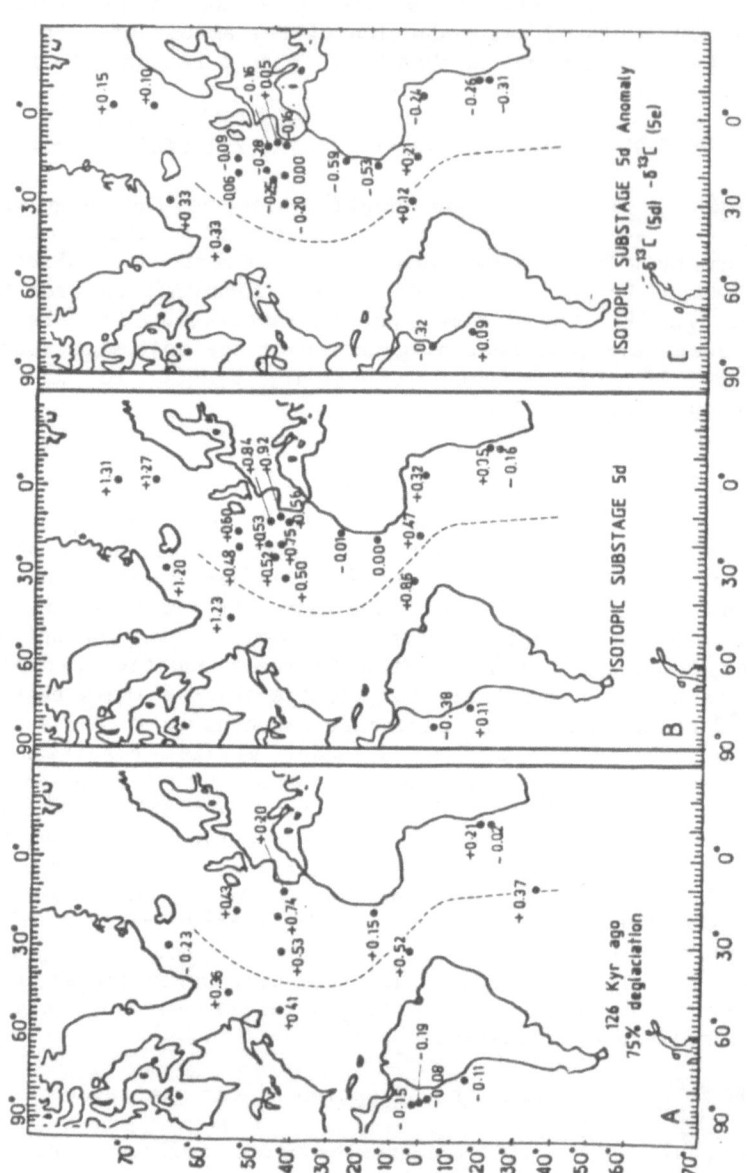

Figure 5: Estimates of $\delta^{13}C$ values of total CO_2 in ocean deep water for:
A) 75% of the penultimate deglaciation
B) the beginning of the glaciation (isotope substage 5d)
C) the 5d anomaly is calculated as $\delta^{13}C$ difference between 5d and interglacial values at the same core site.

deposited over Northern America and Northern Europe. The ocean deep water circulation remained probably similar to that which prevailed during the inception of the glaciation because the δ^{13}C records do not exhibit major changes between the end of isotopic substage 5e and the maximum of 5d (Figure 2).

The δ^{13}C values of benthic calcite during isotopic substage 5d are plotted in Figure 5b and the 5d anomaly (δ^{13}C difference between substage 5d and 5e) in Figure 5c. Isotopically heavy δ^{13}C values are found in the Norwegian Sea and in the Northwestern Atlantic Ocean. A strong positive anomaly in these basins indicates an enhanced flux of the Norwegian Sea Overflow Water into the North Atlantic Ocean. The high production of NADW may be related to the waxing of continental ice sheets that resulted in an increase of the world ocean salinity. This salinity increase was first experienced in surface waters, which must have received less runoff water during a period of strong evaporation (Ruddiman and Mc Intyre, 1979). This process increased the surface density, reduced the density difference between surface and bottom waters and was therefore favorable to deep water formation in the high northern latitudes. A western boundary current stronger than today prevailed in the North Atlantic Ocean and maintained a δ^{13}C difference between the Eastern and the Western basins much larger than the modern one. Low δ^{13}C values and large negative anomalies along the coast of Africa indicate that upwellings were more active than during full interglacial conditions. In the Eastern Atlantic, the surface productivity and the in situ consumption of organic matter at depth were higher than today, resulting in the large δ^{13}C anomalies shown in Figure 5c. By contrast with the pattern of stage 6 however, constant δ^{13}C values were observed between 2200 m and 4000 m, and no strong δ^{13}C minimum developed far from the coast (Fig. 4c).

Cores on the path of the Mediterranean Sea Overflow Water exhibit a positive δ^{13}C anomaly (Fig. 4c). The Mediterranean Sea is the most important concentration basin of the world ocean. Its salinity is high, its productivity is low and the residence time of the deep water is shorter than one century. Consequently, δ^{13}C values of TCO$_2$ are high and similar to those of the Norwegian Sea (Duplessy, 1972) and, today, the Mediterranean Overflow Water serves to increase both the δ^{13}C value and the salinity of the North Atlantic Ocean (Worthington, 1976). We therefore interpret the isotopically heavy δ^{13}C values measured in these cores during isotopic substage 5d as the indication of a strong outflow of Mediterranean water into the Atlantic Ocean. Such a pattern, which provides the North Atlantic Drift with the high salinity required to form deep waters at high latitudes (Reid, 1980), would also serve to increase the production of NADW.

6. CONCLUSION

The variations of the CO$_2$ cycle linked to the Earth's climatic changes are the set of many different processes: decreasing biospheric size over the continents under glacial conditions, deposition of some sediment over the continental shelf during transgression of the sea, phosphate concentration changes in the different oceanic basins. A simple two box model is not sufficient to simulate the reconstructed pCO_2 variations. Large variations of the deep water circulation have been discovered and stronger upwelling has been found along the continental margin of many continents. These variations should be quantified to determine if they may account for the part of the pCO_2 drop which is not explained by simple ocean chemistry changes.

ACKNOWLEDGEMENTS

I wish to thank the participants at the ASI who contributed comments and suggestions to this manuscript, especially C Jeandel, C. Lalou, J. Merril, U. Siegenthaler and A. Spitzy. Thanks are due to CNRS and CEA for financial support.This is CFR contribution n°722.

REFERENCES

Berner, W., H. Oeschger and B. Stauffer, 1980: 'Information on the CO_2 cycle from ice core studies.' Radiocarbon, 22, 2, 227-235.
Boyle, E.A. and L.D. Keigwin, 1985: 'Comparison of Atlantic and Pacific paleochemical records for the last 250,000 years: changes in deep ocean circulation and chemical inventories'. Earth Planet. Sci. Lett., 76, 135-150.
Broecker, W.S., 1982: 'Ocean chemistry during glacial time'. Geochim. Cosmochim. Acta, 46, 1689-1705.
Broecker, W.S. and T.H. Peng, 1982: 'Tracers in the sea'. Eldigio Press. 690 pp.
CLIMAP project members, 1976: 'The Surface of the Ice-Age Earth'. Science, 191, 1131-1135.
Delmas, R.J., J.M. Ascencio and M. Legrand, 1980: 'Polar ice evidence that atmospheric CO_2 20,000 yr BP was 50% of present'. Nature, 284, 155-157.
Duplessy, J.C., 1972: 'La géochimie des isotopes stables du carbone dans la mer'. Note CEA N-1565, CEN Saclay, 197 pp.
Duplessy, J.C. and N.J. Shackleton, 1984: Carbon-13 in the World Ocean during the last interglaciation and the penultimate glacial maximum. Reevaluation of the possible biosphere response to the earth's climatic changes.' Progress in Biometeorology, 3, 48-54.
Duplessy, J.C. and N.J. Shackleton, 1985: 'Response of global deep-water circulation to Earth's climatic change 135,000-107,000 years ago.' Nature, 316, 500-507.
Duplessy, J.C., N.J. Shackleton, R.K. Matthews, W. Prell, W.F. Ruddiman, M. Caralp and C.H. Hendy, 1984: '[13]C record of benthic foraminifera in the last interglacial ocean: Implications for the carbon cycle and the global deep water circulation.' Quat. Res., 21 , 225-243.
Gate, W.L., 1976: 'Modeling the Ice-Age Climate' Science, 191, 1136-1144.
Neftel, A., H. Oeschger, J. Schwander, B. Stauffer and R. Zumbrunn, 1982: 'Ice core sample measurements give atmospheric CO_2 content during the past 40,000 years'. Nature, 295, 220-223.
Raynaud, D. and B. Lebel, 1979: 'Total gas content in polar ice: rheological and climatic implications.' In: International Association of Hydrological Sciences, Pub. n° 118 (General Assembly of Grenoble, 1975 -Isotopes and Impurities in snow and ice), 326-335

Reid, J.L., 1979: 'On the contribution of the Mediterranean sea outflow
    to the Norwegian-Greenland Sea'. Deep Sea Res., 26, 1199-1223.

Ruddiman, W.F. and A. McIntyre, 1979: 'Warmth of the Subpolar North
    Atlantic Ocean during Northern Hemisphere Ice-Sheet Growth'. Science
    204, 173-175.

Ruddiman, W.F., B. Molfino, A. Esmay and E. Pokras, 1980: 'Evidence
    bearing on the mechanism of rapid deglaciation'. Climatic Change, 3,
    65-87.

Shackleton, N.J., 1977: 'Tropical rain forest history and the equatorial
    Pacific carbonate dissolution cycles'. In: The fate of fossil fuel
    CO₂ in the oceans, (N.R. Anderson and A. Malahoff, eds.) Plenum, New
    York, 401-428.

Shackleton, N.J., J. Imbrie and M.A. Hall, 1983: 'Oxygen and carbon
    isotope record of East Pacific core V19-30: implications for the
    formation of deep water in the late Pleistocene North Atlantic.'
    Earth. Planet. Sc. Lett., 65, 233-244.

Siegenthaler, U., 1986: 'Carbon dioxide: its natural cycle and
    anthropogenic perturbation'. This volume.

Siegenthaler, U. and Th. Wenk, 1984: 'Rapid atmospheric CO₂ variations
    and ocean circulation'. Nature, 308, 624-626.

Stauffer, B., H. Hofer, H. Oeschger, J. Schwanfer and U. Siegenthaler,
    1984: 'Atmospheric CO₂ concentrations during the last glaciation'.
    Annals of Glaciology, 5, 160-164

Worthington, L.V., 1976: 'On the North Atlantic circulation' John Hopkins
    Press, Baltimore/London

EXCHANGE OF CO AND H_2 BETWEEN OCEAN AND ATMOSPHERE

Ralf Conrad and Wolfgang Seiler
Max-Planck-Institute for Chemistry
Saarstraße 23
D-6500 Mainz
Fed.Rep.Germany

1. INTRODUCTION

The surface water of the oceans is generally supersaturated with respect to atmospheric CO mixing ratios (Swinnerton and Lamontagne, 1974; Swinnerton et al., 1970; Seiler and Junge, 1970; Seiler and Schmidt, 1974; Seiler, 1978; Conrad et al., 1982). Oceans thus represent a net source in the atmospheric CO budget. With respect to atmospheric H_2 it is less clear, whether oceans represent a net source or not. In tropical, temperate and antarctic regions the ocean surface water seems to be supersaturated with respect to atmospheric H_2 (Seiler and Schmidt, 1974; Herr and Barger, 1978; Setser et al., 1982; Scranton et al., 1982; Williams and Bainbridge, 1973). In arctic regions, on the other hand, the surface water is slightly undersaturated (Herr et al., 1981; Herr, 1984).

In order to be able to calculate quantitatively the total net flux of CO or H_2 between oceans and atmosphere, it is necessary to evaluate data from individual determinations of fluxes conducted at a few stations and during a limited time period. Fluxes are generally determined from the saturation values in the surface water by applying the laminar film model (Liss and Slater, 1974; Broecker and Peng, 1974). However, saturation is influenced by transport, production and destruction processes, which by themselves are influenced by various parameters, such as wind speed, light intensity, microbial activity, etc. To arrive at reliable data of fluxes on a global basis the following points should be considered:

1. Reliability of the measurements of dissolved H_2 or CO in surface water and of the supersaturation factor.
2. Assessment of spatial and temporal changes of H_2 or CO concentrations.
3. Assessment and quantification of processes responsible for changes in concentrations as a function of environmental parameters.
4. Applicability of the laminar film model.

P. Buat-Ménard (ed.), The Role of Air-Sea Exchange in Geochemical Cycling, 269–282.

CO and H_2 are trace gases which exhibit a relatively strong temporal and spatial variability of their concentrations in the water. In the following we will present and discuss data which were obtained during ship cruises in the Atlantic Ocean (see Conrad et al., 1982).

2. DETERMINATION OF THE SUPERSATURATION FACTORS OF CO AND H_2

The supersaturation factor (s) is given by

$$s = C_W / (H\ C_a) \tag{1}$$

(where C_W = concentration of gas dissolved in surface water; H = Henry's law constant of gas as function of water temperature and salinity; C_a = concentration of gas in atmosphere).

CO and H_2 concentrations (C_a) in the atmosphere were monitored by using continuously working analyzers based on the HgO-to-Hg vapor conversion technique (Seiler, 1978; Seiler et al., 1980). CO and H_2 concentrations in water samples were measured by using an extraction technique (Conrad and Seiler, 1982). Water samples were transferred into glass vessels which were filled completely, poisoned with metabolic inhibitors (e.g. $HgCl_2$) and processed as fast as possible in order to minimize changes by microbial or chemical reactions. Part of the water sample was replaced by CO and H_2-free air and the dissolved CO and H_2 extracted into the headspace. The mixing ratios of CO and H_2 in the headspace were analyzed by a discontinuously operating analyzer (Seiler, 1978; Seiler et al., 1980) and used to calculate the dissolved CO and H_2. Since the solubility of CO and H_2 in water is very low, more than 97% of the dissolved gas is extracted into the headspace. The remainder can be calculated from the Henry's law constant (H) at the time of extraction:

$$C_W = C\ (V_g / V_W) + H\ C \tag{2}$$

(C = gas concentration in headspace; Vg = volume of gaseous headspace; V_W = volume of water).

For calculating supersaturation factors the value of H must also be known for the in-situ temperature and salinity. Knowledge of H is not necessary, however, if the equilibrium concentration (E) of the dissolved CO or H_2 is measured directly. Since

$$E = C_W / H \tag{3}$$

the supersaturation factor is given by

$$s = E / C_a \tag{4}$$

The measurement of the equilibrium value (E) is in particular of advantage, if supersaturation is relatively small and the Henry's law constant (H) is large. In this case, inaccuracies due to the value of H that has to be determined by independent laboratory experiments are avoided.

Equilibrium concentrations (E) were determined by a continuously operating extraction technique which has been described by Conrad et al. (1982). The schematic of the apparatus is shown in Fig. 1. This

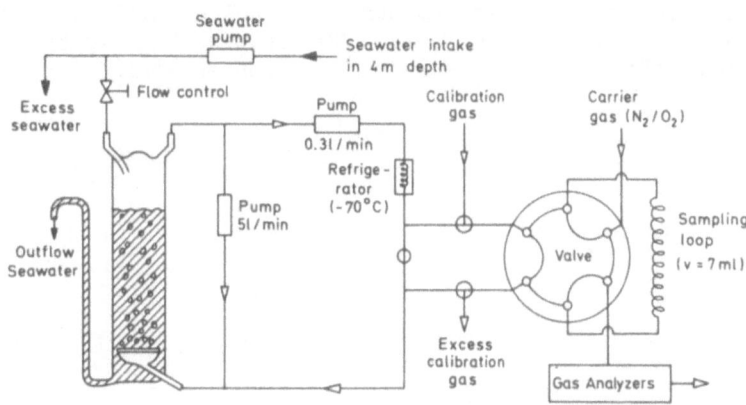

Figure 1. Apparatus for the continuous equilibration and analysis of CO dissolved in surface sea water. After Conrad et al.(1982).

technique allows the frequent analysis of dissolved gases in water which is sampled by a seawater pump. In our case, the seawater intake was in approx. 4 m water depth. Values of dissolved CO obtained by the continuous extraction technique were comparable to those obtained by extraction of distinct water samples taken from the water surface (10-100 cm) using hydrocasts or sampling by hand from a rubber raft.

For H₂, however, the continuous extraction technique could not be applied since the sampled water was contaminated with H₂ produced by water electrolysis due to the electrodes mounted at the ship's wall to prevent corrosion.

Inaccuracies due to contamination or concentration changes by reactions between time of sampling and analysis present a serious problem which is difficult to be assessed. Contaminations by electrolytical H₂ is prevented by plastic and glass materials for sampling and processing. Checks for the inertness of materials is easily done in the laboratory. Concentration changes by microbial or chemical reactions, however, may be a function of the water quality, e.g. the

microbial activity. With respect to CO, it is necessary to utilize
glass materials which are thouroughly cleaned from organic residues
using concentrated chromic sulfuric acid, to keep water samples in
absolute darkness and to inactivate metabolic activities by adding
HgCl$_2$. Applying these precautions, CO concentratiohs in water samples
do not change during storage times of several hours.

3.   SPATIAL AND TEMPORAL CHANGES OF DISSOLVED CO AND H$_2$

    The CO concentrations in surface water show a high degree of
spatial and temporal variability and thus make it difficult to
extrapolate individual measurements to global conditions. Fig. 2 shows
the zonal distribution of dissolved CO between 35° S to 55° N obtained
during three ship expeditions made in autumn or spring. The individual

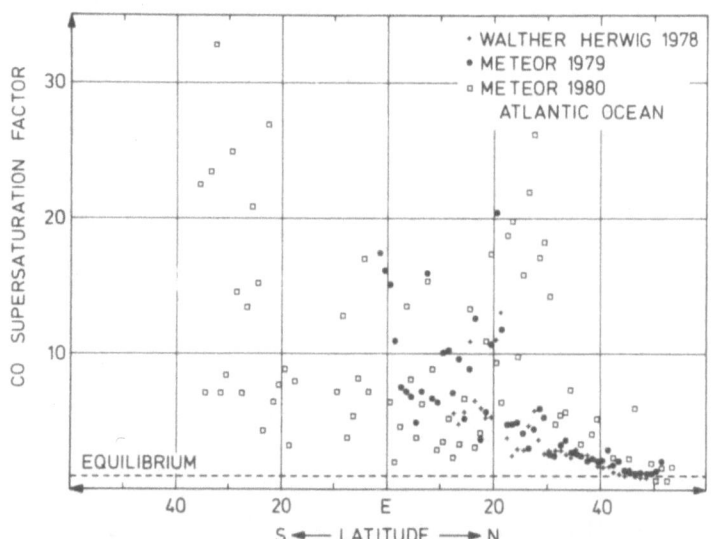

Figure 2.   Latitudinal distribution of dissolved CO in the
surface water of the Atlantic Ocean. Dissolved CO is given
as saturation factor relative to atmospheric CO. A factor of
one indicates equilibrium with atmospheric CO. After Conrad
et al.(1982).

data points are mean values averaged over 1° latitude. The large
scatter of the data is due to the fact that CO concentrations are a
function of different parameters which change with the water quality
and with time. Fig. 3 shows vertical profiles of dissolved CO which
were determined at different daytimes. The data demonstrate that CO
changes with daytime in the upper 80 m of the water column (the
euphotic layer), but stays at a relatively constant and low level in
the deeper water. Fig. 4 shows the equilibrium values of CO dissolved

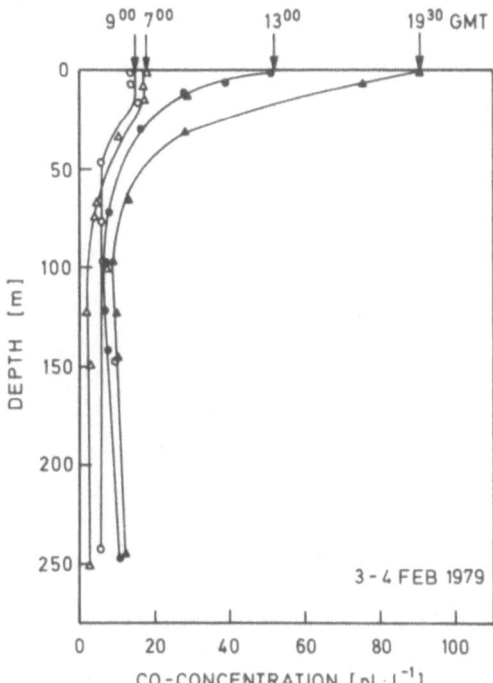

Figure 3. Vertical distribution of dissolved CO in the tropical Atlantic Ocean at different daytimes. After Conrad and Seiler (1985).

in surface water which were obtained by the continuously operating extraction technique during a ship cruise in the tropical Atlantic Ocean. The data clearly show that CO concentrations in surface water follow a diel rhythm with maxima in the afternoon and minima in the early morning. However, the data also show that other factors in addition to daytime influence the magnitude of the CO concentration. It is clear that a reliable extrapolation of the data to global conditions can only be done if the processes and factors influencing CO concentrations are analyzed.

The H₂ concentrations in surface water show a smaller scatter compared to CO. However, there are marked differences in the degree of supersaturation or undersaturation in different regions of the oceans (Scranton et al., 1982; Seiler and Schmidt, 1974; Herr, 1984). Fig. 5 shows the data of vertical profiles of dissolved H₂ analyzed at stations in the tropical Atlantic Ocean. Similar to CO, H₂ concentrations are relatively high within the euphotic layer and approach equilibrium or slight undersaturation to atmospheric mixing ratios in

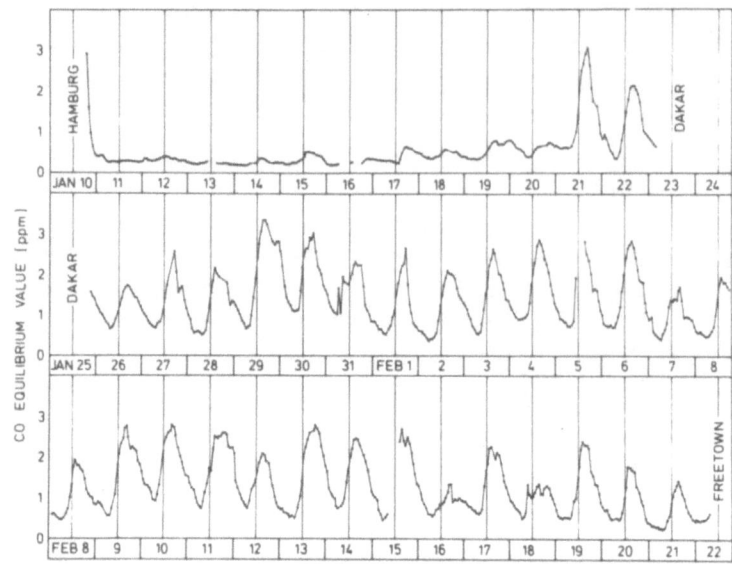

Figure 4. Continuous record of CO concentrations in surface
water as measured by the equilibrium technique during a
cruise with FS Meteor in 1979. After Conrad and Seiler
(1985).

deeper water. Similar profiles were observed by Seiler and Schmidt
(1974), Herr and Barger (1978) and Lilley et al. (1982). In contrast
to CO, however, H_2 concentrations in surface water do not exhibit a
general diel rhythm. Although diurnal changes with maxima in the
afternoon have been observed at some stations, they have not been
observed at others (Bullister et al., 1982; Herr et al., 1984; Setser
et al., 1982; own observations). It seem obvious that H_2 concentrations are affected by processes which are different from those
affecting CO concentrations.

4. PROCESSES SUSTAINING CO AND H_2 CONCENTRATIONS IN SURFACE WATER

The gas concentrations in surface water are sustained and change
due to the following processes: (1) transport into atmosphere, (2)
transport within the ocean, (3) consumption and (4) production.

4.1. Production processes

CO concentrations in surface water change in a diel rhythm
parallel to light intensity and are correlated to phytoplankton,
particulate carbon, primary productivity and chlorophyll a (Conrad and
Seiler, 1980a; Conrad et al., 1982; Swinnerton et al., 1977; Bullister
et al., 1982). CO concentrations in surface water were found to be
proportional to the daily mean light intensity (Conrad et al., 1982).

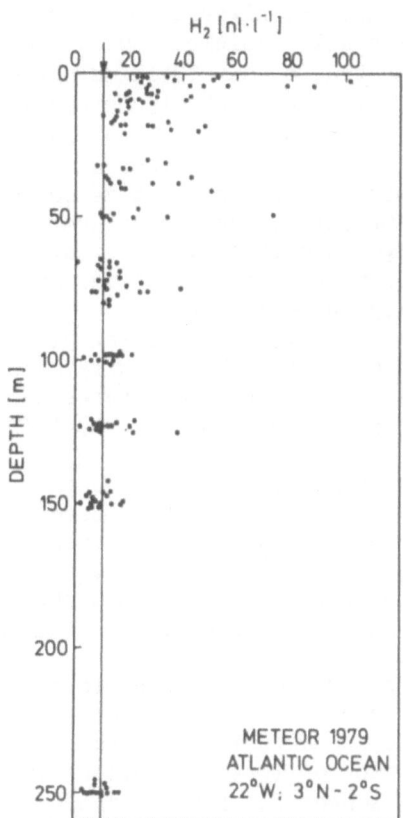

Figure 5. Vertical distribution of dissolved H_2 at different stations in the tropical Atlantic Ocean ($22°W$, $2°S$ -
$4°N$; Jan/Feb 1979). The arrow indicates equilibrium with
atmospheric H_2. After Conrad and Seiler (1980b).

The linear relationship between CO concentration and light intensity
is the reason for the exponential decrease of dissolved CO with depth
within the euphotic water layer (Fig. 3). Hence, it has generally been
believed that CO is produced by photometabolism of phytoplankton.
Indeed, CO is a metabolic product of the synthesis of phycobilins from
heme precursors in algae (Troxler and Dokos, 1973) and it is most
likely also produced during any oxidative degradation of porphyrin-
rings (e.g. chlorophyll). Bauer et al. (1980) demonstrated, on the
other hand, that degradation of chlorophyll could not account for the
CO production rates observed in phototrophic bacteria and green algae.
Their results rather indicated that CO is produced by chemical
photo-oxidation of cell material. Instead of photometabolic CO
production, photooxidative CO production from cell material could as
well explain the observed correlations between CO concentrations in
ocean surface water and biomass parameters such as chlorophyll (Conrad
et al., 1982). In experiments in which unfiltered and filtered

seawater was incubated in the light the highest CO production rates
were observed in water from which all particles > 0.2 μm had been
removed. This experiment demonstrates that CO is mainly produced by
photooxidation of dissolved material and that microorganisms and/or
plankton which generally is larger than 0.2 μm do not enhance CO
production but rather stimulate CO consumption (Conrad et al., 1982).
A likely soluble organic precursor for photooxidative CO formation is
"Gelbstoff" (Ehrhardt, 1984). "Gelbstoff" resembles humic material and
is most probably formed from algal excretions. These materials may
initiate photochemical oxidation reactions (Baxter and Carey, 1983;
Haag et al., 1984) and eventually CO formation. For review of photooxidative processes see Zafiriou et al.(1984) and Zafiriou (this
volume). Photooxidative CO production from "Gelbstoff" materials would
also be consistent with the observed correlations between dissolved CO
and chlorophyll etc. as biomass parameter (Swinnerton, 1977; Conrad et
al., 1982).

Compared to CO, the origin of H_2 dissolved in ocean water is
still unclear. This is mainly due to the fact that observations
obtained in a particular water body often can not be reproduced in
other water bodies. The same is the case with experiments in which
water is incubated to test for H_2 production or consumption activities
(e.g. Bullister et al., 1982; own observations). Presently there are
three major hypotheses being discussed to explain H_2 supersaturation
in ocean surface water: (1) H_2 production by anaerobic bacteria
present in detritus particles and protozoa (Lilley et al., 1982);
(2) H_2 production by N_2-fixing cyanobacteria (Scranton, 1983; 1984);
(3) H_2 production by photooxidation of longchain aldehydes (Herr et
al., 1984). There is definitely a need for further measurements and
experiments under in-situ conditions to elucidate the origin of H_2 in
ocean surface water.

4.2. Consumption processes

CO consumption in ocean surface water was demonstrated by
incubation experiments. It was shown that this activity is due to
microorganisms which are able to consume CO at the low concentrations
present in ocean water (Conrad and Seiler, 1980a; 1982; Conrad et al.,
1982). At the moment, it is unclear which species of microorganisms
are responsible for CO consumption and it is also unclear whether
these microorganisms are bacteria or small eukaryotes in the size
range of 0.2 to 3.0 μm. The ammonium-oxidizing Nitrosomonas species
are possible candidates; they are able to co-oxidize CO and exhibit a
relatively high affinity to this gas (Jones and Morita, 1983).
Experiments with inhibitors specific for ammonium oxidizers indicate,
however, that CO is in addition oxidized by other bacterial groups
which have not yet been characterized (Jones et al., 1984; Conrad and
Seiler, 1982).

Similar to H_2 production, H_2 consumption reactions in ocean water
have not yet been characterized. During two ship expeditions in the
Atlantic Ocean we sometimes have observed H_2 consumption activity but
in most experiments we did not see any change at all in H_2 concentra-

tions during incubation periods of 1 day. In freshwater lakes, on the other hand, H_2 consumption was observed in all water layers (Conrad et al., 1983) and thus, it is very likely that H_2 consumption in ocean water is just a matter of numbers of H_2-oxidizing bacteria. Existence of microbial H_2 consumption in ocean water has been postulated since H_2 production was not balanced by H_2 loss into the atmosphere (Scranton, 1984). In other cases, however, the existence of microbial H_2 consumption was not necessary as H_2 production was balanced just by loss into atmosphere and distribution in the water column (Herr et al., 1984). Microbial H_2 consumption has also been postulated for the arctic ocean water which is apparently undersaturated with respect to atmospheric H_2 (Herr et al., 1981; Herr, 1984). The observed fluxes of H_2 from the atmosphere into the ocean require the presence of a highly efficient microbial population activity throughout the mixed ocean water body (Conrad, 1984).

4.3. Transport processes

Transport of gases from the surface water into the atmosphere is one of the major loss reactions. The transport is a function of the extent of supersaturation in the surface water and of the transfer velocity. The latter increases linearly with wind speed at low velocities and increases to a greater extent at higher wind speeds with the onset of breaking waves (Deacon, 1977; Jähne et al., 1979; Merlivat and Memery, 1983). Fig. 6 shows the data of CO saturation of

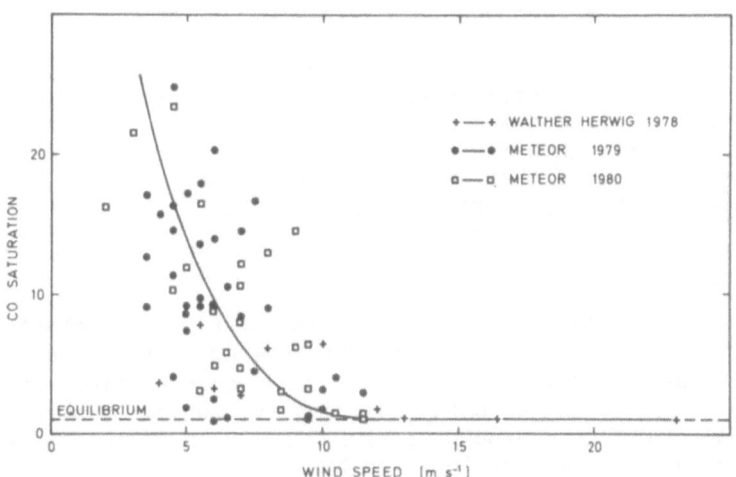

Figure 6. Daily average CO saturation as a function of the daily average wind speed.

surface water as function of wind speed which were obtained during 3 ship cruises in the Atlantic Ocean. The CO concentrations decreased with increasing wind speed and reached values in equilibrium with the atmospheric CO mixing ratio at wind speeds > 13 m s^{-1}. The relatively large scatter of the individual data points is due to the fact that the measurements were carried out in different regions of the Atlantic Ocean and at different times when other factors such as light intensity, biomass content, microbial activity affected the CO concentration as well. The data demonstrate, however, that the CO concentration in surface water is strongly dependent on the transfer velocity between water and air.

Vertical transport of CO within the mixed euphotic water column of approx. 100 m seems to be of minor importance. Vertical profiles of dissolved CO at different daytimes show that CO production occurs as function of light intensity within the entire euphotic zone and that CO consumption in the individual depths is of greater importance than vertical transport (Conrad et al., 1982). In the top 1 m water layer, on the other hand, vertical gradients due to vertical change of light intensity were not observed. Apparently, vertical mixing in this shallow water layer is strong enough to prevent the establishment of a vertical gradient (Conrad et al., 1982).

Very little is known on the relative importance of transport for the concentration of H$_2$ in the water and between water and air. Vertical transport of H$_2$ seems to balance H$_2$ production and to explain the decline of dissolved H$_2$ in those cases when a diurnal change of H$_2$ concentrations has been observed (Herr et al., 1984).

5.   CALCULATION OF FLUXES BY THE "LAMINAR FILM MODEL"

Gas fluxes between ocean and atmosphere are generally determined by using the "Laminar Film Model" (Broecker and Peng, 1974; Liss and Slater, 1974). This model assumes that the flux is limited by molecular diffusion through a laminar boundary layer at the water surface. It is further assumed that the air above and the water below the layer are well mixed. The flux (F) is then given by

$$F = (D / z) \ H \ (E - C_a) \qquad (5)$$

(where D = molecular diffusivity in water; z = thickness of the laminar film; H = Henry's law constant; E = equilibrium value of the dissolved gas (equivalent to C_w / H); C_a = concentration in the air).
In contrast to the transfer velocity (K_w) which is different for each gas species, the hypothetical laminar film thickness (z) is assumed to be the same and is given by

$$z = D / K_w \qquad (6)$$

Hence, the model allows the application of the transfer velocity (K_w) measured for a particular gas species to other gas species by correcting for their molecular diffusivities in water.

There are two major errors which are possibly made by determining fluxes from equations (5) and (6): (1) The equilibrium value E is generally determined in a bulk water sample taken by hydrocasts or water pumps from the upper 1-4 m of the water column. In case of CO or H_2, but possibly also for other gases, there are often vertical concentration gradients with increasing values towards the water surface. It is very questionable whether the concentration measured in bulk water samples really represents the concentration just below the laminar film. (2) The actual value of the thickness (z) of the laminar film is not known. In-situ measurements of transfer velocities from which film thickness can be calculated usually use the measurement of radon deficiencies. It is unclear whether the film thickness determined for the flux of a noble gas may also be applied to the flux of a chemically and biologically reactive gas, such as CO or H_2. Although the laminar film thickness is a purely hypothetical entity, there do exist films on the water surface with particular physical, chemical and biological properties (Wangersky, 1976; Norkrans, 1980; Lion and Leckie, 1981). These actually existing films may affect the transport

Table 1. Budgets of atmospheric CO and H_2 given in Tg
(10^{12}g) per year [a]

	CO	H_2
Sources:		
Oceans	100 ± 90	4 ± 2
Vegetation	75 ± 25	< 0.1
Soils	17 ± 15	3 ± 2
Technological Sources	640 ± 200	20 ± 10
Biomass burning	1000 ± 600	20 ± 10
Oxidation of CH_4	600 ± 300	10 ± 4
Oxidation of NMHC	900 ± 500	18 ± 7
Total sources	3332 ± 1700	75 ± 35
Sinks:		
Soil	390 ± 140	90 ± 20
Oxidation by OH	2000 ± 600	8 ± 3
Stratosphere	110 ± 30	< 0.1
Total sinks	2500 ± 770	98 ± 23

[a] Data taken from Seiler and Conrad (1986)

of particular gas species in a way which is more complex as suggested
by the "Laminar Film Model". Therefore, it is unclear whether the
usually applied film thickness of 30 ± 20 µm (Peng et al., 1979)
reflects reality for gases which might have increased production or
consumption rates in the existing surface layer.

6.   ROLE OF OCEANS IN THE BUDGET OF ATMOSPHERIC CO AND H_2

Oceans are a net source for both, atmospheric CO and H_2. The
source strength for CO has recently been estimated to 10 - 180 Tg
yr^{-1} by Conrad et al. (1982). The oceanic H_2 production of 4 ± 2 Tg
yr^{-1} has been estimated by Seiler and Schmidt (1974). Both data sets
have a rather large degree of uncertainty because of the large range
of supersaturations observed and of the multitude of factors influencing the flux. Table 1 compares the oceanic source strengths of CO
and H_2 with other sources and sinks of the budgets of CO and H_2. The
oceans represent only 3% of the total CO sources, but 52% of the
biogenic sources. In case of H_2, they represent 5% of the total and
57% of the biogenic sources.

7.   REFERENCES

Bauer, K., R. Conrad and W. Seiler, 1980: Photooxidative production of
    carbon monoxide by phototrophic microorganisms, Biochim. Biophys.
    Acta 589, 46-55.
Baxter, R.M. and J.H. Carey, 1983: Evidence for photochemical genera-
    tion of superoxide ion in humic water, Nature 306, 575-576.
Broecker, W.S. and T.H. Peng, 1974: Gas exchange rates between air and
    sea, Tellus 26 , 21-35.
Bullister, J.L., N.L. Guinasso Jr. and D.R. Schink, 1982: Dissolved
    hydrogen, carbon monoxide, and methane at the CEPEX Site,
    J. Geophys. Res. 87, 2022-2034.
Conrad, R., 1984: Capacity of aerobic microorganisms to utilize and
    grow on atmospheric trace gases (H_2, CO, CH_4), In: "Current
    Perspectives in Microbial Ecology" (M.J. Klug & C.A.Reddy, eds),
    American Society for Microbiology, Washington D.C., pp. 461-467.
Conrad, R., M. Aragno and W. Seiler, 1983: Production and consumption
    of hydrogen in a eutrophic lake, Appl. Environ. Microbiol. 45,
    502-510.
Conrad, R., and W. Seiler, 1980a: Photooxidative production and
    microbial consumption of carbon monoxide in seawater, FEMS
    Microbiol. Lett. 9, 61-64
Conrad, R., and W. Seiler, 1980b: Die Bedeutung mikrobiologischer
    Prozesse für den Kreislauf des Wasserstoffs in der Atmosphäre,
    Forum Mikrobiol. 3, 219-225.
Conrad, R. and W. Seiler, 1982: Utilization of traces of carbon
    monoxide by aerobic oligotropic microorganisms in ocean, lake and
    soil, Arch. Microbiol. 132, 41-46.

Conrad, R., W. Seiler, G. Bunse and H. Giehl, 1982: Carbon monoxide in seawater (Atlantic Ocean), J. Geophys. Res. 87, 8839-8852.

Conrad, R., and W. Seiler, 1985: Feldmessung von Emission und Deposition atmosphärischer Spurengase in Boden und Wasser. GIT Supplement Umweltschutz - Umweltanalytik 3, 74-78.

Deacon, E.L., 1977: Gas transfer to and across an air-water interface, Tellus 29, 363-374.

Ehrhardt, M., 1984: Marine gelbstoff, In: "The handbook of environmental chemistry, vol. 1C" (O. Hutzinger, ed.), Springer Verlag, New York, pp. 63-77.

Haag,W.R., J. Hoigne, E. Gassman and A.M. Braun, 1984: Singlet oxygen in surface waters. Quantum yields of its production by some natural humic materials as a function of wavelength ,Chemosphere 13, 641-650.

Herr, F.L., 1984: Dissolved hydrogen in Eurasian arctic waters, Tellus 36 B, 55-66.

Herr, F.L. and W.R. Barger, 1978: Molecular hydrogen in the near surface atmosphere and dissolved in waters of the tropical north Atlantic, J. Geophys. Res. 83, 6199-6205.

Herr, F.L., E.C. Frank, G.J. Leone and M.C. Kennicutt, 1984: Diurnal variability of dissolved molecular hydrogen in the tropical South Atlantic Ocean, Deep Sea Res. A 31, 13-20.

Herr, F.L., M.I. Scranton and W.R. Barger, 1981: Dissolved hydrogen in the Norwegian Sea: Mesoscale surface variability and deep water distribution, Deep-Sea Res. 28, 1001-1016.

Jähne, B., K.O. Münnich and U. Siegenthaler, 1979: Measurements of gas exchange and momentum transfer in a circular wind-water tunnel, Tellus 31, 321-329.

Jones, R.D. and R.Y. Morita, 1983: Carbon monoxide oxidation by chemolithotrophic ammonium oxidizers, Can. J. Microbiol. 29, 1545-1551.

Jones, R.D. , R.Y. Morita and R.P. Griffiths, 1984: Method for estimating in-situ chemolithotrophic ammonium oxidation using carbon monoxide oxidation, Marine Ecol. Progress Series 17, 259-269.

Lilley, M.D., J.A. Baross and L.I. Gordon, 1982: Dissolved hydrogen and methane in Saawich Inlet, British Columbia, Deep-Sea Research 29, 1471-1484.

Lion, L.W. and J.O. Leckie, 1981: The biogeochemistry of the air-sea interface, Ann. Rev. Earth Planet. Sci. 9, 449-486.

Liss, P.S. and P.G. Slater, 1974: Flux of gases across the air-sea interface, Nature 247, 181-184.

Merlivat, L. and L. Memery, 1983: Gas exchange across an air-water interface: Experimental results and modeling of bubble contribution to transfer, J. Geophys. Res. 88, 707-724.

Norkrans, B., 1980: Surface microlayers in aquatic environments, Adv. Microb. Ecol. 4, 51-85.

Peng, T.H., W.S. Broecker, G.G. Mathieu and Y.H. Li, 1979: Radon evasion rates in the Atlantic and Pacific oceans as determined during the GEOSECS program, J. Geophys. Res. 84, 2471-2486.

Scranton, M.I., 1983: The role of the cyanobacterium *Oscillatoria* (Trichodesmium) thiebautii in the marine hydrogen cycle, Mar. Ecol. Progr. Ser. 11, 79-87.

Scranton, M.I., 1984: Hydrogen cycling in the waters near Bermuda: the role of the nitrogen fixer, Oscillatoria thiebau tii, Deep-Sea Res. A 31, 133-144.

Scranton, M.I., M.M. Jones and F.L. Herr, 1982: Distribution and variability of dissolved hydrogen in the Mediterranean Sea, J. Marine Res. 40, 873-891.

Seiler, W., 1978: The influence of the biosphere on the atmospheric CO and H$_2$ cycles, In: "Environmental biogeochemistry and geomicro-biology, Vol. 3: Methods, Metals and Assessment" (W.E. Krumbein, ed.), Ann Arbor Science Publisheres Inc., Ann Arbor (Mich.), pp. 773-819.

Seiler, W. and R. Conrad, 1986: Contribution of tropical ecosystems to the global budgets of trace gases, especially CH$_4$, H$_2$, CO and N$_2$O. In: "The Geophysiology of Amazonia" (R.E. Dickinson, ed.), John Wiley, in press.

Seiler, W., H. Giehl and P. Roggendorf, 1980: Detection of carbon monoxide and hydrogen by conversion of mercury oxide to mercury vapor, Atmos. Technol. 12, 40-45.

Seiler, W. and C. Junge, 1970: Carbon monoxide in the atmosphere, J. Geophys. Res. 75, 2217-2225.

Seiler, W. and U. Schmidt, 1974: Dissolved nonconservative gases in seawater, In: "The Sea, Vol. 5: Marine Chemistry" (E.D. Goldberg, ed.), John Wiley & Sons, New York, pp.219-243.

Setser, P.J., J.L. Bullister, E.C. Frank, N.L. Guinasso Jr and D.R. Schink, 1982: Relationships between reduced gases, nutrients, and fluorescence in surface waters off Baja California, Deep-Sea Res. 29 A, 1203-1215.

Swinnerton, J.W., R.A. Lamontagne and V.J. Linnebom, 1974: Carbon monoxide in the ocean environment, Tellus 26, 136-142.

Swinnerton, J.W., R.A. Lamontagne and J.S. Bunt, 1977: Field study of carbon monoxide and light hydrocarbon production related to natural biological processes, NRL Report 8099, Naval Research Lab. Washington, D.C., pp.1-9.

Swinnerton, J.W., LinnenbomV.J. and R.A. Lamontagne, 1970: Ocean: A natural source of carbon monoxide, Science 167, 984-986.

Troxler, R.F. and J.M. Dokos, 1973: Formation of carbon monoxide and bile pigment in red and blue-green algae, Plant Physiol. 51, 72-75.

Wangersky, P.J., 1976: The surface film as a physical environment, Ann. Rev. Ecol. Syst. 7, 161-176.

Williams, R.T. and A.E. Bainbridge, 1973: Dissolved CO, CH$_4$ and H$_2$ in the southern Ocean, J. Geophys. Res. 78, 2691-2694.

Zafiriou, O.C., 1986: Atmospheric, oceanic, and interfacial photoche-mistry as factors influencing air-sea exchange fluxes and processes, this volume.

Zafiriou, O.C., J. Joussot-Dubien, R.G. Zepp and R.G. Zika, 1984: Photochemistry of natural waters, Environ. Sci. Technol. 18, 358A-371A.

THE AIR-SEA EXCHANGE OF LOW MOLECULAR WEIGHT HALOCARBON GASES

Peter S. Liss
School of Environmental Sciences
University of East Anglia
Norwich   NR4 7TJ
United Kingdom

1. INTRODUCTION

The gases considered in this chapter all contain a halogen-carbon bond
and have up to three carbon atoms in the molecule; they include methyl
halides (CH_3Cl, CH_3Br, CH_3I), haloforms ($CHCl_3$, $CHBr_3$), carbon
tetrachloride (CCl_4), and man-made chlorofluorocarbons, amongst others.
Organo-halogens of high molecular weight are discussed in the article
in this book by Atlas. Inorganic halogen gases aren't dealt with here,
even though air-sea transfer is important in the cycling of several of
them; for example the air-to-sea transfer of SF_6, the possible
emission of I_2 from the oceans by reaction of O_3 with I^- in surface
seawater (Garland and Curtis, 1981), and the evolution of HCl_g from
sea-salt droplets in the atmosphere (see Duce and Hoffman, 1976 for a
review). The only fluorine compounds to be mentioned are the man-made
chlorofluorocarbons (often referred to by the Dupont tradename 'Freons'),
since natural organo-fluorine gases do not appear to have been
detected in the environment.
    The basic equations describing air-sea gas transfer are given in
the chapter by Liss and Merlivat in this volume and will not be repeated
here. With respect to air-sea exchange all the gases included in the
present chapter have rather similar properties. They are all slow to
react in seawater so that the possibility of chemical enhancement of
their air-sea exchange may be ignored (α =1.0). Further, the gases
are generally rather insoluble and tend to partition in favour of the
atmosphere. Henry's Law constants (H) tend to be in the range logH =
0 ± 1; useful compilations are to be found in Glew and Moelwyn-Hughes
(1953), Horvath (1982), and Hunter-Smith et al. (1983). H values of
this magnitude mean that air-sea transfer of these gases will be under
the control of processes in the near-surface seawater, i.e. k_w will be
the dominant transfer velocity (Liss and Merlivat, this volume). Values
of H are also required to calculate the degree of saturation of surface
waters with respect to atmospheric concentrations. In some cases,
particularly for man-made chlorofluorocarbons and CCl_4, where the
surface seawater is close to equilibrium with the atmospheric partial
pressure, H values need to be known accurately in order to assess even

P. Buat-Ménard (ed.), The Role of Air-Sea Exchange in Geochemical Cycling, 283–294.
© *1986 by D. Reidel Publishing Company.*

the net direction of transfer across the interface.

The gases considered here are divided into two classes, i.e. those for which the oceans are a net source for the atmosphere, and those for which the oceans are a net sink. As far as current knowledge allows, each gas is discussed in terms of the following properties: direction and magnitude of air-sea flux, size of other inputs to the atmosphere relative to addition/removal across the sea surface, mode of formation/consumption in the oceans, and importance of the gas in atmospheric chemistry and as a component in the geochemical cycling of the halogen it contains. Finally, the present best estimates of the air-sea fluxes of the gases are gathered together in a summary table.

2. GASES FOR WHICH THE OCEANS ARE A NET SOURCE FOR THE ATMOSPHERE

2.1 Alkyl (mainly Methyl) Halides

2.1.1 Methyl Chloride (CH_3Cl). Three groups of workers have estimated the size of the oceans as a source of methyl chloride for the atmosphere. Singh et al. (1979) and Watson et al. (1980) calculated global sea-to-air fluxes in the range 3-8 x 10^{12} g yr^{-1}. The more recent measurements of Singh et al. (1983) yield a global oceanic source strength in the middle of this range (i.e. 5 x 10^{12} g yr^{-1}), and give some confidence that we can estimate this flux reasonably well. Even though there is some uncertainty over marine tropospheric concentrations of CH_3Cl, Rasmussen et al. (1980) report values (755-815 pptv) significantly higher than those of Singh et al., 1979 and 1983 (613 and 633 pptv, respectively), this doesn't greatly affect the calculated sea-to-air fluxes since surface seawater is generally 200-300% supersaturated with respect to atmospheric levels, with no reports of undersaturation.

Since there is little or no inter-hemispheric difference in air concentrations of CH_3Cl (Khalil and Rasmussen, 1981; Singh et al., 1983; Penkett, 1982), it is generally assumed that the compound is largely of natural origin. Cicerone (1984) states that anthropogenic releases are <5% of natural emissions. Lovelock (1975) gives industrial production in 1973 as 3.5 x 10^{11} g yr^{-1} but since it is almost entirely used as a chemical feedstock, little if any of this reaches the atmosphere. Methyl chloride is also produced during biomass burning, particularly smouldering combustion, and the rate of emission to the atmosphere from this source has been estimated by Crutzen et al. (1979) to be 0.3-0.6 x 10^{12} g yr^{-1} (i.e. an order of magnitude less than the oceanic source), The ability of fungi to produce methyl halides has recently been demonstrated in laboratory culture experiments (Harper, 1985).

The mechanism for the production of CH_3Cl in seawater is not known. It has been suggested that CH_3Cl is formed by reaction between the abundant Cl^- ions in seawater and methyl iodide produced by algae (Zafiriou, 1975). As will be discussed, surface seawater also appears to be a source of CH_3I to the atmosphere ($\sim 10^{12}$ g yr^{-1}), so that if the mechanism proposed by Zafiriou is correct then the algae must be

producing substantially more CH_3I than is required to account for the
sea-to-air flux of this compound alone. However, Singh et al. (1983)
have argued against the production of CH_3Cl by reaction between Cl^- and
CH_3I since in their measurements of several methyl halides in the
Eastern Pacific Ocean there is no apparent relationship between CH_3I
and CH_3Cl concentrations. They reason that since the CH_3I/Cl^- reaction
is relatively rapid, a strong correlation between concentrations of the
two gases in the water is to be expected. However, Zika et al. (1984)
have argued that the kinetics of the reaction vary considerably, from
a half-life of 5-6 days in tropical surface waters to 150 days at high
latitudes, so that correlations between concentrations of CH_3I and its
decomposition products CH_3Cl and CH_3Br may be masked by other processes.
Alternative production routes for CH_3Cl in seawater are biological
reactions analogous to those by which CH_3Br and CH_3I are formed. Such
production can be either by direct emission from organisms or by
reaction between dimethyl sulphonium ions (important intermediates in
the production of dimethyl sulphide - see chapter by Andreae in this
volume) and halide ions in seawater, according to the following
equations (from White, 1982):

$$(CH_3)_2S^+-R + Cl^- = CH_3Cl + CH_3SR$$

$$(CH_3)_2S^+-R + Br^- = CH_3Br + CH_3SR$$

$$(CH_3)_2S^+-R + I^- = CH_3I + CH_3SR$$

Where R = propionate $(CH_2-CH_2-COO^-)$.

The main reason why people are interested in estimating the flux
of methyl chloride to the atmosphere is because it may be (along with
the man-made Freons) a natural source of chlorine atoms in the
stratosphere, and may also be of importance for the behaviour of gases
in the troposphere (Rasmussen et al., 1980). In this context it is
noteworthy that the flux of CH_3Cl out of the oceans (5×10^{12} g yr^{-1},
Singh et al., 1983) is numerically equal to the rate of destruction of
the compound in the troposphere by reaction with $OH^{\cdot}$, as computed by
Logan et al. (1981) - see Table 3 of chapter by Zafiriou in this volume.
If further work confirms these figures then clearly much of the CH_3Cl
released by tropospheric sources is being destroyed there, leaving only
a small fraction to enter the stratosphere.

2.1.2 Methyl Bromide (CH_3Br). The only attempt to assess the air-sea
flux of methyl bromide is that of Singh et al. (1983). From seawater
and air measurements made from 40°N to 32°S in the Eastern Pacific
Ocean they find the water to be always supersaturated with respect to
the air concentrations of CH_3Br and extrapolate their computation to
obtain a global sea-to-air flux of methyl bromide of 3×10^{11} g yr^{-1}.
This is about an order of magnitude less than the evasion flux of the
other major organo-bromine gas in seawater - bromoform (see Section
2.2.2.).
On a unit area basis, CH_3Br fluxes in coastal waters are probably

significantly larger than in open ocean areas, e.g. CH_3Br concentrations
measured in coastal waters by Lovelock (1975) are about an order of
magnitude greater than the oceanic values found by Singh et al. (1983).
There is insufficient data at the present to assess whether elevated
levels in inshore waters can appreciably affect the calculated global
flux.

Methyl bromide is used as an agricultural fumigant. Wofsy et al.
(1975) have estimated the amount released to the atmosphere via
agricultural use at $7.5 \times 10^{10}g \ yr^{-1}$ (figure for 1977), i.e. about 25%
of the flux out of the oceans. Interhemispheric measurements of CH_3Br
concentrations in air tend to support a man-made N-hemisphere source.
Singh et al. (1983) found significantly higher CH_3Br levels in the
Northern versus the Southern Hemisphere (26-30 as opposed to 19 pptv).
A similar size inter-hemispheric increase has been found by Penkett et
al. (1985), although the absolute concentration levels are less (15.4
and 10.6 pptv for the Northern and Southern Hemispheres, respectively).

The mechanism of production of CH_3Br in the sea is even less well
studied than that for CH_3Cl. Reaction of Br^- with CH_3I, analogous to
the reaction of Cl^- with this compound to form CH_3Cl (Section 2.1.1.),
is possible. However, although the reaction rate constant when Br^- is
the nucleophile is significantly faster than with Cl^-, the much lower
concentration of Br^- in seawater relative to Cl^- means that overall
the reaction with Br^- is about 80 times less favourable than with Cl^-
(Zafiriou, 1975). Biological production by algae is a likely source
of CH_3Br, either by direct production or via reaction between Br^- and a
dimethylsulphonium ion (see Section 2.1.1.).

No assessment seems to have been made of the role of methyl bromide
in the geochemical cycling of bromine. It is worth pointing out that
the flux of CH_3Br out of the oceans ($3 \times 10^{11}g \ yr^{-1}$, Singh et al., 1983)
closely matches the calculated rate of destruction of the compound in
the troposphere by reaction with $OH^\cdot$ (Logan et al., 1981)—a situation
similar to that already described for CH_3Cl (Section 2.1.1.). However,
this apparent agreement should be treated with caution since Penkett
et al. (1985) have recently computed that the atmospheric breakdown
rate of CH_3Br by $OH^\cdot$ requires an input of only $0.9 \times 10^{11}g \ yr^{-1}$, which
is significantly less than Singh et al.'s calculated input from the
oceans. Reference should be made to Section 2.2.2 for further discussion
of the atmospheric chemistry of organo-bromine gases.

2.1.3 Methyl Iodide (CH_3I). Using measurements of CH_3I in air and
seawater in the North and South Atlantic (Lovelock et al., 1973), Liss
and Slater (1974) calculated the flux of the gas from the sea to the
atmosphere as $3 \times 10^{11}g \ yr^{-1}$. More recently, Rasmussen et al. (1982)
have done a similar calculation but using their own significantly
larger data set. They obtained a flux from the oceans in low
productivity areas in good agreement with that of Liss and Slater.
However, Rasmussen et al. suggest that if areas of high biological
production are included then the global flux might be four times higher
(i.e. $13 \times 10^{11}g \ yr^{-1}$). Similarly Lovelock (1975) reports substantially
higher water concentrations of CH_3I in coastal areas, particularly in
the vicinity of beds of the large kelp, *Laminaria*. Singh et al. (1983)

from their measurements in the Eastern Pacific compute the global flux to be in the range $3-5 \times 10^{11}$ g yr^{-1}. Given the observed variability in CH_3I concentrations in different waters, the agreement between these three estimates of the global flux is reassuring. The oceans always appear to be a source of CH_3I to the atmosphere, with no reports of surface waters being undersaturated with respect to air concentrations.

The oceans would seem to be the only significant source of methyl iodide for the atmosphere; no other sources have been suggested.

Methyl iodide is clearly produced biologically in seawater. In the open oceans this is probably due to (unidentified) algae, although in coastal waters seaweeds, such as *Laminaria digitata*, are established as prolific producers. Another possible mechanism for CH_3I production is by reaction of I$^-$ with dimethyl sulphonium ions (Section 2.1.1.), although the low and variable concentrations of I$^-$, relative to Cl$^-$ and Br$^-$, found in surface seawaters probably militate against this being an important route.

In order to achieve balance in the geochemical cycle of iodine a flux of some volatile species from the oceans to the atmosphere and thence to the land-based hydrosphere is required. In this respect the geochemical behaviour of iodine is similar to that of sulphur (see chapter by Andreae in this volume). Miyake and Tsunogai (1963) have calculated that a sea-to-air flux of about 5×10^{11} g yr^{-1} is required for geochemical balance. The fluxes of CH_3I out of the oceans calculated from the field measurements discussed above can account for between 50 and 230% of this amount. Clearly ocean emissions are important, if not dominant, in the geochemical cycling of iodine. This cycle has considerable implications for higher terrestrial life forms, including human beings, since the thyroid gland requires iodine to function properly. A significant portion of the necessary iodine appears to be supplied by the sea-to-air flux of CH_3I - a good example of 'Gaia' in action (Lovelock, 1979).

The importance of CH_3I in atmospheric chemistry is a matter of some controversy. Chameides and Davis (1980) have suggested CH_3I may catalytically remove ozone in the troposphere, as well as participating in several other reactions. In contrast, Jenkin et al. (1985) argue that many of the tropospheric reactions of CH_3I are inefficient at destroying O_3. Rasmussen et al. (1982) reason that on a global scale CH_3I is unlikely to be important in atmospheric chemistry due to its low air concentration ($\sim$ 2 pptv), but that in the marine atmosphere over biologically-productive ocean areas, where levels of 10-20 pptv are found, it may play a key role.

2.2 Haloforms

2.2.1 Chloroform (CHCl$_3$).
Khalil et al. (1983) have used the small number of measurements they have made of CHCl$_3$ in surface waters and marine air to estimate the global air-sea flux. From this calculation it appears that the oceans are a net source of CHCl$_3$ to the atmosphere, the magnitude of the flux being approximately 3.6×10^{11} g yr^{-1}.

There is considerable uncertainty concerning the size of the anthropogenic release of CHCl$_3$ (Yung et al., 1975), although several authors have concluded that much of the chloroform in the atmosphere is of natural origin (Penkett, 1982; Khalil et al., 1983). Support for a significant man-made release of $CHCl_3$ in the Northern Hemisphere comes from the approximately 10 pptv higher values found in N. over S. hemisphere air, although the absolute values differ between studies. Singh et al. (1983) found 21 and 11 pptv in the N. and S. hemispheres respectively, whereas the corresponding figures from Khalil et al. (1983) are about 30 and 20 pptv. There is some suggestion that terrestrial ecosystems may be natural sources of $CHCl_3$ to the atmosphere (W. Seiler, personal communication).

2.2.2 Bromoform ($CHBr_3$). There are to date no published estimates of the air-sea flux of bromoform. However, Penkett et al. (1985) have presented air concentrations over the North and South Atlantic which average about 0.7 pptv, and Dryssen and Fogelqvist (1981), and Fogelqvist (1985) give the results of surface seawater measurements in the Eastern Arctic Ocean (approximately 10 ng 1^{-1}). Assuming these concentrations are representative of the surface oceans as a whole, it is straightforward to show that the oceans are supersaturated relative to air concentrations and to compute a global sea-to-air flux of approximately $2 \times 10^{12}g$ yr^{-1}. Even if the much higher air concentrations reported by Berg et al. (1984) for the Arctic atmosphere (15±13 pptv) are used in the calculation, the sea-to-air flux is only reduced to about $1 \times 10^{12}g$ yr^{-1}.

The atmospheric measurements of Penkett et al. (1985) show no significant N-S interhemispheric concentration difference for $CHBr_3$, which argues for a widespread natural source (or sources) dominating input to the atmosphere. These authors calculated from the rate of photochemical breakdown of $CHBr_3$ in the atmosphere that the compound must have an atmospheric lifetime of about 2 weeks. This lifetime in turn implies an input into the atmosphere of approximately $1.0 \times 10^{12}g$ yr^{-1}, close to the sea-to-air fluxes calculated above. It is tempting to conclude from this that the oceans are the sole source of atmospheric $CHBr_3$. However, such a deduction is premature in view of the very small data base of field, particularly surface ocean, measurements for this compound - the Arctic seawater concentration given earlier may not be representative of other regions. However, it is clear that the ocean source must be important and certainly warrants further study.

The mode of formation of $CHBr_3$ in seawater can clearly be biological. Fenical (1981) reports that various species of the red alga *Asparagopsis* are prolific producers of bromoform. Further, Gschwend et al. (1985) show from laboratory studies that $CHBr_3$ is a major compound in the suite of volatile halogenated organic compounds emitted by various species of brown and green, as well as red algae.

In the context of the geochemical cycling of bromine, the sea-to-air flux of bromoform appears to be about an order of magnitude more important than the analogous flux of methyl bromide (see Section 2.1.2). One field observation of interest in this context is that of Berg et al. (1983) who found elevated levels of bromine in both gas and particulate phases in the Arctic troposphere from mid-February to mid-May, compared with the rest of the year. Emission of gaseous bromine compounds, such as CH_3Br and particularly $CHBr_3$, from the oceans could clearly be, at least in part, an explanation for these results. Atmospheric fluxes of bromine gases may also be of importance in the stratosphere since Br acts as a more efficent catalyst than Cl for removing ozone (Wofsy et al., 1975).

2.3 Other Organo-Halides

Compounds discussed so far are the ones for which suffcient field data exist so that an attempt can be made to estimate their air-sea fluxes. However, there are several papers, arising mainly from laboratory studies, which report the existence of a wide range of low molecular weight organo halogen compounds, both within marine macroalgae and in the water in which they are grown. For example, Fenical (1981) lists the following compounds in decreasing order of abundance in the red alga *Asparagopsis*: $CHBr_3$, CH_2Br_2, $CHBr_2Cl$, CBr_4, $CHBr_2I$, CH_2ClI, $CHBrCl_2$, CCl_4, $CHCl_3$, and CH_3I. Gschwend et al. (1985) found a similar suite of compounds in several varieties of macroalgae and also detected the following compounds in the water: $CHBr_3$, $CHBr_2Cl$, CH_2Br_2, and at lower concentrations CH_3I, C_2H_5I, $i-C_3H_7I$, $n-C_3H_7I$, CH_2ClI, CH_2BrI and CH_2I_2. Penkett (1982) also reports the presence of C_2H_5I, as well as $n-$ and $i-C_3H_7I$ in seawater containing different types of seaweed. It is abundantly clear that a large number of organo-halogen (including inter-halogen) compounds are waiting to be 'discovered' in sewater, and particularly in coastal waters with macroalgae present.

3. GASES FOR WHICH THE OCEANS ARE A NET SINK FOR THE ATMOSPHERE

The oceans must be a net sink for a whole host of low molecular weight organo-halogen gases injected into the atmosphere during industrial and other anthropogenic activities. However, the only two for which attempts appear to have been made to calculate their air-to-sea fluxes are carbon tetrachloride (CCl_4) and Freon-11 (CCl_3F).

Using the results of Lovelock et al. (1973) for CCl_4 and Freon-11 concentrations in marine air and surface seawater in the North and South Atlantic Ocean, Liss and Slater (1974) calculated global air-to-sea fluxes for these two gases of 1.4×10^{10} and $5.4 \times 10^9 g \ yr^{-1}$, respectively. The mean undersaturation of the surface seawater with respect to CCl_4 concentrations in the air was about 10%. A decade later Hunter-Smith et al. (1983) sampled on an almost identical cruise track and found the air and seawater to be essentially at equilibrium, implying no net flux. Similarly, more recent measurements of Freon-11 in the North Pacific Ocean (Gammon et al., 1982) indicate almost precise saturation equilibrium between atmosphere and surface ocean. It is possible that the apparent changes in air-sea flux over this period are due to analytical problems and uncertainty over values for the Henry's Law constants for these compounds. However, at least for Freon-11, they could be real. At the beginning of the period atmospheric releases (mainly in the Northern Hemisphere) were still increasing rapidly and this led to a significant difference in air

concentrations between the northern and southern hemispheres. In the
circumstances a net flux into the oceans was to be expected. The
figure calculated by Liss and Slater (1974) corresponds to approximately
2% of the anthropogenic production rate at that time. The more recent
changes in Freon-11 release rates have led to a decrease in the N-S
difference in air levels (Hunter-Smith et al., 1983), and may have
allowed the surface oceans and the atmosphere to come close to
equilibrium.

The main reason for interest in the ocean uptake of compounds like
Freons, which have no natural sources, is for their utility as tracers
of water movements and mixing in the oceans. Examples of Freons being
used as tracers are to be found in the papers by Gammon et al. (1982) –
Upper layers of the N.E. Pacific, Bullister and Weiss (1983) –
Greenland and Norwegian Seas, and Weiss et al. (1985) – Deep Equatorial
Atlantic. A novel reversal of this idea, i.e. to use ocean depth
profiles of halogen gases in order to predict atmospheric histories of
such species, has recently been attempted by Watson and Liddicoat (1985)
for CCl_4 and SF_6.

4. SUMMARY

All the air-sea fluxes discussed in the previous sections are
summarised in Table 1, together with other external atmospheric sources
and sinks for the gases concerned. The fluxes across the sea surface
given in Table 1 are all global values and have been calculated from the
product of the interfacial concentration difference driving the flux and
an appropriate transfer velocity. Of these two, it is uncertainty over
the concentration term which is more likely to limit the accuracy of the
calculated fluxes for the gases considered here. Although our knowledge
of the physics and chemistry controlling air-sea transfer velocities is
far from perfect, studies of radionuclides such as ^{14}C and ^{222}Rn set
rather narrow limits to the possible range of global average values for
this parameter. Measurements in surface seawaters show that for
organo-halogen gases water concentrations vary widely, and for gases
whose net flux is from sea to air it is the water concentration which
essentially determines the magnitude of the concentration driving force.
In general, air concentrations of the gases do not show such wide space
and time variations. Thus it is concentration measurements in surface
waters which are currently the real limitation on the accuracy of air-
sea flux calculations. At such time as the data base of reliable
concentration measurements allows better resolution than the global,
annual averages calculated here, knowledge of transfer velocities and
their temporal and spatial variability may become limiting. In the
meanwhile, the first priority in order to improve air-sea flux
estimates for organo-halogen gases is improved field measurements of
the gases in surface ocean and coastal waters.

TABLE 1. Air-Sea and other External Atmospheric Fluxes of Low Molecular Weight Organo-Halogen Gases

Gas	Air-Sea Flux*	Ref.	Other Atmospheric Sources/*Sinks*	Ref.
CH_3Cl	$+ 3-8 \times 10^{12}$	1, 2	Biomass burning $0.3-0.6 \times 10^{12}$	10
	$+ 5 \times 10^{12}$	3	Industrial, <5% natural emissions	11
CH_3Br	$+ 3 \times 10^{11}$	3	Agriculture 7.5×10^{10}	12
CH_3I	$+ 3 \times 10^{11}$ $+ 13 \times 10^{11}$ $+ 3-5 \times 10^{11}$	4 5 3	None	
$CHCl_3$	$+ 4 \times 10^{11}$	6	Industrial ?	13
$CHBr_3$	$+ 2 \times 10^{12}$	7	None ?	
CCl_4	$- 10^{10}$	4	Industrial 4×10^{10}	14
	~ 0	8	*Stratosphere* 5×10^{10}	14
CCl_3F	$- 5 \times 10^{9}$	4	Industrial 3×10^{11}	14
	~ 0	9	*Stratosphere* 4×10^{11}	14

*   +, Sea → Air; −, Air → Sea. All fluxes are in units of g (of compound) yr^{-1}.

Reference Key: 1-Singh et al. (1979); 2-Watson et al. (1980); 3-Singh et al. (1983); 4-Liss and Slater (1974); 5-Rasmussen et al. (1982); 6-Khalil et al. (1983); 7-This paper; 8-Hunter-Smith et al. (1983); 9-Gammon et al. (1982); 10-Crutzen et al. (1979); 11-Cicerone (1984); 12-Wofsy et al. (1975); 13-Yung et al. (1975); 14-N.A.S. (1976).

ACKNOWLEDGEMENTS

I thank the following ASI participants for providing comments which
have been helpful in preparing the chapter: Patrick Buat-Ménard, Hèlène
Cachier, Ralf Conrad, Bob Duce, Tom Fogg, Véronique Garcon, Catherine
Jeandel, Russell Lang, Horst Meyrahn, Wolfgang Seiler, Rik Wanninkof,
and Ollie Zafiriou. In addition, Stuart Penkett made useful suggestions
for improvements to the text.

REFERENCES

Berg, W.W., P.D. Sperry, K.A. Rahn and E.S. Gladney, 1983:'Atmospheric
    bromine in the Arctic.' *J. Geophys. Res.*, 88, 6719-6736.
Berg, W.W., L.E. Heidt, W. Pollock, P.D. Sperry and R.J. Cicerone, 1984:
    'Brominated organic species in the Arctic atmosphere.' *Geophys. Res.
    Letts.*, 11, 429-432.
Bullister, J.L. and R.F. Weiss, 1983: 'Anthropogenic chlorofluoromethanes
    in the Greenland and Norwegian Seas.' *Science*, 221, 265-268.
Chameides, W.L. and D.D. Davis, 1980: 'Iodine: Its possible role in
    tropospheric photochemistry.' *J. Geophys. Res.*, 85, 7383-7398.
Cicerone, R., 1984: 'Halogens.' In: *Global Tropospheric Chemistry: A
    Plan for Action,* National Academy Press, 128-132.
Crutzen, P.J., L.E. Heidt, J.P. Krasnec, W.H. Pollack and W. Seiler,
    1979: 'Biomass burning as a source of atmospheric gases CO, H_2,
    N_2O, CH_3Cl and COS.' *Nature,* 282, 253-256.
Duce, R.A. and E.J. Hoffman, 1976: 'Chemical fractionation at the air-
    sea interface.' *Ann. Rev. Earth Planet. Scis,* 4, 187-228.
Dryssen, D. and E. Fogelqvist, 1981: 'Bromoform concentrations of the
    Arctic Ocean in the Svalbard area.' *Oceanologica Acta,* 4, 313-317.
Fenical, W., 1981: 'Natural halogenated organics.' In: *Marine Organic
    Chemistry* (E.K. Duursma and R. Dawson, eds.), Elsevier, 375-393.
Fogelqvist, E., 1985. 'Carbon tetrachloride, tetrachloroethylene, 1,1,1-
    trichloroethane and bromoform in Arctic seawater.' *J. Geophys.
    Res.,* 90, 9181-9193.
Gammon, R.H., J. Cline and D. Wisegarver, 1982: 'Chlorofluoromethanes
    in the northeast Pacific Ocean: Measured vertical distributions and
    application as transient tracers of upper ocean mixing.' *J.
    Geophys. Res.,* 87, 9441-9454.
Garland, J.A. and H. Curtis, 1981: 'Emission of iodine from the sea
    surface in the presence of ozone'. *J. Geophys. Res.,* 86, 3183-3186.
Glew, D.N. and E.A. Moelwyn-Hughes, 1953: 'Chemical statics of the
    methyl halides in water.' *Dis. Faraday Soc.,* 15, 150-161.
Gschwend, P.M., J.K. MacFarlane and K.A. Newman, 1985: 'Volatile
    halogenated organic compounds released to seawater from temperate
    marine macroalgae.' *Science,* 227, 1033-1035.
Harper, D.B., 1985: 'Halomethane from halide ion - a highly efficient
    fungal conversion of environmental significance.' *Nature,* 315, 55-
    57.
Horvath, A.L., 1982: *Halogenated Hydrocarbons: Solubility-Miscibility
    with Water,* Dekker, 889pp.

Hunter-Smith, R.J., P.W. Balls and P.S. Liss, 1983: 'Henry's Law constants and the air-sea exchange of various low molecular weight halocarbon gases.' *Tellus*, 35B, 170-176.

Jenkin, M.E., R.A. Cox and D.E. Candeland, 1985: 'Photochemical aspects of tropospheric iodine behaviour.' *J. Atmos. Chem.*, 2, 359-375.

Khalil, M.A.K. and R.A. Rasmussen, 1981: 'Atmospheric methylchloride (CH_3Cl).' *Chemosphere*, 10, 1019-1023.

Khalil, M.A.K., R.A. Rasmussen and S.D. Hoyt, 1983: 'Atmospheric chloroform ($CHCl_3$): ocean-air exchange and global mass balance.' *Tellus*, 35B, 266-274.

Liss, P.S. and P.G. Slater, 1974: 'Flux of gases across the air-sea interface.' *Nature*, 247, 181-184.

Logan, J.A., M.J. Prather, S.C. Wofsy and M.B. McElroy, 1981: 'Tropospheric chemistry: a global perspective.' *J. Geophys. Res.*, 86, 7210-7254.

Lovelock, J.E., 1975: 'Natural halocarbons in the air and in the sea.' *Nature*, 256, 193-194.

Lovelock, J.E., 1979: *Gaia: A New Look at Life on Earth*, Oxford, 157pp.

Lovelock, J.E., R.J. Maggs and R.J. Wade, 1973: 'Halogenated hydrocarbons in and over the Atlantic.' *Nature*, 241, 194-196.

Miyake, Y. and S. Tsunogai, 1963: 'Evaporation of iodine from the ocean'. *J. Geophys. Res.*, 68, 3989-3994.

N.A.S., 1976: *Halocarbons: Effects on Stratospheric Ozone*, National Academy of Sciences, 352pp.

Penkett, S.A., 1982: 'Non-methane organics in the remote troposphere.' In: *Atmospheric Chemistry* (E.D. Goldberg, ed.), Springer-Verlag, 329-355.

Penkett, S.A., B.M.R. Jones, M.J. Rycroft and D.A. Simmons, 1985: 'An interhemispheric comparison of the concentrations of bromine compounds in the atmosphere.' *Nature*, 318, 550-553.

Rasmussen, R.A., L.E. Rasmussen, M.A.K. Khalil and R.W. Dalluge, 1980: 'Concentration distribution of methyl chloride in the atmosphere.' *J. Geophys. Res.*, 85, 7350-7356.

Rasmussen, R.A., M.A.K. Khalil, R. Gunawardena and S.D. Hoyt, 1982: 'Atmospheric methyl iodide (CH_3I).' *J. Geophys. Res.*, 87, 3086-3090.

Singh, H.B., L.J. Salas, H. Shigeishi and E. Scribner, 1979: 'Atmospheric halocarbons, hydrocarbons, and sulfur hexafluoride: Global distributions, sources and sinks.' *Science*, 203, 899-903.

Singh, H.B., L.J. Salas and R.E. Stiles, 1983: 'Methyl halides in and over the Eastern Pacific.' *J. Geophys. Res.*, 88, 3684-3690.

Watson, A.J. and M.I. Liddicoat, 1985: 'Recent history of atmospheric trace gas concentrations deduced from measurements in the deep sea: Application to sulphur hexafluoride and carbon tetrachloride.' *Atmos. Env.*, 19, 1477-1484.

Watson, A.J., J.E. Lovelock and D.H. Stedman, 1980: 'The problem of atmospheric methyl chloride.' In: *Proc. NATO ASI on Atmospheric Ozone* (A.C. Aikin, ed.), U.S. Federal Aviation Administration, 365-372.

Weiss, R.F., J.L. Bullister, R.H. Gammon and M.J. Warner, 1985:
    'Atmospheric chlorofluoromethanes in the deep equatorial Atlantic.'
    Nature, 314, 608-610.
White, R.H., 1982: 'Analysis of dimethyl sulfonium compounds in marine
    algae.' *J. Mar. Res.,* 40, 529-536.
Wofsy, S.C., M.B. McElroy and Y.L. Yung, 1975: 'The chemistry of
    atmospheric bromine.' *Geophys. Res. Letts.,* 2, 215-218.
Yung, Y.L., M.B. McElroy and S.C. Wofsy, 1975: 'Atmospheric halocarbons:
    A discussion with emphasis on chloroform.' *Geophys. Res. Letts.,*
    2, 397-399.
Zafiriou, O.C., 1975: 'Reaction of methyl halides with seawater and
    marine aerosols.' *J. Mar. Res.,* 33, 75-81.
Zika, R.G., L.T. Gidel and D.D. Davis, 1984. 'A comparison of
    photolysis and substitution decomposition rates of methyl iodide
    in the ocean.' *Geophys. Res. Letts.,* 11, 353-356.

SEA-AIR EXCHANGE OF HIGH-MOLECULAR WEIGHT SYNTHETIC ORGANIC COMPOUNDS

Elliot Atlas
Department of Oceanography
Texas A&M University
College Station, TX 77843

C. S. Giam
Graduate School of Public Health
University of Pittsburgh
Pittsburgh, PA

1. INTRODUCTION

Studies over the past two decades have verified that man-made organic chemicals are reaching virtually all areas of the earth. Research on persistent chemical compounds in the environment has uncovered traces of chemical contamination in ocean sediments (Hom et al., 1974; Harvey & Steinhauer, 1976), marine fish and mammals (Tanabe et al., 1983; Zell and Ballschmiter, 1980; and Ballschmiter et al., 1981), surface and deep ocean waters (Giam et al., 1978; Tanabe et al., 1982; Harvey & Steinhauer, 1973; deLappe et al., 1983; and Bidleman & Olney, 1974), Antarctic snow (Peel, 1975; and Tanabe et al., 1983), tropical rains (Atlas & Giam, 1981; Bidleman & Leonard, 1982), and in the atmosphere from the Arctic to New Zealand (Atlas & Giam, 1981; Bidleman & Leonard, 1982; Oehme & Mano, 1984; Bidleman et al., 1981; and Giam et al., 1980). Because of their wide dispersion and possible biological consequences, identification of transport mechanisms, transformations, and fate of these chemical contaminants from their source to their ultimate sink is important.

One significant pathway for dispersion of synthetic organic compounds is through the atmosphere (Atlas & Giam, 1981; Atlas et al., 1985b). Atmospheric deposition has been shown to be a major factor in pollution of coastal waters, of the Great Lakes, and in open ocean areas (Giam et al., 1977; Burns et al., 1985; Eisenreich, 1981). Thus, studies to examine the rates and mechanisms of air-water transfer have been undertaken in order to set limits on a critical segment of the geochemical cycle of organic pollutants.

This chapter will examine the basic processes and rates of air-sea exchange of high-molecular weight organic compounds. An attempt will be made to assess our present understanding of these processes and to identify those areas which require more detailed examination. The approach will be as follows: 1) Establish a reasonable estimate of the concentrations of synthetic organics in the marine and atmospheric environment; 2) Examine the physical and chemical behavior of high-molecular weight organic compounds in air and water; 3) Use available data to model the air-sea exchange rates and mechanisms for selected

P. Buat-Ménard (ed.), The Role of Air-Sea Exchange in Geochemical Cycling, 295–329.

organic compounds; and 4) Compare estimated atmospheric inputs to other
inputs (e.g. riverine) and outputs (e.g. sedimentation).

2. COMPOUNDS OF INTEREST

The total production of synthetic organic compounds and the number of
different compounds produced is almost overwhelming. Literally
thousands of organic chemicals, exceeding a total of 70×10^9 kg, are
produced each year. However, only those compounds which meet certain
criteria are of potential concern in the environment. These criteria
are: a) high production rate ($>25 \times 10^6$ kg/yr), b) distribution and use
pattern which favors environmental dispersion, c) persistence in the
environment, d) bioaccumulation and e) toxicity. In practice, only
several classes of high-molecular weight compounds meet these criteria
and have been examined in marine systems. These are chlorinated
hydrocarbons (CHC) and phthalic acid esters (PAE) (Figure 1). Among
the most widely studied chlorinated hydrocarbons are the
polychlorobiphenyls (PCBs) and DDT compounds. In addition, other
chlorinated hydrocarbons-such as chlorobenzenes-and halogenated
pesticides - including hexachlorocyclohexanes (a-HCH and lindane),
dieldrin, and chlordane - will be discussed in this chapter. The second
compound class, phthalic acid esters, are widely used plasticizers. Two
phthalate esters, di-n-butyl phthalate (DBP) and di-2-ethylhexyl
phthalate (DEHP), are the most commonly found plasticizers in
environmental samples and they will be used as examples of this
compound class.

3. SAMPLING/ANALYTICAL ASPECTS OF TRACE ORGANICS

Advances in analytical organic chemistry have provided a basis for
progress in studying the environmental chemistry of trace organic
compounds. Still, the problems of obtaining and analyzing a sample
from the marine environment without the introduction of artifacts is
difficult. Thus, a brief discussion of the procedures used to sample
and analyze trace organic compounds is included here.
    Because of the very low concentration of most synthetic organics
in ocean waters and the atmosphere, methods have been developed to
concentrate organic compounds from very large volumes of water or air.
(The methods for water and air sampling are similar and may differ only
in the flow rates and material used to concentrate organic compounds
from the sample matrix.) For air sampling, particulate compounds are
trapped on a pre-combusted glass or quartz fiber filter. Typically
2,000 to 5,000 m^3 of air are required to obtain sufficient material for
analysis. Compounds present in a vapor phase (including most CHC) are
trapped on a solid adsorbent such as polyurethane foam (Simon and
Bidleman, 1979; Oehme and Stray, 1982) or Florisil (Giam et al., 1975;
Chang et al., 1985). Usually two to three days of pumping time is
required to obtain sufficient sample from the marine atmosphere.

In the laboratory most samples are treated in a similar fashion, with some details changed depending on sample type. Basically, samples are extracted with appropriate solvents and the solvent extract volume is reduced. The concentrated extracts are then separated into compound classes using column chromatography. Final quantitation and compound identification is performed using high-resolution gas chromatography with electron capture detection (ECD) (for CHC and PAEs) or mass spectrometry (for general detection and compound verification). A typical ECD chromatogram of air from the Texas Gulf Coast (Figure 2) illustrates some of the complexity often encountered in air and other environmental samples.

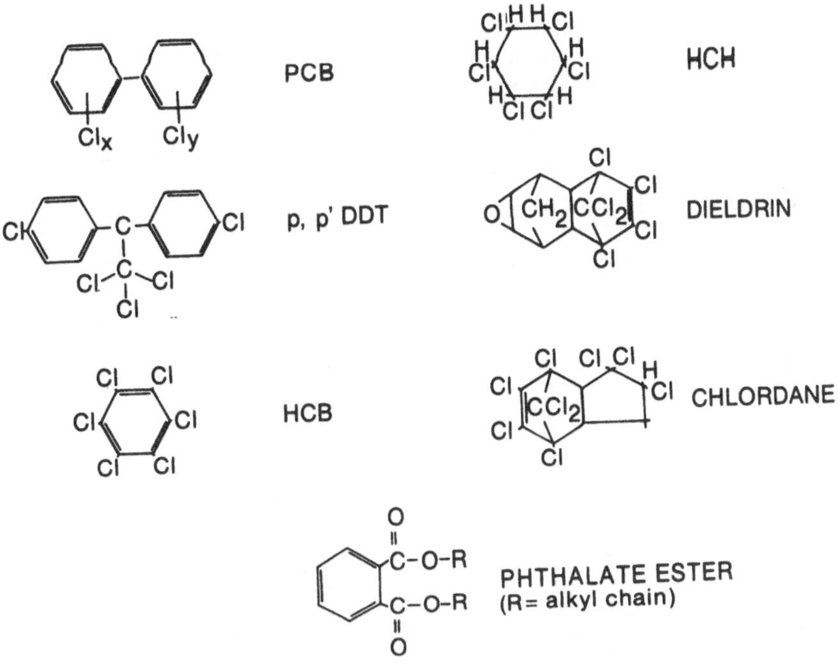

Figure 1.   Structures of selected organic pollutants discussed in text.

4. DISTRIBUTION OF HMW ORGANICS IN THE MARINE ENVIRONMENT

4.1. Water and Organisms

Studies of the distribution of synthetic organic pollutants in the marine environment have been performed by a number of workers. Still, because of difficult analytical problems, the total amount of reliable data, especially for open ocean regions, is small. This section will illustrate some features of the distribution and concentration of synthetic organics rather than critically evaluate the present data

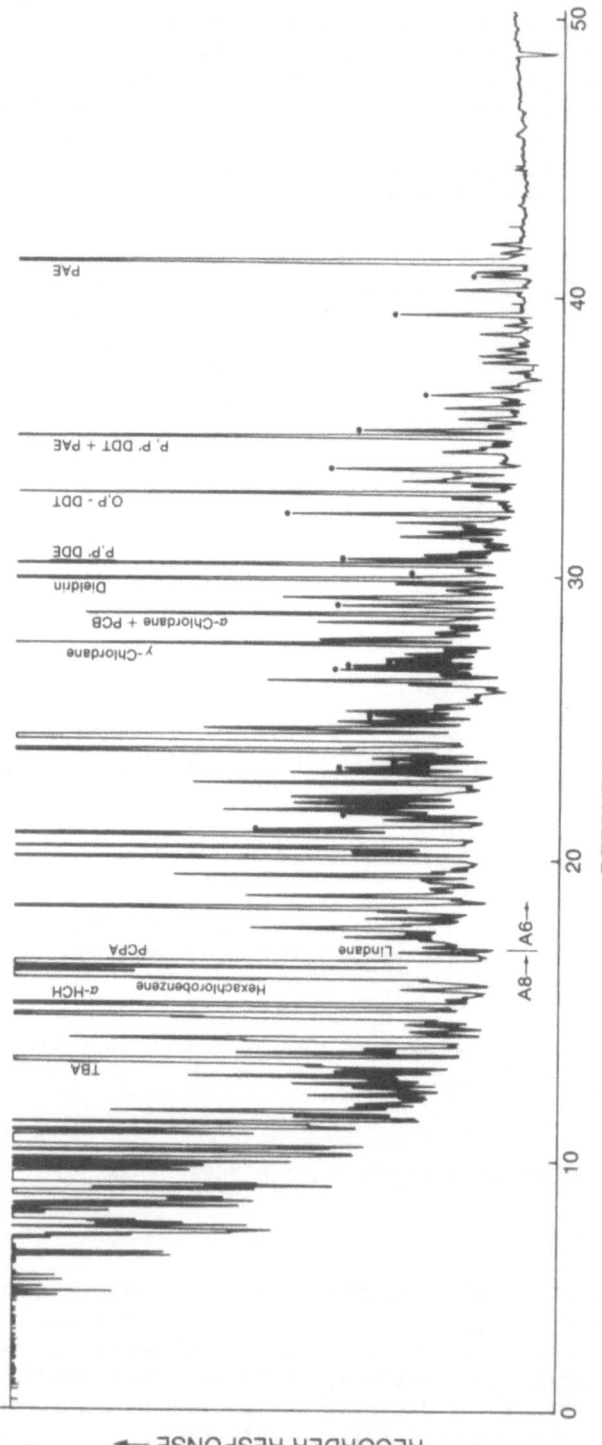

Figure 2. Electron-capture capillary gas chromatogram of an air sample from Ninety-Mile Beach, New Zealand. ● = PCB isomers, TBA = tribromoanisole, PCPA = pentachloroanisole, PAE = phthalic acid ester, other abbreviations as in text.

base.
    Some representative data for surface ocean waters is shown in
Table I. Typical concentrations of PCBs and DDTs are in the sub ng/L
range, while hexachlorocyclohexanes (HCH) and PAEs are found at low
pptr concentration. Concentrations of most organic pollutants increase
in coastal areas and near urban centers (Giam et al., 1978; Tanabe et
al., 1982; deLappe et al., 1983; Bidleman and Olney, 1974).

TABLE I Recent measurements of synthetic organics in surface seawater.

Location	Concentrations (ng/l)					Ref.
	PCBs	DDTs	HCHs	PAEs	Other	
North Atlantic	.1 - .2	~.001	--	5	---	1
North Pacific	<.003-.1	<.001-.03	.3-1	---	0.1(chlordane)	2
	0.28	0.14	2.1	---	---	3
South Pacific	.08	.02	~0.4	---	---	3
Mediterranean	0.8	---	0.5	---	0.4 (HCB)	4
North Sea	~0.6	---	1.9	---	0.1 (HCB)	5

1 Giam et al., 1978   2 deLappe et al., 1983
3 Tanabe et al., 1982b   4 Burns et al., 1985
5 Duinker et al., 1979

    Only a few depth profiles of CHC in the ocean have been measured
(Tanabe and Tatsukawa, 1983). Data indicate only slight differences
between concentrations of DDTs and PCBs in surface and deep (>1000 m)
water. HCH, on the other hand, has a fairly large decrease in
concentration below the surface layer. The differences between the HCH
and PCB/DDT profiles have been attributed to differences in particle
scavenging and subsequent vertical transport (Tanabe and Tatsukawa,
1983). [The different chemical behavior of these compounds is obviously
important in their air-sea transfer, too, and will be discussed more
completely later.]
    Concentrations of CHC in surface water from several North-South
transects in the Western Pacific has been reported by Tanabe and
Tatsukawa (Tanabe et al., 1982). These data clearly show an increase
in concentration of CHC in surface waters between 20 and 45°N. This
distribution pattern suggests that pollutant transport in the
Westerlies can impact the surface water chemistry in the Western
Pacific Ocean. In addition, Tanabe and Tatsukawa note that contaminant
concentrations in marine organisms are highest in the same latitude
zone (Figure 3).
    Other studies have used marine organisms as "pollutant
accumulators" to identify areas of pollutant concentration and to
identify which compounds bioaccumulate (Zell and Ballschmiter, 1980;
Ballschmiter et al., 1981). The main pollutant groups identified in
marine organisms reflect those which have been found in some ocean

water samples. These are the polychlorobenzenes (e.g. hexachlorobenzene (HCB)), HCH, PCBs, DDTs, chlordanes, and polychloroterpenes (toxaphene). Unfortunately, many of compounds seen with electron capture gas chromatography are still unidentified.

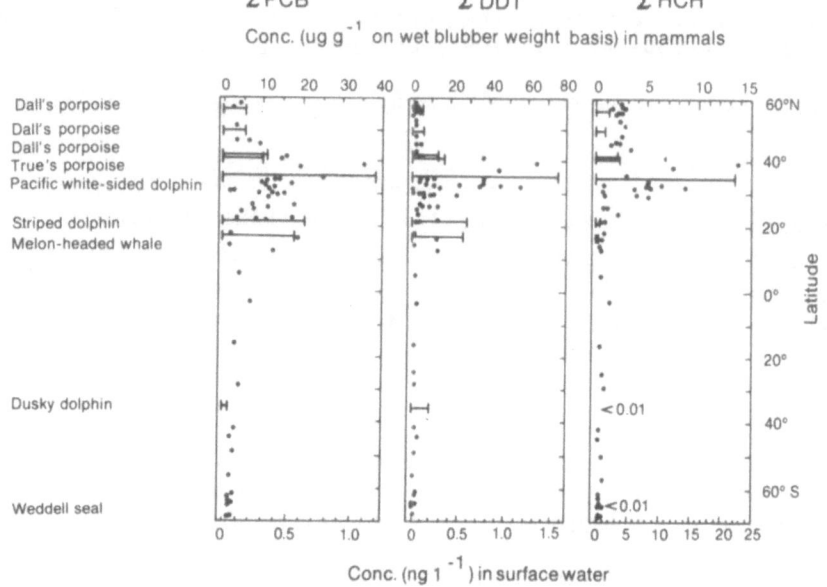

Figure 3. Variation with latitude of organic compounds in water and organisms in the Pacific Ocean, (after Tanabe et al., 1983).

While many details of the distribution of synthetic organic compounds in the ocean are missing, the following general features of their distribution should be noted: a) widespread distribution in surface and deep waters, b) concentrations in the ppb to pptr range, c) accumulation in marine organisms, and d) a distribution pattern suggesting a strong contribution by atmospheric transport.

4.2. Atmospheric Concentrations

Concentrations of synthetic organic compounds have been determined in the atmosphere around the globe. Surprisingly, there are more measurements in remote areas than in urban centers and continental regions (Table II). Synthetic organics are typically measured the pg-ng/m^3 range, with source areas having concentrations one to two orders of magnitude higher than remote areas. For example urban concentrations of PCB are several ng/m^3, while remote areas are usually below 0.1 ng/m^3. Studies in the North and South Pacific (Atlas and Giam, SEAREX Newsletters) have demonstrated a clear interhemispheric difference in atmospheric concentrations of most synthetic organic

TABLE II Average concentration of atmospheric organic compounds at continental and marine sites (ng/m³).

LOCATION	HCB	ΣHCH	ΣDDT	CHLORDANE	DIELDRIN	PCB (A1254)	TOXAPHENE	DBP	DEHP	REFS.
Continental										
College Station, TX	0.21	0.93	0.33	1.05	0.083	0.29	1.80	2.10	2.00	(1)
Lillestrom, Sweden	0.16	1.19	--	--	--	--	--	--	--	(2)
White Sands, NM	0.13	5.40	0.074	0.068	--	0.11	0.54	--	--	(3)
Columbia, SC	0.29	1.10	0.093†	1.30	--	1.50	13.10	--	--	(4)
Denver, CO	0.24	0.30	0.021†	0.063	--	0.45	--	--	--	(5)
New Bedford, MA	0.18	1.00	--	0.24	--	9.30	--	--	--	(5)
Marine/Coastal										
North Inlet, SC	--	--	0.036*	0.15	--	0.25	1.70	--	--	(4)
Gulf Coast, TX	0.13	0.44	0.028	0.036	0.017	0.056	0.57	0.42	0.62	(6)
Barbados	--	--	0.004	0.009	0.005	0.057	<0.10	--	--	(7)
North Atlantic Western Pacific/	0.15	0.39‡	0.006†	0.03	0.02	0.69	--	1.00	2.90	(8)
Eastern Indian	--	0.55	0.28	--	--	0.30**	--	--	--	(9)
Enewetak Atoll	0.10	0.26	≤0.006	0.013	0.010	0.05	≤0.09	0.87	1.40	(8)
American Samoa	0.055	0.032	0.0015	0.001	0.002	0.012	--	<0.20	<0.20	(3)
Bear Island (Arctic)	0.11	0.24	--	--	--	--	--	--	--	(2)
Hopen (Arctic)***	0.12	0.30	--	0.001	--	0.038	--	--	--	(2)

* p, p' - DDT only
† p, p' - DDE only
‡ α-HCH only
*** Concentrations estimated from graphs
** total PCB

1 Atlas et al., 1985a
2 Oehme and Ottar, 1984
3 Atlas and Giam, 1985
4 Bidleman and Christensen, 1979
5 Billings and Bidleman, 1983

6 Chang et al., 1985
7 Bidleman et al., 1981
8 Atlas and Giam, 1981
9 Tanabe et al., 1982

compounds. The differences range from 2 - 15 times (Table III) and reflect the stronger source of synthetic organics in the Northern Hemisphere.

TABLE III Comparison of concentrations of selected chlorinated hydrocarbons in the atmosphere of the North and South Pacific Oceans.

Compounds	Concentration (pg/m^3)		
	North Pacific	South Pacific	Ratio
Pentachlorobenzene	30	16	2
Hexachlorobenzene	100	60	2
α-Hexachlorocyclohexane	253	22	12
Lindane	15	1	15
DDTs	6	2	3
PCBs	49	10	5
Dieldrin	10	3	3
Chlordane	13	1	13

Synthetic organic compounds in the atmosphere are distributed between a particulate (adsorbed) and a vapor phase. Most measurements have found CHC and other synthetic organic compounds almost exclusively in the vapor phase with only a small fraction bound to aerosol particles. Thus, the amount of organic compound in particulate matter accounts for only a small percentage of the total atmospheric concentration. However, as will be shown later, the air-sea flux associated with a particle-bound organic compound can be relatively large. The factors which influence particle/vapor partitioning are, therefore, critical in evaluating mechanisms of air-sea transfer.

Bidleman and co-workers have demonstrated the importance of temperature in vapor/particle partitioning of atmospheric PCB. For example, they used an equation formulated by Yamasaki et al. (1982) to relate the vapor/particle (V/P) ratios of PCB to ambient temperature and total suspended matter. Yamasaki et al. demonstrated that the V/P ratio for polynuclear aromatic hydrocarbons could be described by an equation derived from Langmuir adsorption theory:

$$\log(V/P) = mT^{-1} + b - \log(TSP) \tag{1}$$

where m and b are empirical constants, TSP is the total suspended particle concentration ($\mu g/m^3$), and T is the average absolute sampling temperature. In an attempt to correlate PCB V/P ratios with temperature, Bidleman (1985) evaluated Equation 1 for PCB (Aroclor 1254) in four cities with V/P ratios ranging from 0.2-190 and average temperatures for the sampling periods of -10 to 26°C. For 25 data points, linear regression gave: m = -4583, b = 19.061, and r^2 = 0.836.

Using this relationship, the apparent percentage of particulate PCB
(A-1254) in a city with TSP = 60 $\mu g/m^3$ would be estimated as 2.2% at
20°C and 25% at 0°C. Because of the lower TSP levels in the marine
atmosphere, even lower particulate PCB fractions over the oceans would
be expected. For example, if oceanic TSP is 7 $\mu g/m^3$ (mineral and sea
salt), at 0°C 3.6% of the A-1254 would be on particles. (This
percentage is expected to be an upper limit because the difference in
particle size distributions and chemical characteristics of urban vs.
marine aerosols.) At 20°C this percentage would fall to ~0.3%. Since
lighter PCB components (e.g. A-1242) have a vapor pressure ~5 times
higher than A-1254, the particle-bound A-1242 should be ~1/5 of the
above values. Thus, the low percentage of particulate PCB actually
measured in the marine atmosphere (e.g. Atlas and Giam, 1981; Bidleman
and Leonard, 1982) is consistent with theoretical estimates.

4.3. Atmospheric Deposition

Compared to the number of studies on wet and dry deposition of
inorganic substances, relatively little has been done to investigate
deposition processes for organic compounds. Table IV shows some of the
concentrations of synthetic organics in precipitation samples from non-
urban and marine locations. Not unexpectedly, the concentration
distribution of organics in wet deposition is similar to that observed
in atmospheric samples. High concentrations are found near source
regions where atmospheric concentrations are higher. (Details on
incorporation of organics into precipitation are discussed later.)
    The few fluxes and deposition parameters that have been previously
reported are for PCB, and these are summarized in Tables V and VI.
Table V shows that deposition fluxes to the Great Lakes are in the
range of 10-100 $\mu g/m^2/yr$, while marine deposition may be somewhat
lower. These estimates can be compared to fluxes of CHC derived later
in this chapter. Removal parameters measured by various workers tend to
be higher for the less volatile compounds (Table VI). For example,
washout ratios (W) and V_d are higher for Aroclor 1254 than for Aroclor
1016/1242. The same trend has been found for chlordane and the less
volatile DDT (Bidleman and Christensen, 1979). A likely explanation is
that a greater proportion of the heavier organics are particle-bound
and are removed more efficiently by wet and dry processes.

5. AIR-SEA EXCHANGE MECHANISMS FOR SYNTHETIC ORGANICS

The sections above have provided a general background to illustrate
that, indeed, synthetic organic compounds are present in the atmosphere
and in atmospheric deposition and can impact the chemistry of ocean
waters. In sections to follow we will examine the different mechanisms
of air-sea transfer of synthetic organics. Since these compounds are
partitioned into a gas and a particle phase, we will consider wet and
dry deposition processes of both gaseous and particle-bound material.

TABLE IV   Average concentration of organic pollutants in rainfall from various locations (ng/L).

LOCATION	HCB	α-HCH	LINDANE	DDE	DDT	CHLORDANE	DIELDRIN	PCB	TOXAPHENE	DBP	DEHP	REFS.
Continental/Coastal												
College Station, TX	0.40	11.80	3.3	1.00	2.10	2.30	0.80	2.3	22.2	52.5	210	(1)
North Inlet, SC	---	---	---	-	2.40	1.90	---	10.0	159.0	-	-	(2)
Pigeon Key, FL	---	---	---	0.29	0.49	---	---	7.7	---	-	322	(3)
Great Lakes	<1.00	12.00	5.0	-	3*	tr	1.00	21.0	---	-	-	(4)
	<1.00	28.00	9.0	-	2	tr	1.00	28.0	---	-	-	(4)
Snow	<1.00	1.00	tr	-	1	tr	2.00	29.0	---	-	-	(4)
Great Britain	---	10.40†	---	-	7.9**	---	1.30	15.0	---	-	-	(5)
Beaverton, OR	---	5.90	.45	-	-	-	---	---	---	-	-	(6)
Marine												
Enewetak Atoll	<0.03	3.10	.51	≤0.02	<0.02	<.02	<.02	<0.6	---	31.0	55	(7)
Arabian Sea	---	7.70	1.9	-	0.58	---	---	---	---	-	-	(8)
American Samoa	<0.03	0.32	.03	-	0.02**	<.03	<.03	<0.2	---	<10.0	<65	(3)

* 71% pp'DDT, 9% pp'DDE, 9% op'-DDT, 6% pp'DDD
** ΣDDT
† ΣHCH

1 Atlas et al., 1985a
2 Bidleman and Christensen, 1979
3 Atlas and Giam, 1985
4 Strachan et al., 1980
5 Wells and Johnstone, 1978
6 Pankow et al., 1984
7 Atlas and Giam, 1981
8 Bidleman and Leonard, 1982

TABLE V Precipitation and dry deposition inputs of PCB to water bodies.

Location and Year	Flux $\mu g/m^2/yr$	Reference
L. Michigan 1975-1976	43.0 - 86	Murphy et al., 1977; 1981
L. Huron 1977-1978	18.0	Murphy et al., 1981
L. Superior 1974-1977	80.0 -102	Eisenreich (ed.) 1981
Saginaw Bay, L. Huron, 1977-79	17.0 - 23	Murphy & Schinsky, 1983
North Sea 1975-1976	3.3	Wells & Johnstone, 1978
Five Great Lakes	121.0	Eisenreich et al., 1981
L. Michigan (calculated)	5.4 - 10	Andren, 1983
L. Superior (calculated)	7.3 - 73	Eisenreich et al., 1983

5.1. Dry Deposition

5.1.1. Gas exchange - models and laboratory measurements. The magnitude (and direction) of the gas exchange of synthetic organics across the air/water boundary has been a subject of controversy in recent years, particularly with reference to PCB in the Great Lakes (Eisenreich, 1981; Mackay et al., 1983). It has been shown from mass balance estimates and from water column measurements that the atmosphere must be an important source of PCB to the Great Lakes. However, physicochemical data on PCB vapor pressure, solubility, and Henry's Law constants suggest that the Great Lakes are supersaturated with respect to atmospheric PCB. Therefore, surface water should volatilize excess PCB and be a source, rather than a sink, of PCB. Some have suggested that the physical data are incorrect or have pointed to assumptions regarding mass transfer rates and sorption of PCB as potential sources of miscalculation.

A similar dilemma exists for gas transfer of PCB and other synthetic organics to the ocean surface. Direct measurements are not available to determine the net flux of vapor-phase compounds across the air/water interface. However, some indirect data are available to suggest the magnitude of the gas exchange flux. First, we will examine the equations and parameters used to estimate gas exchange.

The basic parameter which defines the extent of gas exchange of a particular compound is its Henry's Law Constant, H, which may be calculated from vapor pressure and solubility data. Often, though, these data are either not determined at appropriate temperatures or are unreliable. Direct determination of H is possible by measuring the equilibrium distribution of a chemical dissolved in water and its concentration in the vapor phase above the liquid. This technique has also been extended to include multiple equilibrations of the liquid and gas phases. Determinations of this type work well for compounds of relatively high H. MacKay et al. (1979) have devised a dynamic method for determining H for various hydrophobic chemical pollutants. In this method the test compound is stripped from solution and the removal rate, which is proportional to H, is determined.

Atlas et al. (1982) used this method and an independent radiotracer technique (Atlas et al., 1983) to determine H for selected high molecular weight organic pollutants, such as polychlorinated biphenyls (PCB), in distilled water and seawater. Results of these experiments and other estimates are given in Table VII. The stripping method gave results generally higher than the radiotracer technique, which appears to yield move accurate results. Effects of different parameters on H were also examined in our laboratory and by others. These effects are summarized below: a) Salinity--A change in salinity from distilled water to $36\,^{o}/_{oo}$ increases partition coefficients by a factor of 1.5 to 2. Thus, effects of salinity changes within the marine environment on partition equilibria will be relatively small, though changes in gas exchange rates from fresh water to marine systems could be significant. b) Dissolved organic substances--A comparison of gas partitioning in UV-oxidized sea water and lake water and untreated natural waters showed no measurable difference in Henry's Law Constant. Similar conclusions were found by MacKay et al. (1982) and Whitehouse (1985). They found a negligible effect of normal concentrations of DOC on aqueous solubility of PAH and other organics. Caron et al. (1985), however, reported a strong binding of DDT to dissolved organic matter in fresh water. c) Effect of temperature--Burkhard et al. (1985) estimated the temperature dependence of H for PCB mixtures based on theoretical arguments. They calculated a strong temperature dependence in H. The Henry's Law constants they calculate decrease by a factor of 2-3 for a temperature decrease of 10°C. This effect may be quite important in natural systems and will be discussed later. d) Effect of PCB composition--Experiments by Atlas et al. (1982) and Murphy et al. (1982) and the estimates of Burkhard et al. (1985) show that PCB isomer composition makes only a slight difference in Henry's Law constant. This effect is due to vapor pressure and solubility decreasing at approximately the same rate for PCB of increasing degrees of chlorination.

The compounds studied here show a wide range of partition constants which will affect their gas-transfer properties. It has been useful to model gas-transfer processes in terms of diffusion in thin films adjacent to the air-water interface (Whitman, 1923).

The two-film resistance model of Whitman has been widely used to

Table VI Typical atmospheric removal rates and deposition parameters for PCB.

Compound	Location	Dry Deposition Velocity v_d, cm/sec [a]	Washout Factor W [b]	Reference
Aroclor 1242/1016	Columbia, SC	<0.04		Bidleman & Christensen, 1979; Christensen et al., 1979
	North Inlet Estuary, SC	<0.06		Bidleman & Christensen, 1979; Christensen et al., 1979
	Kingston, RI	0.07		Bidleman & Christensen, 1979; Christensen et al., 1979
	Enewetak Atoll		1	Atlas & Giam, 1981
	Lake Michigan		14	Murphy & Rzeszutko, 1977
Aroclor 1254	Columbia, SC	0.43		Bidleman & Christensen, 1979; Christensen et al., 1979
	North Inlet Estuary, SC	0.16	94	Bidleman & Christensen, 1979; Christensen et al., 1979
	Kingston, RI	0.11		Bidleman & Christensen, 1979; Christensen et al., 1979
	LaJolla, CA	1.20		McClure, 1976
	Lake Michigan		103	Murphy & Rzeszutko, 1977
	Pigeon Key, FL		18-32	Giam & Atlas, unpublished data
Total (or unspeci-fied) PCB	Minneapolis, MN	0.13		Eisenreich et al., 1981

[a]Deposition velocity for total PCB (particle & vapor); [b]W = g/kg rain ÷ g/kg air.

examine gas exchange in natural water systems (Liss and Slater, 1974;

TABLE VII   Air-water partition coefficients ($\times 10^3$) for selected organic compounds at 25°C.

Compound	Partition Coefficient		Ref.
	Dist. H_2O	Seawater	
PCB 1242	9	--	1
PCB 1242*	32*	120	2
PCB 1242			
(di, tri chlorobiphenyls)	14	29	3
PCB 1242	8.2-16.8	--	4
PCB 1254	8.4	--	1
PCB 1254			
(tetra, pentachlorobiphenyls)	6.7	14	2
PCB 1254	19.2-29.9	--	4
Hexachlorobenzene	29	41	3
Hexachlorobenzene*	54	70	2
DDE*	50	150	2
γ-chlordane*	55	230	2
α-Chlordane*	36	170	2
α-HCH	0.96	(0.4)	2
Dieldrin	--	1.0	5
Di-n-butylphthalate	0.074	0.11	3

* measured with a stripping method; value probably too high
1 Murphy et al., 1982   2 Atlas et al., 1982
3 Atlas et al., 1983   4 Wescott et al., 1981
5 Slater and Spedding, 1981

Smith et al., 1980). In this model, the transfer of a gas is limited by diffusion across thin stagnant films of air and water adjacent to the interface. The overall transfer coefficient, K_{OL}, is based on the individual transfer coefficients across the liquid and gas films and on the Henry's Law constant of the compound. Thus:

$$1/K_{OL} = 1/K_L + RT/HK_G \qquad (2)$$

where: K_L, K_G are the individual mass transfer coefficients in the liquid (L) and gas (G) film, and (H/RT) = dimensionless air/water partition coefficient, H=Henry's Law constant, R=gas constant, and T=temperature. Depending on the magnitude of (H/RT), one of the two terms in Equation 2 may become large compared to the other and the exchange will be designated as either "gas" or "liquid-phase" controlled. Usually, K_L and K_G are based on estimates of oxygen and water vapor mass transfer rates in the ocean. Typical values of 3000 and 20 cm/hr have been suggested for K_G and K_L, respectively (Liss and Slater, 1974).

Experiments in our laboratory evaluated the use of Equation 2 to predict the behavior of high molecular weight organics (Atlas et al., 1982). This was done by measuring mass-transfer rates of oxygen and organic compounds in a stirred vessel. Reasonable agreement between Equation 2 and measured exchange rates were obtained when differences in diffusion rates of oxygen and the high-molecular weight organic compound were considered. For the compounds discussed here a correction factor of ~0.2 is required to correct exchange coefficients for differences in diffusion constants. Thus, Equation 2 can be re-written, for a compound x, as:

$$1/K_{OL}^{x} = 1/[K_L f_1]^n + 1/[K_G f_2 (H/RT)]^n \qquad (3)$$

where: K_L and K_G = transfer constants for oxygen and water, respectively; (H/RT) = air-water partition coefficient; f_1 and f_2 are the ratios of diffusion constants of compound x to oxygen (1) and water (2); and n = 0.5 to 1. For studies of large, high-molecular weight organics an exponent n = 1 appears appropriate (Smith et al., 1981), while for gases and some lighter organic compounds, e.g. CCl_4, an exponent n of ~0.5 is found to better describe air-water transfer. The reason for this difference in behavior is not yet clear, but may be related to uncertainties in diffusion constants of organic molecules.

Given the mass transfer rate K_{OL}, the net flux F_D^G depends on the disequilibrium between air and water concentrations. Or,

$$F_D^G = K_{OL} (C_{eq} - C) \qquad (4)$$

where C = the concentration of dissolved compound in water, and C_{eq} = the aqueous concentration of compound in equilibrium with the atmosphere. If C > C_{eq}, the flux is from the water to the air, and vice versa. While apparently straightforward, practical application of Equation 4 is not simple. This is because some portion of an organic compound in solution may be associated with particulate matter and/or organic colloids and not be involved in gas-liquid exchange. This aspect is covered in a later section.

5.1.2. Particle deposition. Dry deposition of trace organics occurs by gravitational settling and turbulent impaction of aerosol-bound materials, and adsorption of vapors to the deposition surface. A number of studies have used artificial surfaces to collect the dry deposition of high-molecular weight organics from the atmosphere. Oil-coated plates and nets, ethylene glycol/water-covered pans, and diol-coated filters have been used to estimate dry deposition fluxes (see Table VI).

How well any of these surrogate surfaces approximate dry deposition to surfaces of environmental interest (ocean, lake, plant foliage) is unknown. Removal of particles is strongly influenced by their size, meteorological factors, and surface properties of the collector. Reviews of dry deposition reveal a large number of field experiments carried out over the last 25 years. Particle deposition velocities V_d vary by three orders of magnitude. A discussion of all factors affecting washout and dry deposition is beyond the scope of this chapter and the reader is referred to several articles on the subject (Sehmel, 1980; Slinn et al., 1978; Slinn and Slinn, 1980).

Certainly, particle deposition to a natural water surface is highly complicated process. In a very simple mathematical form which masks many complexities, the flux of particles may be calculated from:

$$F_D^P = V_D C_p \qquad (5)$$

where F_D^P = dry flux of particle-bound compound, V_D = particle deposition velocity, and C_p = concentration of the compound bound to particles. Large differences in particle deposition rates are expected as a function of particle size (Sehmel, 1980; Slinn et al., 1978; Slinn and Slinn, 1980). Thus, measurements of V_D (total) described earlier (Table V) may not be applicable to open-ocean areas. Data of Bidleman and Christensen (1979) and Murphy et al. (1981) suggest that dry deposition rates of PCB measured in continental areas may be dominated by a small percentage of PCB which is associated with large particles. Rapid deposition of these continent-derived large particles near the source will lessen the effective deposition rate of PCB over most ocean areas. Since particulate PCB and other synthetic organic compounds are present in the ocean atmosphere is mainly on small particles (Atlas and Giam, unpublished data), they may behave similarly to other atmospheric compounds associated with small particles, such as Pb. A recent analysis of Pb deposition in the marine atmosphere (Settle and Patterson, 1982) suggests a net dry deposition velocity on the order of 0.1 cm/sec. This value will be used for subsequent calculations. However, larger values might be more appropriate in coastal areas where large-particle fluxes are potentially more important.

5.2. Wet Deposition

5.2.1. Particle scavenging.
Particles are efficiently scavenged from the atmosphere by precipitation. Data on particulate organics in precipitation (Zafiriou et al., 1985; Atlas et al., in preparation) and on trace metal scavenging (Arimoto et al., 1985) suggest a scavenging ratio of 300-700 [ng/kg (rain)/ng/kg(air)]. For subsequent calculations we will use a value of 500. Then, to calculate the flux due to wet deposition of particle-associated organics, F_w^P, one can use the following equation:

$$F_w^P = 500 R C_p \qquad (6)$$

where R = annual rainfall rate over the ocean (~1m/yr).

As noted by Zafiriou et al. (1985), estimates of precipitation scavenging of particles may be biased by selective analysis of (usually large) rainfalls, and a full suite of rainfall samples covering a

variety of conditions is required for accurate assessment of
precipitation fluxes. They reported an inverse correlation of
particulate organic compounds and rainfall amount at Enewetak to follow
the form:

$$C = AV^B \qquad\qquad\qquad (7)$$

where C = concentration in rain and V = rainfall volume. The exponent B
was found to be approximately 2/3. Similar results for particulate CHC
in rural rainfall were found by Atlas and Giam (Atlas et al., in
preparation). These factors are in reasonable agreement with the
square-root dependence reported by others (Barrie and Neustadter, 1983;
Hicks and Shannon, 1979) for sulfate and radionuclide scavenging.
Thus, detailed examination of fluxes due to particle scavenging of
organics should take into account the observed volume dependency,
though we will not apply such a correction for our subsequent analyses
here.

5.2.2. **Gas scavenging by precipitation.** This process involves
equilibrium of rain with vapor phase compounds in the atmosphere.
Essentially this mechanism is the same as discussed previously for gas
exchange with the sea surface. Where actual measurements of synthetic
organics in rain are unavailable, it is assumed that organic vapors
partition into the raindrop according to their Henry's Law constants
(see Equation 4). Data for soluble organic compounds, such as HCH,
suggest this is a reasonable approximation (Atlas et al., in
preparation). The relative importance of this process in precipitation
scavenging of atmospheric organics is illustrated in Figure 4. The
simple equation used to calculate a flux by this process is:

$$F^G_W = RC \qquad\qquad\qquad (8)$$

where: F = the flux of compound from wet deposition associated with
vapor scavenging; R = rain rate, and C = concentration in rain
(directly measured or calculated from atmospheric concentration and
appropriate Henry's Law Constant).

5.3. Adsorption and Partitioning in Surface Waters

The fate of atmospherically-derived organic compounds in surface waters
requires some consideration here. Various solution equilibria,
adsorption processes, and biochemical transformations will affect the
residence time of organic compounds in surface waters and also may
influence the extent and direction of gas exchange. In this section we
will consider some recent data on organic/solid adsorption to evaluate
the partitioning of synthetic organic compounds in surface waters.
        Chiou et al. (1983) and others (Gschwend and Wu, 1985; Hassett and
Anderson, 1982; Means et al., 1983; Wijayaratne and Means, 1984) have
demonstrated that adsorption equilibria between non-polar organics and
solid particles is strongly influenced by the "quality" and "quantity"

of dissolved organic substances in the system. Thus, synthetic
organics can be partitioned into several phases: a) adsorbed on solids,
b) associated with dissolved organic matter, c) bound to colloidal
material, and d) free in solution. Several examples and estimates of
this partitioning follow.

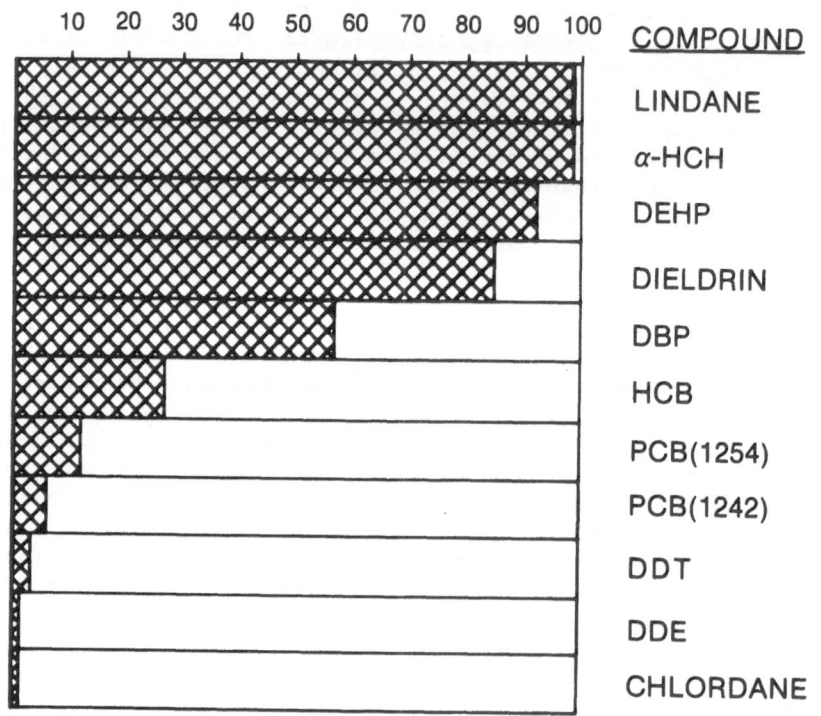

Figure 4. Contribution of vapor scavenging to organic
compounds found in rainfall.

5.3.1. Binding of DDT and Lindane to DOC. Caron et al. (1985) described
the binding of DDT and lindane to humic acids extracted from a
freshwater reservoir. As might be expected based on their different
chemical properties, DDT and lindane were found to have vastly
different affinities for DOC. Association constants were 4.1 x 10^5 for
DDT and 1 x 10^3 for lindane. In normal seawater (DOC = 1 mg/l) 28% of
DDT and 0.1% of lindane would be bound to dissolved organic carbon.
However, studies by Whitehouse (1985) suggest that terrestrial humates
may bind nonpolar organics more strongly than marine humic material.
Thus, organic/DOC associations in seawater may be somewhat less than
that reported by Caron et al. (1985).

5.3.2. <u>Binding of PCB to DOC.</u> Hassett and Milicic (1985) measured the binding of 2, 2', 5, 5' tetrachlorobiphenyl to dissolved humates. The K_{DOC} they measured ($7.1 \pm 2.4 \times 10^4$) suggests that 7% of PCB in surface seawater (1 mg/L DOC) would be bound to DOC.

5.3.3. <u>Binding of PCB to Particles and Colloids.</u> Recent studies by Brownawell and Farrington (1985) have shown that in an environment of high particle content and high DOC, a significant fraction of PCB can be associated with colloidal material. Their measurements of distribution coefficients of PCB on suspended solids in surface waters ranged from ~0.5 - 1 x 10^6 (L/kg). Distribution constants of similar magnitude have been found in the Great Lakes also (Eisenreich, 1981). Distribution coefficients of this magnitude would produce the following partitioning:

Suspended Solids (ng/L)	% Particulate	% Dissolved & Colloidal
0.1	5-10	90-95
1.0	33-50	50-67
10.0	84-91	9-16

5.3.4. <u>"Fugacity" Approach.</u> In an experiment similar to Hassett and Milicic (1985), Murphy (1982) measured the equilibrium vapor of PCB above a spiked natural water (Lake Michigan) and compared the vapor concentration to that expected based on previously measured Henry's Law Constants in pure water. Any binding of PCB to particles or colloids would effectively reduce the "fugacity" of dissolved PCB and lower the equilibrium concentration of the vapor above the liquid, though the exact nature of the interactions in the liquid phase would be undetermined. In this experiment, Murphy found that 60% of PCB was "dissolved" and 40% was somehow "bound" and unable to undergo gas exchange.

5.3.5. <u>Field Measurements of Particle/Dissolved Partitioning.</u> Table VII summarizes some of the few measurements of dissolved and particulate concentrations of PCB and other organics in water. Data from the North Sea, tropical West Pacific, and Lake Michigan show adsorption of PCB close to that predicted by a K'_D of ~1 x 10^6. Higher than predicted amounts of particulate PCB are found in the Eastern Pacific and the Southern Ocean. Reasons for this difference in PCB adsorption are not clear, but possibilities include temperature effects, particle chemistry, and different filtration efficiencies for particulate matter. Data in Table VIII also illustrate the large differences in adsorptive behavior of PCBs and DDTs compared to the more soluble HCH compounds.

5.3.6. <u>Summary.</u> From field data and laboratory experiments it appears that in open-ocean regions ~15 - 50% of PCB and DDT compounds are somehow associated with particulate or dissolved material and do not equilibrate with the atmosphere. However, it is clear there is a large uncertainty here. It is suggested that experiments at sea on the volatility of dissolved synthetic organics would be most valuable in clarifying this problem.

6. AIR-SEA FLUXES IN THE NORTH PACIFIC

In this section we will utilize the equations and processes described above to estimate the rates and mechanisms of air-sea flux of selected synthetic organic compounds in the North Pacific. In doing so, we will consider variations in concentration, temperature, rainfall, wind speed, etc. and will evaluate the effect of these parameters on the air-sea flux of synthetic organic compounds. As examples of different chemical behaviors, we will examine the fluxes of HCH (low Henry's Law Constant, non-particle reactive) and PCB (high Henry's Law Constant, particle reactive), though calculations will be presented for other synthetic organic compounds.

The data base and estimated parameters used for these calculations were as follows: 1) atmospheric concentrations measured at Enewetak (Table III); 2) Henry's Law Constant from laboratory experiments (Table VII) or literature data with temperature dependence estimated from halocarbon data (Lincoff and Gossett, 1984), and salting-out coefficient=2; 3) precipitation (Elliott and Reid, 1984); 4) Wind and temperature (U.S. Navy, 1969); 5) atmospheric vapor/particle partition – Equation 1, adjusted for differences in vapor pressure; 6) effect of wind speed on exchange velocity (Wanninkhof et al., 1984); 7) dry deposition velocity, 0.1 cm/sec.; 8) atmospheric concentration vs. latitude, estimated from data of Tanabe and co-workers; and 9) flux equations as described earlier in this chapter. Some of these parameters are illustrated in Figure 5. Details of these estimates can be found elsewhere (Atlas and Giam, in preparation).

First we tested this approach by comparing estimated precipitation concentrations with actual concentrations measured in rainfall at Enewetak. Figure 6 shows reasonable agreement for most compounds, even though a wide range of concentrations for different compounds was encountered. As suggested earlier differences in precipitation scavenging are related to variations in Henry's Law Constant. Relatively soluble compounds, such as HCH, are efficiently scavenged while poorly soluble compounds, e.g. PCB, are scavenged mainly as particles (Figure 4).

While wet deposition appears to be reasonably well predicted , estimating total deposition is a more difficult problem. The major problem is estimating the saturation state of seawater with respect to gas exchange. Most measured concentrations of CHC in surface ocean waters are apparently near saturation or indicate supersaturated conditions. Under these conditions, gas flux into surface water would be a minimum or possibly negative. If, however, a major portion of CHC is bound to particles or colloidal matter, then surface waters may become undersaturated. With present information, it is estimated that up to 50% of particle reactive compounds (PCB/DDT) may be present in "true" solution, while >95% of more soluble (e.g. HCH) compounds are in a dissolved state in surface seawaters.

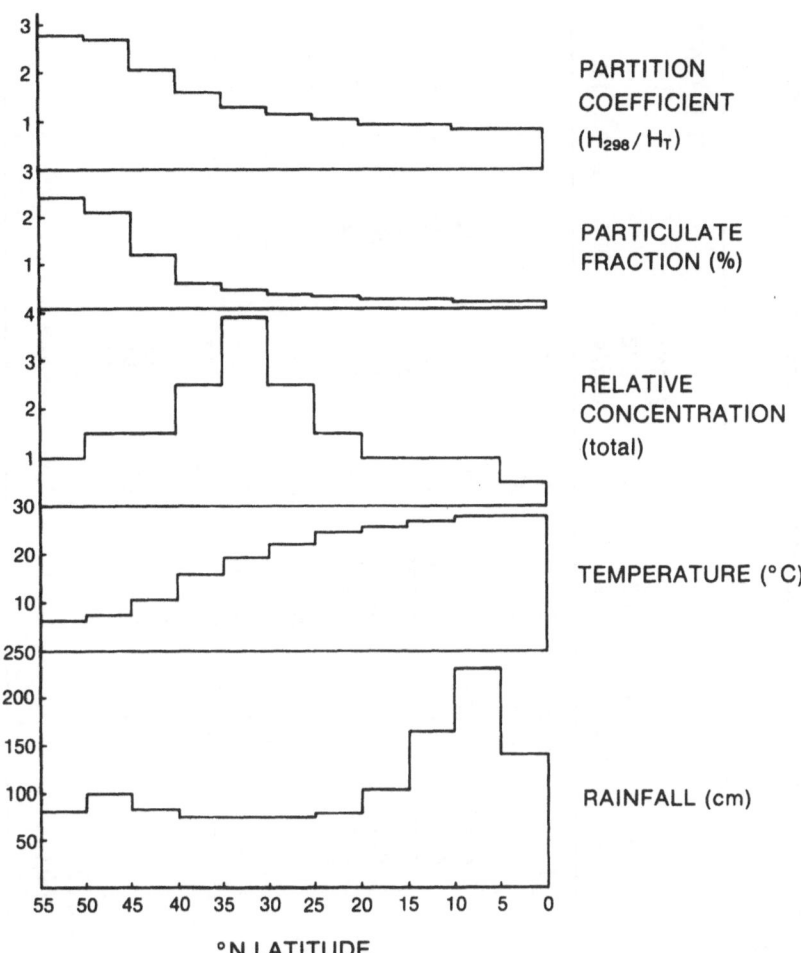

Figure 5. Variation with latitude of selected parameters used in a model of air-sea flux of organics to the North Pacific.

For the purpose of our calculations we will estimate deposition fluxes for several conditions of seawater saturation - including 100% saturated and 0% saturated. The latter condition, used by others for flux estimates (Bidleman et al., 1976), assumes some mechanism to make the sea surface a perfect sink for CHC. While this assumption may be unrealistic, it sets a reasonable upper bound on total fluxes from the atmosphere to the ocean.

TABLE VIII Selected measurements of particle/dissolved concentrations of PCB and other synthetic organics.

LOCATION	SS (mg/l)	PCB DISS. (ng/L)	PCB PART. (ng/L)	PCB % ADS.	PARTIC. PCB (ng/g)	OTHER (%ADS) DDE	OTHER (%ADS) DDT	OTHER (%ADS) HCH	OTHER (%ADS) CHL.	REF.
NORTH SEA	13	0.55	2.34	81	180					1
	0.8	0.28	0.12	30	146					1
W. PACIFIC										
22°N	0.10	0.08	0.006	7	58	24	12	0.4	--	2
16°N	0.11	0.11	0.014	12	130	20	23	0.1	--	2
49°S	0.27	0.07	0.094	58	340	42	58	1.3	--	2
64°S	0.21	0.06	0.027	32	130	78	67	2.1	--	2
64°S	0.52	0.06	0.059	49	110	72	91	2.8	--	2
E. PACIFIC										
Gold. Gate	--	0.16	0.26	62	--	--	--	--	--	3
South SF.	--	0.13	0.15	54	--	--	--	--	--	3
Pacifica	--	0.07	0.10	57	--	--	--	--	--	3
SC Bight	--	0.003	0.002	44	--	24	82	4	9	3
LAKE MICHIGAN										
Microlayer	1.7	9.9	7.7	44	4700					4
Subsurface	1.6	4.8	0.9	17	606					4
NY BIGHT					187					5
BALTIC (sed. trap)					11–161					6
MED. SEA (sed. trap)					x̄=129					7

1 Duinker and Hillebrand, 1983   2 Tanabe and Tatsukawa, 1983
3 deLappe et al., 1983   4 Rice et al., 1982
5 Boehm, 1983   6 Larsson, 1984   7 Burns et al., 1985

TABLE IX Area-weighted averages of air-sea fluxes of synthetic organic compounds to the North Pacific, and estimated riverine inputs.

Compound	Flux (pg/cm²/yr) MIN (gas exchange=0)	MAX	Riverine flux
PCB (1254)	20 (187)*	284 (451)*	10
DDE	0.24 (2.3)*	5.3 (7.4)*	0.5
DDT	1.3 (11.6)*	21 (31)*	0.5
Chlordane	1.0 (9.9)*	23 (32)*	1
Dieldrin	16 (50)*	95 (129)*	1
HCB	4.6	286	1
α-HCH	556	2563	10
Lindane	97	218	10
DBP	5.7	12.6	(ng) 1
DEHP	23	25	(ng) 1

* indicates calculated flux if particulate concentration increased 10 times.

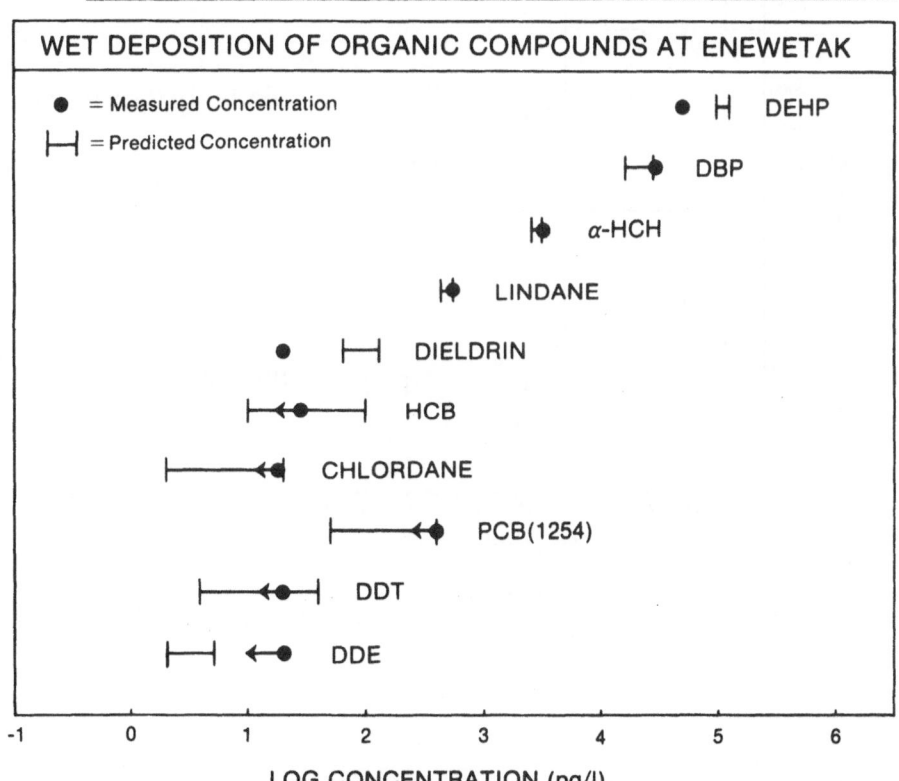

Figure 6. Comparison of measured versus predicted concentrations of organic compounds in rainfall at Enewetak Atoll.

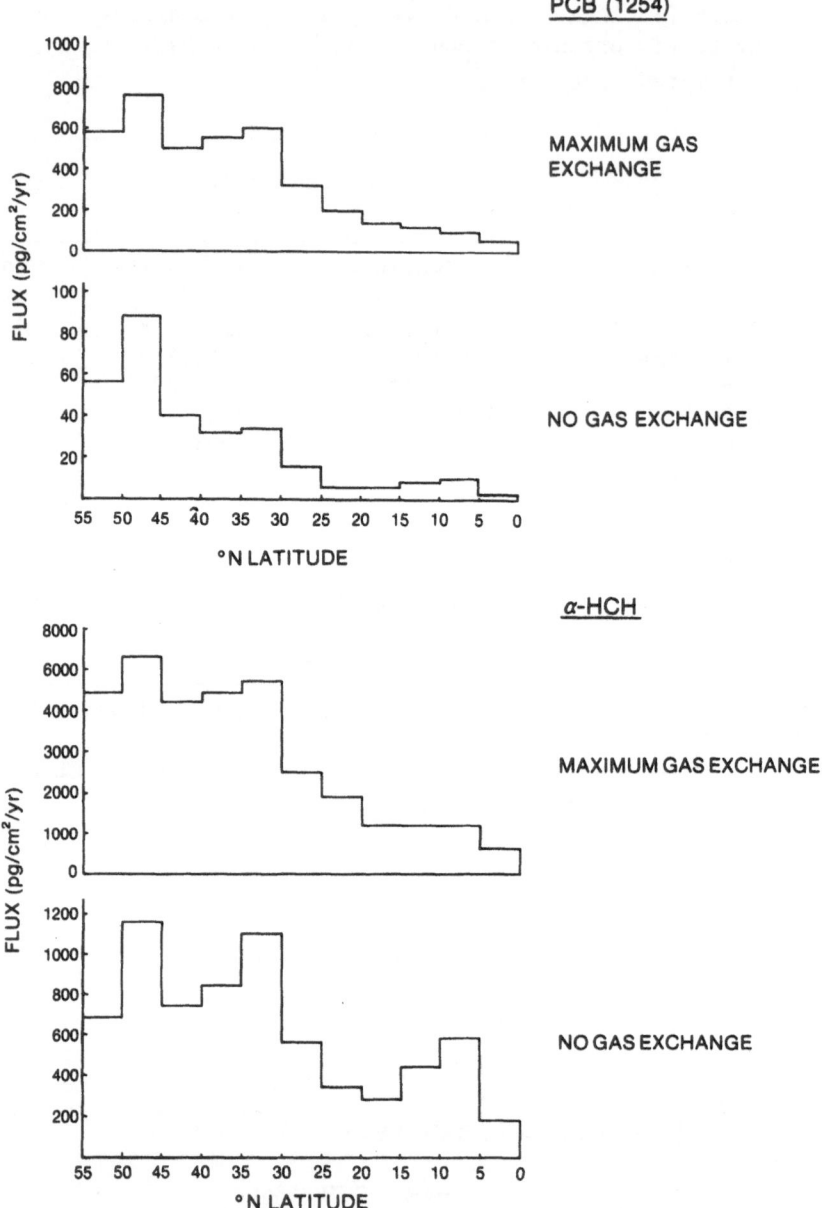

Figure 7. Estimated profiles of air-sea flux of PCB(1254) and α-HCH in the North Pacific for different conditions of gas-exchange flux.

        Results of these calculations are illustrated in Figures 7 and 8.
The profile of air-sea flux of both PCB and HCH (Fig. 7) versus
latitude show similar features. The single most notable feature is the
enhanced flux at high latitudes. The increased flux is the result of
two mechanisms: 1) enhanced particle adsorption and precipitation
scavenging (PCB) and 2) enhanced partition of vapor phase compound
into precipitation. The relative importance of different mechanisms of
air-sea exchange is seen in Figure 8. For conditions of maximum gas
flux, gas exchange (dry deposition) is the major factor for both PCB
and HCH. For saturated seawater (gas exchange=0), precipitation
scavenging of particles (PCB) or vapor (HCH) is the primary mechanism
of air-sea transfer. Dry deposition accounts for ≤10% of the total
deposition under these conditions. A summary of these calculations for
these and other compounds is presented in Table IX.
        Because of the difficulty in obtaining an accurate measurement of
particle-bound material at low atmospheric concentrations, calculations
are also included in Table IX to illustrate the effect of a 10-fold
increase in particle-bound CHC. The increased range is not as large as
it seems. For example, in tropical areas this would increase
particulate PCB from 0.15 to 1.5% of the total. This is not unrealistic
range for aerosol CHC in the atmosphere and may represent a reasonable
uncertainty in the true vapor/particle partitioning in the atmosphere.
As can be seen, this type of increase in particulate CHC will cause
nearly the same increase in flux as the change in gas phase deposition.
Thus, uncertainties in the overall flux are related both to poor
understanding of compound chemistry in natural waters and to somewhat
imprecise estimation of particulate organic compounds in the
atmosphere. In the next section we will compare these estimates to
other deposition measurements, to estimated riverine influx, and to
sedimentation rates of CHC.

7. RELATIVE IMPORTANCE OF ATMOSPHERIC DEPOSITION TO THE CHC CYCLE

The preceeding sections have set limits on the rates and mechanisms of
air-sea transfer of certain organic pollutants, though there are
clearly major areas of uncertainty. In this section we will compare
these estimates to other segments of the geochemical cycle of organic
pollutants.
        First, to estimate the riverine input of organic pollutants to the
global ocean, we could only uncover very limited data on some river
systems, e.g. the Mississippi, and no data on other major rivers, ie.
the Yangtze and Amazon. In addition, since much of the river load of
organic pollutants is associated with particles, the bulk riverine
inputs will be deposited near the river mouth. Thus, our estimate of
riverine input of organic pollutants to the ocean is a reasonable
guess, at best. We chose as upper limits for global average river
concentrations the following: PCBs, HCHs: 1 ng/l; DDTs, HCB, dieldrin,
chlordane; 0.1 ng/l; DBP, DEHP 100 ng/l. The results of this
calculation are shown in Table IX. This rather crude comparison

suggests that maximum estimated riverine fluxes are typically less than
minimum estimated atmospheric fluxes. Thus, for most ocean areas away
from riverine influences it is expected that atmospheric inputs will be
the dominant source of pollutants to the sea.

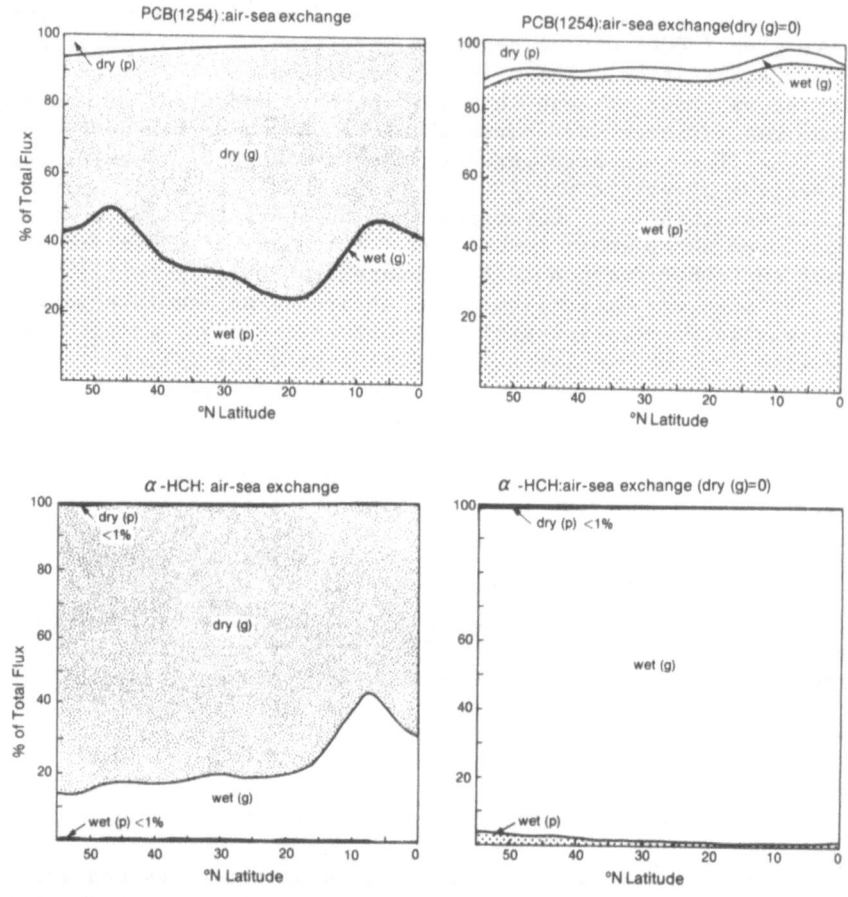

Figure 8. Mechanisms of air-sea exchange of PCB(1254) and
α-HCH in the North Pacific under different assumptions of
gas-exchange flux.

        The second area we wish to compare is atmospheric input with
sedimentation. Only a few measurements are available to make this
comparison. Burns et al. (1985), Larsson (1984), Knap and Binkley
(1986) have used sediment traps to measure the sedimentation of PCB and
a few other chlorinated hydrocarbons in marine areas. These results
are presented in Table X. These data suggest that open ocean
atmospheric fluxes are at least equivalent to present day sedimentation
rates in the open ocean, while sedimentation fluxes in the coastal zone

are possibly higher. We might speculate further and suggest that the less particle reactive compounds, such as chlordane, dieldrin, and lindane, may accumulate in surface waters since atmospheric inputs appear to exceed sedimentation. Such profiles have been reported for HCH, but no data are available to test this speculation for other CHC.

TABLE X   Sedimentation rates of chlorinated hydro-carbons measured in sediment traps.

| | SEDIMENTATION | | | Atmos. |
| | Coastal | | Open Ocean | Deposit. |
Compound	Hano Bight	Mediterranean	North Atlantic	North Pacific
PCB*	5,800	3,650 3,200	160	20-284
DDE	270	97 97	---	0.2-5.3
Lindane	---	85 65	---	97-218
Chlordane	---	---	2.1	1.0-23
Dieldrin	---	---	3.8	16-95
References	1	2	3	4

* as Aroclor 1254

1 Larsson, 1984   2 Burns et al., 1985
3 Knap and Binkley, 1986   4 This work

8. SUMMARY AND CONCLUSIONS

This chapter has presented some data and speculation on the processes of air-sea exchange and the importance of these processes in the cycle of organic pollutants in the marine environment. It is clear from measurement and calculation that atmospheric exchange processes are a critical link in the geochemical cycle of synthetic organic compounds. Furthermore, it was shown that the chemical properties of individual synthetic organic compounds can be quite different, and these differences will affect transfer mechanisms and chemical reactivity in surface ocean waters. However, research to determine appropriate chemical factors such as partition coefficients, distribution coefficients, and association with natural organic compounds still requires more progress, especially as these parameters vary with temperature. In addition, more work is required to accurately quantitate the particulate fraction of synthetic compounds in air (also as a function of temperature) and to examine the partition of these compounds between particle and dissolved forms in precipitation. The lack of quantitative information on the "fugacity" of synthetic organic compounds in natural waters is also apparent. Perhaps, experiments at

sea could be performed to determine air/water equilibria for a range of
different compounds under ambient conditions.

    Finally, one area, thus far ignored, needs to be mentioned: that
is the effect of biological mediation and transformations on pollutant
transfer at the ocean surface. The interaction of non-polar organic
pollutants with both living and non-living organic material has been
well-demonstrated in marine sedimentation and distributions of organic
contaminants in sediments. The possibility of similar interactions
affecting air/sea transfer are high. One possible example of
biological mediation of pollutant transfer would be the bacterial
methylation of halogenated phenols in surface waters, and subsequent
volatilization of the resultant anisoles from the ocean surface. Such
halogenated anisoles have recently been discovered in the marine
atmosphere, though their source is unknown (Atlas et al., 1986). Other
such interactions of synthetic compounds with organisms are likely,
though they have been largely overlooked.

9. REFERENCES

Andren, A. W., 1983, Processes determining the flux of PCBs across air-
    water interfaces, Chapter 8 in: *Physical Behavior of PCBs in the Great
    Lakes*, MacKay, D., Paterson, S., Eisenreich, S. J., Simmons, M.
    (eds.), Ann Arbor Sciences, Ann Arbor, MI, 127.

Arimoto, R., Duce, R. A., Ray, B. J., and Unni, C. K., 1985,
    'Atmospheric trace and dry deposition.' *J. Geophys. Res.* **90**, 2391.

Atlas, E. L. and Giam, C. S., 1981, 'Global transport of organic
    pollutants: ambient concentrations in the remote marine
    atmosphere.' *Science* **211**, 163-165.

Atlas, E. and Giam, C. S., *SEAREX Newsletters* **1-9**.

Atlas, E., and Giam, C. S., 1985. Unpublished data.

Atlas, E., Foster, R. and Giam, C. S., 1982, 'Air-sea exchange of high
    molecular weight organic pollutants: laboratory studies.' *Environ.
    Sci. Technol.* **16**, 283.

Atlas, E., Velasco, A., Sullivan, K., and Giam, C. S., 1983, 'A
    radiotracer study of air-water exchange of synthetic organic
    compounds.' *Chemosphere* **12**, 1251.

Atlas, E., Sullivan, K., Madero, M., and Giam, C. S., 1985a, 'Trace
    organic compounds in rural air and rainfall.' Submitted to Water
    Air and Soil Pollution.

Atlas, E., Bidleman, T. F. and Giam, C. S., 1985b, Atmospheric transport of PCB to the Ocean. In: *PCB and the Environment*, J. S. Waid, ed., CRC Press, Boca Raton (in press).

Atlas, E. L., Sullivan, K., and Giam, C. S., 1986, 'Widespread occurrence of polyhalogenated aromatic ethers in the marine atmosphere.' *Atmos. Environ.* (in press).

Ballschmiter, K., Buchert, H., Bihler, S. and Zell, M., 1981, 'Baseline studies of global pollution: IV. The pattern of pollution by organo-chlorine compounds in the North Atlantic as accumulated by fish.' *Fresenius Z. Anal. Chem.* 306, 323-339.

Barrie, L. and Neustadter, J., 1983, The dependence of sulphate scavenging ratios on meteorological variables. In: *Precipitation Scavenging, Dry Deposition, and Resuspension*, Pruppacher et al. (eds.), Elsevier Science Publishing, New York, 203-215.

Bidleman, T. F., 1985, personal communication.

Bidleman, T. F. and Leonard, R., 1982, 'Aerial transport of pesticides over the Northern Indian Ocean and adjacent seas.' *Atmos. Environ.* 16, 1099-1107.

Bidleman, T. F. and Olney, C. E., 1974, 'Chlorinated hydrocarbons in the Sargasso Sea atmosphere and surface water.' *Science* 183, 516-518.

Bidleman, T. F., Olney, C. E., and Rice, C. P., 1976, High-molecular weight chlorinated hydrocarbons in the air and sea: Rates and mechanisms of air/sea transfer. In: *Marine Pollutant Transfer*, H. L. Windom and R. A. Duce (eds.), D.C. Heath and Co., Lexington, MA, 323-351.

Bidleman, T. F. and Christensen, E. J., 1979, 'Atmospheric removal processes for high-molecular weight organochlorines.' *J. Geophys. Res.* 84, 7857.

Bidleman, T. F., Christensen, E. J., Billings, W. N., and Leonard, R., 1981, 'Atmospheric transport of organochlorines in the North Atlantic gyre.' *J. Mar. Res.* 39, 443-464.

Billings, W. N., and Bidleman, T. F., 1983, 'High volume collection of chlorinated hydorcarbons in urban air using three solid adsorbents.' *Atmos. Environ.* 17, 383-391.

Boehm, P. D., 1983, 'Coupling of particulate organic pollutants between the estuary and continental shelf and the sediments and water column in the New York Bight region.' *Can. J. Fish. Aquatic Sci.* 40 (Suppl 2), 262-276.

Brownawell, B. J., and Farrington, J. W., 1985, 'Biogeochemistry of
    PCBs in interstitial waters of a coastal marine sediment.'
    Geochim. Cosmochim. Acta, (In press).

Burkhard, L. P., Armstrong, D. E., and Andren, A. W., 1985, 'Henry's
    Law Constants for the polychlorinated biphenyls.' *Environ. Sci.
    Technol.* **19**, 590-596.

Burns, K. A., Villeneuve, J. -P., and Fowler, S. W., 1985, 'Fluxes and
    residence times of hydrocarbons in the coastal Mediterranean: How
    important are the biota?' *Est. and Coastal Shelf Sci.* **20**, 313-330.

Caron, G., Suffet, I. H., and Belton, T., 1985, 'Effect of dissolved
    organic carbon on the environmental distribution of nonpolar
    organic compounds.' *Chemosphere* **14**, 993-1000.

Chang, L. W., Atlas, E., and Giam, C. S., 1985, 'Chromatographic
    separation and analysis of chlorinated hydrocarbons and phthalate
    esters from ambient air samples.' *J. Intern. Environ. Anal. Chem.* **19**,
    145.

Chiou, C. T., Porter, P. E., and Schmedding, D. W., 1983, 'Partition
    equilibria of nonionic organic compounds between soil organic
    matter and water.' *Environ. Sci. Technol.* **17**, 227-231.

Christensen, E. J., Olney, C. E., and Bidleman, T. F., 1979,
    'Comparison of dry and wet surfaces for collecting organochlorine
    dry deposition.' *Bull. Environ. Contam. Toxicol.* **23**, 196-202.

deLappe, B. W., Risebrough, R. W., and Walker, W. II, 1983, 'A large-
    volume sampling assembly for the determination of synthetic
    organic and petroleum compounds in the dissolved and particulate
    phase of seawater.' *Can. J. Fish. Aquat. Sci.* **40** (suppl. 2), 322.

Duinker, J. C., and Hillebrand, M. T. J., 1979, 'Behavior of PCB,
    pentachlorobenzene, hexachlorobenzene, α-HCH, β-HCH, γ-HCH,
    dieldrin, endrin, and p,p'-DDD in the Rhine-Meuse estuary and the
    adjacent coastal area.' *Neth. J. Sea Res.* **13**, 256.

Duinker, J. C., and Hillebrand, M. T. J., 1983, 'Composition of PCB
    mixtures in biotic and abiotic marine compartments (Dutch Wadden
    Sea).' *Bull. Environ. Contam. Toxicol.* **31**, 25-32.

Eisenreich, S. (ed), 1981, *Atmospheric Pollutants in Natural Waters*, Ann
    Arbor Science, Ann Arbor, MI, 512 p.

Eisenreich, S. J., Hollod, G. J., and Johnson, T. C., 1981, Atmospheric
    concentrations and deposition of PCB to Lake Superior, Chapter 21
    in: *Atmospheric Pollutants in Natural Waters*, Eisenreich, S. J. (ed.),
    Ann Arbor Science, Ann Arbor, MI, 425-444.

Eisenreich, S. J., Looney, B. B., and Hollod, G. J., 1983, PCBs in the Lake Superior atmosphere 1978-1980, Chapter 7 in: *Physical Behavior of PCBs in the Great Lakes*, MacKay, D., Paterson, S., Eisenreich, S. J., and Simmons, M. (eds.), Ann Arbor Science, Ann Arbor, MI, 115-125.

Elliott, W. P., and Reid, R. K., 1984, 'A climatological estimate of precipitation for the World Ocean.' *J. Clim. and App. Met.* 23, 434-439.

Giam, C. S., Chan, H. S., and Neff, G. S., 1975, 'Rapid and inexpensive method for detection of PCB and phthalates in air.' *Anal. Chem.* 47, 2319-2320.

Giam, C. S., Atlas, E., Chan, H., and Neff, G., 1977, 'Estimation of fluxes of organic pollutants to the marine environment: Phthalate ester concentration and fluxes.' *Rev. Int. Oceanogr. Med.* 47, 79-84.

Giam, C. S., Chan, H. S., Neff, G. S., and Atlas, E. L., 1978, 'Phthalate esters plasticizers: a new class of marine pollutant.' *Science* 199, 419-421.

Giam, C. S., Atlas, E. L., Chan, H. S., and Neff, G. S., 1980, 'Phthalate esters, PCB, and DDT residues in the Gulf of Mexico atmosphere.' *Atmos. Environ.* 14, 65-69.

Gschwend, P. M. and Wu, S., 1985, 'On the constancy of sediment-water partition coefficients of hydrophobic organic pollutants.' *Environ. Sci. Technol.* 19, 90.

Harvey, G. R., and Steinhauer, W. G., 1976, Biogeochemistry of PCB and DDT in the North Atlantic Ocean. In: *Environmental Biogeochemistry*, 1, J. O. Nriagu (ed.), Ann Arbor Science, Ann Arbor, MI, 203-221.

Harvey, G. R., Steinhauer, W. G., and Teal, J. M., 1973, 'Polychlorobiphenyls in North Atlantic Ocean water.' *Science* 180, 643-644.

Hassett, J. P., and Anderson, M. A., 1982, 'Effects of dissolved organic matter on adsorption of hydrophobic organic compounds by river- and sewage-borne particles.' *Water Res.* 16, 681-686.

Hassett, J. P., and Milicic, E., 1985, 'Determination of equilibrium and rate constants for binding of a polychlorinated biphenyl congener by dissolved humic substances.' *Environ. Sci. Technol.* 19, 638-643.

Hicks, B. B., and Shannon, J. P., 1979, 'A method for modeling the deposition of sulfur by precipitation over regional scales.' *J. Appl. Met.* 18, 1415-1420.

Hom, W., Risebrough, R. W., Soutar, A., and Young, D. R., 1974, 'Deposition of DDE and polychlorinated biphenyls in dated sediments of the Santa Barbara Basin.' *Science* 184, 1197-1199.

Knap, A. H., and Binkley, K., 1986, 'The flux of synthetic organic chemicals to the deep Sargasso Sea.' *Nature* (in press).

Larsson, P., 1984, 'Sedimentation of polychlorinated biphenyls (PCBs) in limnic and marine environments.' *Water Res.* 18, 1389-1394.

Lincoff, A. H., and Gossett, J. M., 1984, The determination of Henry's Constants for volatile organics by equilibrium partition in closed systems. In: *Gas Transfer at Water Surfaces*, W. Brutsaert and G. H. Jirka (eds.), D. Reidel Publishing Co., Holland, 17-26.

Liss, P. S., and Slater, P. G., 1974, 'Fluxes of gases across the air-sea interface.' *Nature* 247, 181.

Mackay, D., Shiu, W. Y., and Sutherland, R. P., 1979, 'Determination of air-water Henry's Law constants for hydrophobic pollutants.' *Environ. Sci. Technol.* 13, 333-337.

Mackay, D. M., Shiu, W. Y., Bobra, A., Billington, J., Chau, E., Yeun, A., Ng, C., and Zeto, F. S., 1982. Volatilization of Organic Pollutants from Water. EPA Final Report - EPA 600/3-82-019, 212 pp.

Mackay, D., Paterson, S., Eisenreich, S. J., and Simmons, M., (eds.), 1983. *Physical Behavior of PCBs in the Great Lakes*, Ann Arbor Science, Ann Arbor, MI, 442 pp.

McClure, V. E., 1976, 'Transport of heavy chlorinated hydrocarbons in the atmosphere.' *Bull. Environ. Contam. Toxicol.* 17, 1223-1229.

Means, J. C., Wijayaratne, R. D., and Boynton, W. R., 1983, 'Fate and transport of selected herbicides in an estuarine environment.' *Can. J. Fish. Aquat. Sci.* 40 (suppl. 2), 337.

Murphy, T. J., and Rzeszutko, C. P., 1977, 'Precipitation inputs of PCBs to Lake Michigan.' *J. Great Lakes Res.* 3, 305-312.

Murphy, T. J., and Schinsky, A. W., 1983, 'Net atmospheric inputs of PCBs to the ice cover on Lake Huron.' *J. Great Lakes Res.* 9, 92-96.

Murphy, T. J., Schinsky, A., Paolucci, G., and Rzeszutko, G., 1981, Inputs of polychlorinated biphenyls from the atmosphere to lakes Huron and Michigan. In: *Atmospheric Pollutants in Natural Waters*, S. J. Eisenreich (ed.), Ann Arbor Science, Ann Arbor, MI, 445-458.

Murphy, T. J., Pokojowczyk, J. C., and Mullin, M. D., 1982, Vapor exchange of PCBs with Lake Michigan: The atmosphere as a sink for PCBs. In: *Physical Behavior of PCBs in the Great Lakes*, D. Mackay (ed.), Ann Arbor Science, Ann Arbor, MI, 49-58.

Oehme, M., and Mano, S., 1984, 'The long-range transport of organic pollutants to the Arctic.' *Fresenius Z. Anal. Chem.* **319**, 141.

Oehme, M., and Ottar, B., 1984, 'The long-range transport of polychlorinated hydrocarbons to the Arctic.' *Geophys. Res. Lett.* **11**, 1133-1136.

Oehme, M., and Stray, H., 1982, 'Quantitative determination of ultra-traces of chlorinated compounds in high-volume air samples from the Arctic using polyurethane foam as the collection medium.' *Fresenius Z. Anal. Chem.* **311**, 665-673.

Pankow, J. F., Isabelle, L. M., and Asher, W. E., 1984, 'Trace organic compounds in rain. 1. Sampler design and analysis by adsorption/thermal desorption (ATD.).' *Environ. Sci. Technol.* **18**, 310.

Rice, C. P., Eadie, B. J., and Erstfeld, K. M., 1982, 'Enrichment of PCBs in Lake Michigan surface films.' *J. Great Lakes Res.* **8**, 265.

Peel, D. A., 1975, 'Organochlorine residues in Anarctic snow.' *Nature* **254**, 324-325.

Sehmel, G. A., 1980, 'Particle and gas dry deposition: A review.' *Atmos. Environ.* **14**, 983.

Settle, D. M., and Patterson, C. C., 1982, 'Magnitudes and sources of precipitation and dry deposition fluxes of industrial and natural leads to the North Pacific at Enewetak.' *J. Geophys. Res.* **87**, 8857.

Simon, C. G., and Bidleman, T. F., 1979, 'Quantitative determination of ultra-traces of chlorinated compounds in high volume air samples from the Arctic using polyurethane foam as a collection medium.' *Fresenius Z. Anal. Chem.* **311**, 665-673.

Slater, R. M., and Spedding, D. J., 1981, 'Transport of dieldrin between air and water.' *Arch. Environ. Contam. Toxicol.* **10**, 25-.

Slinn, S. A., and Slinn, W. G. N., 1980, 'Predictions for particle deposition on natural waters.' *Atmos. Environ.* **12**, 2055-.

Slinn, W. G. N., Hasse, L., Hicks, B. B., Hogan, A. W., Lal, D., Liss, P. S., Munnich, K. O., Sehmel, G. A., and Vittori, O., 1978, 'Some aspects of the transfer of atmospheric trace constituents past the air-sea interface.' *Atmos. Environ.* **12**, 2055.

Smith, J. H., Bomberger, D. C., Jr., and Haynes, D. L., 1980,
    'Prediciton of the volatilization rates of high-volatility
    chemicals from natural water bodies.' *Environ. Sci. Technol.* **14**,
    1332.

Smith, J. H., Bomberger, D. C., Jr., and Haynes, D. L., 1981,
    'Volatilization rates of intermediate and low volatility chemicals
    from water.' *Chemosphere* **10**, 281-289.

Strachan, W. M. J., Huneault, H., Schertzer, W. M., and Elder, F. C.,
    1980, 'Organochlorines in precipitation in the Great Lakes Region.
    In: *Hydrocarbons and Halogenated Hydrocarbons in the Aquatic Environment*
    Afghan, B. K. and Mackay, D. (eds.), Plenum Publishing Co., New
    York, 387-396.

Strachan, W. M. J., Huneault, H., 1984, 'Automated rain sampler for
    trace organic substances.' *Environ. Sci. Technol.* **18**, 127-130.

Tanabe, S., and Tatsukawa, R., 1983, 'Vertical transport and residence
    time of chlorinated hydrocarbons in the open ocean water column.'
    J. Oceanogr. Soc. Japan **39**, 53.

Tanabe, S. Tatsukawa, R., Kawano, M., and Hidaka, H., 1982a, 'Global
    distribution and atmospheric transport of chlorinated
    hydrocarbons: HCH (BHC) isomers and DDT compounds in the Western
    Pacific, Eastern Indian, and Antarctic Oceans.' *J. Oceanogr. Soc.
    Japan* **38**, 137-148. Tanabe, S., Kawano, M., and Tatsukawa, R.,
    1982b, 'Chlorinated hydrocarbons in the Antarctic, Western
    Pacific, and Indian Oceans.' *Trans. Tokyo Univ. Fisheries* **5**, 97-109.

Tanabe, S., Hidaka, H., and Tatsukawa, R., 1983a, 'PCBs and chlorinated
    hydrocarbon pesticides in Antarctic atmosphere and hydrosphere.'
    Chemosphere **12**, 277-288.

Tanabe, S., Mori, T., Tatsukawa, R., and Miyazaki, N., 1983b, 'Global
    pollution of marine mammals by PCBs, DDTs, and HCHs (BHCs).'
    Chemosphere **12**, 1269.

U.S. Navy, 1969, 'Marine Climatic Atlas of the World.' vol. 8, NAVAER
    50-IC-535, U.S. Government Printing Office, Washington, D. C.

Wanninkhof, R., Ledwell, J. R., and Broecker, W. S., 1984, 'Gas-
    exchange - wind speed relation measured with sulfur hexafluoride
    on a lake.' *Science* **227**, 1224-1226.

Wells, D. E., and Johnstone, S. J., 1978, 'The occurrence of
    organochlorine residues in rainwater.' *Water, Air, and Soil Pollut.* **9**,
    271-280.

Wescott, J. N., Simon, C. G., and Bidleman, T. F., 1981, 'Determination of polychlorinated biphenyl vapor pressures by a semi-micro gas saturation method.' *Environ. Sci. Technol.* **9**, 1375-1378.

Whitehouse, B., 1985, 'The effects of dissolved organic matter on the aqueous partitioning of polynuclear aromatic hydrocarbons.' *Estuarine, Coastal, and Shelf Science* **20**, 393-402.

Whitman, W. G., 1923, 'Preliminary experimental confirmation of the two-film theory of gas adsorption.' *Chem. Metal. Eng.* **29**, 146.

Wijayaratne, R. D., and Means, J. C., 1984, 'Affinity of hydrophobic pollutants for natural estuarine colloids in aquatic environments.' *Environ. Sci. Technol.* **18**, 121-123.

Yamasaki, H., Kuwata, K., and Miyamoto, H., 1982, 'Effects of ambient temperature on aspects of airborne polycyclic aromatic hydrocarbons.' *Environ. Sci. Technol.* **16**, 189-194.

Zafiriou, D. C., Gagosian, R. B., Peltzer, E. T., Alford, J. B., and Loder, T., 1985, 'Air-to-sea fluxes of lipids at Enewetak Atoll.' *J. Geophys. Res.* **90**, 2409.

Zell, M., and Ballschmiter, K., 1980, 'Baseline studies of global pollution: II. Global occurrence of hexachlorobenzene (HCB) and polychlorocamphenes (Toxaphene) (PCC) in biological samples.' *Fresenius Z. Anal. Chem.* **300**, 387-402.

THE OCEAN AS A SOURCE OF ATMOSPHERIC SULFUR COMPOUNDS

Meinrat O. Andreae
Department of Oceanography
Florida State University
Tallahassee, FL 32306
U.S.A.

1. SOURCES OF SULFUR TO THE ATMOSPHERE: AN OVERVIEW

The impact of human activity on the global atmospheric sulfur cycle is easily seen in densely inhabited, industrialized regions: the degradation of visibility by haze, the acidity of atmospheric precipitation, and the damage to forest vegetation are among the more obvious symptoms of this impact. Atmospheric transport propagates these effects well beyond their source regions. The human perturbation of the atmospheric sulfur cycle results largely from the emission of sulfur dioxide (SO_2) from fossil fuel burning. A number of recent papers have reviewed these emissions and presented a detailed source allocation (e.g. Cullis and Hirschler, 1980; Möller, 1984). The estimates for man-made sulfur emissions fall into a relatively narrow range: about 2.5 ± 0.3 Tmol yr^{-1} (Tmol: 1 Teramole = 10^{12} mol = 32 x 10^{12} g).

A summary of natural sulfur emissions from all sources is given in Table I. This Table presents the best estimates of these fluxes based on current information; it must be emphasized that most of these estimates are rather uncertain. This applies especially to the emissions of particulate sulfur in the form of dust and seaspray and to the emissions from soil and plants on the continents. The main reasons for the uncertainty regarding continental emissions of biogenic sulfur compounds are (1) the difficulty of accurately determining hydrogen sulfide (H_2S) and methylmercaptan (MeSH), (2) the technical problems of measuring emission fluxes from forest and brush ecosystems, and (3) the inadequate geographical coverage of existing data.

In the following sections, I will discuss the production of particulate sulfur by the seaspray process, the principles of biogenic sulfate reduction and synthesis of volatile species, the oceanic emission of dimethylsulfide (DMS) and carbonyl sulfide (COS), and the fate of these compounds in the marine atmosphere. I will not discuss the emissions of sulfur compounds from salt marshes; this topic has been reviewed recently (Andreae, 1985a, and references therein).

331

P. Buat-Ménard (ed.), The Role of Air-Sea Exchange in Geochemical Cycling, 331–362.
© *1986 by D. Reidel Publishing Company.*

Table I. Estimates of natural sulfur emissions (Tmol S yr^{-1})

	SO_2	H_2S	COS	DMS	CS_2	Sulfate	Other	Total
Seaspray						1.2–10		1.2–10
Dust						0.1– 1		0.1– 1
Total particulates						1.3–11		1.3–11
Volcanoes	0.25	0.03	0.0003		0.0003	<0.1	?	0.4
Soils and plants	–	0.1– 1.3	0.006– 0.02	0.006– 0.12	0.02– 0.025	–	0.03	0.15– 1.5
Coastal wetlands	–	0.03	0.004	0.02	0.002	–	0.004	0.06
Biomass burning*	0.2	?	0.003	–	?	?	?	?
Oceans (gases)	–	0– 0.5	0.011	1.1	0.01	–	?	1.1– 1.6
Total gases	0.5	0.2– 2.0	0.03– 0.04	1.1– 1.2	0.03– 0.04	<0.1	0.03	1.9– 3.6

*All gaseous emissions (other than COS) assumed to be SO_2

2. SEASPRAY AND THE PRODUCTION OF AEROSOL SULFATE

The only significant direct source of particulate sulfur to the marine
atmosphere results from the production of seasalt aerosol at the ocean
surface. The process that leads to the production of this aerosol is the
formation of seawater droplets as a result of the bursting of air
bubbles at the sea surface. Under very high wind conditions, droplets
can also be formed when water is torn away from the crest of waves. Two
types of droplets are produced by the bubble bursting process: jet
droplets, which result from the break-up of a water jet rising from the
bottom of the collapsing bubble cavity, and film droplets from the
bursting of the thin film of water that separates the air in the bubble
from the atmosphere (Blanchard, 1983; Monahan, this volume). Partial
evaporation of the water from the seawater droplet produces brine drops
or seasalt particles, depending on the relative humidity and the droplet
diameter.

    The major ion composition in the seaspray aerosol is generally
similar to that of seawater, but some evidence suggests an enrichment of
sulfate ion during the spray process. Garland (1981) observed an enrichment of sulfate relative to sodium by about 20% in model experiments,
but Duce et al. (1982) pointed to the possibility of artefacts in
Garland's experiments. Since secondary atmospheric processes can lead to

an enrichment of seasalt aerosol with sulfate by deposition of sulfuric
acid formed in the atmosphere onto seasalt particles (Andreae, 1982;
Andreae et al., 1985b), sulfate enrichment during the spray formation
process cannot easily be deduced from field studies on the marine
aerosol. Therefore, the question whether at least some of the sulfate
enrichment in marine aerosols is due to sea-surface fractionation
processes remains open (Andreae, 1982).

Varhelyi and Gravenhorst (1983) have attempted to estimate the
seasalt sulfate flux directly from a compilation of sulfate data in
marine aerosols and from wet deposition measurements. They give a range
of 4 to 8 Tmol yr^{-1} for the global seasalt sulfate deposition, implying
a source flux of the same size.

All other source estimates for seasalt sulfate begin with an estimate of the total seasalt flux. This presents some conceptual difficulty, as the distribution of seasalt in the marine atmosphere shows a
steep decrease with altitude (Blanchard et al., 1984). Consequently, the
flux through the atmosphere must also be decreasing with altitude: the
emission rate proper, i.e. the flux through the lowest meter of the
boundary layer must be far in excess of Blanchard's (1983) estimate of
10^4 Tg yr^{-1}, while the amount of seasalt cycled through the upper troposphere should be several orders of magnitude smaller. Monahan (this
volume) has used the global rate of whitecap formation at the sea surface and experimental data on the production of droplets by whitecaps to
estimate seasalt aerosol production: this estimate of 3,500 Tg yr^{-1} for
particles in the size range which can be mixed through the marine boundary layer is reasonably consistent with Blanchard's value.

As the residence time and consequently the ability of the seasalt
particles to undergo reactions and to be transported also vary with the
altitude to which they are able to mix before being re-deposited, no
single value for the seaspray aerosol emission rate can be meaningfully
assigned. Petrenchuk (1980) has estimated a seasalt flux of ca. 1,300 Tg
yr^{-1} though the lower troposphere. Using this value and a 20% enrichment
in sulfate during the spray-formation process, we obtain a sulfate flux
of 1.2 Tmol yr^{-1}. On the other hand, if we use Blanchard's (1983) value
of 10^4 Tg yr^{-1}, the predicted sulfate flux becomes ca. 10 Tmol yr^{-1}.

At this time, we can therefore only state that the sulfate flux by
seaspray aerosol formation probably lies between 1.2 and 10 Tmol yr^{-1}.
Both estimates are based on evaluation of the deposition fluxes of
seasalt by precipitation and dry deposition. The authors were forced to
extrapolate rather limited databases on deposition and rainwater composition to a global estimate, which resulted in large uncertainties. In
recent years, our knowledge of the composition of marine rains and on
the dry deposition of seasalt has increased considerably (McDonald et
al., 1982; Galloway et al., 1982, 1983; Church et al., 1982), and it
should be possible to improve the accuracy of the seasalt flux estimate
considerably on the basis of these data. Most of the seasalt entering
the atmosphere is deposited again to the ocean surface, but about 10% of
the total flux is carried over the continents and deposited on the land
surface.

3.   SULFATE REDUCTION BY GEOLOGICAL AND BIOLOGICAL PROCESSES

In the +6 oxidation state, the chemistry of sulfur is dominated by
sulfuric acid and sulfate, which are rather involatile chemical species.
Since only this oxidation state is stable in the presence of oxygen,
sulfate is the predominant form of sulfur in seawater, fresh waters, and
soils. Therefore, the reduction of sulfate to a more reduced sulfur
species is a necessary prerequisite for the formation of volatile sulfur
compounds which can be emitted to the atmosphere. In the global geochemical cycle, there are three processes which lead to a significant
rate of sulfate reduction: a geological one--the cycling of seawater
through the hydrothermal systems of the submarine ridges--and two
biological ones--assimilatory and dissimilatory sulfate reduction. In
Table II we compare estimates of the rates of sulfate reduction by these
processes in order to obtain a feeling for their magnitude relative to
each other and relative to the fluxes of sulfur through the atmosphere.
        The reduction of seawater sulfate by Fe(II) in basaltic rocks as a
result of ocean-floor hydrothermal circulation results in the production
of H_2S which is emitted from submarine hot springs. The rate given here
is an upper limit, the true rate of H_2S production may be as much as an
order of magnitude less. Since this H_2S is almost immediately reoxidized
by bacteria in the hydrothermal vent environment (and in this way sup-

Table II. Sulfate reduction rates by major biogeochemical processes and
          sulfur emission rates to the atmosphere (Tmol SO_4^{2-} yr^{-1})

Process	Rate	Reference
Reduction by Fe(II) in submarine hydrothermal systems (upper limit)	$\leqslant 4$	Von Damm 1983
Bacterial, dissimilatory sulfate reduction		
Coastal zone	2	
Shelf sediments	6	
Slope sediments	9	
Total	12-20	Ivanov and Freney 1983
Assimilatory sulfate reduction		
Land plants	3- 6	
Marine algae	10-20	
Total	13-26	Ehrlich et al. 1977
Gaseous sulfur emissions to the atmosphere		
Natural	2.5	Andreae 1985a
Anthropogenic	2.5	Cullis and Hirschler 1980
Total	5	

plies the metabolic energy which is responsible for the existence of
life in this environment) it cannot persist in seawater long enough to
contribute to atmospheric emissions.

Biological sulfate reduction has two major objectives: (1) the
biosynthesis of organic sulfur compounds which are used for various
purposes by the cell, e.g. in aminoacids, and (2) the use of sulfate as
a terminal electron acceptor to support respiratory metabolism in the
absence of molecular oxygen. The former process is called assimilatory
sulfate reduction (sulfur is being "assimilated"), the latter dissimilatory sulfate reduction. It is important to understand the ecological and biogeochemical differences between these two mechanisms: inadequate awareness of these differences between the two pathways of sulfate
reduction has led to many of the misinterpretations and false assumptions found in the literature on the atmospheric sulfur cycle, e.g. the
assumption that H_2S is the major reduced sulfur compound emitted from
the oceans.

The major pathway for the formation of H_2S is through dissimilatory
sulfate reduction. In contrast to the assimilatory pathway, the objective of which is the biosynthesis of sulfur compounds, dissimilatory
sulfate reduction is used to obtain thermodynamic energy in an oxygen-
depleted environment. The oxidation of organic matter by available
electron acceptors is the energetic basis for essentially all life
processes. The thermodynamically most favorable electron acceptor is
molecular oxygen, and if available, it will be used preferentially in
any ecosystem. However, if the supply of organic compounds exceeds that
of oxygen, other electron acceptors are used, e.g. nitrate or sulfate,
after oxygen has been depleted. Dissimilatory sulfate reduction is
therefore most commonly observed in marine environments where water
circulation, and consequently oxygen availability, is limited (e.g. in
stratified basins or in sedimentary pore waters) but where sulfate is
easily available due to its relatively high concentration in seawater
(28 mmol kg^{-1}).

Under favorable conditions, the rate of sulfate reduction to H_2S
can be quite high, on the order of hundreds of mmol m^{-2} day^{-1}. However,
since the occurrence of this process is dependent on the existence of a
mixing barrier which prevents oxygen from entering the system, the
escape of H_2S from the system will be limited by the same barrier.
Furthermore, in the presence of oxygen, H_2S provides an excellent substrate for microbial oxidation from which certain bacteria can obtain a
substantial amount of energy. Such microorganisms tend therefore to be
present in high numbers at the oxic/anoxic interface. They are very
efficient in removing H_2S and can completely oxidize this compound in a
layer only a fraction of a millimeter thick. Consequently, the very
large amounts of H_2S which are produced in the coastal and marine environment (Table II) cannot usually be transferred to the atmosphere
(Andreae, 1984, and references therein). Only under exceptional conditions, in shallow water environments, can a fraction of the H_2S escape:
through temperature- or wind-driven turnover in estuaries, through
scouring of muds in tidal channels, through bubbling of gas from anoxic
environments, etc. H_2S emissions from the marine environment are therefore limited to nearshore environments like estuaries and salt marshes.

4. ASSIMILATORY SULFATE REDUCTION

Sulfur in the form of a large variety of organosulfur compounds is an
essential element for biological organisms. Animals and protozoans are
dependent on organosulfur compounds in their food to supply their sulfur
requirement. All other biota---bacteria, blue-green algae, fungi, eucary-
otic algae, and plants---are able to carry out assimilatory sulfate
reduction, i.e. they can synthesize organosulfur compounds from environmental sulfate (Anderson, 1980). The biochemistry of assimilatory sulfate reduction has been studied mostly using the green alga <u>Chlorella</u>,
therefore most of the following discussion refers specifically to this
organism and it is not altogether clear at this time how far these
conclusions can be generalized to other organisms.

The assimilation of sulfate to cysteine, the first organosulfur
metabolite produced, is a complex, multi-step process (Figure 1). Sulfate is taken up into the cell by an active transport mechanism, and
inserted into an energetically activated molecule, APS (adenosine-5'-
phosphosulfate), which can be further activated at the expense of one
more ATP molecule to PAPS (3'-phosphoadenosine-5'-phosphosulfate). It is
then transferred to a thiol carrier (RSH) and reduced to the -2 oxidation state. In contrast to nitrate assimilation, where the various
intermediates are present free in the cytoplasm, sulfur remains attached
to a carrier during the reduction sequence. In a final step, the
carrier-bound sulfide reacts with O-acetyl-serine to form cysteine.
Wilson et al. (1978) have suggested that under conditions when the
availability of this or other endogenous sulfide acceptors is limiting
the rate of cysteine synthesis, the volatilization of H_2S could serve as
a mechanism for removing excess reduced sulfur. Such volatilization has
been observed from plants (e.g. Winner et al., 1981), but its possible
occurrence in marine algae has yet to be investigated.

Cysteine serves then as the source of all other sulfur metabolites,
especially the sulfur-containing aminoacids homocysteine and methionine
(Fig. 1). Cysteine and methionine are the major sulfur aminoacids in
plants and represent usually a very large fraction of the sulfur content
of biological materials (Giovanelli et al., 1980). Glutathione (L-
glutamyl-L-cysteyl-L-glycine) plays a variety of biochemical roles,
including redox transfer reactions and the removal of H_2O_2 in chloro-
plasts. Methionine reacts with ATP to form S-adenosyl-methionine (SAM),
the most important methyl group donor in methyl group transfer reactions
in plants and algae. Transfer of a methyl group from SAM to methionine
yields S-methyl-methionine, the precursor of dimethylsulfide (DMS) in
plants.

The processes discussed so far provide the "raw material" for the
production of biogenic volatile sulfur compounds. The reactions which
lead to volatile species will be discussed in subsequent sections:
production of DMS by algae (Section 5), photodecomposition of organic
sulfur compounds to COS (Section 10) and fermentative decomposition to a
variety of sulfur species (Section 11).

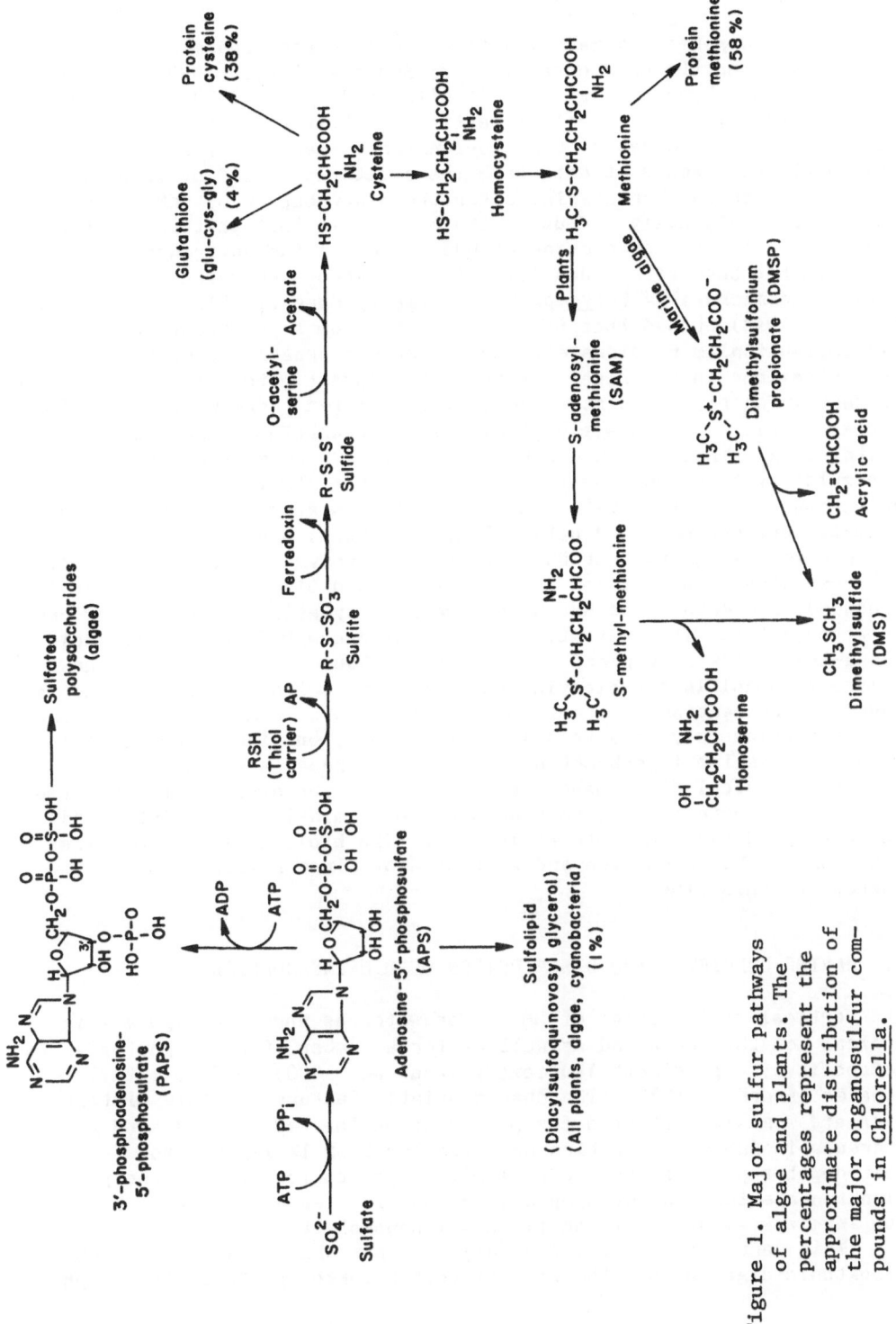

Figure 1. Major sulfur pathways of algae and plants. The percentages represent the approximate distribution of the major organosulfur compounds in Chlorella.

5. BIOSYNTHESIS OF DIMETHYLSULFIDE

In open ocean waters, dimethylsulfide (DMS) is the predominant volatile
sulfur compound (Barnard et al., 1982; Andreae et al., 1983; Andreae and
Barnard, 1984; Nguyen et al., 1984; Cline and Bates, 1983; Bingemer,
1984). Figure 2 shows a typical vertical distribution of DMS and
chlorophyll (an indicator of phytoplankton biomass) in the marine water
column at the example of data from the equatorial Pacific. This type of
profile, which is characteristic for the distribution of DMS throughout
the oceans, originally led us to the hypothesis that phytoplankton are
responsible for the production of DMS in the marine environment.

DMS had been first identified in the gaseous emissions of the
marine red macro-alga Polysiphonia lanosa by Haas (1935). Challenger and
Simpson (1948) showed that DMS was evolved from a precursor substance,
dimethylsulfonium propionate (DMSP) which was present in substantial
concentrations in the algal tissue. Later investigators found DMSP to be
present in nearly all algal species investigated (Ackman et al., 1966;
Tocher et al., 1966; Craigie et al., 1967; Granroth and Hattula, 1976;
White, 1982). This compound appears to have an osmostatic and osmoregulatory function in marine algae (Dickson et al., 1980, 1982;
Vairavamurthy et al., 1985). The similarity in structure and chemical
behavior between DMSP and other plant osmolytes, e.g. glycine betaine
and proline, suggests that DMSP has similar enzyme-protective properties
to these other "compatible" solutes (Brown and Simpson, 1972). DMSP is
produced from methionine by successive S-methylation, deamination, and
decarboxylation. Its enzymatic cleavage produces DMS and acrylic acid on
a one-to-one basis. Cantoni and Anderson (1956) have shown that the
enzyme responsible for cleaving DMSP contains sulfhydryl groups and is
bound to the membrane system. The release of DMS from the DMSP in algae
occurs continuously at a relatively slow rate, but increases manifold
when the organism is subjected to external stress, e.g. salinity
changes, physical disturbance (e.g. stirring), or exposure to the atmosphere. This effect leads to pronounced DMS emissions from intertidal
macro-algae during exposure at low tide. The biological or ecological
function of DMS production and excretion by algae remains, however,
unknown at this time.

6. MARINE CHEMISTRY AND DISTRIBUTION OF DIMETHYLSULFIDE

The vertical distribution of DMS in seawater as shown in Figure 2 is
typical for this compound as well as for a number of other phytoplankton
metabolites, e.g. dimethylsulfoxide (Andreae, 1980) and the methylar-
senates (Andreae, 1979). The characteristic features of this distribution are a maximum at or a few meters below the sea surface, and a sharp
decrease in DMS concentration near the level of 1% light transmission.
This depth represents the lower limit of growth for most of the phytoplankton species. In the deep oceans, DMS is present only at relatively
low levels: ca. 0.03-0.15 nM (1 nM = 1 nanomole L^{-1}).

This vertical distribution suggests that DMS is released by the
planktonic algae within the zone of phytoplankton productivity (euphotic

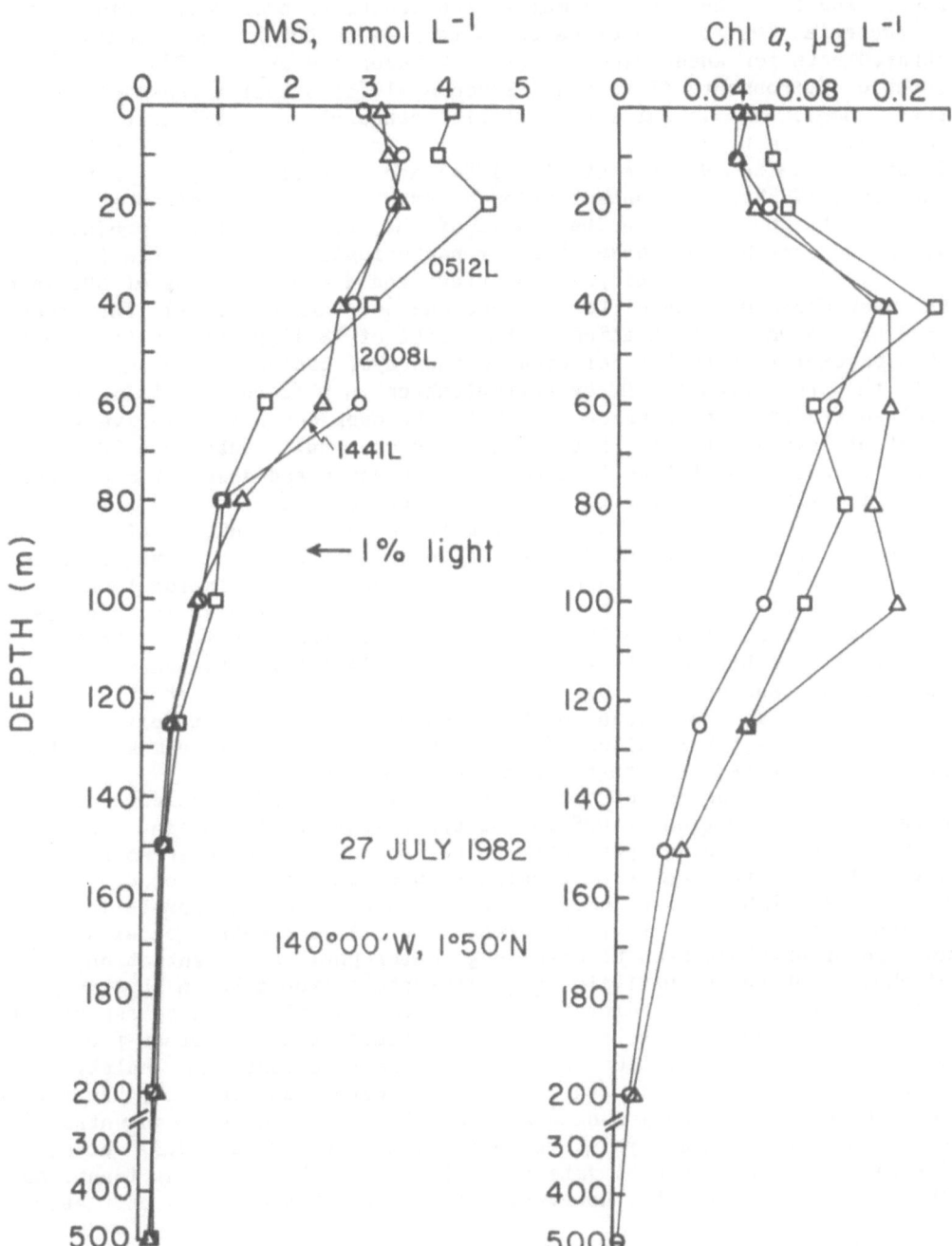

Figure 2. Vertical profiles of DMS concentration and particulate chlorophyll a concentration at a station in the equatorial Pacific. The
samples were collected while the ship was drifting with the current
at 0512 (square), 1441 (triangle), and 2008 (circle) local time.

zone), and that there is a simultaneous consumption of DMS, presumably
by bacteria. The ability of bacteria to grow on DMS has been demonstrated both for anaerobic conditions (Zinder and Brock, 1978) and for
aerobic environments (Sivelä and Sundman, 1975; Scheulderman-Suylen,
pers. comm., 1984). That such bacterial consumption of DMS actually
takes place in the marine environment is also suggested by its behavior
in anoxic basins (Wakeham et al., 1984) and in sedimentary porewaters
(Andreae, 1985b). The concentration of DMS in the water column at any
given depth would then be the result of the interplay of DMS production
by phytoplankton, DMS consumption by bacterioplankton, the volatilization of DMS across the air/sea interface and downward mixing of DMS into
the deep ocean by eddy diffusion (Andreae and Barnard, 1984). The steep
gradient in DMS concentration at the level of 1% light penetration would
then be explained by the relative dominance of bacterial consumption
over the production of DMS by phytoplankton in this region of light-
limited growth. The presence of DMS in the deep ocean at relatively
constant levels suggests that the abiotic chemical breakdown of DMS
under seawater conditions is a very slow process and does not contribute
significantly to the removal of DMS from surface waters.

Based on data on the uptake of sulfate and the concentration of DMS
in the water column of the Peru shelf upwelling region, I have estimated
the relative rates of production, consumption and ventilation loss of
DMS. The results suggest that on the order of 1% of the sulfur assimilated by phytoplankton in this region is converted to DMS, and that
roughly comparable amounts are lost by ventilation and by bacterial
consumption (Andreae, 1985b). These observations are consistent with the
requirement that the release of DMS should be only a relatively small
fraction of the total sulfur assimilated by plankton, since most of the
sulfur is required for other biochemical functions.

For the assessment of the sea-to-air flux of DMS, knowledge of the
oceanwide distribution of DMS in the upper meter of the ocean is required. Since it is not realistic to try to measure DMS everywhere, we
have attempted to find relationships between DMS and other observable
parameters which could be used for the prediction of DMS levels in
regions for which no direct measurements of its concentration exist. A
measure of phytoplankton biomass, e.g. chlorophyll $\underline{a}$ concentration, or
of phytoplankton productivity, e.g. ^{14}C-uptake, would be an obvious
candidate for such a predictor variable. Chlorophyll would be especially
attractive, since it can be estimated by remote sensing either from
aircraft or from satellites. Our attempts to find consistent relationships between chlorophyll and DMS have met with mixed success, however.
When we subject our entire data set on DMS and chlorophyll concentrations to regression analysis, we obtain values of r^2 near 0.3, which,
due to the large number of data (over 1000), are highly significant. As
the value of r^2 suggests, however, this correlation explains only about
30% of the variability.

While such analysis of large data sets (as well as the vertical
distribution of DMS in the marine water column) demonstrates a significant overall relationship between the distributions of DMS and
phytoplankton in the surface ocean, it is difficult to find a clear
correlation between total plankton abundance and DMS concentration

within a given region. This is most likely due to the substantial differences in the DMS output rate between different plankton species
(Andreae et al., 1983; Barnard et al., 1984). In some cases, a single
phytoplankton species can be responsible for most of the DMS in a given
oceanic region, e.g. Phaeocystis poucheti in the Bering Sea shelf region
(Figure 3; Barnard et al., 1984).

     Our data also show that the DMS concentrations in the low-
productivity regions of the oceans, especially the subtropical gyres,
are substantially higher than expected on the basis of the traditionally
accepted primary productivity rates for these areas: when we group our
DMS data according to the biogeographic regions of the oceans defined by
Koblentz-Mishke et al. (1970), and plot the average DMS concentrations
for each group against the average primary productivity for the same

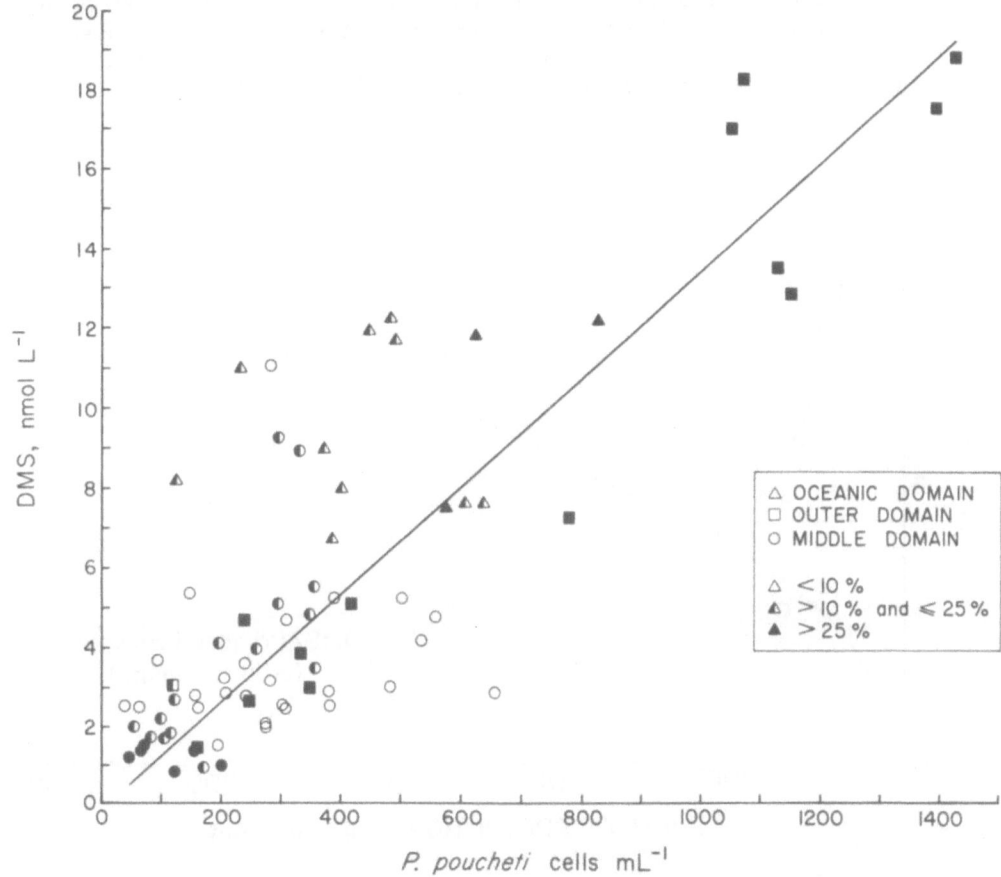

Figure 3. DMS concentration vs. Phaeocystis poucheti cell density. Open,
    half-full, and full symbols represent the percentage of P. poucheti
    cells relative to total phytoplankton cell density. The regression
    line is calculated from the data with a percentage of P. poucheti
    >25%.

group, we observe a close relationship in these highly averaged data
(Figure 4). However, the amount of DMS present per unit primary productivity increases in the regions of lower productivity. This may again be
due to species-related effects, or it may reflect the possibility that
the productivity in these regions has been underestimated, as has been
suggested recently (Shulenberger and Reid, 1981; Jenkins, 1982).
However, there is another speculative, yet tantalizing explanation: I
mentioned above that the DMS precursor, DMSP, functions as an osmolyte
in some algae. In order to achieve the required high internal osmotic
pressure to balance that of the seawater surrounding the cell (osmo-
larity ca. 1.1 mol L^{-1}), the cell needs to produce a substantial amount
of osmoregulatory substances. Many of the preferred osmolytes, however,
contain nitrogen (e.g. proline, betaine). This does not present a
serious problem in the productive regions, where nitrate is present in
the water column in relatively high concentrations. On the other hand,
in the nutrient-depleted regions of low productivity, e.g. the oceanic
gyres, the use of a sulfur osmolyte (DMSP) instead of a nitrogen osmolyte would make all bound nitrogen available for essential uses in

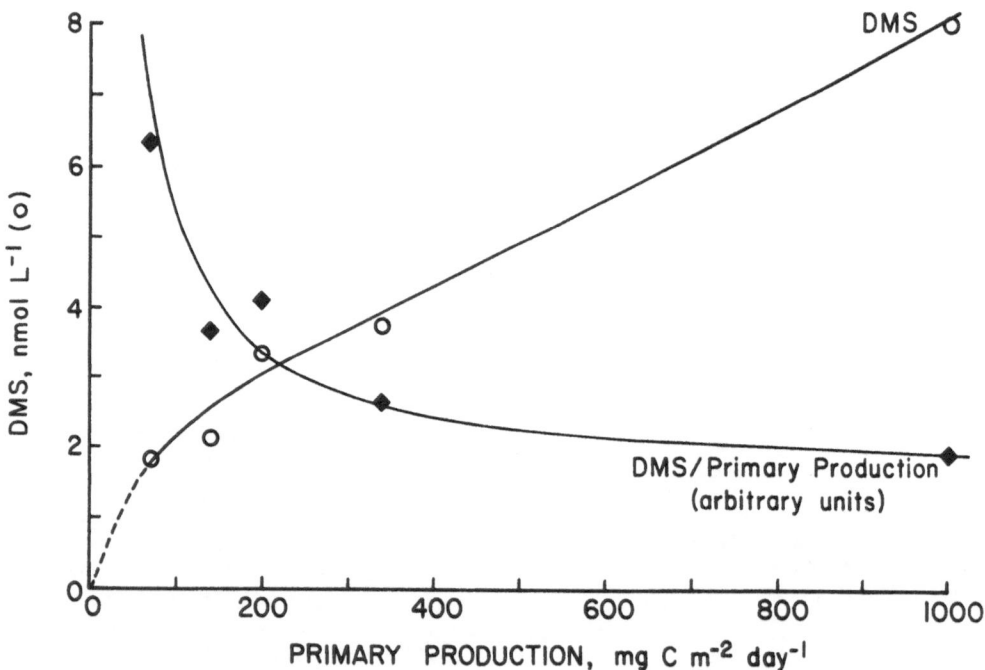

Figure 4. Average DMS concentration for data grouped by primary produc-
    tivity classes according to Koblentz-Mishke et al. (1970) (circles)
    and "specific DMS production" (diamonds). The latter parameter
    represents the amount of DMS released per unit primary produc-
    tivity.

aminoacids, etc. While the thermodynamic energy required to assimilate sulfate is higher than that needed to assimilate nitrate, it is comparable to the energy requirement for nitrogen fixation. Marine blue-green algae solve the problem of nitrogen-limitation by fixing molecular nitrogen. Consistent with this argument, we have found that Synecchococcus sp., a common blue-green alga of oceanic gyres, produces neither DMSP nor DMS. Nitrogen fixation is, however, not available to other algal taxa. These organisms could, therefore, benefit from replacing some of their nitrogen requirement with a molecule which contains sulfur in lieu of nitrogen. While this hypothesis would explain the increased levels of DMSP in species living in nutrient-depleted regions, it does not explain why some of this DMSP is broken down to DMS and excreted.

As a result of numerous cruises conducted by our group at Florida State University and of other groups (Barnard et al., 1982; Andreae et al., 1983; Andreae and Barnard, 1984; Nguyen et al., 1984; Cline and Bates, 1983; Bingemer, 1984) we now have a relatively good picture of the distribution of DMS in the World Oceans. This data is summarized in Table III which forms the basis for our estimate of the flux of DMS from the oceans to the atmosphere.

7. ESTIMATING THE AIR/SEA FLUX OF DIMETHYLSULFIDE

Volatile substances are transferred across the air/sea interface by a combination of molecular and turbulent diffusion processes, which are still poorly understood and for which no entirely satisfactory physical and mathematical models are available. For a discussion of the state of the art in this field see the papers by Liss and Merlivat in this volume.

The sea-to-air flux is proportional to the concentration gradient across the air/sea interface. We therefore need representative data on the DMS concentration in surface seawater and in the air over the sea surface to estimate this gradient. Since most measurements of DMS in seawater are made in samples collected about one meter below the sea surface, we need to exclude the possibility of significant gradients between the sampling depth and the actual surface. In the case of DMS there is no evidence for such gradients. Comparison between the concentration of DMS in bulk seawater from a few meters to a few centimeters depth and its concentration in sea surface microlayer material (a layer of some 10-100 μm thickness) has failed to show significant differences (Barnard et al., 1982; Andreae et al., 1983; Turner and Liss, 1985a). In view of the moderate reactivity of DMS in seawater (Andreae and Barnard, 1984) and the short transit time through the microlayer (on the order of a few seconds) we cannot expect any significant breakdown of DMS to occur in the microlayer.

The atmospheric concentration of DMS which would be in equilibrium with surface seawater can be calculated using the Henry's Law constant for DMS (0.074: Bingemer, 1984) and an average concentration in seawater of about 3 nM. The resulting equilibrium concentration in air (220 nmol m^{-3}) is so much higher than observed concentrations (on the order of 3 nmol m^{-3}) that the atmospheric DMS concentrations can be assumed to be

Table III. DMS concentrations and fluxes from the major biogeographic regions of the World
Oceans (fluxes are subject to uncertainty up to ±50% largely due to the uncertainty inherent in the conversion of concentration gradients to fluxes)

Ocean Region	N	Average DMS (nmol L^{-1})	s.d.[†]	Area (10^6 km^2)	Transfer Velocity (cm hr^{-1})	Flux (Tmol yr^{-1})	Ref.*
Oligotrophic Areas							
Tropical N. Atlantic, 1980	17	3.1	1.4				(1)
Tropical N. Atlantic, 1983	53	2.1	1.0				(2)
Tropical S. Atlantic, 1983	40	2.1	1.4				(2)
Gulf of Mexico, 1981	4	1.6	0.3				(3)
Gulf Stream, Sargasso Sea, 1981	80	2.5	0.7				(4)
Gulf Stream, Sargasso Sea, 1983	30	2.3	0.5				(5)
Brazil Current, 1980	63	1.4	0.5				(1)
Tropical N. Pacific, 1982	25	1.9	0.3				(6)
Tropical N. Pacific, 1982	69	2.9	1.4				(7)
Tropical N. Pacific, 1983	267	2.7	1.0				(7)
Total	648	2.4	0.04	148.3	10.4	0.32	
Transitional Areas							
Temperate N. Atlantic, 1980	44	2.1	0.9				(1)
Temperate N. Atlantic, 1984	29	4.4	2.9				(5)
Temperate N. Pacific, 1983	29	0.9	0.3				(7)
Temperate N. Pacific, 1984	22	0.7	0.1				(7)
Total	124	2.1	0.17	82.8	12.9	0.20	
Upwelling Areas							
Equatorial Zone							
Atlantic, 1980	22	2.2	0.7				(1)
Atlantic, 1983	21	1.7	0.8				(2)
Pacific, 1982	106	3.8	1.4				(6)
Coastal Upwelling							
Peru Shelf, 1982	134	7.2	7.3				(6)
N.W. Africa, 1983	33	3.0	3.6				(2)
S.W. Africa, 1983	5	6.6	1.4				(2)
Frontal Areas							
Bering Sea, 1981	13	4.7	4.7				(8)
Grand Banks, 1984	10	9.4	6.7				(5)
Ushant Front, 1980	17	4.9	1.2				(1)
Rio de la Plata Estuary, 1980	5	17.8	7.4				(1)
Total	366	5.4	0.3	86.5	12.1	0.49	
Coastal and Shelf Zone							
North Sea, Eng. Channel, 1980	27	1.7	0.9				(1)
South America, E. Coast, 1980	39	5.0	0.4				(1)
Ecuador Shelf, 1982	34	5.5	3.8				(6)
South Africa, W. Coast, 1983	8	4.3	2.0				(2)
South America, E. Coast, 1983	2	1.7	0.03				(2)
Florida, Gulf Coast, 1983	18	7.6	4.9				(5)
Florida, Atlantic Coast, 1983	5	2.8	1.3				(5)
Bahamas Coast, 1983	11	3.2	0.7				(5)
U.S., N.E. Coast, 1984	16	2.1	1.0				(5)
U.S., N.W. Coast, May 1983	136	1.4	0.7				(7)
U.S., N.W. Coast, Feb 1984	23	0.7	0.2				(7)
U.S., N.W. Coast, Aug 1984	44	2.3	1.9				(7)
Total	363	2.7	0.12	49.4	12.1	0.14	
World Ocean Total	1501	3.1	0.09	367.1	11.6	1.2	

*(1) Barnard et al., 1982; (2) Bingemer, 1984; (3) Andreae et al., 1983; (4) Andreae and
Barnard, 1984; (5) Andreae, unpublished; (6) Andreae and Raemdonck, 1983; (7) T. Bates,
pers. comm., 1985; (8) Barnard et al., 1984.

†Standard error of the mean is shown for area totals instead of standard deviation.

zero for the purpose of estimating of the sea-to-air flux of DMS. Consequently, we can estimate the oceanic source strength for DMS based on the concentration of DMS in surface seawater and the transfer velocities derived by the radon deficit and ^{14}C budget methods (Merlivat, this volume).

In Table III, I have summarized the currently available information on DMS concentrations in surface seawater. It contains the data collected by our group, as well as the work of H. Bingemer and T. Bates. The data of Turner and Liss (1985a) and of Nguyen et al. (1984) are not in a format which permitted their inclusion in this table, but are generally consistent with the values presented here. In Table III, the data are organized by biogeographical regions as defined by Koblentz-Mishke et al. (1970); averages for each of these regions are used together with an estimate of their areal extent for the prediction of the flux of DMS from each region.

The only measurements of the transfer velocities for any of the sulfur gases were made by Bingemer (1984) in a chamber system using the DMS naturally present in seawater. For DMS concentrations of 1.2 to 30 nM, he found fluxes that compared reasonably well with the predictions from theory. In the flux calculations in Table III, I use the data of Peng et al. (1979) and of Smethie et al. (1985) to estimate the transfer velocities. (To estimate the transfer velocity for DMS from that for radon, it was assumed that the transfer velocity is proportional to the square root of the diffusivity.) If we use the global average $^{14}CO_2$ transfer velocity (~21 cm hr^{-1}: Merlivat, this volume), we obtain a significantly higher flux: 1.6 instead of 1.2 Tmol DMS yr^{-1}. This is probably due to the fact that the $^{14}CO_2$ transfer velocity integrates over the whole year, while the radon transfer velocity is based almost entirely on summer data, when wind speeds are lower. In view of the extensive data on DMS concentrations in the surface ocean as presented in Table III, I feel that the major uncertainty about the sea-to-air flux of DMS now rests in the uncertainties associated with the use of the "stagnant-film" model, and in particular with the estimation of the transfer velocities. This uncertainty may be as large as a factor of two.

From Table III we can reach some interesting conclusions. First, there is a surprisingly small difference in the average DMS concentrations for the different regions. The average for the oligotrophic areas is essentially the same as for the transition areas of the temperate oceans, and both types of open ocean regimes have DMS concentrations quite similar to the coastal waters. Only in upwelling areas do we observe a substantially higher average concentration, but even here the difference is only a factor of two. One reason for these relatively small differences is that the tropical regimes have relatively high DMS concentrations year-round, whereas in temperate regions, especially the coastal temperate regions, there is a pronounced seasonality with quite low values during the cold season. Second, the large areas of low and moderate biological productivity contribute amounts of DMS to the atmosphere comparable to those from the relatively small regions of high productivity in the upwelling regions and the coastal areas. This is in contrast to earlier views which had assumed that the biogenic sulfur

flux from the oceans would be dominated by localized "hot-spots" of
biological productivity. Finally, we find that the estimate for the
global flux has by now become very robust relative to the addition of
new data (even including data from a number of different groups). While
the number of data points in Table III is about 2.5 times greater than
in the comparable table in Andreae and Raemdonck (1983), the global mean
DMS concentration has only changed from 3.2 to 3.1 nM, and the global
flux remains unchanged at ca. 1.2 Tmol yr^{-1}. The major remaining uncertainties in this estimate (besides those inherent in using the
"stagnant-film" model) are the concentrations of DMS in the Southern
Ocean, where no data exist at this time. The high wind speeds and correspondingly high transfer velocities in this region may make this
region an important source during the spring and summer seasons.

In view of the large uncertainties associated with the "stagnant-
film" model, it seems very important that independent methods be
developed to test the predictions based on this model. However, alternative methods to determine the flux, e.g. the eddy correlation or
gradient techniques, still face large experimental difficulties. No
rapid-response sensor which would make the eddy-correlation technique
possible is available for DMS, or in fact any of the sulfur gases. The
gradient method has been used on shipboard by Bingemer (1984) and by
Nguyen et al. (1984) by sampling at different levels above the water-
line. While the results compare well to the predictions from the gas
transfer calculations, they may contain substantial error due to the
influence of the ship on the air flow characteristics. Due to the difficulty of simulating a realistic wave climate inside a flux chamber,
direct measurements of sulfur gas fluxes across the air/sea interface by
the chamber technique have not been attempted.

8.   CHEMICAL REACTIONS AND TRANSFORMATIONS OF DIMETHYLSULFIDE IN THE MARINE ATMOSPHERE

After its transition from the ocean into the atmosphere, DMS can react
with a variety of oxidizing atmospheric trace species. While not all
possible reactants have been investigated, currently available information suggests that the reactions with hydroxyl radical (OH) and nitrate
radical (NO_3) are the most important oxidation processes for DMS in the
atmosphere (Atkinson et al., 1984; Baulch et al., 1982; Cox and Sheppard, 1980; Grosjean, 1984; Hatakeyama et al., 1982; Niki et al., 1983;
Winer et al., 1984). The direct reaction between DMS and ozone seems to
be of negligible importance in the atmosphere (Winer et al., 1984, and
references therein).

A considerable amount of work has been done to determine the rate
of the reaction between DMS and these radicals; however, the actual
reaction sequences and products are still quite uncertain. Figure 5
shows possible reaction schemes, based largely on the work of Grosjean
(1984). It must be understood that these pathways are largely hypothetical, however. DMS can react with OH radical either by the addition of OH
to the DMS molecule, or by abstraction of a H atom. The latter process
appears to be the more significant one, at least in laboratory experi-

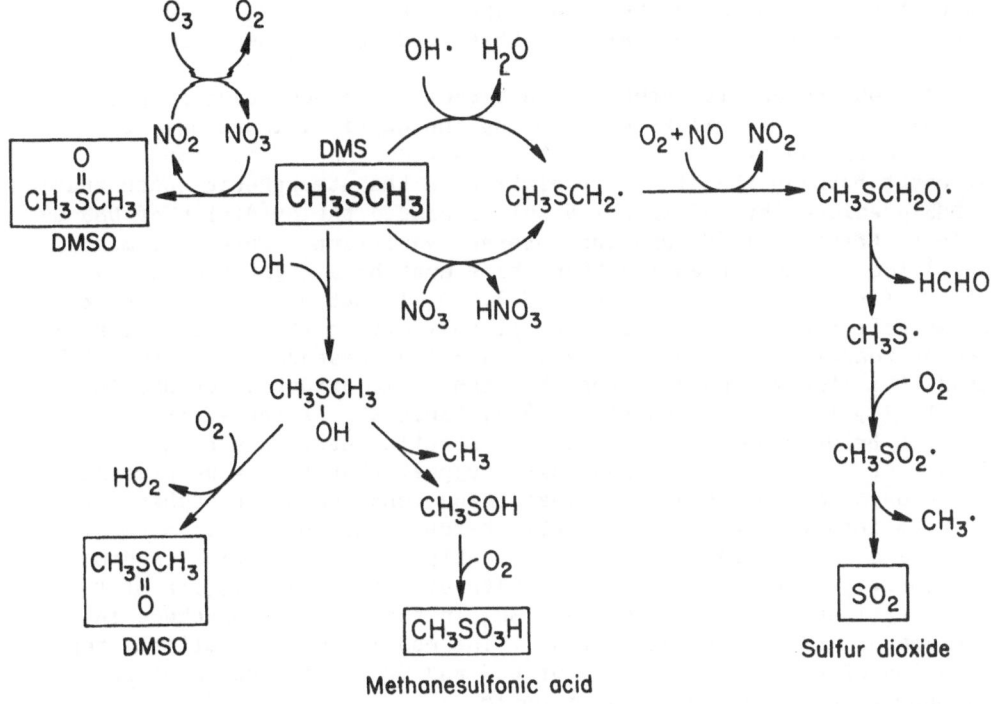

Figure 5. Suggested reaction pathways for the oxidation of DMS by OH and NO₃ radicals (based largely on Grosjean (1984) and Atkinson et al. (1984)).

ments. The CH₃SCH₂· radical formed by H-abstraction can then react with O₂ and NO, beginning a reaction chain that eventually leads to SO₂. The addition of the OH radical to DMS would lead to the formation of an intermediate [CH₃S(OH)CH₃], which could react with O₂ to form dimethyl-sulfoxide (DMSO). Alternatively, this intermediate could first expel a methyl radical and then react with O₂ to produce methanesulfonic acid.

DMS could react with the nitrate radical, NO₃, either by hydrogen abstraction or by oxygen addition. The product of hydrogen abstraction would be the same intermediate as that formed as a result of hydrogen abstraction by OH, and the reactions would follow the same subsequent pathway to SO₂. On the other hand, oxygen addition would produce DMSO in one reaction step. DMSO is relatively stable and would probably not be oxidized further in the atmosphere before being removed by deposition processes. The question whether NO₃ reacts by the addition or abstraction mechanism is an important one for the atmospheric nitrogen cycle in the marine atmosphere. In the case of the addition mechanism, NO₂ is formed as reaction product, which can again react with ozone to form NO₃. This reaction is in fact a form of NO₂ catalysis for the reaction between DMS and O₃, and does not represent a sink for NOₓ. The hydrogen abstraction mechanism, on the other hand, produces HNO_3, which is removed by wet and dry deposition; this reaction is therefore a NO_x sink.

The OH radical is formed by photochemical processes and is so short-lived that it exists only during the daytime. Therefore, the oxidation of DMS by OH can take place only during the daytime. The disagreement between the diurnal variation in the atmospheric concentration of DMS predicted by models which only included the oxidation of DMS by OH (e.g. Graedel, 1979) and the observed variations (which are much lower) led to the suggestion that there must be a night-time oxidation process (Andreae and Raemdonck, 1983). NO_3 is subject to rapid photodecomposition and can therefore exist only during the night. The rapid rate of reaction between DMS and NO_3 and its presence during the night makes it a likely candidate for the night-time oxidation of DMS (Atkinson et al., 1984; Winer et al., 1984). Since NO_3 is formed from the reaction between NO_2 and O_3, its concentration depends strongly on the amounts of NO_x present. Current data suggest that it may be present in high enough concentrations in coastal and continental atmospheres to be the most important oxidant for DMS, whereas over the remote oceans it probably does not contribute significantly to DMS oxidation (Andreae et al., 1985a). Figure 6 shows the results of a model simulation of the concentrations of DMS, OH, NO_3 and SO_2 in the marine atmosphere in the presence of moderate amounts of NO_x. The model gives a realistic representation of the concentrations of DMS and SO_2 in the clean marine atmosphere, e.g. the tropical Atlantic.

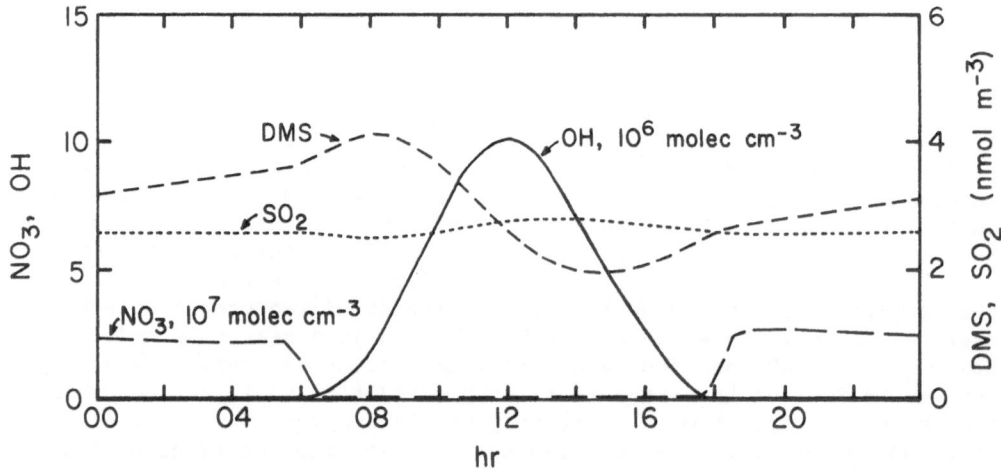

Figure 6. Diurnal cycle of the concentrations of DMS, SO_2, OH, and NO_3 in the marine atmosphere as predicted by the model of Ferek et al. (1985). The model assumes clear sky conditions, a CO concentration of 130 ppbv, and a NO_x concentration of 100 pptv.

Observations on the relative abundances of SO_2 and the other possible DMS oxidation products (DMSO, methanesulfonic acid) in the marine
atmosphere suggest that SO_2 is the major product (Saltzman et al., 1983;
Andreae, unpublished data). This would imply the the H abstraction
mechanism is dominant. However, the reactions following the abstraction
step require significant amounts of NO (Figure 5) which is present only
at very low concentrations in the remote marine atmosphere. It is therefore likely that additional pathways for the reaction of the $CH_3SCH_2 \cdot$
radical to SO_2 exist which become important at low NO_x concentrations.
The information on the atmospheric abundance of DMSO is currently
limited to a few measurements in marine rain (Andreae, 1980). These
measurements are not sufficient to assess the role of DMSO as a product
of DMS oxidation in the marine atmosphere, and futher studies on the
abundance of this compound should be conducted.

9. A MODEL OF THE CYCLE OF BIOGENIC SULFUR OVER THE OCEANS

Based on intensive field studies during the last few years, we now have
a reasonably good idea of the concentrations of DMS in the atmospheric
boundary layer over most of the major ocean regions (Andreae and Raemdonck, 1983; Andreae et al., 1985a; Nguyen et al., 1984). This information is summarized in Table IV. Together with available information on
the concentrations of the products of DMS oxidation in the marine atmosphere, and the estimates of DMS emission from the oceans and the
deposition fluxes of the oxidation products, we can construct a simple
model of the cycle of biogenic sulfur in the marine atmosphere

Table IV. DMS concentrations (nmol m^{-3}) in the remote marine atmosphere

	Time-weighted Mean	Range	Quartile Range
Equatorial Pacific	5.0	2.1 −13	3.6− 6.1
Cape Grim	5.2	1.1 −15	3.8− 7.0
Bahamas	3.8	0.2 −21	2.1− 3.6
North Atlantic: all points	3.0	0.1 −13	1.1− 3.8
"marine"	5.6		
"continental"	1.8		
Sargasso Sea	7.3	0.03−32	1.5−12
North Pacific	0.9*	0.3 − 1.8	
Tropical South Pacific (1)	3.1*	1.8 − 6.2	

*Arithmetic mean
(1) Nguyen et al., 1984

(Figure 7). In this model, I have ignored the contributions from gases other than DMS, which probably do not make a major contribution to the sulfur cycle over the oceans.

This model represents conditions characteristic of the remote tropical or subtropical marine atmosphere. The flux of DMS across the air/sea interface was based on the data in Table III. After it has entered the atmosphere, DMS is distributed through the marine troposphere. I selected a scale height of 2.5 km for DMS mixing in the troposphere based on our data from the tropical Atlantic near Barbados (Ferek et al., 1985). The sea-to-air transfer of 0.34 μmol S m^{-2} hr^{-1} corresponds to a DMS input into this 2.5 km column at a rate of 0.14 nmol S m^{-3} hr^{-1}. The concentration of DMS in the tropical marine atmosphere is

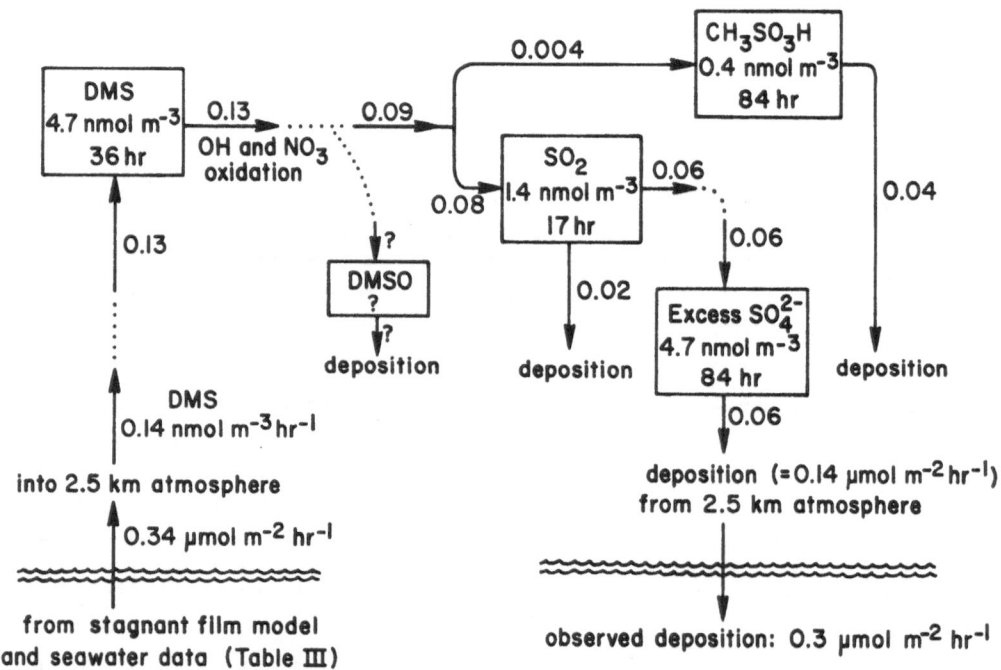

Figure 7. Box model of the atmospheric cycle of sulfur compounds derived from marine DMS. The concentrations and lifetimes for the species in the boxes are based on the best currently available data (see text). The fluxes (indicated in units of nmol m^{-3} hr^{-1}) are calculated independently for each species; their match between species is a test for the consistency of independent observations. For calculating exchange rates between the atmosphere and the surface, an atmospheric column height of 2.5 km has been assumed.

about 4.7 nmol m^{-3}* (Andreae et al., 1985a); when combined with the current best estimate of the lifetime of DMS in the remote marine atmosphere (ca. 36 hr), this corresponds to a DMS oxidation rate of 0.13 nmol m^{-3} hr^{-1}. Note that the estimate of the DMS input rate into the atmosphere and the estimate of its removal rate by oxidation were derived independently of one another; we can therefore use a comparison of these rates to test if the information from the air-sea transfer model, the concentration of DMS in the atmosphere, and the DMS oxidation rate constants are mutually consistent.

The known oxidation products of DMS are SO_2, methanesulfonic acid (MSA), and DMSO. We do not know the atmospheric concentrations and lifetime of DMSO; current information suggests that for the conditions of our model it is a minor product. The concentration of SO_2 used in the model is from a number of sources (Herrmann and Jaeschke, 1984; Bonsang et al., 1980; Maroulis et al., 1980; Nguyen et al., 1983), the SO_2 lifetimes relative to dry deposition and oxidation to sulfate are from Nguyen et al. (1983). The concentration of MSA is based on the data of Saltzman et al. (1983) and our own unpublished data. Assuming that there is no significant photochemical removal of MSA, its removal is dominated by wet and dry deposition (Saltzman et al., 1984) and I have assigned it a residence time typical of submicron aerosols (84 hr). The same lifetime was used for the excess sulfate aerosol. The concentration of excess sulfate is again based on a number of different authors (Andreae, 1982; Savoie, 1984; Maenhaut et al., 1981, 1983; Raemdonck et al., 1985; Bonsang et al., 1980).

Figure 7 shows that the fluxes through the MSA, SO_2 and sulfate compartments are mutually consistent, and that their sum is consistent with the DMS oxidation rate. Finally, if we strip the excess sulfate back out of the 2.5 km air column of our model, we obtain a sulfate deposition flux of ca. 0.14 µmol m^{-2} hr^{-1}, which is somewhat smaller than the rate of deposition of excess sulfate in the remote tropical regions (ca. 0.3 µmol m^{-2} hr^{-1}) (Galloway, 1985, and references therein). While an agreement within a factor of two is normally considered satisfactory for this type of calculation, the observed difference leaves room for contributions from anthropogenic sources, volcanic emissions and other possible biogenic sources of sulfur compounds.

In conclusion, we find that we can construct a reasonably consistent cycle of biogenic sulfur for the marine atmosphere. This suggests that DMS is indeed the major source of non-seasalt sulfur in the marine atmosphere, and that its flux is on the order of 1.0-1.2 Tmol yr^{-1} as estimated on the basis of its concentration in surface seawater. While the preceding discussion shows that DMS can certainly provide the required amounts of SO_2 to the marine boundary layer, the question of the origin of SO_2 in the marine free troposphere has not yet been addressed. The atmospheric reaction/transport models which used eddy diffusion for the parameterization of vertical transport (Logan et al., 1979; Rodhe and Isaksen, 1980) predicted that gases with relatively short atmospheric lifetimes, like DMS, could not be transported to the middle and

---

*All atmospheric concentrations are expressed in units of mol m^{-3} at standard temperature and pressure.

upper troposphere. Recent models, which include intermittent, rapid
transport of boundary layer air into the upper troposphere by large
convective cloud systems, predict that DMS can make a significant contribution to free tropospheric SO_2 as well (Gidel, 1984; Chatfield and
Crutzen, 1984). These predictions were substantiated by our measurements
in the marine troposphere near Barbados, which showed that convective
transport increased the free tropospheric DMS levels by a factor of ten
over the values found during non-convective conditions. The resulting
rate of SO_2 production can account for much, if not all, of the SO_2 in
the free troposphere at least in tropical regions (Ferek et al., 1985).

10. CARBONYL SULFIDE

Carbonyl sulfide (COS) is the most abundant atmospheric sulfur species
in the remote troposphere with an average concentration near 20 nmol
m^{-3}. Due to its low reactivity in the troposphere, and its correspondingly long residence time (on the order of one year), it is the only
sulfur compound which can enter the stratosphere (with the exception of
SO_2 injections during violent volcanic eruptions). The input of COS is
considered responsible for the maintenance of the sulfate aerosol layer
in the stratosphere during volcanically quiescent periods. Therefore
even a relatively small COS source flux can be of considerable importance in atmospheric chemistry.
      COS is present in surface seawater at concentrations of about 0.03
to 1.0 nM (Rasmussen et al., 1982; Ferek and Andreae, 1983, 1984; Turner
and Liss, 1985a). The observed concentrations are almost always higher
than the equilibrium concentration relative to the overlying atmosphere,
so that a net sea-to-air flux exists essentially from the entire ocean
surface. Johnson (1981) has speculated that the ocean should be a sink
for COS due to its hydrolysis at the slightly alkaline pH of seawater.
This suggestion is clearly not supported by the measured COS supersaturation ratios across the air/sea interface. In the following sections I will discuss the formation process of COS in seawater and
present an estimate of the oceanic source of COS.

10.1. Photochemical Production of COS

In the upper few meters of the ocean, carbonyl sulfide is produced by
photochemical reactions from dissolved organic sulfur compounds (Ferek
and Andreae, 1984), resulting in strong diurnal variations of the COS
concentration in surface seawater (Figure 8). Laboratory experiments
with seawater and with solutions of organosulfur compounds in distilled
water showed that seawater sulfate did not participate in the reaction,
and that only the presence of dissolved organic sulfur compounds, dissolved O_2 and light were necessary to produce COS. Carbonyl sulfide was
formed by irradiation of a variety of organic sulfur compounds commonly
found in biological materials, e.g. cysteine, methionine, glutathione
and dimethylsulfonium propionate. On the other hand, the presence or
absence of living microorganisms—planktonic algae or bacteria—had no

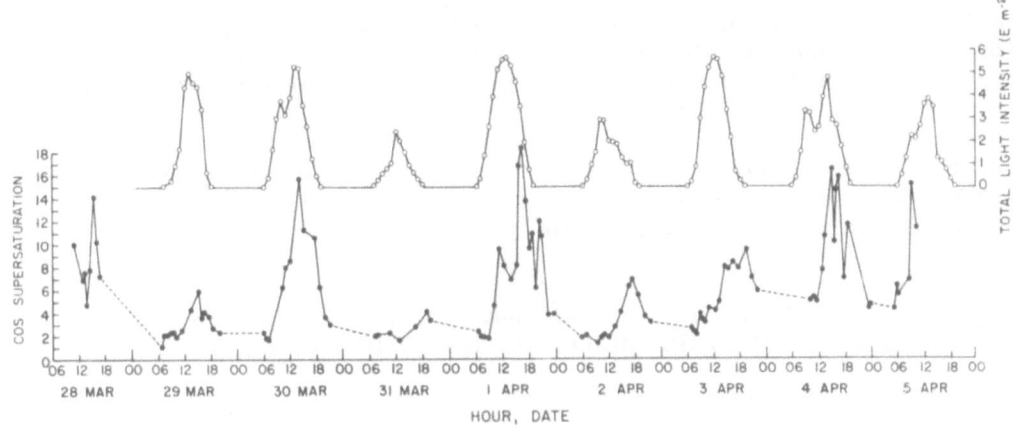

Figure 8. COS supersaturation and light intensity measured in surface
    waters during cruise of R/V Cape Hatteras in Chesapeake and
    Delaware Bays, spring 1983. (Reprinted by permission from Nature,
    vol. 307, no. 5947, p. 149, Copyright (c) 1984 Macmillan Journals
    Ltd.)

influence on the rate of formation of COS in seawater. It appears that
the role of organisms in the production of COS in seawater is limited to
the synthesis of dissolved organic sulfur compounds which are then
abiotically photolyzed to COS. The mechanism of this reaction is not
known at this time, but it is likely that short-lived, photochemically
produced radicals (e.g. OH) are involved.

10.2. Air/Sea Exchange of COS

In order to obtain a realistic estimate of the flux of COS from the
oceans to the atmosphere, we have to integrate the supersaturation over
the diurnal cycle and over the different light fluxes, dissolved organic
sulfur concentrations, and wind conditions found in different ocean
regions. Previous estimates clearly do not satisfy this requirement, and
can therefore be only considered rough guesses. On the basis of samples
collected without consideration of the diurnal cycle and only separated
into coastal and open ocean values, Rasmussen et al. (1982) estimate a
global COS flux from the oceans of ca. 10 Gmol yr^{-1}. Based on largely
coastal data and interpolation from an observed relationship of DMS and
COS, Ferek and Andreae (1983) estimate a global flux of ca. 16 Gmol
yr^{-1}.
       An attempt to obtain a more representative estimate of the sea-to-
air flux of COS is presented in Table V. In this table, I have broken
down the ocean surface into the same biogeographic regions as used in
Table III. Then, based on our (diurnally averaged) data on the supersaturation of COS in surface seawater relative to the overlying atmo-

sphere, and the average temperature of the surface ocean in these
regions, I have calculated the flux of COS across the air/sea interface
for these regions (the piston velocities for COS are a factor of 1.3
higher than for DMS, due to the higher diffusivity of COS). We see that,
in contrast to DMS, the flux of COS is dominated by the high productivity regions, especially the coastal and shelf areas. Due to the low
levels of COS in oligotrophic areas, they contribute little to the
global flux, which I estimate to be ca. 11 Gmol yr[-1], similar to the
previous flux estimates given above.

11. FORMATION AND EMISSION OF OTHER SULFUR SPECIES: HYDROGEN SULFIDE, CARBON DISULFIDE, METHYLMERCAPTAN, DIMETHYLDISULFIDE, ETC.

11.1. Hydrogen Sulfide

H_2S from "biological decomposition" was assumed to be responsible for
essentially all of the flux of reduced sulfur from the oceans to the
atmosphere in models of the atmospheric sulfur cycle published up to the
last few years (e.g. Rodhe and Isaksen, 1980). Following the discovery
of the important role of DMS in the sea-to-air transfer of biogenic
sulfur, there is now a tendency to ignore H_2S altogether. Since there
are no data on the concentration of dissolved H_2S in surface seawater,
and few reliable measurements of H_2S in the marine atmosphere, the
assumption that the H_2S flux is negligible is usually based on the fact
that it is oxidized rapidly in oxygenated seawater. Halflives for H_2S on
the order of a few hours are reported (Almgren and Hagström, 1974);
however, other workers have found values as high as 50 hours (Chen and
Morris, 1972). It must be remembered that biological processes can
result in the production and release of substantial amounts of H_2S even
in the presence of oxygen (see Section 4). This is especially true in
the presence of high ambient sulfate concentrations, as is the case in

Table V. Average COS supersaturation, concentration, and flux for the World Oceans (based
on Ferek and Andreae 1983, 1984, and unpublished data by Ferek and Andreae)

Region	Area (10^6 km^2)	Temp. ($^{\circ}$C)	Saturation Ratio*	Seawater Concentration (pmol L^{-1})	Transfer Velocity (cm hr^{-1})	Flux (μmol m^{-2} day^{-1})	Flux (Gmol yr^{-1})
Oligotrophic	148.3	25	1.5	11	13.8	14	0.7
Transition	82.8	15	1.9	20	17.1	45	1.2
Upwelling	86.5	20	2.6	24	16.0	64	1.8
Coastal/Shelf	49.4	15	10.5	112	16.0	440	7.0
		Mean:	3.1	30		Total:	10.7

*COS partial pressure in equilibrium with seawater divided by the atmospheric partial
pressure

seawater. It appears that attempts should be made to determine H_2S directly in seawater, even though the low concentrations that must be expected to be present make this a difficult analytical problem.

H_2S has been observed in the marine atmosphere at levels of 1 μmol m^{-3} or less (Slatt et al., 1978; Delmas and Servant, 1982; Herrmann and Jaeschke, 1984). It is not altogether clear from these papers how much of these values could be due to a DMS interference. Graedel (1979) predicted on the basis of a photochemical model that a source of 0.15 Tmol H_2S per year would be required to support concentrations of the magnitude indicated above. The model of Rodhe and Isaksen (1980) introduces 1.2 Tmol H_2S yr^{-1} from the oceans into the atmosphere, and predicts atmospheric H_2S concentrations which are at least an order of magnitude too high. This suggests that a flux of ~0.1 Tmol yr^{-1} may be a more realistic estimate for the marine H_2S source. If we simply use an average H_2S concentration of 0.4 nmol m^{-3} with a scale height of 2.5 km, a diurnally averaged OH concentration of 2×10^6 molec cm^{-3} and the measured reaction rate for the oxidation of H_2S by OH (5×10^{-12} cm^3 $molec^{-1}$ sec^{-1} (Cox and Sheppard, 1980), we obtain an estimate for the H_2S flux of only 0.08 Tmol yr^{-1}. A flux of this magnitude appears to be the most reasonable estimate at the current state of knowledge.

It is not clear, however, if the source of this H_2S is necessarily the ocean surface or if other processes could be responsible for its presence. For example, advection from coastal regions, where H_2S is emitted from salt marshes, may supply some of the H_2S found in the marine atmosphere. On the other hand, McElroy et al. (1980) have speculated that reactions of COS and CS_2 with OH radical could produce the necessary amounts of H_2S. However, this suggestion has not yet been experimentally verified.

In view of the unsatisfactory state of knowledge on the role of H_2S in the global sulfur cycle, it appears essential that attempts should be made to determine directly the concentration of H_2S in surface seawater and in the overlying atmosphere, so that the magnitude (and direction, since the sea surface may well even be a sink for H_2S!) of the exchange flux can be estimated.

11.2. Carbon Disulfide

The presence of CS_2 in seawater was first observed by Lovelock (1974) who measured an average concentration of 6.8 pM in 35 samples taken in the open Atlantic. Inshore values were about an order of magnitude higher. Turner and Liss (1985a) also report on the presence of CS_2 in coastal waters off England, but due to a chromatographic separation problem between DMS and CS_2, they present quantitative information only for a few samples with values near 0.15 nM. They found substantially higher concentrations in the low salinity region of an estuary (up to ca. 1 nM: Turner and Liss, 1985b). It is possible that much of the CS_2 found in coastal waters is the result of the diffusion of this substance from the porewaters of the underlying sediments. This would be consistent with the relatively high concentrations and fluxes of CS_2 observed in coastal marsh environments (Adams et al., 1981; Steudler and Peterson, 1984). CS_2 could be formed there either by fermentation reactions

of organosulfur compounds or by "pulp—mill" type reactions of terrigenic
plant matter with dissolved polysulfides originating from bacterial
dissimilatory sulfate reduction.

We have recently determined CS_2 in in coastal and open ocean
seawater from the temperate and tropical North Atlantic and have obtained a mean value of 7.4 pM, consistent with Lovelock's results. Using
this concentration as an estimate for the global average concentration
of CS_2 in surface seawater, we estimate a sulfur flux of ca. 0.005 Tmol
S yr^{-1} in the form of CS_2 from the world ocean, about one percent of the
DMS flux.

11.3. Methylmercaptan, Dimethyldisulfide, and Other Sulfur Compounds

In coastal waters, we have observed peaks for methylmercaptan and its
oxidation product, dimethyldisulfide. Due to calibration problems, we
have not been able to report quantitative data, but the concentrations
of these compounds are always much smaller than those of DMS. The
presence of dimethyldisulfide and a compound tentatively identified (on
the basis of its retention time) as ethyl propyl sulfide in coastal
waters off England was reported by Turner and Liss (1985a). Like CS_2,
these compounds may originate in the sediment porewaters and coastal
marshes, rather than in the marine water column. Due to their consis-
tently low concentrations even in coastal waters, and their absence in
open ocean waters, they probably do not make a significant contribution
to the global sulfur cycle.

12. CONCLUSION

The oceans are a major source of sulfur compounds to Earth's atmosphere.
Sulfate is introduced into the atmosphere as a component of the seasalt
aerosol. A variety of biological processes leads to the formation of
reduced, volatile sulfur compounds: dimethylsulfide, carbonyl sulfide,
carbon disulfide, hydrogen sulfide and others. The pathway for the
formation of dimethylsulfide, the most important of these compounds,
starts with assimilatory sulfate reduction by marine algae and proceeds
through dimethylsulfonium propionate, a major algal metabolite. The
release of DMS is related to phytoplankton ecology in complex ways,
which are still inadequately understood. About 1.2 Tmol (40 Tg) of
sulfur enter the atmosphere annually in the form of DMS from the oceans.
Such a flux is consistent with our current knowledge of the concentrations and behavior of DMS and its oxidation products in the atmosphere.
Carbonyl sulfide is produced by photochemical reactions from dissolved
organic sulfur compounds in seawater. While its sea—to—air flux is much
smaller than that of DMS, the tropospheric stability of COS makes it a
source of sulfur to the stratosphere. The importance of the flux of H_2S
from the ocean surface remains still uncertain, while the flux of the
minor sulfur compounds CS_2, methylmercaptan, etc. is almost certainly
negligible.

13. ACKNOWLEDGMENTS

Our research on the biogenic production and air-sea transfer of sulfur compounds has been supported by the National Science Foundation under grants ATM-8017574, ATM-8407137, and OCE-8315733.

14. REFERENCES

Ackman, R. G., C. S. Tocher, and J. McLachlan. 1966. 'Occurrence of dimethyl-β-propiothetin in marine phytoplankton'. J. Fish. Res. Bd. Can. 23:357-364.

Adams, D. F., S. O. Farwell, E. Robinson, M. R. Pack, and W. L. Bamesberger. 1981. 'Biogenic sulfur source strengths'. Environ. Sci. Technol. 15:1493-1498.

Almgren, T., and I. Hagström. 1974. 'The oxidation rate of sulphide in sea water'. Water Res. 8:395-400.

Anderson, J. W. 1980. 'Assimilation of inorganic sulfate into cysteine'. In The Biochemistry of Plants, Vol. 5 (P.K. Stumpf and E.E. Conn, eds.) New York: Academic Press, 203-223.

Andreae, M. O. 1979. 'Arsenic speciation in seawater and interstitial waters: the influence of biological-chemical interactions on the chemistry of a trace element'. Limnol. Oceanogr. 24:440-452.

Andreae, M. O. 1980. 'Dimethylsulfoxide in marine and fresh waters'. Limnol. Oceanogr. 25:1054-1063.

Andreae, M. O. 1982. 'Marine aerosol chemistry at Cape Grim, Tasmania and Townsville, Queensland'. J. Geophys. Res. 87:8875-8885.

Andreae, M. O. 1984. 'Atmospheric effects of microbial mats'. In Microbial Mats: Stromatolites (Y. Cohen, R.W. Castenholz and H.O. Halverson, eds.) New York: Alan R. Liss, 455-466.

Andreae, M. O. 1985a. 'The emission of sulfur to the remote atmosphere'. In The Biogeochemical Cycling of Sulfur and Nitrogen in the Remote Atmosphere (J.N. Galloway, R.J. Charlson, M.O. Andreae and H. Rodhe, eds.) Dordrecht: Reidel, 5-25.

Andreae, M. O. 1985b. 'Dimethylsulfide in the water column and the sediment pore waters of the Peru upwelling area'. Limnol. Oceanogr. 30:1208-1218.

Andreae, M. O., and W. R. Barnard. 1984. 'The marine chemistry of dimethylsulfide'. Marine Chem. 14:267-279.

Andreae, M. O., and H. Raemdonck. 1983. 'Dimethyl sulfide in the surface ocean and the marine atmosphere: a global view'. Science 221:744-747.

Andreae, M. O., W. R. Barnard, and J. M. Ammons. 1983. 'The biological production of dimethylsulfide in the ocean and its role in the global atmospheric sulfur budget'. Ecol. Bull. 35:167-177.

Andreae, M. O., R. J. Ferek, F. Bermond, K. P. Byrd, R. T. Engstrom, S. Hardin, P. D. Houmere, F. LeMarrec, H. Raemdonck, and R. B. Chatfield. 1985a. 'Dimethyl sulfide in the marine atmosphere'. J. Geophys. Res., in press.

Andreae, M. O., R. J. Charlson, F. Bruynseels, H. Storms, R. E. Van Grieken, and W. Maenhaut. 1985b. 'Salts, silicates, and sulfates: internal mixture in marine aerosols'. Science, submitted.

Atkinson, R., J. N. Pitts, Jr., and S. M. Aschmann. 1984. 'Tropospheric reactions of dimethyl sulfide with NO_3 and OH radicals'. <u>J. Phys. Chem</u>. **88**:1584-1587.

Barnard, W. R., M. O. Andreae, W. E. Watkins, H. Bingemer, and H. W. Georgii. 1982. 'The flux of dimethylsulfide from the oceans to the atmosphere'. <u>J. Geophys. Res</u>. **87**:8787-8793.

Barnard, W. R., M. O. Andreae, and R. L. Iverson. 1984. 'Dimethylsulfide and <u>Phaeocystis poucheti</u> in the southeastern Bering Sea'. <u>Cont. Shelf Res</u>. 3:103-113.

Baulch, D. L., R. A. Cox, P. J. Crutzen, R. F. Hampson, J. A. Kerr, J. Troe, and R. T. Watson. 1982. 'Evaluated kinetics and photochemical data for atmospheric chemistry'. <u>J. Phys. Chem. Ref. Data</u>, 11:327-496.

Bingemer, H. 1984. 'Dimethylsulfid in Ozean und mariner Atmosphäre—Experimentelle Untersuchung einer natürlichen Schwefelquelle für die Atmosphäre'. Ph.D. Dissertation, J.W. Goethe Universität, Frankfurt am Main.

Blanchard, D. C. 1983. 'The production, distribution, and bacterial enrichment of the sea-salt aerosol'. In <u>Air-Sea Exchange of Gases and Particles</u> (P.S. Liss and W.G.N. Slinn, eds.) Boston: Reidel, 407-454.

Blanchard, D. C., A. H. Woodcock, and R. J. Cipriano. 1984. 'The vertical distribution of the concentration of sea salt in the marine atmosphere near Hawaii'. <u>Tellus</u> **36B**:118-125.

Bonsang, B., B. C. Nguyen, A. Gaudry, and G. Lambert. 1980. 'Sulfate enrichment in marine aerosols owing to biogenic gaseous sulfur compounds'. <u>J. Geophys. Res</u>. **85**:7410-7416.

Broecker, W. S., and T.-H. Peng. 1974. 'Gas exchange rates between air and sea'. <u>Tellus</u> **26**:21-35.

Brown, A. D., and J. R. Simpson. 1972. 'Water relations of sugar-tolerant yeasts: the role of intracellular polyols'. <u>J. Gen. Microbiol</u>. **72**:589-591.

Cantoni, G. L., and D. G. Anderson. 1956. 'Enzymatic cleavage of dimethylpropiothetin by <u>Polysiphonia lanosa</u>'. <u>J. Biol. Chem</u>. **222**:171-177.

Challenger, F., and M. I. Simpson. 1948. 'Studies on biological methylation. Part XII. A precursor of the dimethyl sulphide evolved by Polysiphonia fastigiata. Dimethyl-2-carboxyethylsulphonium hydroxide and its salts. <u>J. Chem. Soc</u>. 3:1591-1597.

Chatfield, R. B., and P. J. Crutzen. 1984. 'Sulfur dioxide in remote oceanic air: cloud transport of reactive precursors'. <u>J. Geophys. Res</u>. **89**:7111-7132.

Chen, K. Y., and J. C. Morris. 1972. 'Kinetics of oxidation of aqueous sulfide by O_2'. <u>Environ. Sci. Technol</u>. **6**:529-537.

Church, T. M., J. N. Galloway, T. D. Jickells, and A. H. Knap. 1982. 'The chemistry of western Atlantic precipitation at the mid-Atlantic coast and on Bermuda'. <u>J. Geophys. Res</u>. **87**:11,013-11,018.

Cline, J. D., and T. S. Bates. 1983. 'Dimethyl sulfide in the equatorial Pacific Ocean: a natural source of sulfur to the atmosphere'. <u>Geophys. Res. Lett</u>. **10**:949-952.

Cox, R. A., and D. Sheppard. 1980. 'Reactions of OH radicals with gaseous sulphur compounds'. <u>Nature</u> **284**:330-331.

Craigie, J. S., J. McLachlan, R. G. Ackman, and C. S. Tocher. 1967.

'Photosynthesis in algae. III. Distribution of soluble carbohydrates and dimethyl-β-propiothetin in marine unicellular chlorophyceae and prasinophyceae'. Can. J. Bot. 45:1327-1334.

Cullis, C. F. and M. M. Hirschler. 1980. 'Atmospheric sulfur: natural and man-made sources'. Atmos. Environ. 14:1263-1278.

Delmas, R., and J. Servant. 1982. 'The origins of sulfur compounds in the atmosphere of a zone of high productivity (Gulf of Guinea)'. J. Geophys. Res. 87:11,019-11,026.

Dickson, D. M., R. G. Wyn Jones, and J. Davenport. 1980. 'Steady state osmotic adaptation in Ulva lactuca'. Planta 150:158-165.

Dickson, D. M., R. G. Wyn Jones, and J. Davenport. 1982. 'Osmotic adaptation in Ulva lactuca under fluctuating salinity regimes'. Planta 155:409-415.

Duce, R. A., F. MacIntyre, and B. Bonsang. 1982. 'Enrichment of sulfate in maritime aerosols' (Discussion). Atmos. Environ. 16:2025-2034.

Ehrlich, P. R., A. H. Ehrlich, and J. P. Holdren. 1977. Ecoscience. Population, resources, environment. San Francisco: Freeman, 1051 p.

Ferek, R. J., and M. O. Andreae. 1983. 'The supersaturation of carbonyl sulfide in surface waters of the Pacific Ocean off Peru'. Geophys. Res. Lett. 10:393-396.

Ferek, R. J., and M. O. Andreae. 1984. 'Photochemical production of carbonyl sulfide in marine surface waters'. Nature 307:148-150.

Ferek, R. J., R. B. Chatfield, and M. O. Andreae. 1985. 'Vertical distribution of dimethylsulfide in the marine atmosphere: implications for the atmospheric sulfur cycle'. Nature, submitted.

Galloway, J. N. 1985. 'The deposition of sulfur and nitrogen from the remote atmosphere'. In The Biogeochemical Cycles of Sulfur and Nitrogen in the Remote Atmosphere (J.N. Galloway, R.J. Charlson, M.O. Andreae and H. Rodhe, eds.) Dordrecht: Reidel, 143-175.

Galloway, J. N., G. E. Likens, W. C. Keene, and J. M. Miller. 1982. 'The composition of precipitation in remote areas of the world'. J. Geophys. Res. 87:8771-8786.

Galloway, J. N., A. H. Knap, and T. M. Church. 1983. 'The composition of western Atlantic precipitation using shipboard collectors'. J. Geophys. Res. 88:10,859-10,864.

Garland, J. A. 1981. 'Enrichment of sulphate in maritime aerosols'. Atmos. Environ. 15:787-791.

Gidel, L. T. 1984. 'The role of clouds in micro and macro-scale transport of atmospheric constituents'. In Gas-Liquid Chemistry of Natural Waters (L. Newman, ed.) Upton, N.Y.: Brookhaven Natl. Lab., (6)1-8.

Giovanelli, J., S. H. Mudd, and A. H. Datko. 1980. 'Sulfur aminoacids in plants'. In The Biochemistry of Plants, Vol. 5 (P.K. Stumpf and E.E. Conn, eds.) New York: Academic Press, 453-505.

Graedel, T. E. 1979. 'Reduced sulfur emission from the open oceans'. Geophys. Res. Lett. 6:329-331.

Granroth, B., and T. Hattula. 1976. 'Formation of dimethyl sulfide by brackish water algae and its possible implication for the flavor of Baltic herring'. Finn. Chem. Lett., 148-150.

Grosjean, D. 1984. 'Photooxidation of methyl sulfide, ethyl sulfide, and methanethiol'. Environ. Sci. Technol. 18:460-468.

Haas, P. 1935. 'The liberation of methyl sulfide in seaweed'. Biochem.
    J. 29:1297–1299.
Hatakeyama, S., M. Okuda, and H. Akimoto. 1982. 'Formation of sulfur
    dioxide and methane sulfonic acid in the photooxidation of dimethyl
    sulfide in the air'. Geophys. Res. Lett. 9:583–586.
Herrmann, J., and W. Jaeschke. 1984. 'Measurements of H_2S and SO_2 over
    the Atlantic Ocean'. J. Atmos. Chem. 1:111–123.
Ivanov, M. V., and J. R. Freney. 1983. 'The global biogeochemical sulfur
    cycle'. New York: Wiley, 470 p.
Jenkins, W. J. 1982. 'Oxygen utilization rates in North Atlantic sub-
    tropical gyre and primary production in oligotrophic systems'. Nature
    300:246–248.
Johnson, J. E. 1981. 'The lifetime of carbonyl sulfide in the tropo-
    sphere'. Geophys. Res. Lett. 8:938–940.
Koblentz-Mishke, O. J., V. V. Volkovinsky, and J. G. Kabanova. 1970.
    'Plankton primary production of the world ocean'. In Scientific Ex-
    ploration of the South Pacific (W.S. Wooster, ed.) Washington, D.C.:
    Natl. Acad. Sci., 183–193.
Liss, P. S. 1983. 'Gas transfer: experiments and geochemical
    implications'. In Air-Sea Exchange of Gases and Particles (P.S. Liss
    and W.G.N. Slinn, eds.) Boston: Reidel, 241–298.
Logan, J. A., M. B. McElroy, S. C. Wofsy, and M. J. Prather. 1979.
    'Oxidation of CS_2 and COS: sources for atmospheric SO_2'. Nature
    281:185–188.
Lovelock, J. E. 1974. 'CS_2 and the natural sulphur cycle'. Nature
    248:625–626.
Maenhaut, W., M. Darzi, and J. W. Winchester. 1981. 'Seawater and non-
    seawater aerosol components in the marine atmosphere of Samoa'. J.
    Geophys. Res. 86:3187–3193.
Maenhaut, W., H. Raemdonck, A. Selen, R. Van Grieken, and J. W.
    Winchester. 1983. 'Characterization of the atmospheric aerosol over
    the eastern equatorial Pacific'. J. Geophys. Res. 88:5353–5364.
Maroulis, P. J., A. L. Torres, A. B. Goldberg, and A. R. Bandy. 1980.
    'Atmospheric SO_2 measurements on project GAMETAG'. J. Geophys. Res.,
    85:7345–7349.
McDonald, R. L., C. K. Unni, and R. A. Duce. 1982. 'Estimation of atmo-
    spheric sea salt dry deposition: wind speed and particle size
    dependence'. J. Geophys. Res. 87:1246–1250.
McElroy, M. B., S. C. Wofsy, and N. D. Sze. 1980. 'Photochemical sources
    for atmospheric H_2S'. Atmos. Environ. 14:159–163.
Möller, D. 1984. 'Estimation of the global man-made sulphur emission'.
    Atmos. Environ. 18:19–27.
Nguyen, B. C., B. Bonsang, and A. Gaudry. 1983. 'The role of the ocean
    in the global atmospheric sulfur cycle'. J. Geophys. Res. 88:10,903–
    10,914.
Nguyen, B. C., C. Bergeret, and G. Lambert. 1984. 'Exchange rates of
    dimethyl sulfide between ocean and atmosphere'. In Gas Transfer at
    Water Surfaces (W. Brunsaert and G.H. Jirka, eds.) Dordrecht: Reidel,
    539–545.
Niki, H., P. D. Maker, C. M. Savage, and L. P. Breitenbach. 1983. 'An
    FTIR study of the mechanism of the reaction HO + CH_3SCH_3'. Int. J.

Chem. Kinet. 15:647–654, 1983.

Peng, T.-H., W. S. Broecker, G. G. Mathieu, and Y.-H. Li. 1979. 'Radon evasion rates in the Atlantic and Pacific Oceans as determined during the Geosecs program'. J. Geophys. Res. 84:2471–2486.

Petrenchuk, O. P. 1980. 'On the budget of sea salts and sulfur in the atmosphere'. J. Geophys. Res. 85:7439–7444.

Raemdonck, H., W. Maenhaut, and M. O. Andreae. 1985. 'Chemistry of the marine aerosol over the tropical and equatorial Pacific'. J. Geophys. Res., in press.

Rasmussen, R. A., M. A. K. Khalil, and S. D. Hoyt. 1982. 'The oceanic source of carbonyl sulfide (OCS)'. Atmos. Environ. 16:1591–1594.

Rodhe, H., and I. Isaksen. 1980. 'Global distribution of sulfur compounds in the troposphere estimated in a height/latitude transport model'. J. Geophys. Res. 85:7401–7409.

Saltzman, E. S., D. L. Savoie, R. G. Zika, and J. M. Prospero. 1983. 'Methane sulfonic acid in the marine atmosphere'. J. Geophys. Res. 88:10,897–10,902.

Saltzman, E. S., L. T. Gidel, R. G. Zika, P. J. Milne, J. M. Prospero, D. L. Savoie, and W. J. Cooper. 1984. 'Aerosol chemistry of methane sulfonic acid'. In Gas–Liquid Chemistry of Natural Waters (L. Newman, ed.) Upton, N.Y.: Brookhaven Natl. Lab., (53)1–8.

Savoie, D. L. 1984. 'Nitrate and non-sea-salt sulfate aerosols over major regions of the world ocean: concentrations, sources, and fluxes'. Ph.D. Dissertation, University of Miami, Florida.

Shulenberger, E., and J. L. Reid. 1981. 'The Pacific shallow oxygen maximum, deep chlorophyll maximum, and primary productivity, reconsidered'. Deep–Sea Res. 28:901–919.

Sivelä, S., and V. Sundman. 1975. 'Demonstration of Thiobacillus-type bacteria, which utilize methyl sulphides'. Arch. Microbiol. 103:303–304.

Slatt, B. J., D. F. S. Natusch, J. M. Prospero, and D. L. Savoie. 1978. 'Hydrogen sulfide in the atmosphere of the northern equatorial Atlantic Ocean and its relation to the global sulfur cycle'. Atmos. Environ. 12:981–991.

Smethie, W. M., Jr., T. Takahashi, D. W. Chipman, and J. R. Ledwell. 1985. 'Gas exchange and CO_2 flux in the tropical Atlantic Ocean determined from $222Rn$ and pCO_2 measurements'. J. Geophys. Res. 90:7005–7022.

Steudler, P. A., and B. J. Peterson. 1984. 'Contribution of the sulfur from salt marshes to the global sulfur cycle'. Nature, in press.

Tocher, C. S., R. G. Ackman, and J. McLachlan. 1966. 'The identification of dimethyl-β-propiothetin in the algae Syracosphaera carterae and Ulva lactuca'. Can. J. Biochem. 44:519–522.

Turner, S. M., and P. S. Liss. 1985a. 'Measurements of various sulphur gases in a coastal marine environment'. J. Atmos. Chem. 2:223–232.

Turner, S. M., and P. S. Liss. 1985b. 'Measurements of sulphur gases in coastal marine environments'. Searex Newsletter, January 1985, 12–17.

Vairavamurthy, A., M. O. Andreae, and R. L. Iverson. 1985. 'Biosynthesis of dimethylsulfide and dimethylpropiothetin by Hymenomonas carterae in relation to sulfur source and salinity variations'. Limnol. Oceanogr. 30:59–70.

Varhelyi, G., and G. Gravenhorst. 1983. 'Production rate of airborne seasalt sulfur deduced from chemical analysis of marine aerosols and precipitation'. J. Geophys. Res. 88:6737–6751.

Von Damm, K. L. 1983. 'Chemistry of submarine hydrothermal solutions at 21° North, East Pacific Rise, and Guaymas Basin, Gulf of California'. Ph.D. Dissertation, Massachusetts Institute of Technology, Cambridge, 241 p.

Wakeham, S. G., B. L. Howes, and J. W. H. Dacey. 1984. 'Dimethylsulfide in a coastal stratified salt pond'. Nature 310:770–772.

White, R. H. 1982. 'Analysis of dimethyl sulfonium compounds in marine algae'. J. Marine Res. 40:529–536.

Wilson, L. G., R. A. Bressan, and P. Filner. 1978. 'Light–dependent emission of hydrogen sulfide from plants'. Plant Physiology 61:184–189.

Winer, A. M., R. Atkinson, and J. N. Pitts, Jr. 1984. 'Gaseous nitrate radical: possible nighttime atmospheric sink for biogenic organic compounds'. Science 224:156–159.

Winner, W. E., C. L. Smith, G. W. Koch, H. A. Mooney, J. D. Bewley, and H. R. Krouse. 1981. 'Rates of emission of H_2S from plants and patterns of stable sulphur isotope fractionation'. Nature 289:672–673.

Zinder, S. H., and T. D. Brock. 1978. 'Production of methane and carbon dioxide from methane thiol and dimethyl sulphide by anaerobic lake sediments'. Nature 273:226–228.

CYCLING OF MERCURY BETWEEN THE ATMOSPHERE AND OCEANS

William F. Fitzgerald
Marine Sciences Department and Institute
The University of Connecticut
Groton, CT., 06340 U.S.A.

1. Introduction

It is well known that mercury and many of its compounds are volatile, and readily cycled at the earth's surface. Moreover, the major transfer of Hg from terrestrial sources to the oceans occurs by exchange at the sea surface. Thus, the atmospheric cycle of Hg over the oceans and the processes affecting Hg exchange between the air and the sea are a predominant feature of the global Hg cycle. Our understanding of the behavior and fate of Hg in the marine environment is improving. Elucidation of the current state of knowledge regarding the cycling of Hg between the atmosphere and oceans makes a very fine case study. There are sufficient reliable data and geochemically consistent results to provide a working framework. There also appears to be much room to question current dogma and for speculation.

Mercury is one of the most challenging and frequently measured trace constituents in the environment. Yet, it is an insidiously difficult element to measure accurately. Thus, reliable and sufficient observations for critical parts of the global geochemical cycle have been lacking. Historically, global scale estimates for Hg flows, source strengths, and reservoir contents have been widely divergent. Corresponding mass balance estimates of environmental impact from Hg emissions have been poorly constrained. Examples of the limited and often unreliable data base include sparse information for Hg fluxes in precipitation. The historical record of atmospheric Hg deposition is poor and conflicting. There are relatively few atmospheric Hg measurements in remote regions, particularly in the southern hemisphere. Little is known about the chemical reactivity of Hg in the atmosphere. On a broader basis, it is not yet clear how atmospheric chemical processes are affecting the global cycle of Hg, nor has the role of the oceans as a source and a sink for atmospheric Hg been well defined.

I have often viewed my studies of Hg in the environment as akin to a pursuit of an elusive silvery comet whose trail is full of tortuous twists and turns leading frequently to surprising results, challenging experiments, remote and romantic places, and some false starts. I wish to share aspects of this environmental chase as we address the geochemical cycling of Hg between the oceans and the atmosphere within a global framework. It is my intention to examine the atmospheric cycle of Hg over the oceans and to carefully evaluate the role of air-sea exchange in the global movement and distribution of Hg. An experimentally constrained picture of the the atmospheric mobilization of Hg and its exchange with the marine environment can

P. Buat-Ménard (ed.), The Role of Air-Sea Exchange in Geochemical Cycling, 363–408.
© *1986 by D. Reidel Publishing Company.*

be developed. Such studies have been an important part of our efforts
concerned with the global cycling of Hg and its behavior and fate in
the marine environment.

2.   Global Models

     Valuable insight into the geochemical behavior of an element
such as Hg can be derived from consideration of relatively simple
global models. Mass balance geochemical formulations can provide
estimates for the contents of the major reservoirs, and identify
principal sources and sinks. Source strengths can be evaluated and
geochemical pathways elucidated. The global cycle framework is often
useful in finding deficiencies and/or inconsistencies associated with
a data base. Thereby, serving as a guide for future research.
Further, and quite importantly, the degree to which human-related
emissions are interfering with natural flows can be assessed from an
examination of the global cycle for an element.

     Hg additions to the environment are a continuing worry because
of the toxic nature of certain Hg compounds (e.g., alkylated mercurials)
and the potential for Hg species to be transformed chemically and
biologically to more toxic forms. (Jensen and Jernelov, 1969; Wood,
et al., 1968; Wood, 1974; Ridley, et al., 1977). The Minimata Bay
disaster is a tragic reminder that Hg pollution in the marine
environment can have debilitating and lethal consequences. Moreover,
such concerns are reflected in the extraordinary number of published
studies dealing with the environmental behavior of Hg and by the
large scale assessments of environmental impact that have been derived
from global modeling of the geochemical cycle of Hg.

     The major features of global geochemical modelling are presented
in R. Wollast's chapter entitled "Basic Concepts in Geochemical
Modelling". Here the general approach for studying the environmental
behavior of Hg will be outlined. The basic mass balance model for Hg
is patterned after V. M. Goldschmidt's (1954) geochemical budgets and
deals with major exchanges among the atmospheric, oceanic, and
terrestrial reservoirs. Such a simplified representation of the
geochemical cycle of Hg and its principal pathways is presented in
Figure 1.

     This approach, while actually a conceptual framework or a
broadbrush view, has been rigorously applied to model the preindustrial
and present global cycle of Hg. In 1975, for example, Garrels et
al., using a steady-state assumption for the "pre-man" Hg cycle and
estimates of the prevailing anthropogenic contributions to rivers and
the atmosphere, concluded that "man's contributions to the mercury
cycle rival the natural fluxes". These authors, estimated the then
present day Hg flux to the atmosphere ($\sim 4 \times 10^{10}$g y^{-1}) at about 1.6
times the "pre-man" rate. They indicated as likely "that mercury is
currently being stored on land and in the oceans". Further speculation
included the suggestion that the average concentration of Hg in the

oceanic mixed layer (taken as 100m) might show a 30% increase by the year 2000 if Hg use continued at a linearly increasing rate. Perhaps the most extraordinary result from the Garrels et al., modelling efforts was the indication that the oceans were the most significant source of Hg emissions (~2.2 x 10^{10}g y^{-1}) to the atmosphere. Wollast et al., (1975) also offered a similar analysis for Hg flows in natural systems.

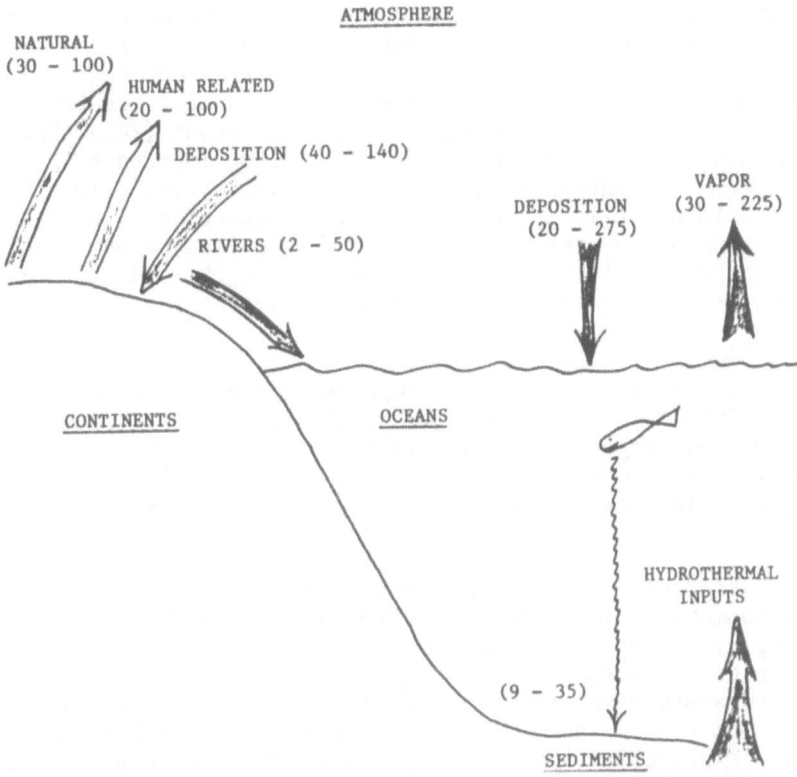

Figure 1. Simplified geochemical cycle of Hg. Ranges of reported fluxes in units of 10^8 g Hg y^{-1} given for exchanges between the continents, atmosphere, oceans, and sediments.

MacKenzie and Wollast (1977) presented a slightly refined version of the Garrels et al. model for the present-day global cycle of Hg. While Hg fluxes were in general identical, this newer formulation predicted that the major chemical forms of Hg in the atmosphere were gaseous; that natural volcanic contributions of Hg to the atmosphere and oceans would be insignificant when compared to other emissions, and that about 50% of the riverine input to the oceans would be in the particulate load. Implicitly, atmospheric gaseous Hg was assumed to be principally the elemental species, and an average residence for Hg in the atmosphere was estimated to be about 60 days. It is pertinent to note here that a year earlier Matsunaga and Goto (1976) estimated an atmospheric residence time for Hg at 5.7 years from rainfall data and atmospheric Hg measurements at a coastal site in Hakodate City.

In 1979, "An Assessment of Mercury in the Environment" was published by the U. S. National Academy of Sciences. This report included an updated version of the Garrels et al. model. Apparent refinements were made by the addition of more boxes and compartments for the formulation of a "present-day global cycle for mercury". The model results were similar to the MacKenzie and Wollast work. However, the oceanic degassing flux was found to be lower by about 60%, while the natural continental degassing flux was assessed at twice the earlier estimate. The oceans remained as a significant (24%) though not dominant source of Hg to the atmosphere. Other important conclusions from this analysis included a very low estimate for the atmospheric residence time of Hg @ 11 days; a suggestion that "from 25 to 30 per cent of the total atmospheric burden is anthropogenic", and a strong appeal that in terms of global Hg budgets "substantial evidence remains to be collected before definitive appraisals are possible".

Millward (1982) designed a time dependent five reservoir global model to specifically examine the effects of anthropogenic mobilization on the geochemical cycle of Hg. Simulations using a variety of initial conditions showed that both the atmosphere and the ocean surface layer could increase in Hg content for at least 500-800 years.

In 1984, Lindqvist et al., authored a report on "Mercury in the Swedish Environment: Global and Local Sources" in which aspects of the global Hg cycle were evaluated but not modeled. Some relevant major features that emerged were: 1) present Hg emissions associated with human endeavors rival emissions from natural processes; 2) greater than 90% of the total Hg in the atmosphere is in the vapor state and it is probably elemental Hg; 3) the residence time of this "elemental Hg" is at least a few months, or maybe 1 or 2 years, and 4) the continents are more likely to be important sources of Hg to the global atmosphere than the sea. Much of this information is also contained in a review article by Lindqvist and Rodhe which was published in 1985.

This brief summary of principal geochemical conclusions derived from global mass balance modelling suggests that little improvement in our first-order understanding of the global cycling of Hg has occurred during the past decade. There has been general but nonquantitative agreement that: (1) as a consequence of the high volatility of Hg and many of its compounds, there must be a considerable flux of Hg from natural geological, biological, and anthropogenic sources to the atmosphere; (2) atmospheric processes must significantly affect the global distribution of Hg and provide a major means for Hg transfer between the continents and oceans, and (3) the oceans are a significant source of Hg to the atmosphere. However, specific details regarding large scale geochemical questions appear as yet unanswered. For example, how significant is the interference of human activities on the global Hg cycle?; are ocean sources of atmospheric Hg important?; what is "continental degassing" and does it produce as large a flux of Hg as the models suggest?; what role does biological mobilization play in the atmospheric Hg cycle?; and does the assumption of elemental Hg (Hg^0) as the primary focus distort the models?

3.   Physico-Chemical Models

There are additional and fundamental questions that should be considered within the global framework. These queries are concerned with specific chemical behavior and reactivity which ultimately determine the fate of Hg in the environment. One may wonder, for example, about the role of organo-Hg species in the global mobilization of Hg. We should be curious about the types of atmospheric chemical reactions affecting Hg and about the kinds of Hg compounds that may be involved. We may ask: what are the important Hg species that are present in the atmosphere and oceans?; are particulate forms of Hg significant in atmospheric processes?; or why does Hg accumulate in large pelagic fish such as tuna and swordfish? These represent but a few of the many questions related to the speciation and reactivity of Hg in the environment. Thus, it is appropriate to broaden the mass balance geochemical view of the Hg cycle to the possibility that significant rate and fate determining reactions may involve Hg compounds other than Hg^0. Such species could be present in low abundances and as yet are not identified.

A rigorous and generalized chemical speciation model for the generation and exchange of Hg in the environment is not possible. However, a limited but plausible speciation view can be presented. It is reasonable to assume that there are only three potentially important representative chemical forms associated with the geochemical cycling of Hg between the oceans and atmosphere. These would be elemental Hg (Hg^0), divalent Hg (Hg^{++}), and methylated Hg (CH_3Hg^+). Elemental Hg is mobilized by both high and low temperature processes. It has a high vapor pressure, a low solubility in water, and is thought to be the principal component of the total gaseous Hg in the atmosphere. Divalent Hg should be the principal form of Hg in most natural waters. It will be found in association with inorganic and organic ligands. Divalent Hg displays a significant attraction for sulfur. Methyl Hg is the predominant form of Hg found in fish (Westoo, 1966). Methyl Hg compounds can be produced by a variety of microorganisms (Jensen and Jernelov, 1969; Wood et al., 1968; Bisogni, 1979) and there are some reports of abiotic production (Beijer and Jernelov, 1979). Biological enrichment of CH_3Hg^+ is evident in aqueous systems, where uptake may occur directly from the aqueous phase or during consumption as one proceeds through food webs. Methyl Hg compounds are more toxic than either Hg⁰ or Hg⁺⁺, and present a more serious environmental hazard.

    Physico-chemical representations of the Hg cycle in aqueous systems have been prepared by a number of authors such as Wood, (1974), Beijer and Jernelov, (1979), Stumm and Morgan, (1981), and Brosset, (1981). Thermodynamic modeling permits the inorganic speciation of Hg in natural waters to be defined for a system at equilibrium. Organo-Hg complexes, while less well defined, appear to be quite important in chemical and physical transformations of Hg ranging from biological uptake and incorporation to air-sea exchange. Biological mediation (e.g. bacteria; phytoplankton) is an integral part of the organo-Hg system and thus the assumption of an equilibrium state for a physico-chemical model is less reliable. Analytical data for the amounts and distribution of organo-Hg species in the aqueous phase are lacking. This problem is exacerbated by the myriad of potentially important organic substances in natural waters that are as yet insufficiently identified. Thus, any physico-chemical model of the Hg cycle in aqueous systems is quite idealized. Given this caveat, I have produced the physico-chemical view of the Hg cycle in sea water that appears in Figure 2.

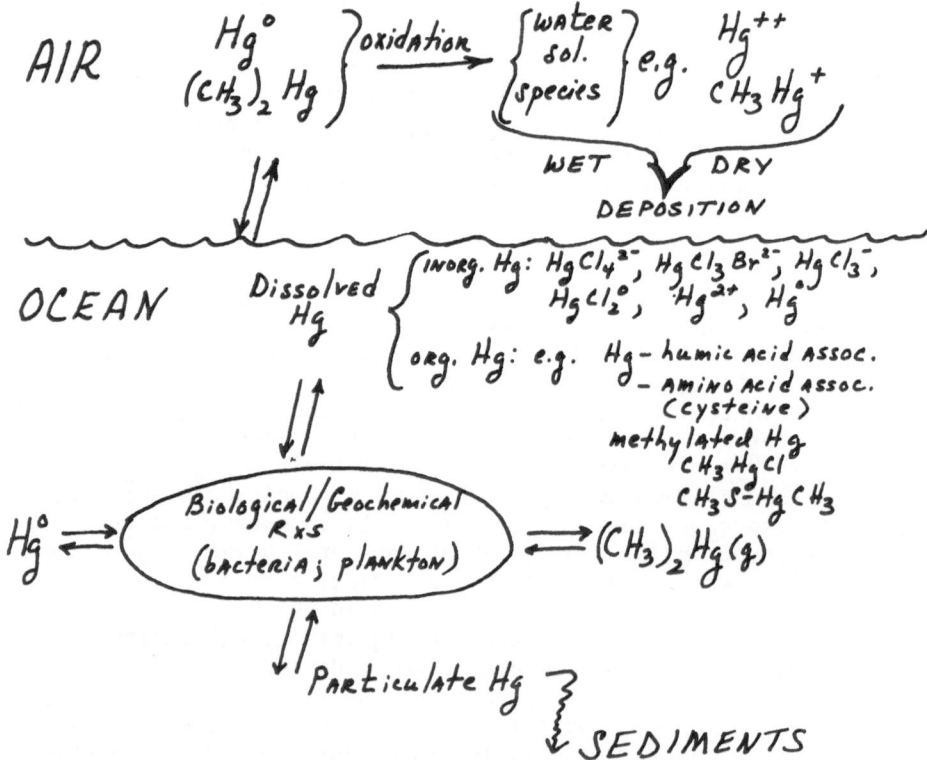

Figure 2. Physico-chemical view of the Hg cycle at the air/sea interface.

The mass balance models demonstrate the importance of atmospheric mobilization of Hg , and indicate that exchange of Hg species at the sea surface is a preeminent pathway in the global Hg budget. Figure 2 illustrates a proposed chemical composition, as well as examples of biological and chemical interactions, and phase changes that may occur in one dimensional three phase formulation of Hg cycle in the marine environment. Thermodynamic chemical models for dissolved Hg in sea water predict that chloride and mixed halide complexes (e.g. $HgCl_4^{2-}$ > $HgCl_3Br^{2-}$ > $HgCl_3^-$ > $HgCl_2^0$) will predominate (Dyrssen and Wedborg 1980); complexation of Hg by organic ligands such as free amino acids or "dissolved humic material" and carbon-Hg bonded species such as monomethyl mercury may add significantly to the dissolved component. Only a very small amount of Hg^0 would be predicted to be in equilibrium with the dissolved inorganic phase (<10^{-6} of total inorganic Hg concentration under oxidizing conditions).

The high chemical activity of Hg in biological processes is reflected by the interconversion of dissolved Hg species into elemental Hg, methylated Hg compounds, and other kinds of organo-Hg complexes in biologically mediated reactions. Part of the total Hg will remain associated with solid phases (biological and nonbiological) to be eventually carried to the bottom with fecal material or other settling detritus. A fraction of the Hg entering the sediments, reenters the water as the organic carrier phases are consumed through benthic biological activity. The Hg component that eventually becomes immobilized and buried in sediment can be associated with a variety of substrates. In the near shore regime, for example, Hg will be primarily combined with sulfide phases, while in open ocean surficial sediments the association may involve residual organic matter, hydrous oxides of Fe and Mn, carbonaceous oozes, or siliceous remains.

As shown in Figure 2, volatile species of Hg produced in surface waters can evade to the atmosphere. Thus, we can speculate, that in certain ocean regions, particularly those characterized by high biological production, there may be a flux of volatile Hg species (e.g.Hg^0; $(CH_3)_2Hg$, etc.) to the atmosphere. In other oceanic regimes such as the cold high latitude waters, there may be a net uptake of gaseous Hg species at the sea surface. We can speculate further by suggesting that evasion and invasion of volatile Hg compounds across the sea surface will show seasonal variability.

What types of reactions affect the volatile Hg species (whether land or oceanically derived) in the air over the oceans? The suggested scheme proposes photochemical oxidation processes to convert the presumed prominent but sparingly water soluble atmospheric volatile chemical forms of Hg such as Hg^0 and $(CH_3)_2Hg$ to more soluble species (e.g. Hg^0 -> Hg^{++}; $(CH_3)_2Hg$ -> CH_3Hg+). Niki et al., (1984), for example, reported laboratory results suggesting that HO-radical initiated oxidation of $(CH_3)2Hg$ could occur in the atmosphere and at a rapid rate. A rate constant of $1.9(\pm13\%)$ x10^{-11} cm^3 molecule $^{-1}$ sec^{-2} was derived for the HO-radical reaction with $(CH_3)_2Hg$.

One envisions a natural cycle between the atmosphere and ocean where dissolved Hg compounds are converted via biological activity to Hg species which can volatilize from the sea. These species once airborne can undergo photochemically mediated oxidation reactions which convert them to more water soluble forms that reenter the oceans by dry and wet deposition. This process can be repeated as part of the Hg cycling between the atmosphere and oceans. In "steady-state" preindustrial times the burial flux of Hg to the sediments would balance the Hg entering the ocean from the continental runoff and atmospheric transport. Correspondingly, modern emissions related to world-wide human endeavors may be producing an increase in the preindustrial background levels of Hg in the atmosphere and surface ocean.

Given these two views of the Hg cycle, one broad and qualitative; the other specific but speculative, we are ready to examine the state of knowledge concerning the role of air-sea exchange processes in the geochemical cycle of Hg. The first consideration is analytical. That is, what types of observations are made and how reliable are the observations.

4.   Atmospheric Hg Determinations

4.1 Total Gaseous Hg

Geochemical consistent results are being obtained for Hg in the marine atmosphere, particularly over open ocean areas of the Atlantic and Pacific Oceans (Fitzgerald et al., 1983; Fitzgerald et al., 1984; Slemr et al., 1981). The field technique that was developed in my laboratory for the collection of particulate and vapor phase Hg in both the coastal and remote marine atmosphere has been described (Fitzgerald and Gill, 1979). The procedures used by Slemr et al., 1981 are similar and can be found in Slemr et al., 1979. In these studies, the volatile atmospheric Hg species are collected by trapping on gold or on a gold-coated substate such as quartz sand or wool. It appears that Au effectively captures all the major gaseous Hg species thought to be present in the marine boundary layer. This collection has been operational defined as total gaseous Hg (TGM). The particulate Hg fraction is usually defined as the Hg collected on filters having a pore size of between about 0.3 and 0.45μ (Fitzgerald and Gill, 1979 and Millard and Griffin, 1980). Thus, the TGM is defined operationally by passage through such filters.

Our mercury analyses are conducted using a two-stage Au amalgamation gas sampling train with detection of the eluting elemental Hg^0 by either flameless atomic absorption or direct current (d.c.) plasma emission. The Hg collected on a gilded quartz sand tube in the field is transferred by controlled heating to a standardized analytical gold coated quartz sand column using Hg free air or He as the carrier gas. Following this step, the Hg is eluted from the analytical column by controlled heating, and the gas phase absorption or emission of elemental Hg is determined at 253.7 nm. A standard curve is prepared for the analytical column using known injections of Hg vapor. The coefficient of variation for the determination of 1.5 ng Hg is 1.7% and 0.06 ng of Hg can be confidently measured. At present, we estimate that the overall variability associated with the collection, volume determination, and analysis of TGM at 1.0 ng/m3 in the marine atmosphere is less than 10%. Slemr et al., 1981 report their overall precision at 14% for Hg concentrations between 1 and 2 ng/m^3, using a 400ℓ sample.

The particulate Hg collected on a Gelman Type A-E 47 mm quartz fiber filter is volatilized by pyrolysis and then collected on a gilded quartz column. Following this step, the Hg is measured in a manner identical to the volatile Hg determinations.

4.2 Volatile Hg Species

The TGM phase has been successfully partitioned into two fractions using a stacked gas sampling train consisting of a Ag collection column followed by a Au collection column. In this tandem pattern, elemental Hg and other volatile inorganic forms of Hg (e.g. $HgCl_2^0$) are collected on Ag while organo-Hg species such as $(CH_3)_2Hg$ are passed and subsequently trapped on the Au stage. Operationally, the Ag/Au sequence defines an inorganic and organic Hg fraction with the inorganic component usually considered to be elemental Hg. This separation scheme is based on early work by Braman and Johnson (1974) and has been used in the open ocean studies cited.

5. Hg Analysis in Seawater and Rainwater

The accurate determination of Hg in seawater and in rainwater is a most demanding task. Ultra-clean laboratory protocol is required to obtain uncompromised shipboard collections of such natural waters. Indeed, the variety of problems that have plagued trace metal studies in the marine environment is now well known and significant improvements have taken place (e.g. Schaule and Patterson, 1981; Boyle et al., 1977, and Bruland, 1980). The NATO ASI volume entitled Trace Metals in Seawater (1983) is recommended as a source of much useful information and insight into marine trace metal geochemistry and analytical practice.

Mercury analyses in seawater are generally conducted using a two stage reduction-aeration preconcentration procedure. Dissolved Hg species are converted to elemental Hg with the addition of a reducing agent such as $SnCl_2$ or $NaBH_4$. In the next step, the generated elemental Hg is purged from the reaction vessel and trapped on a suitable absorbent which is usually Au. The Hg collected in this manner can be subsequently analyzed using a procedure similar to the atmospheric Hg determinations and with a variety of detection techniques such as gas phase atomic absorption, plasma emission, or atomic fluorescence. A useful summary of analytical practices and

procedures can be found in the 1982 report from the ICES International
Intercalibration Study for Hg in Seawater (Olafsson, 1982).

We usually carry out part of the Hg analyses of sea water on
board the research vessel. We employ a two-stage Au amalgamation
$SnCl_2$ reduction-aeration preconcentration procedure with the
determination of Hg as Hg^0 by gas phase absorption at 253.7nm. Our
entire methodology from sea water collections to analysis has been
recently published (Gill and Fitzgerald, 1985) and the reader is
referred to this work for analytical details. As we note, the
precision of analysis for the determination of 1.0ng Hg in a 500ml
sample (10 pM) is 10% and about 0.12ng/1 (0.6pM) can be confidently
measured. Increased sensitivity is readily achieved by using larger
volumes for the reduction-aeration step.

5.1 Reactive and Total Hg

Mercury measurements in sea water are often experimently divided
into two fractions-reactive and total Hg (Fitzgerald and Hunt, 1974;
Olafsson, 1983; Gill and Fitzgerald, 1985). Reactive Hg measurements
are broadly defined as representing those species of Hg that are
readily reducible (e.g. with $SnCl_2$) in acidified sea water. In our
work, the reactive Hg fraction represents the amount of Hg measured
(within 24 h of collection) in unfiltered sea water acidified to a pH
of 1.2. Thus, a reactive Hg determination will include dissolved
inorganic species, labile organo-Hg associations, and Hg easily
leached from particulate matter in the sample. Operationally, the
difference between a reactive and total Hg determination is a measure
of the presence of reasonably stable organo-Hg associations, e.g.,
methyl and dimethyl mercury, which require photochemical or prolonged
acid digestion to cleave covalent C-Hg bonds and liberate inorganic Hg
for reduction and detection (Fitzgerald and Lyons, 1973; Baker, 1977;
Olafsson, 1978).

5.2 Volatile Hg

Dissolved gaseous Hg constituents in sea water can be measured
using purge and trap techniques on "raw" freshly collected sea water
samples (Slemr et al., 1981; Iverfeldt and Lindqvist, 1982; and Kim
and Fitzgerald, 1984 and 1985). Moreover, an indication of the
inorganic and organic speciation of these volatile phases can be
obtained with the Ag/Au trapping sequence that was described for
atmospheric Hg speciation studies.

5.3 Determinations of Hg in Rain

Procedures used to determine Hg fluxes associated with rainfall
are similar to those employed in sea water studies. As emphasized,
the uncompromised collection of rain is a most demanding analytical/
field operation. In our investigations, at sea and at land-based
stations, rain is collected in large diameter glass (38 cm), or thin

walled, all Teflon (45 cm) funnels, supported by anodized Al frames. Rainwater feeds directly into Teflon storage bottles (1 or 2l) through a threaded Teflon transfer block joining bottle to funnel. The collection apparatus and storage containers are exhaustively and progressively cleaned at 55°C over a period of 10 days (Gill and Fitzgerald 1985). Rainfall collections are acidified with concentrated high purity subboiling distilled HNO_3 acid to yield a pH of 1.0. In addition, 1 ml of $AuCl_3$ (15ppm) is added per liter to ensure no loss of Hg during storage (Moody et al., 1976). The rain collector and Teflon bottles are stored before and after use in acid washed Teflon bags.

Mercury analyses in rainwater are conducted using a two stage Au amalgamation modification of the cold-trap $SnCl_2$ reduction-aeration flameless atomic absorption method we initially employed (Fogg and Fitzgerald, 1979). The analysis procedures are identical to our sea water determinations of Hg.

6. Air-Sea Exchange of Hg

As part of our general interests and efforts concerned with the global cycling of Hg and with its behavior and fate in the marine environment, we have been investigating the atmospheric cycle of Hg over the oceans. Much of this work has been conducted within the SEAREX (Sea-Air Exchange) Program. SEAREX is a multidisciplinary, coordinated research endeavor that is investigating the sources, fluxes, and air-sea exchange behavior of a large number of trace metals and organic substances in the Pacific Ocean,. The geographic coverage of important marine regions and oceanic phenomena has been broad and includes measurements of Hg in the near surface troposphere, in sea water, and in rainwater from both hemispheres. New information for Hg has been obtained at remote tradewind sites in each hemisphere (i.e. Enewetak Atoll, Marshall Islands and at Tutuilla Island, American Samoa and during cruises to equatorial regions. Shipboard investigations have also taken place in the northwest Atlantic Ocean and in the nearshore environs of the Peru Upwelling. Major experiments took place in the South Pacific westerlies at a land-based sampling facility on Ninety Mile Beach, northern North Island, New Zealand, and during a joint U.S./N.Z. oceanographic cruise to the Tasman Sea. The North Pacific westerlies will be examined on SEAREX cruises between May and August of 1986. Additional work includes ongoing studies of Hg associated with our local Long Island Sound coastal/urban environment.

6.1 Preliminary Studies

6.1.1 Hg: A Trace Atmospheric Gas

Most of the atmospheric Hg species in the marine boundry layer over the open ocean are in the vapor phase (>99%). Even in a coastal/urban region such as Long Island Sound, greater than 97% of the total Hg in the near surface in the gaseous state. This physical

speciation has been established experimentally by our laboratory
(Fitzgerald and Gill, 1979; Fitzgerald et al., 1983). A quantitative
indication of the partitioning observed from our early investigations
in the northern hemisphere is presented in Table I.

Table I   Mercury Distribution in the Marine Atmosphere
          at Searex Sampling Locations

Collection Site	No. of Samples	Gaseous Hg Distribution Range ng/scm*	Arithmetic Mean & Std. Dev. ng/scm	Particulate Hg Distribution No. of Samples	pg/scm
			Oceanic Regions		
Enewetak, Dry Season (4/27-5/21/79)	27	0.8 - 2.9	1.6 ± 0.6	2	0.4 and 0.7
Enewetak, Wet Season (6/28-8/6/79)	67	1.1 - 3.2	1.7 ± 0.5	2	0.5 and 2
Northwest Atlantic Ocean (7/10-7/24/79)	7	1.0 - 1.9	1.6 ± 0.4	1	0.7
			Long Island Sound		
Avery Pt., Groton, Ct. (1979-1980)	130	1.6 - 9.9	3.1 ± 1.0	18	20 ± 20 (8-60 range)

*scm = standard cubic meter

     Three important features of the atmospheric cycle of Hg were
quite evident in this early work. Firstly, gaseous Hg concentrations
were in the ng m^{-3} range while atmospheric particulate Hg was in pg m^{-3}
quantities, about a 1000 times smaller. Secondly, the means and
standard deviations for the gaseous Hg distributions were similar for
open ocean areas of the Atlantic and Pacific Oceans. Thus, total
gaseous Hg as defined by collection on Au was characterized by a
relatively stable concentration in the open ocean marine boundry
layer. This lack of variability suggested that atmospheric gaseous Hg
could have a long residence time. This result was surprising since
it did not support predictions from most global models (Section 2).
Moreover, it should be emphasized that about 1/2 of the variation
found in these early oceanic studies of vapor phase Hg was due to

analytical imprecision. The third significant feature was the increase of both concentrations and variability for particulate and gaseous Hg at the nearshore Long Island Sound station. This pattern supported the view that the continents were the principal sources of atmospheric Hg. In 1981, Brosset reported yearly means for TGM at different places in Sweden to range between 2.7 and 4.0 ng m^{-3}.

During the SEAREX wet season investigations at Enewetak, A. Hewitt conducted five speciation experiments. The results indicated that vapor phase Hg could be completely trapped on a Ag substrate. While the analytical uncertainty was at least $^+_-$20%, this work suggested that the gas phase Hg at this remote tropical Pacific Ocean site consisted principally of inorganic Hg, with elemental Hg as the probable major component. In comparable experiments at our Long Island Sound station, an organo-Hg fraction was found and it ranged from 27 to 82% (average = 50%; n=13) of the total gaseous Hg (Fitzgerald, et al., 1983). These data were consistent with the hypothesis that the major source of atmospheric Hg was continental. The speciation contrast between the open ocean and coastal stations suggested that while mercury is emitted from natural and anthropogenic sources principally in gaseous forms (Hg^0, (CH$_3$)$_2$Hg, HgX, etc.) only Hg^0 was long lived in the atmosphere.

This work agreed quite satisfactorily with speciation studies conducted by Slemr et al., 1981 in the Atlantic Ocean. In clean marine air masses they found the organo-Hg fraction (e.g. (CH$_3$)$_2$Hg) to be lower than their detection limit of 0.1 ng m^{-3}. In continental air masses, the organo-Hg concentrations ranged between 0.2 and 2.2 ng m^{-3}.

The results from these initial experiments encouraged us to model the air-sea exchange behavior of Hg in a preliminary manner by treating Hg as a trace tropospheric gas (Fitzgerald, et al., 1981). We suggested that Hg transfer and exchange with the oceans was analogous to a trace gas whose primary sources were continental and included both natural and anthropogenic processes (Fitzgerald, 1982; Fitzgerald, et al., 1983). The trace gas modelling for Hg yielded an average tropospheric residence time of total gaseous Hg, assumed to be Hg^0, of about a year. This long residence time was corroborated by our estimates of annual Hg deposition to the earth's surface using a steady state model for the global Hg cycle and by the work of Slemr, et al., 1981 (Section 9).

6.2 Present Status

6.2.1 South Pacific Westerlies

We will begin our consideration into the present state of understanding regarding the atmospheric cycle of Hg over the oceans with the South Pacific westerlies. The SEAREX Ninety Mile Beach site on the northern North Island, New Zealand is shown in Figure 3. A

20m aluminum air sampling tower faced into westerlies. The samplers
used for total gaseous Hg collections, speciation studies and
particulate Hg determinations were mounted at the top level. Sampling
was controlled automatically and occurred only if the wind speed
exceeded 8 km/hr, condensation nuclei counts were <500 cm^{-3}, the wind
was from a selected open ocean sector (200°-290°), and it was not
raining. A further and important criterion was the desire to sample
air arriving at Ninety Mile Beach whose trajectory history was known.
We were interested, firstly, in determining the Hg distribution in
purely oceanic air representative of the South Pacific westerlies,
and secondly, if possible, to evaluate the contributions from the
Australian continent by analyses of Hg in air whose back trajectories
intercepted land areas of Australia. The New Zealand Meterological
Service provided extraordinarily valuable air trajectory forecasts
during the experiments. This information has now been enhanced by
10 day isentropic trajectories prepared by J. Merrill.

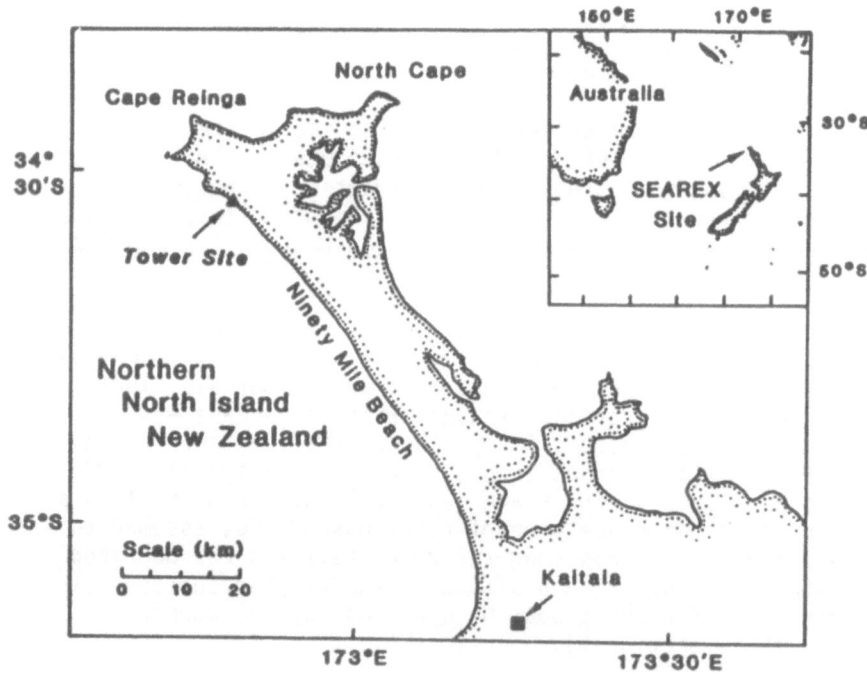

Figure 3. Sea-Air Exchange (SEAREX) Program
atmospheric study site in New Zealand.
(Courtesy of R. Arimoto)

The results from the New Zealand experiment are illustrated in Figure 4, and summarized in Table II. There were only 3 ideal air mass trajectories sampled during the approximately 60 day investigation. These trajectories have been classified as oceanic and represent air arriving from the Southern Ocean with no interference from land. Ten measurements were made in air whose ten day history indicated

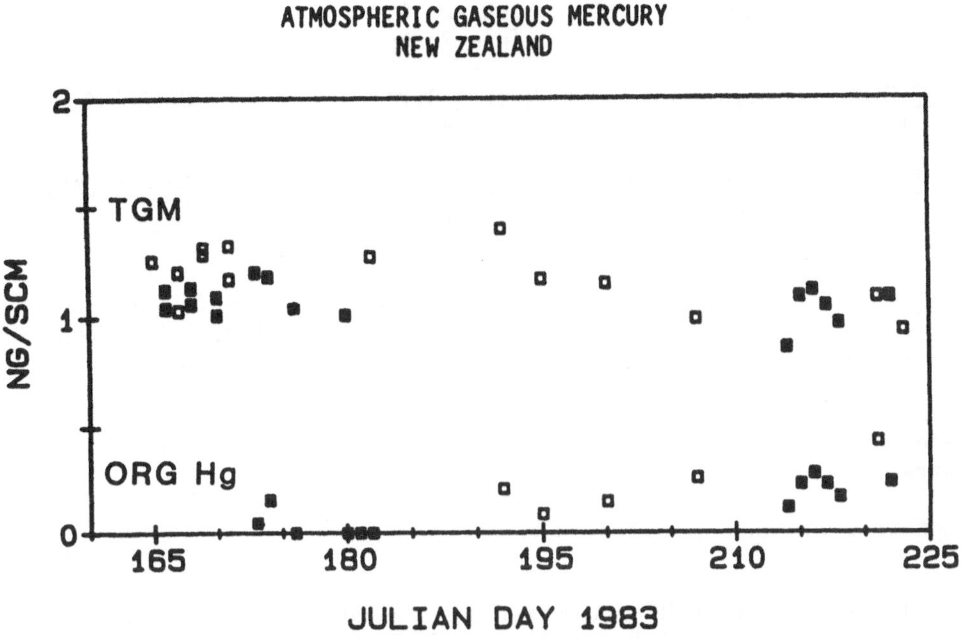

ATMOSPHERIC GASEOUS MERCURY
NEW ZEALAND

Figure 4. Atmospheric gaseous Hg distribution at Ninety-Mile Beach, New Zealand (July - August, 1983) Filled squares represent results from ocean + oceanic* sector.

passage over southern regions of Australia and/or Tasmania. These trajectories have been called oceanic*. Additional measurements were made in air which had passed over either the North or South Island or intercepted local features along the beach. These trajectories are indicated as "out of the preferred sampling sector" and the relevant data are included in Table II.

Table II. Atmospheric Gaseous Mercury Distribution at Ninety-Mile
           Beach, Northern North Island, New Zealand,
           (June-August, 1983)

Atmospheric Mercury Concentration mean and std. dev. (ng/scm)			Trajectory
Total Hg	Org. Hg	Inorg. Hg	
1.08 $\pm$ 0.04 (n = 4)	0.00 and 0.24 (n = 2)	1.05 and 0.85 (n = 2)	OCEANIC
1.06 $\pm$ 0.09 (n = 12)	0.16 $\pm$ 0.10 (n = 8)	0.91 $\pm$ 0.14 (n = 8)	OCEANIC*
1.07 $\pm$ 0.08 (n = 16) [0.86 - 1.12]#	0.15 $\pm$ 0.11 (n = 10) [0.05 - 0.28]	0.92 $\pm$ 0.13 (n = 10) [0.74 - 1.15]	OCEANIC + OCEANIC*
1.18 $\pm$ 0.13 (n = 15) [0.94 - 1.40]	0.16 $\pm$ 0.15 (n = 7) [0.00 - 0.43]	1.00 $\pm$ 0.28 (n = 7) [0.63 - 1.27]	OUT OF PREFERRED SAMPLING SECTOR (OSS)

*Trajectories passed over either Tasmania or southern regions of
 Australia.

#Concentration range (ng/scm) in brackets.

       The total gaseous Hg concentrations showed very small variations,
with a mean and standard deviation of 1.12 ± 0.10 ng/scm for all
measurements (n=31), and a range between 0.86-1.40 ng/scm. This
variability is comparable to our estimates of overall experimental
error. There was no significant statistical difference between the
oceanic and oceanic* determinations and it is appropriate to consider
the combined measurements as representative of air associated with
the westerlies over open ocean regions of the southern hemisphere. A
comparison of the mean value for total gaseous Hg in the out of
sector category with the mean ocean + oceanic* determination shows
only a slight influence from passage of air over various parts of New
Zealand. This result was somewhat surprising because the North
Island contains potentially large sources of gaseous Hg emissions
such as the large urban center of Auckland, and in the central area,
the active volcanic region with associated geothermal activity. The
Hg emissions must be readily dispersed. Moreover, an analytical
precision of < 5% would appear required to confidently observe concentration differences among air masses with different origins and
trajectories.

   The gaseous organic component might be a better indicator of the
influence of terrestrial Hg emissions on atmospheric Hg concentrations.
However, as illustrated in Figure 4, the organo-Hg fraction did not
increase in the air at Ninety Mile Beach for trajectories that passed
over New Zealand (OSS). Some oceanic mobilization is suggested by
the presence of gaseous organic Hg in the oceanic trajectory. An
oceanic source of atmospheric Hg may also be responsible for the
organic Hg signal associated with the oceanic* trajectories.

   In summary, the open ocean trajectories in the South Pacific
westerlies yielded on the average: TGM @ 1.07 ± 0.08 ng/scm; org.
Hg. @ 0.15 ± 0.11 ng/scm; inorg. Hg @ 0.92 ± 0.13 ng/scm, and
part. Hg @ 1.5 pg/scm. The Australian continental influence on
atmospheric gaseous Hg concentrations was not evident. Nor were
localized contributions to the atmospheric gaseous Hg concentrations
pronounced. Most of the gaseous Hg in the open ocean atmosphere
appears to be Hg^0. However, the organic Hg fraction, while small,
may be providing evidence for oceanic Hg emissions.

6.2.2   Interhemispheric Atmospheric Hg Distribution

   During May - June, 1984 J. P. Kim and I participated in an air/sea
study of Hg in the equatorial Pacific Ocean. The cruise track of the
R.V. Researcher appears in Figure 5. This cruise is described in

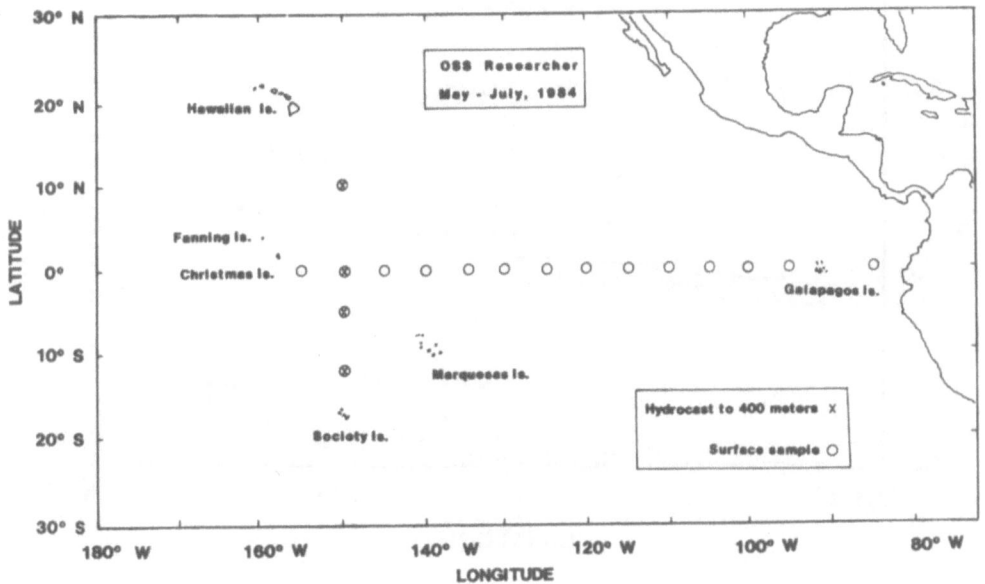

Figure 5. Cruise track of the O.S.S. Researcher
in the equatorial Pacific Ocean (May-June, 1984).
Water sampling stations are indicated.

Section 7.2 and appropriate acknowledgements given. Here I wish to
focus on atmospheric Hg measurements and interhemispheric aspects of
the expedition. The latitudinal distributions of total and organic
gaseous Hg are plotted for the interhemispheric transect at 150° W,
and appear in Figure 6. The Hg collections were made at bow level
(~10m) while the ship was underway. Thus, the bars refer to the
latitudinal sampling interval over which a collection was made.

There are three important features illustrated by the plots in
Figure 6. Firstly, the southern hemispheric end members (> 6°S) show
excellent agreement with the data obtained at Ninety Mile Beach. The
mean and standard deviation for the 5 TGM measurements between 6°S
and 17°S was 1.00 ± 0.08 ng/scm, which can be compared to the value
of 1.07 ± 0.08 ng/scm in the South Pacific Westerlies (Table II).
Moreover, the chemical speciation for these tropical latitudes
indicates that the total gaseous Hg consists almost entirely of Hg⁰
as defined by the Ag/Au fractionation sequence (Section 4.2). This
would imply, as suggested by the New Zealand studies, that the organo
Hg contribution would be superimposed on a southern hemispheric
background of Hg⁰ at about 1.0 ng/scm.

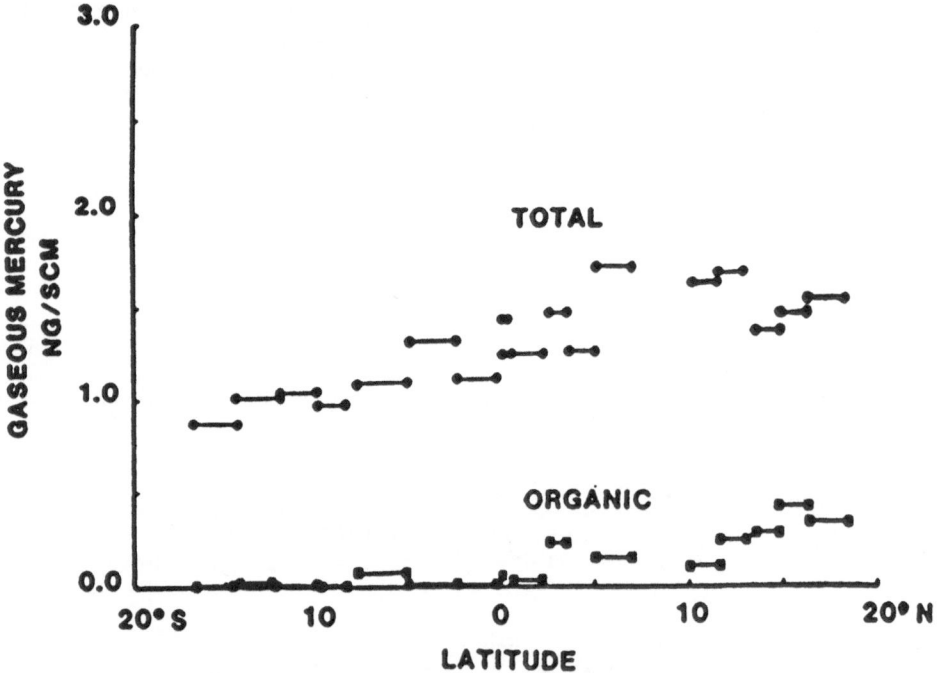

Figure 6. Interhemispheric shipboard study of Hg in near surface
marine atmosphere during May–June, 1984 at 150°W. Total gaseous
Hg in filled circles, and total gaseous organic Hg in filled
squares (the bars refer to latitudinal sampling intervals).

The second significant finding is the increase in TGM that occurs going from the southern to the northern hemisphere. This pattern is similar to a trace atmospheric gas such as CO whose primary sources have been established to be continental (Seiler, 1974), and whose atmospheric residence time is sufficiently short so as to allow the interhemispheric open ocean marine boundary layer concentrations to reflect the time required to mix air between the hemispheres. These results for TGM are very much like the interhemispheric Hg distribution reported for the near surface atmosphere by Slemr et al., (1981) and their later results, all of which were summarized in a figure for the Swedish Report (Lindqvist, et al., 1984) and reproduced here as Figure 7.

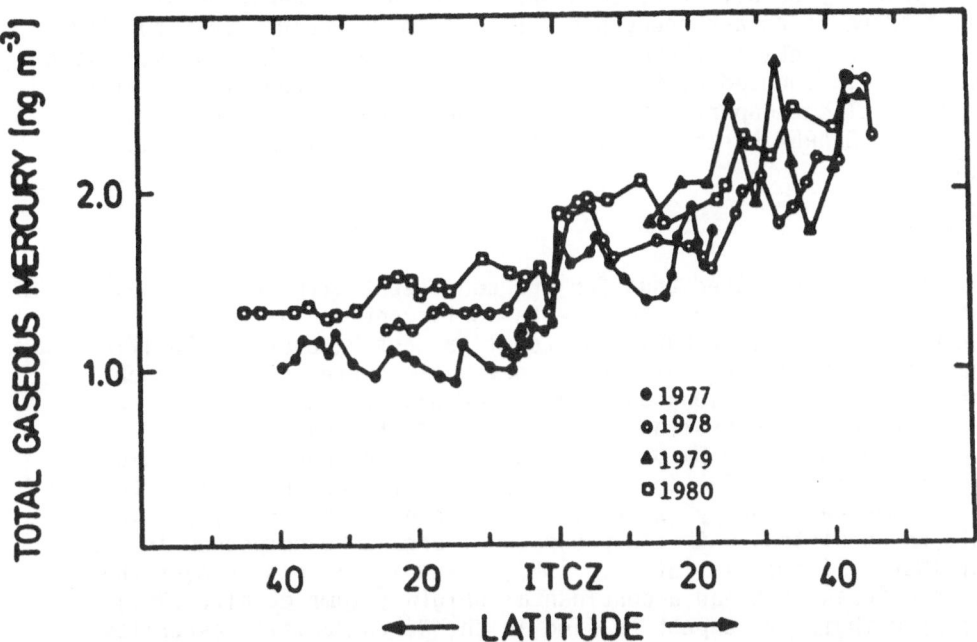

Figure 7. Latitudinal distribution of total gaseous mercury in surface air over the Atlantic Ocean. The mid point represents the position of the Intertropical Convergence Zone (ITCZ) and the Northern Hemisphere is to the right. (After W. Seiler in Lindqvist et al., 1984.)

It should be noted that during the cruise we passed through the ITCZ at about 4^0-6^0 N according to the bow level particle distribution (A. Clarke, personal communication). The total gaseous mercury concentrations for the latitudes (> 5^0 N) have a mean and standard deviation of 1.55 ± 0.17 ng/scm (n=4) that agrees satisfactorily with our early results from Enewetak (summarized in Table I).

The third important result from our shipboard investigation was the presence of organo-Hg species in the northern hemispheric collections. The concentrations were as large as 0.42 ng/scm. Moreover, there is a suggestion that this organic fraction may increase toward higher latitudes. In constrast to the New Zealand results, we suspect that the organo-Hg species are land derived, since the ocean regions north of 10⁰ N are oligiotrophic and similar to the ocean areas covered during the southern hemispheric leg of the cruise track. Very small quantities were found in the latter regime.

It is appealing to speculate on the possibility that terrestrially produced organo-Hg compounds are being observed at these distant oceanic locations. We can suggest, for example, that the atmospheric lifetimes of these species would at least be days; the chemical forms would probably be different than species with an oceanic origin; and if there is an anthropogenic cause, then we would expect to find increasing amounts of these substances in regions of the North Pacific westerlies influenced by the Asian continent. Fortunately, we will have an exellent opportunity to investigate these phenomena further during the SEAREX North Pacific westerlies cruise in May - August, 1986.

6.3 Summary

The data presented thus far are consistent with the hypothesis that the major source of atmospheric Hg is continental. The interhemispheric distribution of total gaseous Hg over the Pacific Ocean is characteristic of a trace gas whose terrestrial emissions exceed oceanic emanations on a unit area basis. This is supported by the increased concentrations of Hg observed, for example, by us over Long Island Sound and by Brosset (1981) in near-shore regions along the Swedish coast. The organic Hg fraction in the tropical North Pacific marine boundary layer may have a continental origin, which would be consistent with the presence of organo-Hg compounds in the near shore (Fitzgerald, et al., 1983), and in air masses over the Atlantic Ocean that had a continental origin (Slemr et al., 1981). Correspondingly, we expect to find in the North Pacific westerlies increased total gaseous Hg concentrations and perhaps a larger organic Hg fraction that results from emissions on the Asian continent. The New Zealand experiments suggest albeit weakly that oceanic sources of Hg exist. Let us now consider the more substantial evidence for oceanic Hg evasion.

7.   Ocean Sources of Hg

7.1 Hg Evasion from the Equatorial Pacific Ocean: 1980

The initial evidence for Hg evasion from the surface ocean was obtained on an interhemispheric research cruise of the R/V T. G. THOMPSON (University of Washington) in the tropical Pacific Ocean during October, 1980, (Fitzgerald et al., 1984). The meridional cruise track at 160°W and the air sampling locations are summarized in Figure 8. Shipboard determinations were made of total gaseous mercury (TGM) in the marine atmosphere at bow level (~10 m). No speciation experiments were conducted. In addition, surface water was collected and analyzed on shipboard for reactive Hg. Nutrient concentrations were measured and standard hydrographic determinations of salinity and temperature were also obtained (Strickland and Parsons, 1972). This information was provided by K. Bruland, University of California at Santa Cruz, who was chief scientist. We were grateful cruise participants. The observed atmospheric TGM concentrations (ng/scm), sea surface mixed layer water temperature (°C), nitrate + nitrite concentrations (µmoles/1), and reactive Hg concentrations (log pmoles/ℓ) are plotted as a function of latitude in Figure 9.

We found pronounced increases (>0.5 ng/SCM) of TGM in the equatorial region defined approximately by the latitudes between 4°N and 10°S (Fig. 9). Moreover, these maxima appear to be superimposed on a background of TGM which decreases interhemispherically (N → S). A relatively smooth transition from higher concentrations of TGM in the northern hemisphere to lower concentrations in the southern hemisphere had been anticipated (Section 6.2.2). We found such a pattern during the 1984 Researcher study of the equatorial Pacific Ocean (Figure 6), and Figure 7 illustrates such interhemispheric distributions as reported by Slemr et al., 1981 for TGM in the near surface atmosphere over the Atlantic Ocean.

We suggested that the localized enhancement of TGM in the atmosphere over the equatorial Pacific was caused by evasion of Hg species from the oceans to the atmosphere. The increases in TGM were defined by a few measurements in which we had a high degree of confidence. The overall analytical uncertainty in any one measurement was about 10% (Section 4.1), and the end-member TGM determinations agree extraordinarily well with other results. For example, the TGM concentrations (1.47 ± 0.04 ng/scm) at latitudes 7°-20°N are similar to our earlier observations at Enewetak Atoll, Marshall Islands, and in the northwest Atlantic Ocean (Table I) as well as to the average (TGM) concentration of 1.55 ng/scm found on the interhemispheric leg of the 1984 Researcher cruise (Section 6.2.2). At latitudes >10°S smaller concentrations of TGM (mean = 1.05 ± 0.09 ng/scm) were observed and these results correspond well to the average concentration of TGM at 0.98 ± 0.07 ng/scm from comparable latitudes during the Researcher work.

This end-member agreement conicides satisfactorily with the Intertropical Convergence Zone (ITCZ) as defined by cloudiness at or near 160°W. During the cruise period (October 1-21) this transition region which serves to inhibit interhemispheric exchange in the troposphere varied between 5°N and 15°N (O. Garcia, ERL/NOAA Boulder, CO, personal communication). Thus, southern hemispheric chemical distributions in the troposphere would be expected south of 5°N.

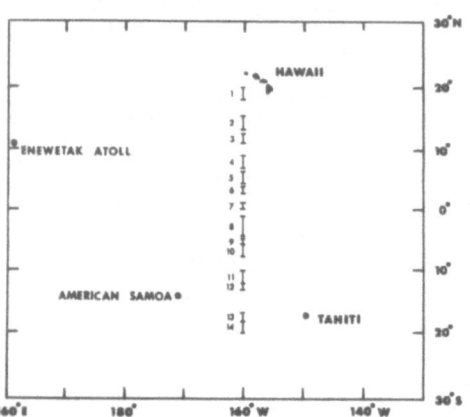

Figure 8. Cruise track of the R.V.T.G. Thompson (1 to 21 October 1980). Each bar represents the latitudinal distance traversed during a period when atmospheric Hg was collected. (After Fitzgerald et al., 1984.)

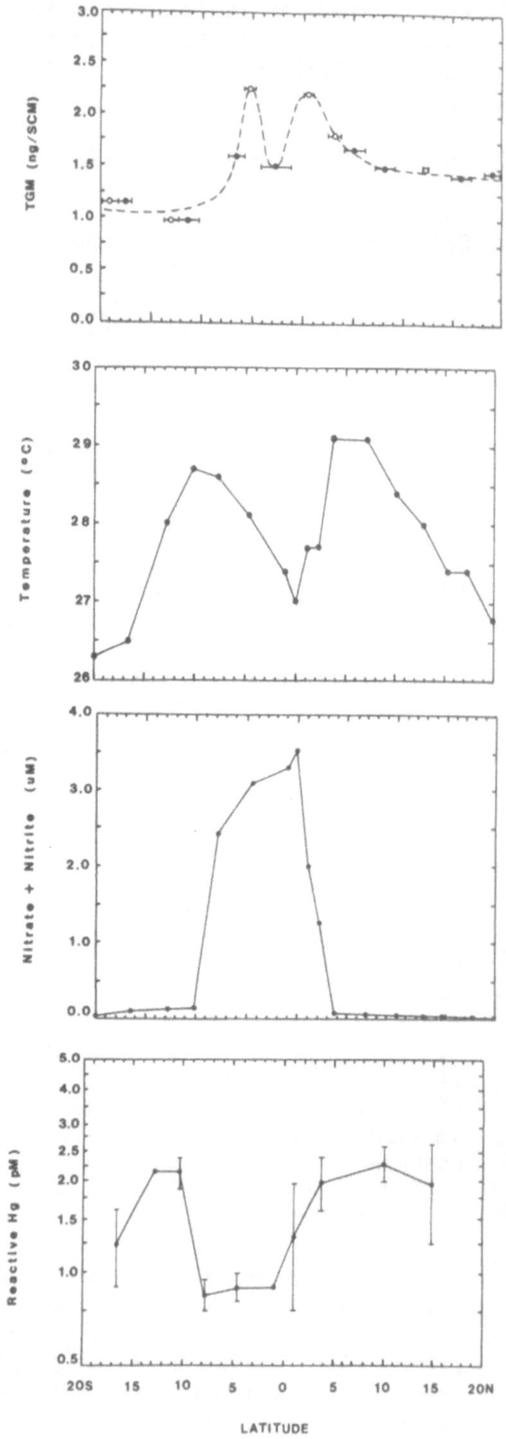

Figure 9. Interhemispheric shipboard study of Hg in near surface marine atmosphere and in surface sea water at 160°W. Latitudinal distributions plotted are: Total gaseous Hg, where open circles represent collections during daylight hours and filled circles represent night collections (the bars refer to latitudinal sampling intervals as indicated in Figure 8; temperature of the mixed layer at 5 m below the sea surface; nitrate plus nitrite concentration in the sea-surface mixed layer; reactive Hg concentrations in the seasurface mixed layer with the 2 sigma experimental error limits indicated for replicate samples. (After Fitzgerald et al., 1984.)

The equatorial Pacific regime is hydrographically and biologically quite dynamic. Westward flowing north and south equatorial currents are intersected by an eastward directed equatorial counter current located near the equator, resulting in a strong but variable zonal water flow (Wyrtki et al., 1981). Significant wind-driven equatorial upwelling (Ekman divergence) occurs in the South Equatorial Current that flows between ~4°N to ~10°S. The diverging surface waters are replaced by cooler, nutrient rich water from depths of about 100 to 200 meters, supporting an enhancement of primary production and biological activity at the surface (Wyrtki, 1981).

The observed TGM fluctuations are distributed over the same latitudinal range as the oceanographic features. Indeed, the elevated levels of TGM in the atmosphere coincide closely with sea surface manifestations of equatorial divergence and upwelling as reflected in measured changes of temperature and nutrient concentrations (Fig. 9). The sea surface mixed layer water temperature can be used as an indicator of upwelling along the equator and the nitrate + nitrite concentrations show that the cooler waters reaching the surface are enriched in nutrients. This reasonably good correlation between the near surface atmospheric changes in Hg and hydrographic variability suggests, firstly, that Hg evasion is enhanced within the equatorial upwelling region and, secondly, biological mediation is probably involved in the process.

The day-night variability observed for TGM as a function of latitude (Fig. 9) provides additional support for an equatorial oceanic source of gaseous Hg and further evidence for a role of biota in the process. In the upwelling region (4°N - 10°S) the TGM concentrations are about 0.7 ng/scm higher during the day as compared to the night periods. Although, we do not have diel data from the same site, we assume this difference to reflect temporal rather than spatial variability. No significant diel difference in TGM was observed at the higher latitude stations in either hemisphere. This pattern is consistent with the distribution of biological productivity in the region. That is, primary production in upwelling zones is considerably higher than in the oligotrophic open ocean ocean regimes to the north and south (Broecker and Peng, 1982). Laboratory studies indicate that both algae (Ben-Bassat and Mager, 1975) and bacteria (Jensen and Jernelöv, 1969; Wood et al., 1968) can convert inorganic dissolved Hg to volatile vaporous forms.

In summary, we proposed the existence of an oceanic source of atmospheric Hg in the equatorial Pacific Ocean. We suggested that the oceanic evasion process was probably related to upwelling of nutrient rich water and to the enhanced biological productivity associated with the equatorial Pacific Ocean.

We were very excited with these findings regarding the marine biogeochemical behavior of Hg and the concomitant implications to the global Hg cycle. However, this enthusiasm was tempered by recognition of the limitations of the experiment and the need for confirming results. For example, the coincidental reduction in reactive Hg concentrations in surface water between 4ºN and 10ºS (Figure 9) was offered as supporting evidence for Hg evasion from the equatorial zone. A decrease in reactive Hg of about 65% occurred in this equatorial region. While we had not yet measured dissolved gaseous Hg species directly, we suggested that part of the decrease in reactive Hg may represent conversion of dissolved species to volatile gaseous forms which can evade from the sea surface.

We were anxious to return to the equatorial Pacific Ocean upwelling regions for a more thorough study of the proposed evasion process. This follow up investigation would include simultaneous measurements of gaseous Hg in the near surface atmosphere and in surface sea water. Dissolved gaseous Hg concentrations would be compared with the solubility predicted for a sea water solution in equilibrium with the atmosphere containing the measured gaseous Hg concentration. If our hypothesis was correct, then substantial evidence of dissolved gaseous Hg at supersaturated concentrations would be obtained.

7.2 Hg Evasion from the Equatorial Pacific Ocean: 1984

In 1984, we were given the opportunity to return to the equatorial Pacific Ocean as part of a NOAA sponsored oceanographic expedition on the O.S.S. Researcher. This cruise enabled us to carry out a more complete examination of the Hg distribution in both the air and surface water of this marine region where potentially important Hg emissions were indicated. This work represents part of J. P. Kim's doctoral dissertation research at the University of Connecticut. Some of the principal findings have been reported (Kim and Fitzgerald, 1984) and a more complete paper has been submitted for publication (Kim and Fitzgerald, 1985).

The Hg experiments were quite successful, and I wish to express our appreciation prior to discussion of the results which are relevant to this present paper. We acknowledge with much gratitude the professional assistance and competent support provided by the Captain, officers, crew and survey technicians of the Researcher. Our scientific colleagues were most helpful. V. Thayer and R. Barber provided the chlorophyll a data. We are indebted to Steve Piotrowicz, who served as Chief Scientist during the first leg of the cruise. Steve invited us, and ensured the success of the cruise and our new venture in the marine biogeochemical world of Hg. His generous aid included a completely outfitted Class 100 trace metal clean laboratory for our use.

The cruise track along the equator appears in Figure 5. It extended eastward from a longitude of 155ºW to 93ºW. A short interhemispheric leg was run between 4º30' S and 4º30' N at 85ºW. Total gaseous and organic gaseous Hg measurements were made from bow level atmospheric collections. These determinations were carried out as described in Section 4. Surface seawater was collected in acid-cleaned 30 l Teflon-lined Go Flo bottles using a synthetic fiber hydrographic cable (Kelvar) made taut with plastic-bagged concrete weight. Aliquots were acidified and stored in acid-cleaned 2l Teflon bottles for later (> 2 months) analysis of total reactive Hg in our laboratory (see Section 5 and Gill and Fitzgerald, 1985). Standard hydrographic determinations of temperature, salinity, and oxygen were made. Nutrient concentrations (dissolved nitrate + nitrite and phosphate) concentrations were stored frozen for later laboratory analyses by T. Loder using a Technicon autoanalyzer. Chlorophyll a was measured fluorometrically.

Dissolved gaseous Hg in sea water was determined by a purge and trap procedure with subsequent detection of Hg by gas phase atomic absorption. The general features of the analytical determination of Hg are similar to the other Hg measurements in air and in sea water which have been described (Section 4 & 5). The dissolved gaseous Hg was measured in sea water collected with the acid-cleaned 30 l Go Flo samplers. Thirty to 60 l samples were processed by purging with Hg free air in a 4 l customized gas sparging vessel under continuous flow conditions. The flow rates for the sea water and purging gas were, respectively: 0.5 l/min and 1 l/min. The stripping efficiency for elemental Hg was between 90-95% as inferred from calibration experiments with dissolved oxygen as the monitoring gas. The dissolved gaseous Hg stripped from sea water was collected on Au and Ag/Au traps in the manner previously described. Thus, a measure of the total dissolved and organic gaseous Hg concentrations in surface sea water was obtained. The samples were processed in a Class 100 portable clean laboratory.

The atmospheric measurements of total gaseous Hg along the equatorial track are displayed in Figure 10. The concentrations range from 0.85 to 1.12 ng/scm, with a mean and standard deviation of 1.02 ± 0.08 ng/scm (n = 19). This variation is comparable to our estimates of the total experimental error associated with a measurement at these atmospheric Hg concentrations (Section 4.1). The predominant form making up the total gaseous Hg in the atmosphere was elemental Hg (80-96%) as defined by the Ag/Au separation technique.

These results compare extraordinarily well with the other southern hemisphere determinations of total and organic gaseous Hg made during the 1980 and 1984 shipboard studies and with the 1983 South Pacific westerlies investigation at New Zealand (Sections 6.2.1, 6.2.2, and 7.1). It should be clear to the reader that we have temporal information on background total gaseous mercury concentrations in the remote southern hemispheric ocean atmosphere. This important aspect of the atmospheric cycling of Hg over the oceans will be discussed fully in Section 9 which is entitled "Atmospheric Cycling of Hg Over the Oceans: Global Perspectives".

The total dissolved Hg in the surface sea water generally increases along the equator from relatively constant levels of about 60 femtomolar (fM) between 155 and 125 W to 225 fM near the Galapagos Islands (90 W). The Ag/Au speciation experiments show that all the dissolved gaseous Hg is Hg^0. The increasing trend of dissolved gaseous Hg coincides with decreasing surface sea temperature and increasing concentrations of nitrate + nitrite that occurs east of about 130 W (Figure 10). The cooler, nutrient rich surface water distribution is an indicator of upwelling of deeper waters.

The increases of dissolved gaseous Hg show a significant correlation (for n=21, r=0.426, and p=0.054) with increases in phytoplankton biomass as reflected by chlorophyll a concentrations. The chlorophyll a distribution generally increases towards the Galapagos Islands in a variable fashion that reflects patchiness in the phytoplankton population (Figure 10). Data from the short interhemispheric leg at 85°W which are not shown in Figure 10 have been included in the regression analysis (Kim and Fitzgerald, 1985). The correspondence between dissolved gaseous Hg and chlorophyll a is sufficient to suggest that dissolved Hg in the surface waters of the equatorial Pacific Ocean may be converted in association with phytoplankton activity to a volatile form that is most likely elemental Hg.

Emission of gaseous Hg species from the equatorial Pacific Ocean does not appear to be an unusual marine biogeochemical process. For example, enhanced concentrations of the trace gases, ethylene, propylene and nitrous oxide (N_2O) in association with increased biological productivity have also been observed in the Pacific equatorial upwelling region. A source of ethylene and propylene was observed between 9°N and 20°S on a cruise track from Hawaii to New Zealand (Lamontagne et al., 1974). The excess levels of these unsaturated hydrocarbons in the surface seawater were assumed to have biological origin. A source of N_2O has been identified between 4°N and 6°S in the surface seawater of the eastern equatorial Pacific Ocean (Pierotti and Rasmussen, 1980). It was also suggested that this N_2O source was associated with primary production (nitrate reduction) in the upwelling zone. Photochemically induced nitric oxide production was observed in samples of nutrient rich equatorial surface water from the central Pacific Ocean near 170°W (Zafiriou and McFarland, 1981).

In 1983, Andreae and Raemdonck reported finding evasion of dimethyl sulfide (DMS) from equatorial Pacific water as well as a diurnal variation in the atmospheric DMS concentrations due to photooxidative destruction. Moreover, the supersaturated values of DMS in equatorial surface waters were related to high levels of chlorophyll a. They concluded that phytoplankton were important in producing DMS in surface sea water, thereby making the oceans a significant source of DMS to the atmosphere. Dissolved gaseous Hg appears to be behaving in an analogous manner.

Figure 10. Experimental results from the equatorial Pacific Ocean are plotted versus longitude from 155°W to 93°W along the equator. The parameters in descending order are: total and organic gaseous mercury concentrations in the atmosphere in nanograms per standard cubic meter (ng/SCM); dissolved gaseous gaseous mercury in surface seawater in femtomoles per liter (fm); sea surface temperature in degrees Celsius (°C); nitrate plus nitrite in micromoles per liter (μM); and chlorophyll a, contoured every 100 micrograms per cubic meter. Superimposed on the chlorophyll a field are the sampling depths of the dissolved gaseous mercury. A circle denotes one sample, while a circle with a line indicates that two samples taken at 15 and 30 meters were combined for the dissolved gaseous mercury measurement. (After Kim and Fitzgerald, 1985.)

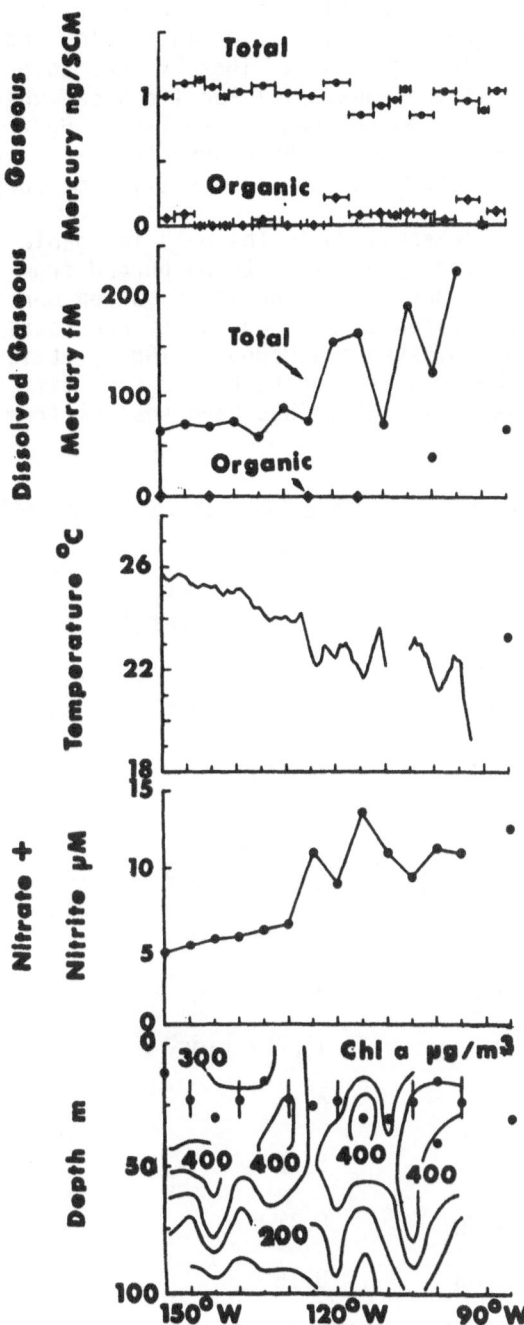

Total reactive mercury concentrations in the surface water of the equatorial Pacific Ocean are low and ranged from 1.0 to 3.5 picomoles Hg/l (pM). These values compare well with our previous work in the equatorial Pacific Ocean where reactive Hg in surface sea water varied between 0.7 to 2.0 pM Hg (Fitzgerald et al., 1983). The concentrations of DGM ranges from 28 to 225 fM Hg and represents a significant fraction of dissolved mercury (2 to 12%).

Elemental Hg is the only inorganic gaseous form of Hg that is sufficiently volatile to be purged from sea water under the experimental conditions. While dissolved mercury chloride gas ($HgCl_2$), is predicted to be about 3% of the total dissolved mercury in sea water (Dyrssen and Wedborg, 1980), it is too soluble to be removed. The Henry's law constants for some different Hg compounds are listed in Table III which has been adapted from Lindqvist et al., 1984. The

Table III.   Henry's Law constants for different mercury compounds

Species	$H=[HgX_{(g)}]/[HgX_{(aq)}]$ (dimensionless)	Temp (^oC)	$I=[Cl^-]$ (M)	Ref
CH_3HgCl	$1.9^+_-0.2 \times 10^{-5}$	25	0.7	1
CH_3HgCl	$1.6^+_-0.2 \times 10^{-5}$	15	1.0	1
CH_3HgCl	$0.9 \quad \times 10^{-5}$	10	0.2×10^{-3}	1
$HgCl_2$	$2.9 \quad \times 10^{-8}$	25	0.2×10^{-3}	2
$HgCl_2$	$1.2 \quad \times 10^{-8}$	10	0.2×10^{-3}	2
$Hg(OH)_2$	$3.2 \quad \times 10^{-6}$	25	0.2×10^{-3}	2
$Hg(OH)_2$	$1.6 \quad \times 10^{-6}$	10	0.2×10^{-3}	2
Hg^o	0.29	20	0	3
Hg^o	0.32	20	sea water	3
$(CH_3)_2Hg$	0.31	25	0	4
$(CH_3)_2Hg$	0.15	0	0	4

1 Iverfeldt and Lindqvist (1982)
2 Iverfeldt and Lindqvist (1980)
3 Sanemasa (1975)
4 Talmi and Mesmer (1975)

only inorganic volatile mercury species with a sufficiently large value of H to be a viable candidate for DGM is Hg^0 (H=0.32 at 20^0 C). If organoHg were found, then $(CH_3)_2Hg$ would be a prime candidate.

It is reasonable to assume that all the DGM in the surface sea water and all the inorganic gaseous mercury in the overlying air is Hg^0. The degree of saturation of mercury in the sea water relative to air can be calculated using the following equation (after Hoyt and Rassmussen, 1985):

$$S=[((C_w \times H)/C_a)] \times 100 \qquad (1)$$

where Cw and Ca are the concentrations of Hg^0 in sea water and air, respectively, and H is the Henry's law constant for Hg^0. Values of S that are >100 indicate supersaturation in the water, while values of S that are <100 indicate undersaturation in the water. Using the measured values of Cw and Ca obtained on the cruise, and values of H as a function of temperature in sea water for Hg^0 (Sanemasa, 1975), we calculate a range for S from +179 to +1769. The surface sea water is greatly supersaturated with Hg^0 for all stations. This indicates that the equatorial Pacific Ocean is a source of mercury to the atmosphere.

A flux of Hg^0 from the ocean to the atmosphere can be estimated following the approach outlined by P. Liss in the chapter entitled "Air-Sea Gas Exchange Rates: Field Measurements". The governing equation for the air-sea gas exchange of Hg^0 is:

$$F=K(C_a H^{-1} - C_w) \qquad (2)$$

where F is the flux of a Hg^0 into (+) or out of (-) the ocean, K is the total transfer velocity for Hg^0. As discussed above, C_a is the average concentration of Hg^0 in the overlying air; C_w is the bulk water concentration of Hg^0 and H is the Henry's Law constant for Hg^0.

An average transfer velocity of 15 cm hr^{-1} was estimated for Hg^0 at 24°C (the average sea surface temperature of all the stations) using the surface renewal model and the relation:

$$K_{Hg^0} = K_{Rn} \frac{(D_{Hg^0})^{1/2}}{(D_{Rn})} \qquad (3)$$

where K_{Hg^0} = average transfer velocity for Hg^0

K_{Rn} = 10 cm hr^{-1}, the average transfer velocity for Rn in the equatorial Pacific Ocean (Broecker and Peng, 1982)

D_{Hg^0} = diffusivity of Hg^0; 2.9×10^{-5} cm^2 sec^{-1} at 24°C (Othmer and Thakar, 1953)

D_{Rn} = diffusitivity of Rn; 1.4×10^{-5} cm^2 sec^{-1}
at 24°C (Broecker and Peng, 1982).

The calculated fluxes of Hg^0 normalized to an annual basis
range from 3.0 to 54 µg Hg/m^2 y^{-1}, with an average of 20±13 µg
Hg/m^2 y^{-1}. The evasion pattern along the equator is illustrated in
Figure 11. Taking the area of the equatorial Pacific Ocean (length
from 82º to 172º W and width from 5º N to 5º S) to be 1.1×10^{13} m^2
(Wyrtki, 1981), an average annual Hg flux from the equatorial Pacific
Ocean to the atmosphere will be about 2.2×10^8g or 0.2Kt.

It is always tempting to go a step further and estimate a flux
of Hg from the world ocean to the atmosphere by assuming that the
volatilization of Hg is linearly proportional to primary production.
Primary production in the equatorial zone has been estimated at 2×10^{15} g C y^{-1} (Chavez et al., 1984) and for the world ocean at 1.8×10^{16} g C y^{-1} (Ryther, 1969). Thus, about 12% of the total primary
production occurs in the equatorial Pacific Ocean. An annual mercury
efflux of 2.3×10^9 g (2.3 kt) is predicted for the world ocean.
This is comparable to the estimate of annual anthropogenic Hg emissions at 2.4×10^9 g Hg y^{-1} (2.4 kt y^{-1}) that was prepared by Watson
in 1979.

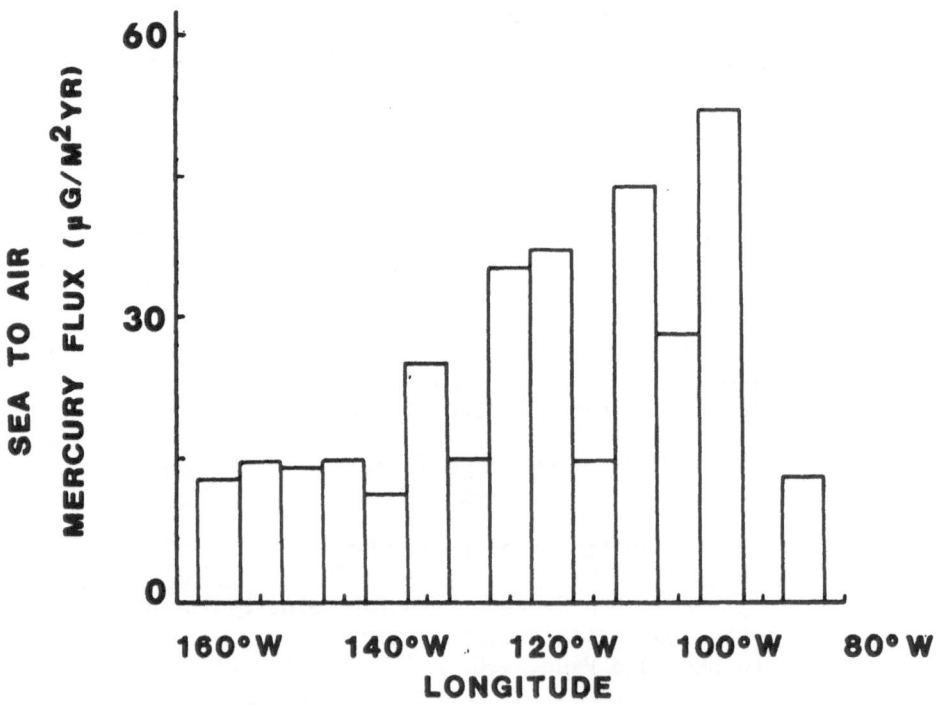

Figure 11. Estimates: Sea to air fluxes of elemental Hg
along an equatorial track in the Pacific Ocean.

In summary, there were significant amounts of volatile Hg found in the surface waters of the equatorial Pacific Ocean. The dissolved volatile Hg was of an inorganic form, most likely Hg^0. Increased levels of volatile Hg occurred in cooler nutrient-rich waters which are characteristic of upwelling or out-cropping of deeper water at the sea surface. Also, higher levels of volatile Hg corresponded with higher chlorophyll a levels, which suggests that dissolved Hg is probably volatilized by phytoplankton in surface sea water. Mercury was supersaturated in sea water, and there was a significant flux of Hg from the equatorial Pacific Ocean to the overlying air. When normalized to primary productivity on a global basis, ocean effluxes of Hg could rival anthropogenic emissions to the atmosphere.

8.   Hg Deposition to the Sea Surface

The efflux of Hg from the tropical Pacific Ocean does not appear to be caused by unique oceanic phenomena. That is, the probable marine recipe for conversion of dissolved Hg to volatile Hg species is nutrient rich water, sufficient sunlight and the right type of biota (phytoplankton such as diatoms; bacteria?). These conditions are present in most oceanic regions. Coastal and estuarine regimes, for example, would readily satisfy these suggested requirements. Moreover, ocean sources of Hg may be more widely distributed and include less productive areas such as the expansive but oligiotrophic subtropical gyres. As noted, (Section 7.2) evasion of dimethylsulfide has been found in central North Pacific Ocean as well as in the equatorial zone. (Andrae and Raemdonck, 198). Perhaps, gaseous Hg will show a similar pattern.

Marine sources of atmospheric Hg must be balanced by atmospheric, riverine, and hydrothermal inputs to the oceans. As we have seen, however, mass balance models for the global cycling of Hg, in general, indicate that the primary mass transfer of Hg to the oceans is through the atmosphere (Section 2). Thus, oceanically produced gaseous outflows of Hg from the sea surface will be countered by atmospheric Hg deposition to the sea surface. Let us examine available information regarding depositional fluxes of Hg to the marine environment, and quantitatively consider the geochemical implications of precipitational and dry depositional Hg flows to the air-sea exchange cycle of Hg.

8.1 Precipitation

There are not many determinations of Hg fluxes associated with oceanic precipitation. Most studies have been carried out at terrestrial locations and inshore regions. A summary of much of the available information can be found in Linquist et al., 1984. The details of our experimental procedures for the collection and analyses of Hg in rain has been described (Section 5.3). As noted, we have made Hg measurements in rains collected at both land-based stations and aboard ship.

Measurements of Hg in rains from the now familiar oceanic sites associated with our investigations of the atmospheric cycle of Hg over the oceans are given in Table IV. In addition, annual Hg precipitational fluxes have been calculated using average Hg concentrations in rain and estimates of the annual rainfall depth for an appropriate geographic region. This work has been recently reported (Fitzgerald and Gill, 1985).

There are several important features to these results. Firstly, the yearly pluvial fluxes of Hg are about a factor of ten greater than the estimated fluvial inputs. An annual riverine flux of about 0.5 μg/m2 was obtained by using an average dissolved Hg concentration of 25pM (5ng/1) as representative of unpolluted rivers. As discussed previously, (Section 2), most models of the global Hg cycle predict that atmospheric processes must significantly affect the global distribution of Hg and provide a major avenue for Hg transfer between the continents and the oceans. This is a global scale demonstration of the atmophilic nature of Hg. Moreover, atmospheric deposition to the open ocean is even more significant because much of the river-borne Hg appears to be removed in the estuary and coastal zone (Gill, 1980).

Table IV. Mercury Measurements in Rainfall and Estimated Oceanic Mercury Precipitation Fluxes

Geographic Region	Hg conc. (ng/1)[1]	No. of Collections	Mean Annual Rainfall Depth[2] (m)	Est. Hg Flux (μg/m2/yr)
Rainfall Measurements				
S. New England[3]	10 ± 5	11	1.1	11
NW. Atlantic (37°N; 69°W) (34°N; 66°W)	9 26	1 1	1.0	18
Enewetak Atoll (11°N; 165°E)	2.8 ± 1.6	3	1.5	4.2
American Samoa (13°S; 170°W)	4.4 ± 2.4	4	1.6	7.0
Tasman Sea (35°S; 170°E)	3.8 ± 0.8	6	1.0*	3.8

[1] 1 ng Hg/liter = 5 pM Hg
[2] Rainfall depths taken from Dorman and Bourke (1979) and (1981).
[3] From Fogg and Fitzgerald (1979).
* New Zealand Meterological Service for Cape Reinga, North Island.

Notice that the estimated pluvial fluxes of Hg to the N. W. Atlantic Ocean are larger than those calculated for the other open ocean locations. Though the data are limited, this pattern is consistent with the hypothesis that the major sources of atmospheric Hg are continental and include natural geological, biological, and anthropogenic processes. Thus, one would expect the contribution of Hg mobilized over the North American continent to be found in N. Atlantic Ocean precipitation.

The third and perhaps most interesting result of this work is the relatively small range for the predicted depositional fluxes. They vary by about a factor of 4, while the average atmospheric particulate Hg concentration decreases from an average of 20 pg/scm at Long Island Sound to 0.4pg/scm at American Samoa. If rainfall scavenging involved principally atmospheric particulate Hg, then a corresponding decrease in Hg precipitational fluxes would be expected from coastal to open ocean regions.

The contribution due to Hg^0 would be very small. For example, the rainwater Hg^0 concentration at equilibrium with 2 ng m^{-3} Hg^0 in air would be 6.7 x 10^{-3} ng/l at 20^0C. Thus, this distribution suggests that water soluble gaseous Hg species may be contributing to the amounts of Hg found in precipitation. The contribution from the water soluble gaseous components would become more important and evident in open ocean regions as the atmospheric particulate Hg concentration decreases. The limitations of the data restrict further speculation.

8.1.1 Hg Deposition in Rainfall to the Equatorial Pacific Ocean

A Hg flux in rainfall to the Equatorial Pacific Ocean can be estimated from our experimental results at Enewetak Atoll and at American Samoa. The data for these locations which are summarized in Table IV, indicate that a precipitational flux between about 4 and 7 μg m^{-2} y^{-1} might be expected for the most western regions (155^0W) of the equatorial coverage during the 1984 O.S.S. Researcher cruise (Figure 5).

The oceanic flux of elemental Hg to the atmosphere near 155^0W is about 14 μg m^{-2} y^{-1} (Figure 11). This value is about 2-4 times the average input of Hg in rainfall that was noted above. The annual Hg inputs to the equatorial Pacific Ocean zone should be higher than the average range determined for the tradewind sites. For example, the mean annual rainfall depth at 5^0N is 3 meters which is about twice the precipitation depth at Enewetak and American Samoa (Dorman and Burke, 1979). We can complete a local material budget for Hg by including an estimate for the depositional flux of Hg to the sea surface due to dry deposition.

8.2 Dry Depositional Hg Flux to the Equatorial Pacific Ocean

The particulate Hg concentrations at Enewetak Atoll and American Samoa ranged between 0.4 and 2.0 pg m^{-3}. The dry depositional flux of Hg can be estimated using the following relation:

$$\text{FLUX} = [\text{Part. Hg Conc.}] \times \text{Depositional Velocity}$$

where a depositional velocity of about 0.1 cm s^{-1} for particle sizes <0.5μ is assumed (Arimoto and Duce, 1985).

The dry depositional flux of Hg at the tradewind sites ranges from 0.01 to 0.06 μg m^{-2} y^{-1}. These quantities represent a rather small fraction of the estimates for Hg flows associated with either precipitation or oceanic evasion in the equatorial Pacific Ocean. Moreover, increasing the depositional velocity to an unrealistically high value of 1.0 cm s^{-1} doesn't alter the relative balance for the exchange of Hg between the atmosphere and the sea surface in open ocean regions. A summary of these calculations and comparisons is presented in Table V.

8.3 Air-Sea Exchange in the Equatorial Pacific Ocean

A rigorous mass balance budget for the equatorial Pacific Ocean study region cannot be prepared. Rainfall data for Hg are not available for the areas where very significant oceanic fluxes of Hg are evident. As shown in Table V, there is a surprisingly reasonable

Table V.   Estimated annual Hg fluxes across the sea surface in the Equatorial Pacific Ocean

Process	Flux (μg Hg/m^2 y^{-1})
Input	
Rain (155°W)*	8 - 14
Dry Deposition	<.1
	8 - 14
Output	
Evasion (155° - 135°W)	~14
Evasion (135° - 95°W)	16 - 54
Average Evasion (All stations)	20 ± 13

*Estimated at 2x the Hg flux for precipitation at Enewetak Atoll and American Samoa.

balance between the estimates for average rainfall Hg input (8-14 $\mu g/m^2y^{-1}$) and the average outflows (20 $\pm$ 13 $\mu g/m^2y^{-1}$) of Hg from the oceans to the atmosphere. Additional insight into the air-sea cycling of Hg can be gained by a geochemical scaling exercise. For example, in the less productive western regions of our equatorial study zone, Hg evasion is nearly balanced by Hg deposition (Table VI). In the biologically productive eastern regime, the evasional fluxes were locally often ~10 x the rainfall fluxes from Enewetak and American Samoa. This is a clear demonstration of the potential for geochemically significant evasion in the equatorial Pacific Ocean.

The concentrations of Hg in the surface ocean mixed layer must be maintained in support of the evasional process. The larger ocean sources of Hg require either a precipitational supply significantly larger than measured at other oceanic locations (Section 8.1), or there must be an enhanced renewal of the surface layer by upwelling and/or horizontal advection of cooler, nutrient rich water containing elevated Hg concentrations. It is probable that both inputs are significantly increased. For example, increases in the Hg content of rain may occur over strong ocean sources of Hg, if photochemical oxidation of gaseous Hg to water soluble forms is enhanced. While we have demonstrated that oceanic Hg evasion is an integral part of the atmospheric cycle of Hg over the oceans, more specific questions relating to ocean sources of Hg are emerging.

8.4 Physico-Chemical Aspects

A physio-chemical model of the Hg cycle was outlined in Figure 2 and discussed in Section 3. Many features of this formulation are supported by our work. Ocean sources of atmospheric Hg were found and a recycling of oceanic derived Hg species via precipitation was indicated. The association of dissolved gaseous Hg and chlorophyll a (plankton biomass) concentrations supports the involvement of biota in the conversion of dissolved Hg species to volatile gaseous forms. Biological activity is also required to remove Hg from surface waters by incorporation or absorption on settling biological debris and fecal material. We found little evidence for evasion of organo-Hg species. This process, however, may occur under more specialized conditions. For example, we did find dissolved gaseous organic Hg at lower depths (400 meters) during the one vertical speciation study conducted (Kim and Fitzgerald, 1984). Unusual hydrographic activity with periodic intrusions of deeper waters may be required to transfer dissolved organo-Hg species from such depths to the surface. These ideas are quite speculative, and must await more complete speciation studies.

9. Atmospheric Cycling of Hg Over the Oceans: Global Perspectives

A refined and experimentally constrained global model would be a valuable contribution to current knowledge of the geochemical cycling of Hg. As the brief review of global models (Section 2)

demonstrated, there is much room for refinements in the mass balance
formulations for the mobilization of Hg in the environment. A rigorous
mass balance simulation is not possible because there are as yet gaps
in the data base. Geochemical compensation for these weaknesses
inevitably leads to order of magnitude estimates in the mass balance
assessments. This situation is illustrated by the budget for the
global atmosphere that appears in Table VI. These estimates of
annual Hg fluxes to and from the atmosphere were carefully prepared
by a scientific working group as part of the report on Hg in the
Swedish environment (Linqvist, et al., 1984).

Table VI. Estimates of fluxes of mercury to and from
the global atmosphere.
Units: 10^9 g Hg a^{-1} (From Lindqvist et al., 1984)

Process	Flux	Method of Estimate
Present anthropogenic emissions	2-10	Watson (1979), Mackenzie and Wollast (1977)
Present background emissions	<15	By balance
Total present emissions	2-17	Assuming balance between emission and deposition
Wet deposition	2-10	From estimates of Hg conc. in rainwater
Dry deposition	<7	Average conc. of soluble mercury <0.1ng m^{-3} Deposition velocity <0.5 cm s^{-1}
Total present deposition	2-17	
Pre-industrial deposition (and emission)	2-10	Data from Danish peat bogs and Greenland ice cores

The global modelling approach will probably not yield a very
clear view of the Hg cycle given the range of estimates for Hg flows.
It is also most disheartening for experimentalists to find their hard
won data in a "back of the envelope calculation" framework. Nevertheless, it is quite important to place the air/sea exchange results
for Hg within a global perspective. Therefore, we will not focus on
budgets, but rather on global perspectives that are gained from

geochemical scaling and comparisons. This approach does provide a realistic basis for making judgements and generating questions concerning the processes which mobilize Hg in nature.

Precipitation plays a major role in the deposition of Hg to the oceans. Our results indicate that rainfall scavenging is the primary removal process for Hg over the oceans. It is also probable that atmospheric deposition of Hg over the continents occurs primarily through precipitation. At present, the latter suggestion cannot be confirmed. The contribution of Hg via dry particle depositon and gaseous uptake in terrestrial systems cannot be estimated.

Our precipitation data are summarized in Table VII along with deposition data reported by the group from the Danish Isotope Center. An average value for wet deposition over continental regions is included from the compilation in Lindqvist, et al., 1984. It is clear that the oceanic data fit quite well into a distributional pattern where the lowest deposition occurs over the open ocean and the largest depositional fluxes are associated with impacted coastal regions such as those affected by coal burning in western Europe (peat bog data). The Long Island Sound measurements which reflect a coastal urban environment and the average rain water value for terrestrial regions also appear quite reasonable.

Table VII. Estimated Annual Mercury Depositional and Precipitation Fluxes for Different Geographic Locations

Location	Flux
Peat Bog, Denmark[1]	42 $\mu g/m^2/yr$
Sweden, Italy, North Sea, & U.K.[2]	~18
Southern New England Coastal Area	11
NW Atlantic Ocean	~18
Samoa	7
Tasman Sea	4
Enewetak Atoll	4
Site Crete Greenland Ice Sheet[3]	3

1. Madsen, P.O., 1981.
2. Lindqvist, O., et al., 1984.
3. Appelqvist, H., et al., 1978.

There are two geochemically important features of these results that deserve special mention. Firstly, the estimated annual Pb deposition at Enewetak Atoll (Settle and Patterson, 1983) and northern Greenland (Murozumi et al., 1969; Davidson et al., 1981) are about the same at ~70 $\mu g/m^2$. Thus, the estimates for Hg deposition at Enewetak and Site Crete, Greenland (~3 $\mu g/m^2$ y^{-1}) are consistent with recent Pb deposition in these remote regions. This correlation gives added and independent support for the observed oceanic depositional Hg fluxes. Secondly, it also suggests that Hg in precipitation from there is behaving in a manner physically similar to Pb and other trace particulate metals in the atmosphere. That is, Hg in rainwater would consist principally of Hg associated with particles. Any contribution of water soluble gaseous Hg species would be small.

However, the relatively small concentration range for Hg in rain does not in general correlate with the substantial decrease in particulate Hg concentrations occurring from the nearshore to the open ocean. This inconsistency probably indicates that water soluble gaseous Hg species are contributing to the amounts of Hg found in precipitation (see Section 8). Brosset (personal communication) has reported finding water soluble gaseous Hg components in the Swedish atmosphere. However, little process oriented information is available concerning the rainfall scavenging of Hg and the removal of specific physical and chemical forms of Hg from the atmosphere. Thus, the relative role of gaseous and particulate Hg contributions to wet deposition is for the present an open and intriguing question.

Mercury deposition to the oceans can be estimated in the following manner. An average annual value of 4 μg Hg/m^2 would be reasonable for remote open ocean regions other than the northern hemispheric westerlies. Deposition in the westerlies including associated coastal regions should range between 10 and 20 μg Hg/m^2 y^{-1}. The northern hemispheric westerlies (30°N - 60°N) encompass about 13% of the total ocean area. Thus, ocean-wide deposition would range between 1.7-2.2 x $10^9 g$ y^{-1} with approximately 25-50% of the input delivered to mid-latitudes in the northern hemisphere. This range for Hg fluxes into the ocean through precipitation also brackets the ocean wide estimate for Hg emissions from the oceans at 1.8 x $10^9 g$ y^{-1} that we put forth speculatively in Section 7.2.

While a rigorous budgeting of the global cycle of Hg will not be attempted, an estimate of the overall Hg deposition to the earth's provides an important mass balance constraint. Using the distribution of fluxes tabulated in Table VII, we infer that the representative range for continental Hg deposition will be between 10 and 40 $\mu g/m^2$ y^{-1}. Moreover, the peat bog results represent an upper limit because these ombrotrophic bogs accumulate Hg from wet and dry deposition and appear to be tracking Hg use in western Europe. If 25 $\mu g/m^2$ y^{-1} is taken as a reasonable average flux, then the total average flux to terrestrial regions will be about 3.7 x 10^9 g y^{-1} Correspondingly, the total flow to the earth's surface would be somewhere between 5 and 6 x 10^9 g y^{-1}.

This estimate for Hg deposition to the earth's surface agrees quite well with a value of 6 x 10^9 Hg yr^{-1} that Slemr et al (1981) found for the global source and sink strengths required to maintain a steady state atmospheric burden of 5 x 10^9 g. This mass balance exercise is summarized in Table VIII. The qualitative nature of such computations is indicated by rounding off the various figures. Thus, Watson's 1979 estimate for world wide anthropogenic Hg emissions is 2 x 10^9 g y^{-1}; our value for potential ocean-wide Hg emissions is given as 2 x 10^9 g y^{-1} and a recent estimate for volcanic Hg emissions at 0.06 x 10^9 g y^{-1} (Fitzgerald, 1985) is included. No values are given for other potentially important natural sources of Hg to the atmosphere such as "crustal degassing," biological mobilization, forest fires, or for remobilization of anthropogenic deposits.

Table VIII.	Global Hg "Budget"	
Hg Deposition	~5-6 X 10^9 g/y^{-1}	This work
	~6	Slemr et al. (1981)
Hg Emissions		
Anthropogenic	~2	Watson (1979)
Volcanic	~0.06	Fitzgerald (1985)
Other Continental Sources		
e.g. Crustal degassing		
Forest fires	~1-2	By difference
Biological mobilization		
Oceanic Sources		
Equatorial Pacific (~0.2)		This work
World ocean	~2	

There are a variety of interesting geochemical permutations suggested by the compilation of Hg flows in Table VIII. For example, if total yearly Hg deposition is assumed to be equivalent to total

annual Hg emissions (steady state approximation) then the interference
of human related activities on the natural cycle would be quite
significant. About 30-40% of the total flow of Hg to the atmosphere
could come from human related activities.

The steady state view would also suggest that the Hg contributions
from the ocean could range from about 30-40% of total emissions.
Moreover, in preindustrial times, the strength of oceanic and terrestrial sources would be nearly the same. In this context, the observed
interhemispheric differences in atmospheric gaseous Hg concentrations
do not necessarily indicate that the enhanced northern hemispheric
concentrations are due to anthropogenic emissions. This pattern, for
example, could be present in preindustrial times because most of the
land sources are in the northern hemisphere and the Hg flux per unit
area is greater from terrestrial regions.

Finally, our southern hemispheric data from the New Zealand
westerlies study in 1983, and the two equatorial cruises (1980 and
1984) show the following:

Time	TGM (ng/scm)	
1980	1.05 ± 0.09	(n = 4)
1983	1.07 ± 0.08	(n = 16)
1984	1.02 ± 0.08	(n = 25)

Clearly, the TGM concentrations appear to be from the same population.
While the measurements were made at different times and widely spaced
geographic regions, the observed variations can be explained solely
by experimental error. A large influence of anthropogenic emissions
on the atmospheric Hg concentrations over open ocean regions of the
southern hemisphere is not apparent. On the other hand, small but
significant increases in the atmospheric Hg burden may be lossed to
analytical error limits.

In summary, air-sea exchange has been shown to play a prominent
role in the global geochemical cycle of Hg. Moreover, recent data
suggest that Hg flows through the atmosphere are smaller than many
previous estimates. Thus, impact from ocean sources of Hg and the
interference of human-related emissions on the atmospheric cycle of
Hg could be quite substantial. It should be clear, however, that only
the framework for the global mobilization of Hg appears to be satisfactorily constrained. Fundamental and more specific questions
regarding the behavior and fate of Hg in the atmosphere and ocean have
not been addressed. The challenge remains.

Acknowledgments:

I am most grateful to the graduate students who have shared in
these studies of the atmospheric cycle of Hg. Messrs. J.P. Kim, G.A.
Gill, A.D. Hewitt, and T.R. Fogg contributed substantially to the
work herein described. The masters, officers, crews, and survey

technicians of the research vessels, R/V Endeavor (U. of Rhode Island), R/V T.G. Thompson (U. of Washington), OSS Researcher (NOAA), and R/V Tangaroa (New Zealand Oceanographic Institute) were especially skilled and cooperative. Shipboard experiments depended on the efforts of K. Bruland, R. Franks, A. Ng, J. Hunt, S. Piotrowicz, and K. Hunter.

M. DeGruy and staff from the Mid-Pacific Marine Laboratory (U. of Hawaii) at Enewetak, provided much physical aid and logistical support to our field program. Additional support for the Enewetak experiments was provided by the Dept. of Energy, Honolulu, Holmes and Narver, Inc., and the University of Hawaii at Manoa. D. Nelson, former station chief of the NOAA Observatory at American Samoa helped ensure the success of our experiments at Cape Matatula.

At New Zealand, I wish to acknowledge the extraordinary and cheerful aid provided by the late J. Spedding and his colleagues at the U. of Auckland. The New Zealand Meteorological Service provided air trajectory and weather forecasts during the study period. The assistance of various personnel associated with the New Zealand Oceanographic Institute is gratefully acknowledged.

I thank D. Hedler for assisting in the preparation of this manuscript. As usual, I am indebted to my colleagues in the SEAREX Program for their full assistance in the field. Finally, this manuscript benefited from the constructive criticism provided by several of the ASI participants. I found the critical review of H. Bingemer to be most helpful. This work was supported by National Science Foundation grants as part of the Sea-Air Exchange Program.

References:

Andreae, M.O. and H. Raemdonck, 1983: 'Dimethyl sulfide in the surface ocean and the marine atmosphere: A global view.' Science, 221, 744-748.

Appelquist, H., K.O. Jensen, T. Sevel, and C. Hammer, 1978: 'Mercury in the Greenland ice sheet.' Nature, 273, 657-659.

Arimoto, R. and R.A. Duce, 1985: 'Dry deposition models and the air/sea exchange of trace elements.' (In press)

Baker, C.W., 1977: 'Mercury in surface waters around the United Kingdom.' Nature, 270, 230-232.

Beijer, K. and A. Jernelöv, 1979: 'Methylation of mercury in aquatic environments.' In: The Biogeochemistry of Mercury in the Environment, (J.O. Nraigu ed.), Elsevier/North Holland Biomedical Press, 203-210.

Ben-Bassat, D. and A. Mayer, 1975: 'Volatilization of mercury by algae.' Physiol. Plant., 33, 128-132.

Bisogni, J.J., 1979: 'Kinetics of methylmercury formation and decomposition in aquatic environment.' In: The Biogeochemistry of Mercury in the Environment, (J.O. Nraigu ed.), Elsevier/North Holland Biomedical Press, 211-230.

Boyle, E.A., F.R. Sclater, and J.M. Edmond, 1977: 'The distribution of dissolved copper in the Pacific.' Earth and Planetary Science Letters, 37, 38-54.

Braman, R.S. and D.L. Johnson, 1974: 'Selective absorption tubes and emission technique for determination of ambient forms of mercury in air.' Environ. Sci. Tech., 8, 996-1003.

Broecker, W.S. and T.H. Peng, 1982: Tracers in the Sea, Eldigio Press, Palisades, N.Y., 690 p.

Brosset, C., 1981: 'The mercury cycle.' Water, Air and Soil Pollut., 16, 253-255.

Brosset, C., 1982: 'Total airborne mercury and its possible origin.' Water, Air and Soil Pollut., 17, 37-50.

Bruland, K.W., 1980: 'Oceanographic distributions of cadmium, zinc, nickel, and copper in the North Pacific.' Earth and Planetary Science Letters, 47, 176-198.

Chavez, F., R. Barber, J. Kogelschatz, V. Thayer and B. Cai, 1984: 'El Nino and primary productivity: potential effects on atmospheric carbon dioxide and fish production.' Tropical Ocean Atmosphere Newsletter, 28, 1-2.

Davidson, C.I., Chu, L., Grimm, T.C., Nasta, M.A. and Qamoos, M.P., 1981: 'Wet and dry deposition of trace elements into the Greenland ice sheet.' Atmos. Env., 15, 1429-1437.

Dorman, C.E. and R.H. Bourke, 1979: 'Precipitation over the Pacific Ocean, 30°S to 60°N.' Month Weath. Rev., 107, 896-910.

Dorman, C.E. and R.H. Bourke, 1981: 'Precipitation over the Atlantic Ocean, 30°S to 70°N.' Month Weath. Rev., 109, 554-563.

Dyrssen, D. and M. Wedborg, 1980: 'Major and minor elements, chemical speciation in estuarine waters.' In: Chemistry and Biogeochemistry of Estuaries (E. Olausson and I. Cato, eds.), John Wiley, New York, 71-119.

Fitzgerald, W.F. and W.B. Lyons, 1973: 'Organic mercury compounds in coastal waters.' Nature, 242, 452-453.

Fitzgerald, W.F. and C.D. Hunt, 1974: 'Distribution of Hg in surface microlayer and in subsurface waters of the northwest Atlantic.' Journal de Recherches Atmospheriques, 8, 629-637.

Fitzgerald, W.F. and G.A. Gill, 1979: 'Subnanogram determination of mercury by two-stage gold amalgamation and gas phase detection applied to atmospheric analysis.' Anal. Chem., 51, 1714-1720.

Fitzgerald, W.F., G.A. Gill, and A.D. Hewitt, 1981: 'Mercury, a trace atmospheric gas.' Presented at the Symposium on the Role of the Ocean in Atmospheric Chemistry, IAMAP Third Scientific Assembly, Hamburg, Federal Republic of Germany, 17-28 August 1981.

Fitzgerald, W.F., 1982: 'Evidence for anthropogenic mercury inputs to the oceans.' EOS Trans. Amer. Geophys. U., 63, #41A-9, 77.

Fitzgerald, W.F., G.A. Gill, and A.D. Hewitt, 1983: 'Air-sea exchange of mercury.' In: Trace Metals in Sea Water, (C.S. Wong, E. Boyle, K.W. Bruland, and J.D. Burton, E.D. Goldberg, eds.), Plenum Press, New York, 297-315.

Fitzgerald, W.F., G.A. Gill, and J.P. Kim, 1984: 'An Equatorial Pacific Ocean Source of Atmospheric Mercury.' Science, 224: 597-599.

Fitzgerald, W.F. and G.A. Gill, 1985: 'Depositional Fluxes of Mercury to the Oceans.' Intern. Conf. on Heavy Metals in the Environ., Athens Sept. 10-13, 1, 79-81.

Fitzgerald, W.F., 1985: 'Volcanic Mercury Emissions and the Global Cycle of Mercury.' (In preparation).

Fogg, T.R. and W.F. Fitzgerald, 1979: 'Mercury in southern New England coastal rains.' J. Geophys Res., 84, C11, 6987-6989.

Garrels, R., F. Mackenzie, and C. Hunt, 1975: Chemical Cycles and the Global Environment: Assessing Human Influences. W. Kaufman, Inc., Los Altos, California, 206 pp.

Gill, G.A., 1980: 'Geochemistry of mercury in Long Island Sound: Analytical and Field Study.' Master's Thesis, Univ. of Connecticut.

Gill, G.A. and W.F. Fitzgerald, 1985: 'Mercury sampling of open ocean waters at the picomolar level.' Deep-Sea Research, 32, 287-297.

Goldschmidt, V.M., 1954: Geochemistry, Clarendon Press, Oxford 730 p.

Hoyt, S. and R.A. Rassussen, 1985: 'Determining trace gases in air and seawater.' In: Mapping Strategies in Chemical Oceanography, (A. Zirino, ed.). American Chemical Society, Washington, pp. 31-56.

Iverfeldt, A. and O. Lindqvist, 1980: 'Determination of distribution equilibria between water and air.' Report No. 415 (in Swedish with summary in English), Project Coal, Health and Environment, The Swedish State Power Board, S-16287 Vallingby, Sweden.

Iverfeldt, A. and O. Lindqvist, 1982: 'Distribution equilibrium of methyl mercury chloride between water and air.' Atmos. Environ., 16, 2917-2925.

Iverfeldt, A. and O. Lindqvist, 1984: 'The transfer of mercury at the air/water interface.' Proc. Int. Symp. Gas Transfer at Water Surfaces, Reidel Publishing Co. 533-538.

Jensen, S. and A. Jernelöv, 1969: 'Biological methylation of mercury
    in aquatic organisms.' Nature, 223, 753-754.
Kim, J.P. and W.F. Fitzgerald, 1984: 'Volatile mercury in the
    equatorial Pacific Ocean.' EOS Trans. Amer. Geophys. U., 65
    84D.
Kim, J.P. and W.F. Fitzgerald, 1985: 'Sea-air Partitioning of
    Mercury in the Equatorial Pacific Ocean.' (Submitted)
Lamontagne, R., J. Swinnerton, and V. Linnenbom, 1974: 'C$_1$ - C$_4$
    hydrocarbons in the North and South Pacific.' Tellus, 16, 71-77.
Lindqvist, O., A. Jernelöv, K. Johansson and H. Rodhe, 1984:
    Mercury in the Swedish Environment: Global and Local Sources
    National Swedish Environment Protection Board, 105 p.
Lindqvist, O. and H. Rodhe, 1985: 'Atmospheric mercury - a review.'
    Tellus, 378, 136-159.
Mackenzie, F.T. and R. Wollast, 1977: 'Sedimentary cycling models of
    global processes.' In: The Sea, (E.D. Goldberg, I.N. McCave,
    J.J. O'Brien and J.H. Steele, eds.). J. Wiley and Sons, New
    York. 739-785.
Madsen, P.O., 1981: 'Peat bog records of tropospheric mercury
    deposition.' Nature, 293, 127-130.
Matsunaga, K. and T. Goto, 1976" 'Mercury in the air and precipi-
    tation.' Geochemical Journal, 10, 107-109.
Millward, G.E., 1982: 'Nonsteady state simulations of the global
    mercury cycle.' J. Geophys. Res., 87, 8891-8897.
Millward, G.E. and J.H. Griffin, 1980: 'Concentrations of partic-
    ulate mercury in the Atlantic marine atmosphere.' Sci. Total
    Environ., 16, 239-248.
Moody, J.R., P.J. Paulsen, T.C. Rains, and H.L. Rook, 1976: 'The
    preparation and certification of trace mercury in Water Standard
    Reference Materials.' NBS Special Publication 422, Washington,
    D.C., Vol II, 267-275.
Murozumi, M., T.J. Chow, and C.C. Patterson, 1969: 'Chemical con-
    centrations of pollutant lead aerosols, terrestrial dusts, and
    sea salts in Greenland and Antarctic snow strata.' Geochim.
    Cosmochim. Acta, 33, 1247-1294.
National Academy of Sciences (NAS), 1978: An assessment of mercury
    in the environment. Washington, D.C., 185 p.
Niki, H., P.D. Maker, C.M. Savage, and L.P. Breltenbach, 1983: 'A
    long-path fourier transform infrared study of the kinetics and
    mechanism for HO-radical initiated oxidation of dimethylmercury.'
    J. Phys. Chem., 87, 4978-4981.
Olafsson, J., 1978: 'Report of the ICES international intercalibration
    of mercury in seawater.' Marine Chem., 6, 87-95.
Olafsson, J., 1982: 'An International Intercalibration For Mercury
    in Seawater.' Marine Chem., 11, 129-142.
Olafsson, J., 1983: 'Mercury concentrations in the North Atlantic
    in relation to cadmium, aluminum, and oceanographic parameters.'
    In: Trace Metals in Seawater, (C. S. Wong, E. Boyle, K.W.
    Bruland, J.D. Burton, and E.D. Goldberg, eds.), Plenum Press,
    New York, 475-485.
Othmer, D. and M. Thakar, 1953: 'Correlating diffusion coefficients
    in liquids.' Ind. Eng. Chem., 45 589-593.

Pierotti, D. and R. Rasmussen, 1980: 'Nitrous oxide measurements in the eastern Pacific Ocean.' Tellus, 32, 56-72.

Ridley, W.P., L.J. Dizikes and J.M. Wood, 1977: 'Biomethylation of toxic elements in the environment.' Science, 197, 329-332.

Ryther, J., 1969: 'Photosynthesis and fish production in the sea.' Science, 166, 72-77.

Sanemasa, I., 1975: 'The solubility of elemental mercury vapor in water.' Bull. Chem. Soc. Jap., 48, 1795-1798.

Schaule, B.K. and C.C. Patterson, 1981: 'Lead concentrations in the Northeast Pacific: evidence for global anthropogenic perturbations.' Earth and Planetary Science Letters, 54, 97-116.

Seiler, W., 1974: 'The cycle of atmospheric CO.' Tellus, 26, 116-135.

Settle, D.M. and C.C. Patterson, 1982: 'Magnitudes and sources of precipitation and dry deposition fluxes of industrial and natural leads to the North Pacific at Enewetak.' J. Geophys. Res., 87, 8857-8869.

Slemr, F., W. Seiler, C. Eberling, and P. Roggendorf, 1979: 'The determination of total gaseous mercury in air at background levels.' Anal. Chim. Acta, 110, 35-47.

Slemr, F., W. Seiler, and G. Schuster, 1981: 'Latitudinal distribution of mercury over the Atlantic Ocean.' J. Geophys. Res., 86, 1159-1166.

Strickland, J.D.H. and T.R. Parsons, 1972: A Practical Handbook of Seawater Analysis (Second Edition), Fisheries Research Board of Canada, Ottawa, 309 p.

Stumm, W. and J.J. Morgan, 1981: Aquatic Chemistry (Second Edition), John Wiley, New York 780 p.

Talmi, Y. and R.E. Mesmer, 1975: 'Studies on vaporization and halogen decomposition of methyl mercury compounds using GC with a microwave detector.' Water Res., 9, 547-552.

Trace Metals in Sea Water, 1983: C.S. Wong, E. Boyle, K.W. Bruland, J.D. Burton and E.D. Goldberg, eds. NATO Conference Series Plenum Press, New York, 920 p.

Watson, W.D., 1979: 'Economic considerations in controlling mercury pollution.' In: The Biogeochemistry of Mercury in the Environment, (J.O. Nriagu, ed.), Elsevier/ North-Holland Biomedical Press 42-77.

Westöö, G., 1966: 'Determination of methyl mercury compounds in food stuffs I. Methylmercury compounds in fish, identification and determination.' Acta Chem. Scand., 20, 2131-2137.

Wollast, R., F. Billen, and F.T. MacKenzie, 1975: 'Behavior of Mercury in Natural Systems and Its Global Cycle.' In: Ecological Toxicology Research, (A.D. McIntyre and C.F. Mills, eds.), Plenum, New York, 145-166.

Wood, J.M., 1974: 'Reaction of methyl mercury with plasmalogens suggests a mechanism for neurotoxicity of metal-alkyls.' Nature 248, 456-458.

Wood, J.M., F.S. Kennedy, and C.G. Rosen, 1968: 'Synthesis of methyl mercury compounds by extracts of a methanogenic bacterium.' Nature, 220, 173-174.

Wyrtki, K., 1981: 'An estimate of equatorial upwelling in the
    Pacific.' J. of Phys. Oceanog., 11, 1205-1214.
Wyrtki, K., E. Firing, D. Halpern, R. Knox, G.J. McNally, W.C.
    Patzert, E.D. Stroup, B.A. Taft, and R. Williams, 1981:
    'The Hawaii to Tahiti shuttle experiment.' Science, 211,
    22-28.

THE AIR-SEA EXCHANGE OF PARTICULATE ORGANIC MATTER: THE SOURCES AND
LONG-RANGE TRANSPORT OF LIPIDS IN AEROSOLS.

Robert B. Gagosian
Department of Chemistry
Woods Hole Oceanographic Institution
Woods Hole, Massachusetts 02543
U.S.A.

1. INTRODUCTION

The sources, composition, transport pathways and fluxes of organic
components in the atmosphere are just beginning to be understood.
Only recently have studies been undertaken to ascertain the processes
controlling the spatial distributions of these substances. However,
aside from methane and some halocarbons, relatively few measurements
of vapor phase or particulate organic material have been made outside
urban areas (for recent reviews see Duce, 1978; Simoneit and Mazurek,
1981; Duce and Gagosian, 1982; Buat-Menard, 1983; Duce et al., 1983a;
Gagosian, 1983). In general these studies have utilized the biomarker
approach to assign terrestrial, marine, or anthropogenic origins to
the organic components of the aerosols. This approach takes advantage
of the unique molecular signatures or source "fingerprints" for various lipid compound classes of marine and terrestrial plants and animals to identify the sources of the organic substances in aerosols.
Although these compounds represent a small portion of the organic
matter in aerosols, they are excellent tracers for the sources and
transformations of this material. This "tracer" approach for lipid
class compounds is somewhat analogous to the use of natural and bomb
fallout radionuclides in the study of oceanic and atmospheric
processes.
    In this chapter I present case studies from the multidisciplinary
experiments of the Sea-Air Exchange Program (SEAREX) as examples of
how the biomarker approach can be utilized to understand atmospheric
processes controlling the distributions, sources, transport and deposition of organic material in aerosols. Throughout the text I have
attempted to emphasize what we have learned from these studies and
what will be needed to answer questions concerning the processes of
interest.

1.1. Background

Most of the research to date in remote environments utilizing the bio-

P. Buat-Ménard (ed.), The Role of Air-Sea Exchange in Geochemical Cycling, 409–442.

marker approach has focussed on five lipid compound classes; fatty
acid esters, fatty alcohols, sterols, aliphatic hydrocarbons and fatty
acid salts. Examples of their specific source marker information can
be found in Table 1 and will be discussed in detail below. In addition polynuclear aromatic hydrocarbons have been used in long-range
transport studies (Bjorseth and Lunde, 1979; Broddin et al., 1980;
Daisey et al., 1981; Marty et al., 1984; Ohta and Handa, 1985). Soot
carbon has also been used to trace long-range transport of combustion-
derived aerosols (Rosen et al., 1981; Andreae, 1983).

a. Fatty Acid Esters

Long chain esters of fatty acids (wax esters) have proven to be most
useful in differentiating between the marine versus terrestrial
origin of oceanic particulate material (Sargent, 1976; Sargent et
al., 1976, 1977; Wakeham, 1982; Wakeham et al., 1983).
    Wax esters are formed naturally from long-chain fatty alcohols
and long-chain fatty acids. Both sides of the molecule reflect the
homolog distribution of the source. Marine wax esters are generally
26-42 carbon atoms long (Sargent et al., 1976); terrestrially-derived
wax esters are 32-64 carbon atoms long (Tulloch, 1976). These compounds are easily hydrolyzed into their fatty alcohol and fatty acid
fragments by saponification or transesterification. Unfortunately,
this obscures some of the source marker information in the free fatty
alcohol and fatty acid fractions. For this reason, specifically
designed analytical methods to separate and identify wax esters intact
must be used. This preserves the source marker information they contain. Because of their high carbon numbers (26-42) marine wax esters
are sufficiently non-volatile that they will not evaporate. Therefore, they are potentially important as marine source markers in the
aerosol phase. There are only two sets of aerosol samples that have
been analyzed for wax esters: Enewetak (Peltzer et al., 1981) and
coastal Peru (Schneider and Gagosian, 1985). A strong terrestrial
signature (homolog distribution: $C_{42}-C_{62}$, high even/odd carbon preference index) was found in both sets.
    Triglycerides are another class of fatty acid esters that have
recently found use as a source marker (Wakeham, 1982; Wakeham et al.,
1983). Like the wax esters these compounds are relatively large
(marine triglycerides: $C_{42}-C_{58}$, Wakeham, 1982) and consequently non-
volatile. Triglycerides are rarely present in the epicuticular waxes
of higher plants (Tulloch, 1976); consequently, they are potentially
important indicators of marine derived material. Because there may be
subtle differences of the triglyceride homolog distributions among the
various marine sources (e.g. phytoplankton vs. zooplankton, Wakeham,
1982), it is important that these compounds be analyzed intact to pre-
serve this information and not lose it among the more abundant fatty
acid salts (Gagosian et al., 1981, 1982; Schneider and Gagosian,
1985).

b. Fatty Alcohols and Sterols

Table 1. Source Marker Information

HYDROCARBONS	FATTY ALCOHOLS AND STEROLS	FATTY ACIDS
Terrestrial Plant-wax hydrocarbons: homologous series: $nC_{23}-nC_{35}$ strong odd/even carbon number predominance	**Terrestrial** Plant-wax alkanols: homologous series: $nC_{21}-nC_{36}$ strong even/odd carbon number predominance β-sitosterol	**Terrestrial** Plant-wax acids (<10% of wax) homologous series: $nC_{14}-nC_{36}$ strong even/odd carbon number predominance frequently there is no major component
Marine Phytoplankton and Zooplankton hydrocarbons: nC_{15}, nC_{17} or nC_{19}, $C_{15:1}$, $C_{17:1}$, $C_{19:1}$, $C_{19:4}$, $C_{19:5}$, $C_{19:6}$, $C_{21:4}$, $C_{21:5}$, $C_{21:6}$ Pristane	**Marine** Phytoplankton and Zooplankton alcohols: generally in the range of $C_{12}-C_{22}$ especially nC_{14}, nC_{16}, $C_{20:1}$, $C_{22:1}$ Dinosterol Diatomsterol	**Marine** Phytoplankton and Zooplankton fatty acids: generally in the range of $C_{12}-C_{24}$ especially: nC_{14}, nC_{16}, nC_{18} and $C_{18:1}$ polyunsaturated acids are also common: $C_{20:4}$, $C_{20:5}$, $C_{22:5}$, $C_{22:6}$
Anthropogenic Petroleum hydrocarbons: $nC_{15}-nC_{40}$ homologous series: no odd/even carbon number predominance unresolved complex mixture common, α-hopane and extended tricyclic terpane series Squalene	**Anthropogenic** nC_{12} and nC_{16}	**Ubiquitous** nC_{16}, $C_{16:1}$, nC_{18}, and $C_{18:1}$

Like the fatty acids, the fatty alcohols and steroidal alcohols are
structurally varied and surface active. Specific compounds in these
classes have terrestrial or marine sources that may serve as useful
markers (Table 1) (Gagosian et al., 1983). Garrett (1967) found the
12-, 16-, and 18-carbon fatty alcohols in sea surface microlayer
samples. They have no major terrestrial natural source. The even-
numbered series is also produced to some extent industrially. The
higher molecular weight fatty alcohol distributions in eolian dusts
have a strong even:odd carbon preference index and have been used as
markers of land plant waxes (Simoneit, 1977, 1979, 1980, 1984b; Cox
et al., 1982). The sterols have been particularly useful as source-
marker compounds in that they have been widely studied. Many specific
sterols are unique to the organism involved, so that there are land-
derived sterols (e.g. ß-sitosterol) as well as specifically marine-
derived sterols (e.g. dinosterol from dinoflagellates and diatomsterol
from certain species of diatoms, Gagosian et al., 1983). The sterols,
like the fatty acid salts and esters, are essentially involatile and
should be confined to the particulate phase. Barbier et al. (1981)
reported the presence of marine derived sterols in atmospheric particles from the west African upwelling. Other triterpenoid compounds
such as α- and ß-amyrin have also been used as specific source
markers (Simoneit et al., 1983). However, these compounds have not
been detected in the remote marine environment.

c. Aliphatic Hydrocarbons

Alkanes are among the least reactive organic compounds and are one of
the most widely studied classes of lipids. Here we cannot review in
depth the voluminous literature on their environmental distribution.
Because they have been so widely studied, we shall stress the analysis
of selected alkanes of non-petroleum origin (e.g. not the "unresolved
complex mixture"). The C_{27}-C_{33} plant wax alkanes serve as useful
terrestrial markers (Simoneit, 1977, 1979, 1980, 1984a,b; Weschler,
1981; Cox et al., 1982; Marty and Saliot, 1982; Simoneit and Mazurek,
1982; Ohta and Handa, 1985). The n-alkanes are sufficiently volatile
that they can also appear in the gas phase (Cautreels and Van Cauwenberghe, 1978; Eichmann et al., 1979, 1980). Some alkanes have source
specificity (e.g. marine-derived pristane, n-C_{15} and n-C_{17}; odd
carbon numbered higher hydrocarbon plant waxes); most have a multipli-
city of origins, including land-derived, marine-derived, and petro-
leum-derived material (e.g. α-hopane and extended tricyclic terpane
series) (Duce and Gagosian, 1982; Gagosian et al., 1982; Schneider et
al., 1983; Simoneit, 1984a). Although members of this class are
certain to be detected in virtually all samples, the alkane data
interpretation in terms of processes benefits greatly from information derived from the behavior of the other compound classes, such as
alcohols, acids, and wax esters, for which some of these ambiguities
are minimal or absent.

d. Fatty Acid Salts

Fatty acids are ubiquitous in nature. They have also been found previously in surface microlayer samples (Garrett, 1967), in eolian dusts
(Simoneit, 1977, 1979; Cox et al., 1982) and in aerosol samples from
marine (Barger and Garrett, 1976; Marty et al., 1979; Gagosian et al.,
1981, 1982; Schneider and Gagosian, 1985) and remote and urban land
sites (Cautreels et al., 1977). This compound class is rich in marker
compounds. Simoneit (1977, 1979) and Cox et al. (1982) found higher
plant (land-derived) fatty acids in eolian dusts as the upper peak of
the bimodal distribution. These higher carbon numbered compounds with
strong even:odd carbon preference peaking at approximately C_{26} have no
major marine sources. The lower ($C_{14}-C_{24}$) acids were present as well;
they were tentatively assigned a lacustrine source. However, in other
locations a marine component dominates the lower molecular weight
fraction (Gagosian et al., 1981, 1982; Peltzer and Gagosian, 1984;
Peltzer et al., 1985; Schneider and Gagosian, 1985). Most marine
fatty acids are in the $C_{14}-C_{22}$ range and display a wide range of un-
saturation. Garrett (1967) reported 17 fatty acids, with from 1-4
double bonds, in oceanic surface microlayer samples.

    Marine-derived fatty acids are present strictly in the particulate
phase in remote regions, as non-volatile salts and wax or glyceryl
esters (Gagosian et al., 1981, 1982; Schneider and Gagosian, 1985).
The high surface activity of these components may subject them to
particularly intense cyclic reinjection into the air from the sea
surface. In contrast, Cautreels and Van Cauwenberghe (1978) found in
land-derived samples that some of the fatty acids, especially the
lower molecular weight ones, appear in the gas-phase. These must
represent free acids or short-chain esters released to the air by
thermal processes or by acidification of the aerosol by pollutant
SO_2 uptake. In remote oceanic sites such processes are unlikely.

2. SAMPLING AND ANALYTICAL METHODOLOGY

Characterization of trace amounts of organic compounds in the marine
atmosphere is a problem distinctly different from the monitoring of
organic pollutants in urban, suburban or rural continental areas.
Typically, aliphatic and polycyclic aromatic hydrocarbons derived from
petroleum and fossil fuel combustion are measured in analyses of atmospheric samples from urban and suburban areas (Lamb et al., 1980). On
the other hand, biogenic lipids, in a variety of compound classes, may
predominate in background aerosol samples from rural continental, seashore and coastal marine locations (e.g. Simoneit, 1980; Simoneit and
Mazurek, 1982). It is the wide variety of compound classes contained
in these lipid materials, and their orders of magnitude lower concentrations at remote marine locations which presents unique analytical
challenges (Peltzer et al., 1984).

    Sampling of atmospheric particulate matter for organic compound
analysis has been done primarily through the use of hi-volume air
samplers with glass fiber filters (GFF). Nylon nets to sample large
(>2 μm) (Simoneit, 1977, 1979) particles have also been used.

Recently, size-fractionated aerosol samples have been collected
(Hoffman and Duce, 1974, 1977; Van Vaeck and Van Cauwenberghe, 1978;
Van Vaeck et al., 1979; Broddin et al., 1980; Chesselet et al., 1981;
Marty and Saliot, 1982; Schneider et al., 1983; Ohta and Handa, 1985;
Schneider and Gagosian, 1985). The concentration of particulate
organic matter may also be affected by processes affecting the partitioning between the particulate and vapor phase. This may be especially important for aliphatic and aromatic hydrocarbons. Hahn (1980)
and Zafiriou et al. (1985) have discussed this issue in depth, with
special emphasis on the gas and aerosol distribution of the n-alkanes.

     Due to the high probability of contamination, it is critical that
several precautionary measures be taken during sampling. For instance, the collection system should be controlled automatically
according to real-time meteorological parameters to sample only when
"clean air" conditions prevail; and therefore, only the uncontaminated
background aerosol collected (Duce, 1981; Peltzer et al., 1984).
Relevant meteorological and chemical conditions such as wind speed,
direction, condensation nuclei counts, precipitation, ozone and radon
should be monitored continuously.

     Like many analytical methods, certain procedures may alter the
physical and chemical state of the analyte. The collection of atmospheric aerosols on glass fiber filters is one such procedure. The
possibilities for altering the physical and chemical state of the
analyte are several.

     First, a filter may not be 100% efficient at collecting the parti-
cles of interest. The GFFs are rated as >98% efficient for the collection of particles with radius >0.015 μm (Butcher and Charlson,
1972). Typically, double filters have been used by several investigators (for example, see Duce et al. (1983b)) for the collection of
aerosols. For n-alkanes and fatty alcohols, >95% of the material was
collected by the first filter (Gagosian et al., 1982). Since these
compound classes have the highest vapor pressure of all the compounds
discussed in this report, we feel that single GFFs are acceptable for
collection of atmospheric aerosols for the lipid compound classes
discussed here.

     A second problem in using GFFs for the collection of aerosols
deals with the loss of material from the filter. This can result in
lower estimates for the "particulate" phase, as well as increased
estimates for the gas phase if a gas phase trapping device is installed serially behind the filter. Some of the problems associated with
the use of serial particle and gas phase samplers have been discussed
by Junk and Jerome (1983). For example, one of the major processes
they cite is the sublimation loss of loosely bound compounds from
organic particles and their subsequent trapping by the serially
installed gas phase collector.

     A third process affecting the collection of aerosols with GFFs
deals with the adsorption of gas phase compounds onto the surface of
the filter. For remote marine locations, the gas phase concentrations
are so low that except for the most reactive of compounds this process
is probably negligible. Comparison of gas phase and aerosol lipid

compositions suggests that these are two distinct phases, and adsorption of the gas phase by the GFF is minimal (Gagosian et al., 1982;
Zafiriou et al., 1985). Indeed, we believe the process proceeds in
the other direction based upon our conclusions about hydrocarbons in
rain (Zafiriou et al., 1985).

A fourth process deals with the interaction of trapped compounds
with reactive gases in the sampling stream. Not only are compounds
lost by this process, but new ones are formed as the product of these
reactions. This process is expected to be especially acute for the
relatively reactive compounds containing one or more double bonds.
Thus, the polyunsaturated fatty acids are especially good candidates
for this type of interaction with ozone, hydroxyl radical, etc. Their
lack of detection in aerosol samples may be evidence of this effect.
Indeed we have evidence that this hypothesis is correct from initial
results of samples collected only during daylight or night-time hours
when the levels of OH are significantly different. Higher levels of
polyunsaturated fatty acids were found at night. No significant
difference was observed for the saturated compounds.

A fifth sampling artifact deals solely with the comparison of two
different types of aerosol samples. Because of the ability of rain to
scavenge aerosols, one might be tempted to consider rain samples as a
different type of "aerosol sample." Indeed, this is exactly what is
done whenever a washout ratio (the ratio of the concentration of a
particular substance in rain divided by its concentration in air) is
calculated. However, there are significant differences between the
temporal and spatial sampling characteristics of the two methods.
Aerosol samples are typically collected over a period of several days
at the base of the marine boundary layer. Rain samples are collected
on a discrete event basis lasting from tens of minutes to a few hours
in duration while the rain scavenges aerosols from cloud base to
ground level. Thus, we are already dealing with two distinctly different "aerosol" samples. An additional factor to be considered is
the possibility that gas phase compounds can be adsorbed by rain
drops. Clearly, the comparison of aerosol and rain samples is a very
complex situation and the reader is referred to Zafiriou et al. (1985)
where the subject is discussed in detail.

3. SOURCE AND LONG-RANGE TRANSPORT STUDIES

3.1. Introduction

In our laboratory our approach has been to analyze for several related
classes of organic compounds with fairly well-known, unique biogeochemical source signatures to allow differentiation between natural
terrestrial, anthropogenic, and marine sources. Under favorable
conditions, atmospheric reactions (primarily those with OH radical
and O_3) might also produce recognizable reaction products from these
components. Such a multi-component analysis has many advantages. The
fact that all the lipid classes can independently lead to the same

major conclusions demonstrates that, in this difficult application,
the source-marker concept is valid and useful. Furthermore, to a
first approximation the various lipid-class compounds consist of
straight hydrocarbon chains with various terminal functional groups
(e.g. -H, -OH, -COO⁻, -COOR). These functional groups result in
markedly different physical properties (volatility, water solubility,
and surface activity, in particular). These differences greatly
facilitate interpretation concerning transport and transformation mechanisms.

There is very little data in the literature concerning the total
lipid content of remote marine aerosols, and these concentrations vary
considerably as a function of the technique used and the location of
aerosol collection. Barger and Garrett (1970, 1976) reported aerosol
concentrations for chloroform extractable organic compounds of 700 -
6300 ng/m^3 at Hawaii and 160-4000 ng/m^3 in the eastern equatorial
Pacific. Simoneit (1977, 1979) reported total neutral lipids for
atmospheric aerosols off the west coast of Africa in the range 0.02 -
10 ng/m^3 with the highest values, 5-10 ng/m^3, for Saharan dusts.
However, the nets used for collecting dust samples clearly discrimin-
ate against the smaller aerosol particles as collection efficiencies
for particles $\leq$2 µm are $\leq$50% (Simoneit, 1977). These concentrations
are thus best regarded as a lower limit. Indeed, Cox et al. (1982)
found 660-6000 ng/m^3 total solvent extractable lipids in Harmattan
aerosols from Nigeria which are a possible source for some of the
samples analyzed by Simoneit (1977). Eichmann et al. (1980) reported
total neutral lipids (as EEOM: ether extractable organic matter) on
the order of 800 ng/m^3 for the tropical North Atlantic; 700-800 ng/m^3
for the Irish west coast (Mace Head and Loop Head); 500 ng/m^3 for onshore only winds at Loop Head, Irish west coast; and 500 ng/m^3 for
Cape Grim, Tasmania. Gagosian et al. (1982) found total neutral and
acidic lipid values of 0.7-1.2 ng/m^3 and 1.6-2.0 ng/m^3 in aerosols
collected at Enewetak atoll (Table 2). N-alkanes, n-alkanols and
fatty acid esters made up approximately 10-25% of the neutral extract.
Greater than 50% of the acidic extract were comprised of fatty acids.
Clearly, the total neutral lipid concentrations for aerosols collected
at Enewetak are several orders of magnitude lower than most of the
other reported values and are at the lower end of the range reported
by Simoneit (1977, 1979).

3.2. North Pacific Trades: Enewetak

The quantitative results from the analysis of the Enewetak sample set
are summarized in Table 3. Column 1 lists the dry season mean aerosol
concentrations; column 2 lists the wet season mean aerosol concentrations (Gagosian et al., 1981, 1982); column 3 lists the mean gas phase
concentrations for both seasons; and column 4 lists the mean concentrations for lipids in rain (Zafiriou et al., 1985). In general, the
levels of total identifiable organic compounds reported (1-4 ng/m^3)
are lower than any previously reported in the literature. Much higher
levels of organic compounds have been reported for aerosols collected

Table 2.   Extract weights for ENAS–24 and ENAS–33.

	Sample No.			
	ENAS–24		ENAS–33	
Mid–date	.14		7.29	
Volume (M^3)	8500		9200	
	pg/m^3	%	pg/m^3	%
Methylene chloride extract	710		1200	
$C_{21}-C_{36}$ n–alkanes	23	(3%)	20	(2%)
$C_{14}-C_{32}$ fatty alcohols	83	(12%)	72	(6%)
$C_{13}-C_{32}$ fatty acid esters	69	(10%)	40	(3%)
SUM	175	(25%)	132	(11%)
Acidic Methanol/Hexane Extract	1600		2000	
Fatty acid salts				
$C_{13}-C_{18}$	730	(46%)	1400	(70%)
$C_{19}-C_{32}$	36	(2%)	53	(3%)
SUM	770	(48%)	1450	(73%)

Table 3.   Summary of Enewetak sample analyses.

| | Aerosol Samples | | Gas | Rain |
	D.S. *1 pg/m^3	W.S. *2 pg/m^3	Phase *3 pg/m^3	Samples *4 ng/L
n–Alkanes				
Sum $C_{17} - C_{20}$	1.0	0.6	28	27
Sum $C_{21} - C_{36}$	99	32	320	104
CPI $C_{21} - C_{36}$	2.8	2.0	1.2	1.1
UCM	ND	ND	1300	1500
Wax esters				
Sum $C_{39} - C_{62}$	59	NA		NA
CPI $C_{39} - C_{62}$	8.5	NA		NA
Fatty alcohols				
Sum $C_{14} - C_{20}$	3.3	7.4		68
Sum $C_{21} - C_{32}$	140	65		20
CPI $C_{21} - C_{32}$	10	14		>4
Sterols				
Cholesterol	3.3	2.9		34
ß–sitosterol	ND	ND		10
Fatty acid salts				
Sum $C_{13} - C_{18}$	2900	960		1430
Sum $C_{19} - C_{32}$	380	70		130
CPI $C_{19} - C_{32}$	2.4	4.8		4.4
Total *5	3600	1100	350	1800

NA = not analyzed; ND = not detected.
*1:   D.S. = dry season mean concentrations.
*2:   W.S. = wet season mean concentrations.
*3:   mean of 3 samples.
*4:   mean of 6 samples.
*5:   excluding UCM (unresolved complex mixture).

off the coast of Africa. Simoneit (1980) found up to 10 ng/m³ of
n-alkanes, fatty acids and fatty alcohols; while Marty and Saliot
(1982) found 6-13 ng/m³ of n-alkanes alone. In both cases, the
organic material in these aerosols was dominated by waxes of terrestrial origin. Van Vaeck et al. (1979) also found naturally derived
organic compounds (approximately 30 ng/m³) in aerosols collected at
coastal North Atlantic sites. They reported the presence of anthropogenic compounds (approximately 5 ng/m³) in these samples as well,
but the n-alkane and fatty acid distributions were still dominated by
terrestrial plant waxes. Ketseridis et al. (1976) and Eichmann et al.
(1979, 1980) found quite a different distribution for n-alkanes in
aerosol samples collected in the North Atlantic and Indian oceans.
They reported very high levels (approximately 5-50 ng/m³) of predominantly anthropogenic n-alkanes. Ohta and Handa (1985) reported n-
alkane concentrations in the range of 4-29 ng/m³ from samples
collected on a cruise from Japan into the south central North
Pacific. Samples collected from stations nearest to Enewetak gave
the lowest values. Barger and Garrett (1976) found levels of fatty
acids (3-300 ng/m³) in aerosols collected over the eastern equatorial Pacific ocean which were much greater than the levels we have
reported for aerosols collected at Enewetak atoll. Finally, Schneider
et al. (1983) and Schneider and Gagosian (1985) found terrestrial
plant waxes in aerosols collected off the coast of Peru at levels
approximately 3-5 fold higher than the Enewetak results.

We feel that the differences between the Enewetak results and
those of other workers are due to differences in sampling locations
and conditions and proximity to sources, especially with regards to
the presence of anthropogenic compounds. However, we find the great
similarity between the n-alkane compositions and concentrations
reported by Eichmann et al. (1979, 1980) for the North Atlantic and
Indian Oceans perplexing, since one would have expected the Indian
Ocean samples to represent a more remote marine location. Comparisons
are difficult because very little local meteorological and no long-
range transport information was collected during most of these
studies.

It is interesting to note that while the authors listed above all
report n-alkane concentrations, and some report fatty alcohol, fatty
acid and polycyclic aromatic hydrocarbon concentrations, none have
reported fatty acid salt concentrations. All the fatty acid concentrations reported by other authors were for fatty acids obtained
following hydrolysis/saponification of an organic solvent filter extract and thus represent esterified fatty acids, i.e. triglycerides,
wax and steryl esters, etc., and not the free fatty acids which exist
in the salt form on the GFFs. Thus, the Enewetak fatty acid salt
concentrations are not directly comparable to the reported fatty acid
ester concentrations of others. For the Enewetak samples, we have
found that while the fatty acid salts have a composition similar to
the fatty acid esters, these two compound classes have important
differences. Most important of these differences is the fact that
the fatty acid salt concentrations are as much as 10 times higher;

and for some samples, the identifiable fatty acid salts represent
>50% of the total solvent extractable organic matter (Gagosian et
al., 1982). This makes them the singly most abundant compound class
yet found.

The compositions of the four major compound classes are shown in
Fig. 1 for a typical Enewetak aerosol sample. This sample was collected at the height of a dust event and thus represents the strongest
terrestrial source signature. Based on the carbon number ranges and
the carbon preference index (CPI) of the n-alkanes, wax esters and
fatty alcohols found, the source of the lipids is clearly the epicuticular waxes of terrestrial vascular plants (see Table 1); strong odd/
even CPI for n-alkanes dominated by C_{27}, C_{29} and C_{31}; and strong even/
odd CPI for fatty alcohols and fatty acids, C_{28}, C_{26}, and C_{30} being
dominant. The wax esters comprise of high carbon numbered compounds.
Whether this material is emitted directly to the atmosphere or co-
transported with soil following senescence and decay of the plants is
difficult to ascertain at this time. Quite possibly both mechanisms
are important: with wind-blown soil being the major source during
the drier months, and the direct wax emission mechanism contributing
significantly at other times of the year.

The higher molecular weight fatty acid salts also show this terrestrial plant wax source, but the fatty acid salt fraction is clearly
dominated by the C_{13}-C_{18} fatty acids. These fatty acids appear to be
of marine origin although they have other sources as well including
terrestrial plants and soil (Peltzer and Gagosian, 1983). The unequi-
vocably marine source polyunsaturated fatty acids were not detected,
but these compounds are known to undergo rapid degradation and photooxidation. Hence, they would not be expected to survive atmospheric
transport or aerosol collection. We attribute a major proportion of
the C_{13} to C_{18} fatty acids to a marine source due to their strong correlation with concentrations of Na (from sea-salt). In addition, the
same aerosol concentrations of these compounds were observed at Samoa
where the terrestrial dust source is much weaker (Table 4; Peltzer and
Gagosian, 1984).

The major feature of the Enewetak aerosol data is a clear terrestrial source signature, which accounts for most or all of the material
for carbon numbers greater than C_{20} in every compound class analyzed.
A strong temporal trend was also observed for these compound classes
(Table 3, Fig. 2). Neither the presence of a strong terrestrial
source marker signature so far from a continental source nor the
temporal variation of the concentration levels were anticipated at
Enewetak. However, both of these trends are completely consistent
with the results of the inorganic component analyses at Enewetak (Duce
et al., 1983b). The obvious decrease in hydrocarbon, wax ester, fatty
alcohol and high molecular weight fatty acid salts (Table 3, Fig. 2)
has been interpreted as a reflection of the seasonal nature of dust
storm activity in China and changes in the wind fields over the North
Pacific Ocean (Duce et al., 1980, 1983b; Uematsu et al., 1983, 1985).
Climatological trends and air-mass trajectory analyses show that dust
transport from well within Asia occurred in the upper level (700 mbar

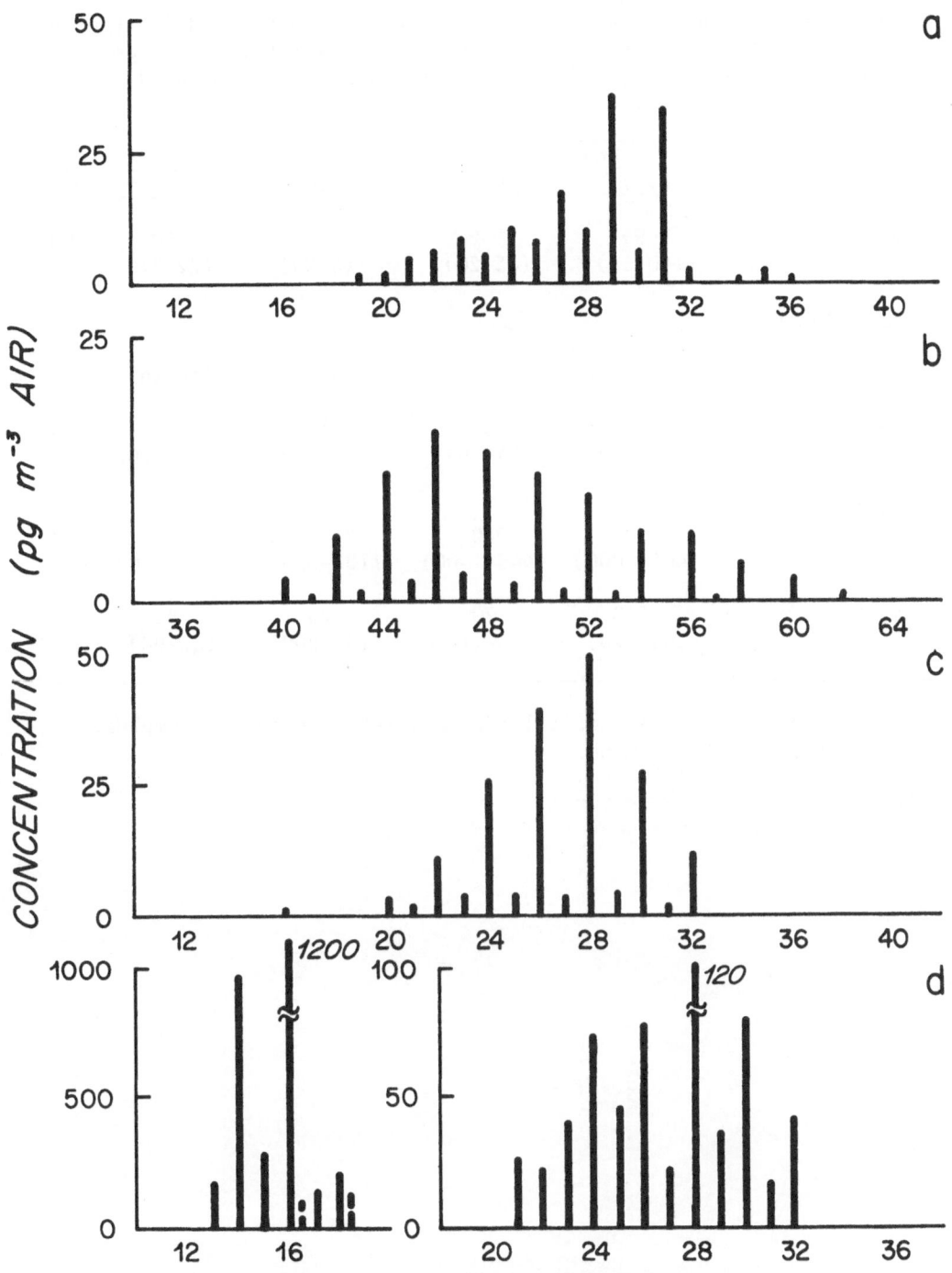

Figure 1. Lipid composition of Enewetak aerosol. (a) n-alkanes, (b) wax esters, (c) fatty alcohols, (d) fatty acid salts.

Table 4. Means and ranges for n–alkane, fatty alcohol, and fatty acid salt concentrations in aerosols collected at Enewetak (1979), Samoa (1981), and New Zealand (1983). (Concentrations are pg/m^3 air).

	Enewetak		Samoa	New Zealand	
	D.S.	W.S.	A.M.	ISS	OSS
n–alkanes	99 (160–63)	32 (45–20)	18 (6–35)	160 (22–270)	60
fatty alcohols					
C_{14}–C_{20}	5 (2–7)		3 (0.5–6)	15 (14–16)	10
C_{21}–C_{36}	140 (72–170)	65 (54–78)	18 (6–43)	270 (32–470)	76
fatty acid salts					
C_{13}–C_{18}	2900 (1400–4000)	960 (645–1400)	2100 (1100–3800)	510 (330–620)	640
C_{19}–C_{32}	380 (200–670)	70 (46–110)	76 (32–200)	81 (65–98)	86

A.M. = Annual mean: mean of all (dry and wet season) ISS samples.
D.S. = Dry season ISS mean.
W.S. = Wet season ISS mean.
ISS = Inside–sample–sector.
OSS = Outside–sample–sector.

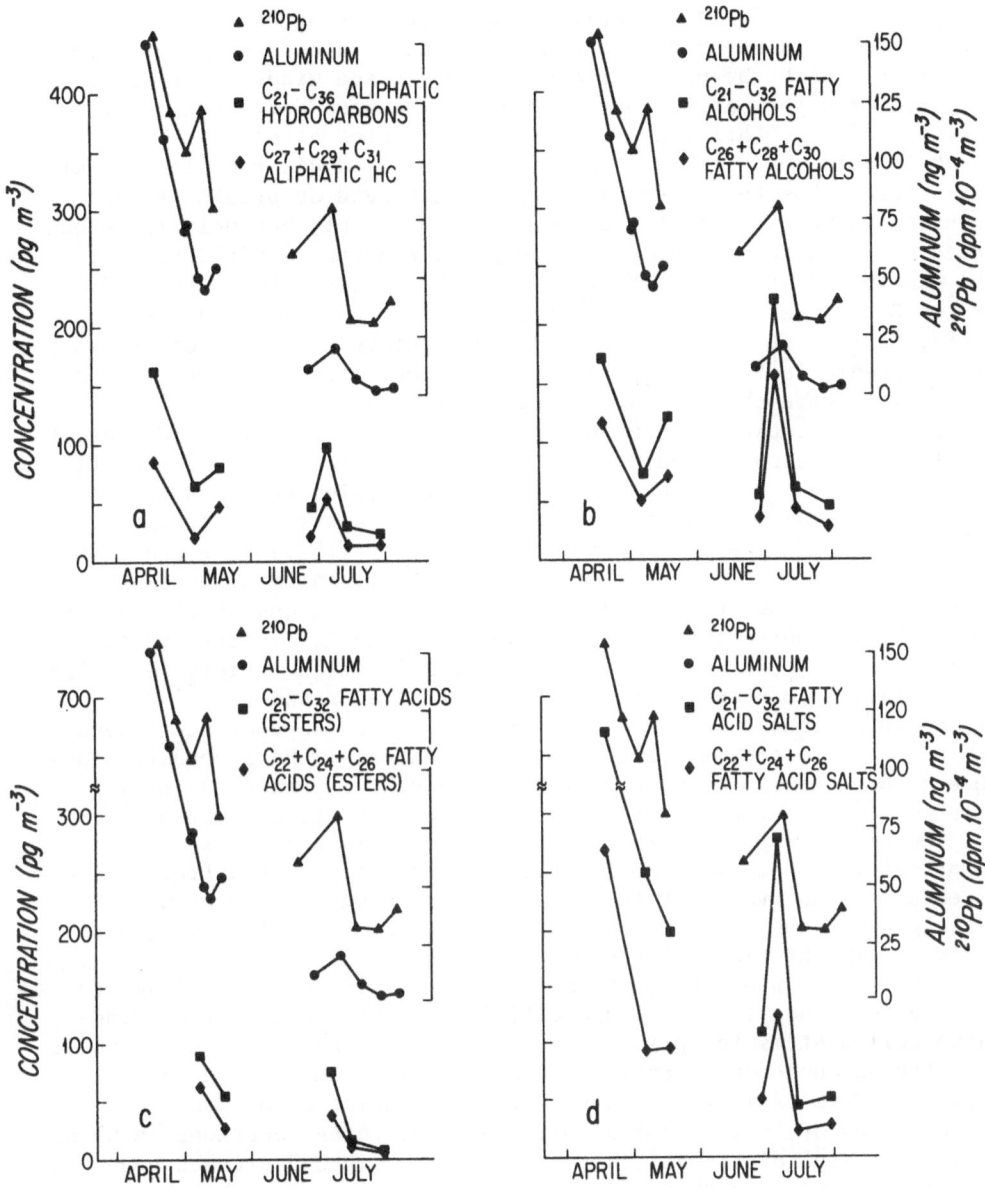

Figure 2.   (a) Aliphatic hydrocarbon, (b) fatty alcohol, (c) fatty
acid ester, and (d) fatty acid salt concentrations from glass fiber
filter samples as a function of time during the Enewetak field experiment. Particulate aluminum (Duce et al., 1980) and ^{210}Pb (Turekian
and Cochran, 1981a,b) are plotted for each compound class (from
Gagosian et al., 1982).

or approximately 3 km) westerlies, which are strong early in the year
and weaken in late spring and summer. Gradual subsidence and mixing
of this air with the near-surface easterly trade winds northeast of
Enewetak transport Asian dust to Enewetak causing decreasing concentrations of Al, other trace metals, and organic material with time.

Meteorological analyses of the frequency and distribution of Asian
dust storms show that several large and extended outbreaks of dust
occurred in the spring of the Enewetak experiment, but only isolated
cases occurred in the summer. Widespread reports of airborne dust
and numerous sustained dust or sandstorms, some severe, were recorded
in the Gobi desert areas during April and May (Merrill and Bleck,
1985a; Merrill et al., 1983, 1986), just before the times of terrestrial organic isotopic signatures and highest inorganic aerosol concentrations were observed at Enewetak (Fig. 2).

Carbon isotopic studies during the Enewetak experiment (Chesselet
et al., 1981) also showed a continental source for small size particles (<0.5 μm). The $\delta^{13}C$ ratios for three samples were -26 to
-28°/oo, typical land (non-marine) organic C values. By using $^{206}Pb/$
^{207}Pb ratios of aerosols, Settle and Patterson (1982) have shown that
the origin of these same air masses was predominantly Asian in April,
shifting to include north and central American regions in the summer
when dust concentrations had decreased. Furthermore, Turekian and
Cochran (1981a,b) have reported that a similar trend exists for ^{210}Pb
concentration in Enewetak air. They have interpreted the ^{210}Pb trend
as due to gaseous ^{222}Rn emanating from the Eurasian continent, decaying en-route to the mid-Pacific (thereby labeling the secondary fine-
mode aerosol particles with nonvolatile daughter products), mixing
with low ^{222}Rn marine air from the east, and arriving at Enewetak.

The similarity of ^{210}Pb and Al behavior (Fig. 2) leads to the
important inference that continental sources with very different modes
of emission and transport (dust suspension versus gas emission and
scavenging of decay products in the troposphere) can lead to quantitatively similar temporal trends at Enewetak.

For most sample dates, the strong correlation between Al and the
organic compounds (regression coefficient >0.9) suggests that wind-
blown soil dust is the principle source, while the anomalous increase
in fatty alcohol concentration in early summer (July 5 mid-date
sample) may be explained by a direct wax emission mechanism.

Analyses of loess samples collected from three locations in China
yield identical distributions for the lipid compound classes found in
Enewetak aerosols further supporting a soil source for the organic
components in the aerosols. The molecular composition for the lipids
found in the Enewetak samples shows no evidence for biomass burning,
anthropogenic or evaporation/condensation sources (Gagosian et al.,
1982).

3.3. South Pacific Westerlies: New Zealand

Cape Reinga lies at the northern most tip of the North Island of New
Zealand. A long stretch of open ocean beach, Ninety Mile Beach, lies

just to the south and is an excellent location for measuring the input of particulate material carried by the Southern Hemisphere westerlies across the Pacific Ocean. This site was the location for the SEAREX South Pacific Westerlies Experiment.

The means and ranges of the concentrations for the three major lipid compound classes found in aerosols at the New Zealand site are summarized in Table 4. Corresponding means and ranges for the Enewetak and Samoa sample sets are provided for comparison. The mean n-alkane concentration measured for clean air in New Zealand was 160 pg/m^3. This is about 1.6 times the mean in the Enewetak dry season, five times the mean for the Enewetak wet season, and almost an order of magnitude greater than the annual mean at Samoa. The low molecular weight ($C_{14}-C_{20}$) fatty alcohols collected in New Zealand were three to five times higher than the annual means for Enewetak and Samoa. The mean concentration in New Zealand for the higher molecular weight ($C_{21}-C_{36}$) fatty alcohols (which have a terrestrial plant wax source) was almost twice the mean observed for the Enewetak dry season and 15 times the annual mean measured at Samoa. In contrast to these high concentrations, the fatty acid salts measured in New Zealand were among the lowest values yet reported. Enewetak dry and wet season mean low molecular weight ($C_{13}-C_{18}$) fatty acid salts were six and two times greater than the New Zealand mean and the Samoa mean was over four times larger. The results are a clear indication that the low molecular weight fatty acid salts have a source other than terrestrial plant waxes or soils, most probably marine. The higher molecular weight ($C_{19}-C_{32}$) fatty acid salt concentrations measured in New Zealand were very comparable to the levels seen in Enewetak during the wet season and at Samoa. Enewetak dry season higher molecular weight fatty acid salts were almost five times the mean observed in New Zealand.

Several conclusions concerning the source of the aerosols can be made by examining individual compound distributions. For all four samples analyzed, the major source signature for the hydrocarbons was indicative of terrestrial plant waxes (see Fig. 3 and 4). A strong odd/even carbon preference was observed, the major homologs being C_{29} and C_{31}. For the first time, however, we also observed the lighter n-alkanes in the range of $C_{17}-C_{20}$. This is most likely due to the lower ambient temperatures in New Zealand vs. the tropical sampling sites. For the three in sector, marine (ISS) samples, C_{31} was the major homolog, while C_{29} was the major component of the n-alkanes in aerosols coming from across New Zealand (NZAS-16). It is interesting to note that for many of the Samoa samples, C_{31} was the major n-alkane. The lighter source signature for NZAS-16 suggests a more temperate source in comparison to the ISS samples whose origin is more tropical in nature.

The fatty alcohols are strong indicators of terrestrial plant wax sources. In all samples, a strong even/odd carbon predominance was observed, and for the first time significant amounts of the lighter ($C_{14}-C_{20}$) homologs were found. For the New Zealand-ISS samples, the mean carbon number was 27.5 - 28. At Enewetak, the mean carbon number

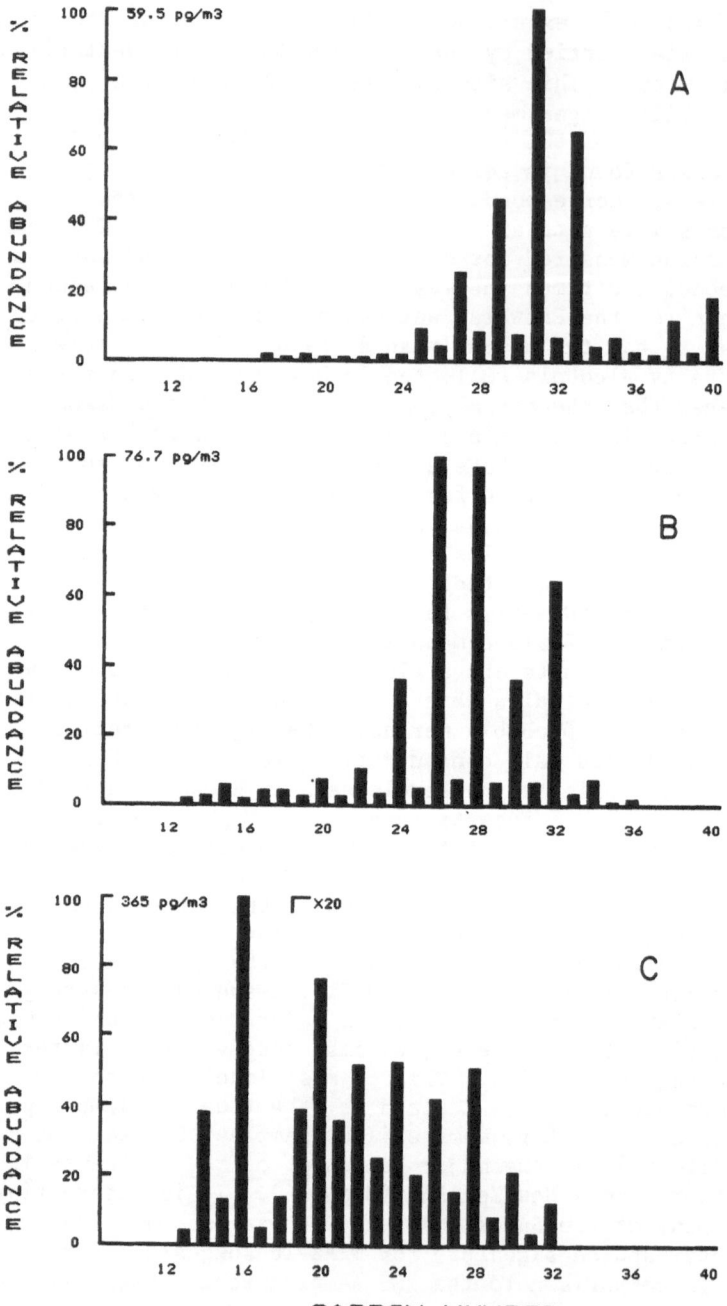

Figure 3.   n-alkane (A), fatty alcohol (B), and fatty acid salt (C) compositions for NZAS-20.

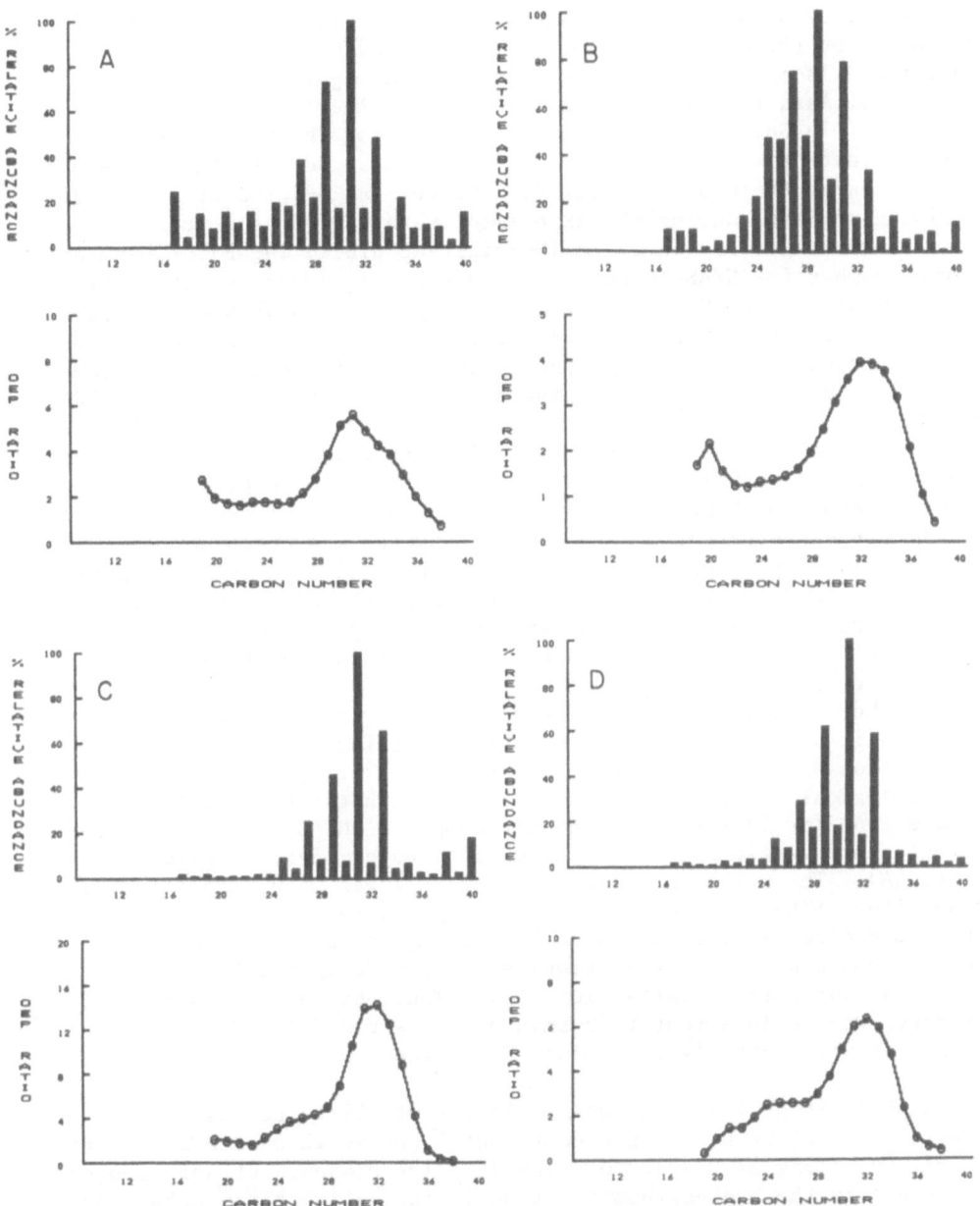

Figure 4 (A-D). n-Alkane distribution patterns and odd/even predominance (OEP)-ratio plots for aerosols collected at Ninety Mile Beach,
New Zealand, June-August, 1983. (A) NZAS-5. (B) NZAS-16. (C) NZAS-20.
(D) NZAS-22. Total hydrocarbon concentrations are 22, 61, 186 and
270 pg/m³, respectively. For air mass trajectories see Figure 5 A-D.

$$OEP(i) = \left[\frac{C(i-2) + 6C(i) + C(i+2)}{4C(i-1) + 4C(i+1)} \right]^{(-1)^i}$$

where C(i) is the observed concentration of the n-alkane that is i
carbon atoms long. The OEP ratio is then plotted as a function of
carbon number (i) (Scalan and Smith, 1970).

for the fatty alcohols showed a steady progression from 27.1 at the beginning of the dry season to 28.1 at the end of the wet season. This trend is indicative of the shift from a major temperate source in central Asia to a more "tropical" Central American source later in the year.   In Samoa, the mean carbon number for the fatty alcohols varied substantially from a low of 27 to almost 30.   A wide variety of sources is therefore indicated.   Indeed, the origins of air masses arriving at Samoa during the experiment were varied as determined by long-range trajectory analysis (Merrill and Bleck, 1985b).   The mean carbon number for NZAS-16 fatty alcohols was 26.8, and unlike most other samples where C_{26} and C_{28} are nearly equal in abundance, C_{26} comprised almost 42% of the fatty alcohols and was almost three times more abundant than C_{28}.   This indicates that NZAS-16 has a substantially different, and a more temperate source than the three NZ–ISS samples.

The low molecular weight (C_{13}–C_{18}) fatty acid salts appear to have a strong marine origin.   In all of the New Zealand samples, C_{16} was the most abundant homolog followed by C_{14} and C_{18}.   This is comparable to the abundance patterns seen in the Enewetak and Samoa samples.   The higher molecular weight (C_{19}–C_{32}) fatty acid salts in contrast show a pronounced land plant source as evidenced by the strong even/odd carbon predominance pattern and the lack of any real major homolog in the range C_{21}–C_{32}.   The mean carbon number of 24 ± 0.5 is typical of this source as well.

Further information can be obtained concerning the source of the atmospheric aerosols by examining the New Zealand aerosol data in conjunction with the long range transport and air parcel trajectory meteorological data.   Figure 4 shows the aliphatic hydrocarbon concentrations for four New Zealand air samples.   NZAS-5 (ISS) had the lowest concentrations of all the New Zealand samples, 22 pg/m^3 (Fig. 4A).   The n-alkane composition had a lower mean carbon number than either NZAS-20 or NZAS-22 suggesting a more temperate source. The fatty alcohol composition of NZAS-5 suggested a mixture of several sources including a "tropical" and a "temperate" source (32 pg/m^3 for terrestrial fatty alcohols).   Thus, based on chemical evidence we conclude that this sample represents "clean" southern ocean air with possibly some minor Tasmanian or South Australian influence.

NZAS-16 (Fig. 4B) had moderate levels of all three compound classes suggesting a stronger continental source than NZAS-5 but not nearly as strong as NZAS-20 or NZAS-22.   The presence of anthropogenic n-alkanes in the hydrocarbon fraction is indicative of an urban source while the very low mean carbon number for the high molecular weight fatty alcohols suggests that this sample had the most temperate source of all the samples analyzed.   Thus, we would predict that this sample represents an urban source in New Zealand or possibly Tasmania.   The relatively low n-alkane and terrestrial fatty alcohol concentrations, 61 and 76 pg/m^3 respectively, suggest that the transport was from high level winds.

Samples NZAS-20 (4C) and NZAS-22 (4D) had very high and similar concentrations of all the compound classes discussed (186 and 270 pg/m^3 for n-alkanes and 291 and 474 pg/m^3 for terrestrial fatty alcohols respectively). The n-alkanes, the high molecular weight fatty alcohols and the high molecular weight fatty acid salts all indicate a major terrestrial source. Both samples exhibited a moderately tropical character for the source, NZAS-20 being the more tropical. The levels of n-alkanes and fatty alcohols were greater than any of the Enewetak samples suggesting a close continental source, probably Australia.

In Figure 5, we have presented four air mass trajectories which are typical of the trajectories followed by the air sampled in New Zealand for each of these samples. In each case, these trajectories are consistent with the predicted source of the aerosol based on the trace organic compound analyses. Figure 5 (A) clearly shows a southern ocean origin for NZAS-5 which is essentially free of all continental influence for many thousands of miles. Some trajectories during the period of sampling approach or cross the far south-western part of the South Island of New Zealand. Most, however, are similar to the case shown, where the "origin" – the position 10 days before arrival at the sampling site, is in the open ocean or near southwestern Australia. Figure 5 (B) shows that NZAS-16 originated in New Zealand and that there was urban and natural terrestrial influence from the south of the North Island and Christchurch. Likewise, NZAS-20 (Fig. 5 (C)) and NZAS-22 (Fig. 5 (D)) clearly originated in southern Australia and Tasmania supporting the strong terrestrial biomarker aerosol signal.

These examples of trace organic compound and air parcel trajectory correlations clearly show the benefits of this approach in defining aerosol sources and transport pathways.

3.4. Short-Range Transport – Coastal Peru

The coastal area of Peru is an excellent location for studying atmospheric source strength processes. The coastline is marked by high sand cliffs up to 1000 m for several miles inland where the topography gives way to the foothills of the Andes. The winds blow strongly from the southeast most of the year transporting aerosol material originating on land into the marine atmosphere. The coast of Peru is also a strong upwelling zone. The area has very high productivity, thereby producing high concentrations of marine biological source markers. In addition, the anthropogenic contribution to atmospheric organic matter is very low. Hence, this area was chosen to undertake particle size studies concerning aerosols with strong marine and continental source strengths.

The concentration range of the terrestrial source marker compound classes found off the coast of Peru is 10-20 times higher than that from a remote area of the southern hemisphere (e.g. Samoa, Table 4) and 2-5 times higher than terrestrially-derived lipids found in aerosols from the tropical North Pacific at Enewetak (Table 4). The predominance of fatty alcohols, wax esters, and fatty acid salts with an

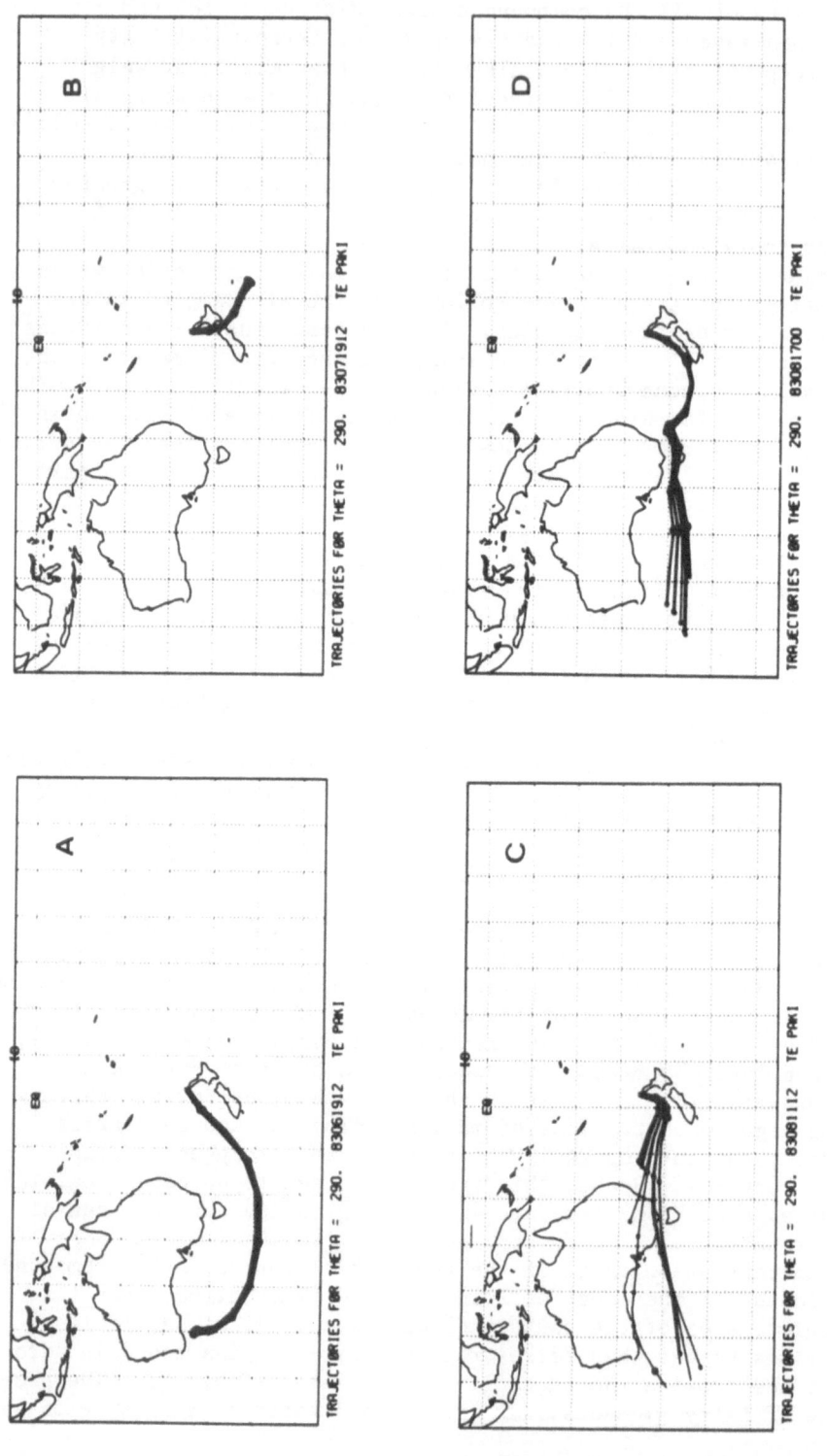

Figure 5. Air mass trajectories for NZAS-5 (A), NZAS-16 (B), NZAS-20 (C), and NZAS-22 (D).

even number of carbon atoms (e.g. C_{26}, C_{28}, C_{30} for fatty alcohols; C_{44}, C_{46}, C_{48} for wax esters; and C_{28}, C_{30}, C_{32} for fatty acid salts) indicating land plant wax (Eglinton and Hamilton, 1963; Plouvier, 1963; Shoreland, 1963; Tulloch, 1976) or soil (Gagosian et al., 1982) origin is clearly present in all stages (Figs. 6b-6d). The same conclusion is reached for total hydrocarbons, but the predominance of compounds with odd carbon atoms (e.g. C_{27}, C_{29}, C_{31}) is more pronounced in the large particle size fractions (Figs. 6a and 7a) (Schneider et al., 1983). Marty and Saliot (1982) observed the same phenomenon from cascade impactor samples taken in the Gulf of Guinea.

Schneider et al. (1983) concluded that the n-alkanes had two sources: direct emission of land plant epicuticular wax or suspension of soil, and anthropogenic origins. The anthropogenic n-alkanes (CPI = 1) which correlate with ^{210}Pb show increasing concentration as a function of decreasing particle size (Fig. 7b,c). The relatively even distribution of the terrestrially derived n-alkanes (CPI = 5) (Fig. 7a), however, is very similar to the fatty alcohol and wax ester distributions (Schneider and Gagosian, 1985). Thus, this result is consistent with the hypothesis of all three compound classes having the same source. These results strongly suggest different emission and transport mechanisms for the terrestrially derived compound classes relative to the anthropogenic n-alkanes.

The even distribution of lipid concentrations over different particle sizes is probably due to the wide spectrum of particle sizes generated by epicuticular plant wax and soil aerosol sources, little modified by short range transport offshore. However, the particle size distribution of compounds with a terrestrial source would be expected to change as a function of distance from land as observed by Chesselet et al. (1981) for particulate organic matter. Indeed, in cascade impactor samples collected offshore, Ohta and Handa (1985) observed that terrestrial derived n-alkanes were present mostly on the small particle fractions suggesting the loss of larger wax and soil derived aerosols with air parcel transport offshore.

Earlier we discussed several possible emission mechanisms for these terrestrially-derived lipids with relation to the Enewetak data. The fact that the terrestrial compounds are evenly distributed over the whole particle size range suggests emission mechanisms which do not include gas-particle partitioning. If this partitioning were a major process, it would lead to a distribution similar to the anthropogenic n-alkanes, i.e. increasing concentrations with decreasing particle size. Furthermore, Zafiriou et al. (1985) conclude that the vapor pressure for fatty alcohols (>C_{24}) and wax esters are too low for a significant fraction of these compounds to exist in the gas phase. This is strongly supported by the consideration that the wax esters would probably not chemically survive any evaporation processes.

There are two possible explanations for the similar trends of ^{210}Pb and the anthropogenic n-alkanes as a function of particle size (Fig. 7b,c). First, the distribution pattern of the anthropogenic n-alkanes could be due to emission of small particles con-

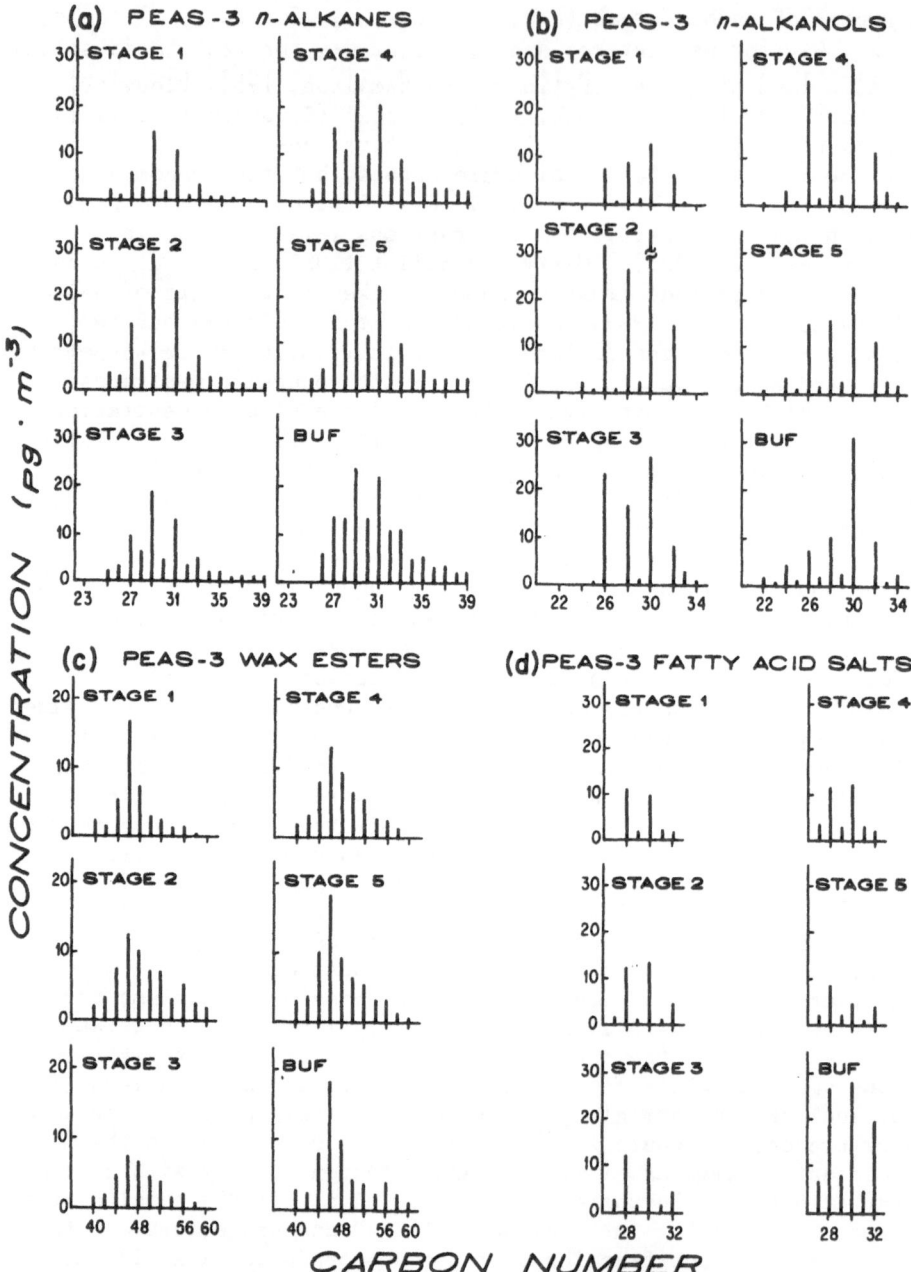

Figure 6. Concentrations of individual terrestrial compounds as a function of carbon number in the different stages of the cascade impactor sample PEAS 3: (a) n-alkanes, (b) n-alkanols, (c) wax esters, and (d) fatty acid salts (equivalent aerodynamic diameter: stage 1, >7.2 μm; stage 2, 3.0-7.2 μm; stage 3, 1.5-3.0 μm; stage 4, 0.96-1.5 μm; stage 5, 0.50-0.96 μm; and backup filter, <0.50 μm) (from Schneider and Gagosian, 1985).

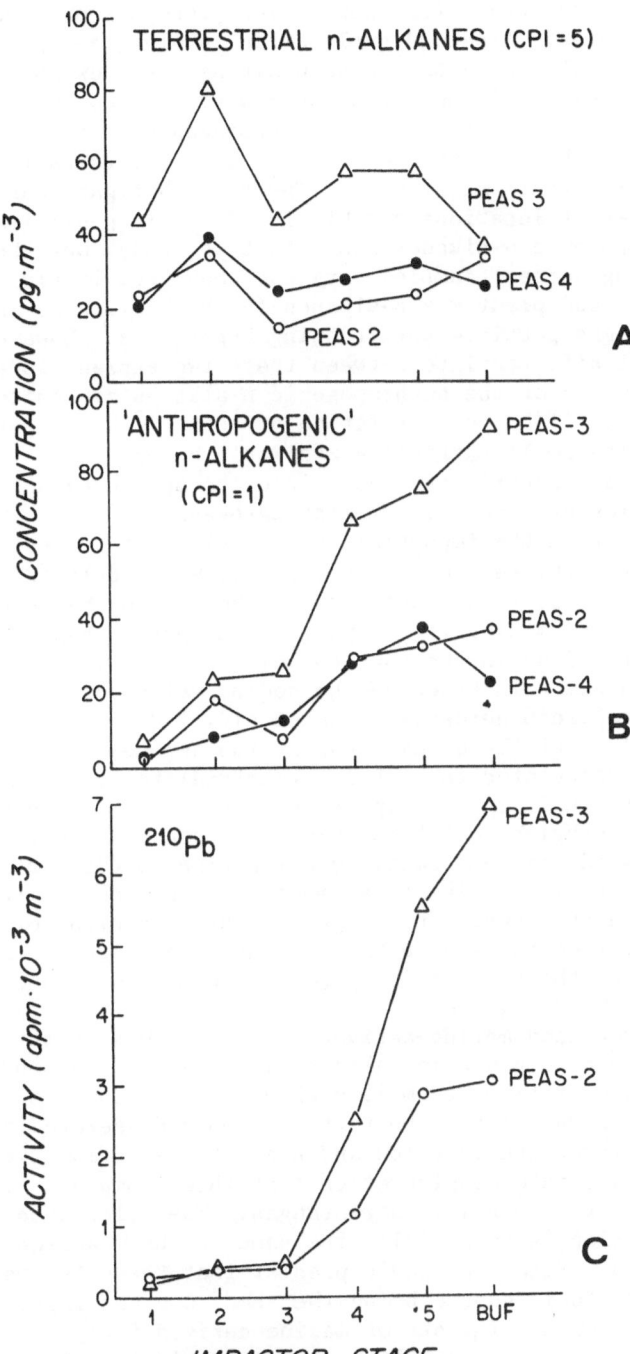

Figure 7. Calculated concentrations for two proposed n-alkane contributions and ²¹⁰Pb as a function of particle size for the cascade impactor samples PEAS-2, PEAS-3 and PEAS-4; a, terrestrial n-alkanes; b, 'anthropogenic' n-alkanes; c, ²¹⁰Pb (from Schneider et al., 1983).

taining n-alkanes during fossil fuel combustion processes or other
anthropogenic activities. A second explanation involves gas-particle
partitioning of the n-alkanes. Gas phase n-alkanes (for example,
evaporated during fossil fuel combustion) can adsorb or condense onto
the small carbon particles produced from anthropogenic activities.
These alkanes can also be incorporated into or attached to particles
covered with natural organic substances. The fact that gas phase
n-alkanes from different locations exhibit low CPI values similar to
those of the anthropogenic n-alkanes found in this study, but the CPI
for the corresponding aerosol samples vary and show from little to
strong influence of land plant wax n-alkanes (high CPI values) is
consistent with the gas-particle partitioning hypothesis. However,
at present we cannot differentiate between these two explanations.

Although the sources of the anthropogenic n-alkanes are quite
different from that of ^{210}Pb, a correlation of both ^{210}Pb and anthropogenic n-alkanes with small particle sizes would be expected. This
suggests that their atmospheric transport into the open ocean on
similar size particles proceeds via similar pathways.

In order to ascertain the importance of recycling of terrestrial
organic compounds from the sea to the atmosphere, we calculated the
expected concentrations of these compounds in the aerosol based on
data from a 1 μm surface microlayer sampler and compared these with
the measured concentrations in the corresponding air sample, PEAS-3
(Table 5). The calculation is based on the sodium values of 4-7 μg
Na m^{-3} for Enewetak and Peru aerosols (Duce et al., 1980). The
expected concentrations of these compounds in the aerosol based on
recycling have been calculated for total terrestrial fatty alcohols
and fatty acid salts, their major components, and for the terrestrially-derived sterol 24-ethylcholest-5-en-3ß-ol. As can be seen from
Table 5, the calculated maximum oceanic contribution to the aerosol
from recycled terrestrial material is at most a few percent of the
measured PEAS-3 concentrations. This suggests that recycling of
terrestrial material from the sea surface to the atmosphere is not a
significant source for the terrestrially-derived compounds found in
aerosols.

The calculated recycled marine-derived fatty alcohols, fatty acid
salts and sterols are also given in Table 5 with their corresponding
measured aerosol concentrations from 15 m above the sea surface. The
calculated atmospheric values for the fatty alcohols by aerosolization
of the surface microlayer are very low and near the detection limit,
especially when we take into consideration that these compounds could
be distributed over 3 to 6 particle size ranges. The calculated value
of the recycled fraction is essentially the same as the measured
aerosol concentration. Since one would predict that these two values
would be equal due to their source being the same, one concludes that
there has been little or no blow-off of marine derived fatty alcohols
from the aerosol filters during sampling. The same is true for the
diatom-derived sterol 6 and for cholesterol.

Table 5. Recycled terrestrial and marine marker compounds in aerosols 15 m above the sea surface.

	Aerosol[a] (pg m^{-3})	Recycled fraction[b] (pg m^{-3})	Max. recycled portion (%)
Terrestrial			
Fatty Alcohols			
Σ C_{21}-C_{32}	484	7-14	< 3
C_{26}	112	1-2	< 2
C_{28}	97	< 1	~ 1
C_{30}	163	1-2	< 2
Sterols			
24-ethylcholest-5-en-3ß-ol (8)	< 6[c]	1-2	--
Fatty Acid Salts			
Σ C_{24} - C_{32}	256	1-2	< 1
C_{28}	84	< 1	< 1
C_{30}	81	< 1	< 1
Marine			
Fatty Alcohols			
Σ C_{14} - C_{20}	< 40[c]	19-38	~95
C_{16}	< 6[c]	4-8	100
C_{18}	< 6[c]	5-10	100
Sterols			
Cholesterol (3)[d]	32	11-22	33-66
24-methylcholesta-5,24(28)-dien-3ß-ol (6)	< 6[c]	3-6	50-100
Fatty Acid Salts			
Σ C_{12} - C_{20}	NM[e]	350-700	
C_{16}	NM	200-400	
C_{18}	NM	85-170	

[a] PEAS-3 data.
[b] Calculated from the microlayer data.
[c] The detection limit for a single compound is 1 pg m^{-3} for a single stage. The values given are the sum over the six stages, since a compound could be distributed over the whole particle size range.
[d] Although cholesterol has several marine sources, it also has terrestrial sources as well and should be considered ubiquitous.
[e] NM - Not measured.

4. CONCLUSIONS

In the previous sections examples were given of the use of the biomarker approach in understanding the processes responsible for the
distribution of organic substances in aerosols in remote environments.
These studies clearly show the need for further temporal and particle
size separation experiments. They also point out the importance of
undertaking future experiments as collaborative investigations. It
is critical that these studies involve several researchers each
contributing their expertise concerning the inorganic, isotopic and
organic tracers of interest. This allows for maximum interpretation
of a single tracer data base since a broad base includes compounds
and elements of varying sources, functional groups, stabilities,
transformation reaction rates, transport pathways, deposition rates,
etc. The section on sampling and analysis was included to impress
upon the reader the importance of contamination free sampling conditions and or establishing the criteria for these conditions. The
necessity of close cooperation between chemists and meteorologists
cannot be overstated; the New Zealand case study is a clear example
of the importance of this collaboration. Although air may be sampled
within a clean air "sector", the parcel of air sampled could have
passed over "contaminated" areas days before. Hence, meteorological
long-range transport analysis information is essential both for planning sampling strategy and interpreting chemical data.

Studies concerned with the spatial distributions of the chemical
composition of aerosols must include a temporal component. The author
feels that the next logical step in these studies is to establish a
"network" of sites in an ocean basin framework with yearly sampling.
In this way the important processes responsible for the long range
transport in space and time of chemical substances in aerosols can be
determined. (The reader is referred to the recent NAS Global Tropospheric Chemistry Report (1985) for further discussion.) This approach
also allows us to accumulate the appropriate data base (both chemical
and meteorological) to determine the boundary conditions necessary
for predicting source strengths, transit times and pathways of the
chemical constituents of interest in the aerosols.

In addition to future applications of the biomarker approach,
research needs to be continued to identify new biomarkers for use as
specific source markers in aerosols. As pointed out in the case
studies, we can differentiate between natural continental, anthropogenic and marine sources by using the lipid classes: aliphatic and
aromatic hydrocarbons, fatty alcohols and acids, sterols and fatty
acid esters (wax esters and triglycerides). However, we cannot differentiate between soil or epicuticular plant wax sources using these
compound classes. Extending our biomarker "arsenal" to include
hydroxy fatty acids, long and mid-chain ketones and lignin material
may allow us to separate these sources more easily.

Studies concerned with regional sources for terrestrial input
determinations are just beginning. Early results are quite promising
for differentiating between temperate and tropical region plant

sources as shown by the New Zealand case study. Biomass burning produces a completely different suite of organic compounds than those observed from plants and soils (e.g. polycyclic aromatic hydrocarbons). Hence, this source can be differentiated quite clearly from others. Identifying the gas to particle conversion process for aerosol sources is not as straightforward. Low molecular weight compounds such as organic acids and diacids may play an important role in our understanding of this process.

Future research on the organic compound classes outlined above in conjunction with organic, inorganic, isotopic and meteorological field data will allow us to put limits on source strengths and source regions for aerosols found in remote areas.

ACKNOWLEDGEMENTS

This work was supported by National Science Foundation grants OCE77-12914, OCE81-11947 and OCE84-06666. We thank J. B. Alford and N. M. Frew for assistance in the analytical work and Eliot Atlas, Hèlène Cachier, Philippe Garrigues and Patrick Buat-Ménard for discussion and critical comments on the manuscript. Special thanks to Edward Peltzer for his major analytical and scientific efforts which have led to several SEAREX publications. This is Woods Hole Oceanographic Institution Contribution No. 6094.

REFERENCES

Andreae, M. O., 1983: 'Soot carbon and excess fine potassium: long range transport of combustion-derived aerosols', Science, 220, 1148-1151.

Barbier, M., D. Tusseau, J. C. Marty and A. Saliot, 1981: 'Sterols in aerosols, surface microlayer and subsurface water in the northeastern tropical Atlantic', Oceanol. Acta, 4, 77-84.

Barger, W. R. and W. D. Garrett, 1970: 'Surface active organic material in the marine atmosphere', J. Geophys. Reas., 75, 4561-4566.

Barger, W. R. and W. D. Garrett, 1976: 'Surface active organic material in air over the Mediterranean and over the eastern equatorial Pacific', J. Geophys. Res., 81, 3151-3157.

Björseth, A. and G. Lunde, 1979: 'Long-range transport of polycyclic aromatic hydrocarbons', Atmos. Environ., 13, 45-53.

Broddin, G., W. Cautreels, and K. Van Cauwenberghe, 1980: 'On the aliphatic and polyaromatic hydrocarbon levels in urban and background aerosols from Belgium and the Netherlands', Atmos. Environ., 14, 895-910.

Buat-Ménard, P., 1983: 'Particle goechemistry in the atmosphere and oceans. In: NATO Advanced Study Institute in Air-Sea Exchange of Gases and Particles, eds. P. S. Liss and G. Slinn. D. Reidel, Hingham, MA, 455-532

Butcher, S. S. and R. J. Charlson, 1972: An Introduction to Air
    Chemistry, Academic Press, New York, 49.
Cautreels, W., K. Van Cauwenberghe and L. A. Guzman, 1977: 'Comparison
    between the organic fraction of suspended matter at a background
    and an urban station', Sci. Total Environ., 8, 79–88.
Cautreels, W. and K. Van Cauwenberghe, 1978: 'Experiments on the
    distribution of organic pollutants between airborne particulate
    matter and the corresponding gas phase', Atmos. Environ., 12,
    1133–1141.
Chesselet, R., M. Fontugne, P. Buat-Ménard, U. Ezat and C. E. Lambert,
    1981: 'The origin of particulate organic carbon in the marine
    atmosphere as indicated by its stable carbon isotopic composi-
    tion', Geophys. Res. Lett., 8, 345–348.
Cox, R. E., M. A. Mazurek, and B. R. T. Simoneit, 1982: 'Lipids in
    Harmattan aerosols of Nigeria', Nature, 296, 848–849.
Daisey, J. M., R. J. McCaffrey, and R. A. Gallagher, 1981: 'Poly-
    cyclic aromatic hydrocarbons and total extractable particulate
    organic matter in the Arctic aerosol', Atmos. Environ., 15(8),
    1353–1363.
Duce, R. A., 1978: 'Speculations on the budget of particulate and
    vapor phase non-methane organic carbon in the global tropo-
    sphere', Pageoph., 116, 244–273.
Duce, R.A., C.K. Unni, B.J. Ray, J.M. Prospero and J.T. Merrill,
    1980: 'Long-range atmospheric transport of soil dust from Asia to
    the tropical North Pacific: temporal variability', Science,
    209, 1522–1524.
Duce, R. A., 1981: 'SEAREX: A multi-institutional investigation of
    the sea/air exchange of pollutants and natural substances', In:
    Marine Pollutant Transfer Processes, eds. M. Waldichuk, G.
    Kullenburg and M. Orren. Elsevier, New York.
Duce, R. A. and R. B. Gagosian, 1982: 'The input of atmospheric
    n-C$_{10}$ to n-C$_{30}$ alkanes to the ocean', J. Geophys. Res., 87,
    7192–7200.
Duce, R. A., V. A. Mohnen, P. R. Zimmerman, D. Grosjean, W. Cautreels,
    R. Chatfield, R. Jaenicke, J. A. Ogren, E. D. Pellizzari and G.
    T. Wallace, 1983a: 'Organic material in the global troposphere',
    Review of Geophysics and Space Physics, 21, 921–952.
Duce, R. A., R. Arimoto, B. J. Ray, C. K. Unni and P. J. Harder,
    1983b: 'Atmospheric trace elements at Enewetak Atoll: 1. Concen-
    trations, sources, and temporal variability', J. Geophys. Res.,
    88, 5321–5342.
Eglinton, G. and R. J. Hamilton, 1963: 'The distribution of alkanes',
    In: Chemical Plant Taxonomy, ed. T. Swain. Academic, Orlando,
    FL, 187–217
Eichmann, R., P. Neuling, G. Ketseridis, J. Hahn, R. Jaenicke and C.
    Junge, 1979: 'n-Alkane studies in the troposphere. I. Gas and
    particulate concentrations in North Atlantic air', Atmos.
    Environ., 13, 587–599.
Eichmann, R., G. Ketseridis, G. Schebeske, R. Jaenicke, J. Hahn, J.
    Warneck and C. Junge, 1980: 'n-Alkane studies in the troposphere.

II. Gas and particulate concentrations in Indian Ocean air',
Atmos. Environ., 14, 695-703.

Gagosian, R. B., O. C. Zafiriou, E. T. Peltzer and J. B. Alford,
1982: 'Lipids in aerosols from the tropical North Pacific:
Temporal variability', J. Geophys. Res. 87(C), 11133-11144.

Gagosian, R. B., 1983: 'Review of Marine Organic Geochemistry', Rev.
Geophysics and Space Physics, 5, 1245-1258.

Gagosian, R. B., E. T. Peltzer, and O. C. Zafiriou, 1981: 'Atmospheric
transport of continentally-derived lipids to the tropical North
Pacific', Nature, 291, 312-314.

Gagosian, R. B., G. E. Nigrelli, and J. K. Volkman, 1983: 'Vertical
transport and transformation of biogenic organic compounds from a
sediment trap experiment off the coast of Peru', In: NATO Advanced
Research Institute on Coastal Upwelling: Its Sediment Record, ed.
E. Suess and J. Thiede. Plenum Press, New York, 241-272.

Garrett, W. D., 1967: 'The organic chemical composition of the ocean
surface', Deep-Sea Res., 14, 221-227.

Hahn, J., 1980: 'Organic constituents of natural aerosols', In:
Anthropogenic and Natural Sources and Transport, T. J. Kneip and
P.J. Lioy (eds.). Annals N. Y. Acad. Sciences, 338, 359-378.

Hoffman, E. J. and R. A. Duce, 1974: 'The organic carbon content of
marine aerosols collected on Bermuda', J. Geophys. Res., 79,
4474-4477.

Hoffman, E. J. and R. A. Duce, 1977: 'Organic carbon in marine atmospheric particulate matter: concentrations and particle size distributions', Geophys. Res. Lett., 4, 449-452.

Junk, G. A. and B. A. Jerome, 1983: 'Sampling methods for organic
compounds in stacks', American Laboratory (December), 16-29.

Ketseridis, G., J. Hahn, R. Jaenicke and C. Junge, 1976: 'The organic
constituents of atmospheric particulate matter', Atmos. Environ.,
10, 603-610.

Lamb, S. I., C. Petrowski, I. R. Kaplan and B.R.T. Simoneit, 1980:
'Organic compounds in urban atmospheres: a review of distribution,
collection and analysis', J. Air Poll. Control Assoc., 30,
1098-1115.

Marty, J. C., A. Saliot, P. Buat-Menard, R. Chesselet and K. A.
Hunter, 1979: 'Relationship between the lipid compositions of
marine aerosols, the sea surface microlayer and subsurface water',
J. Geophys. Res., 84, 5707-5716.

Marty, J. C., and A. Saliot, 1982: 'Aerosols in equatorial Atlantic
air: n-alkanes as a function of particle size', Nature, 298,
312-314.

Marty, J. C., M. J. Tissier and A. Saliot, 1984: 'Gaseous and particulate polycyclic aromatic hydrocarbons (PAH) from the marine atmosphere', Atmos. Environ., 18, 2183-2190.

Merrill, J., R. Bleck, and L. Avila, 1983: 'Transport to Enewetak in
1979', SEAREX Newsletter, 6 (1), 6-11.

Merrill, J. T. and R. Bleck, 1985a: 'Trajectory results for the North
Pacific network', SEAREX Newsletter, 8 (2), 2-8.

Merrill, J. T. and R. Bleck, 1985b: 'Revised Samoa trajectories and

trajectories from TePaki, New Zealand', SEAREX Newsletter, 8
    (1), 42–45.
Merrill, J. T., R. Bleck and L. Avila, 1986: 'Modeling atmospheric
    transport to the Marshall Islands', J. Geophys. Res., in press.
NAS, 1978: The Tropospheric Transport of Pollutants and Other
    Substances to the Oceans, Ocean Sciences Board, National Academy
    of Sciences, Washington D.C., 243.
NAS, 1984: 'Global Tropospheric Chemistry', National Research Council,
    Washington, 194 pp.
Ohta, K. and N. Handa, 1985: 'Organic components in size-separated
    aerosols from the western North Pacific', J. Ocean. Soc. Japan,
    41, 25–32.
Peltzer, E. T., S. G. Wakeham, and R. B. Gagosian, 1981: 'Wax ester
    composition of Enewetak particulate samples', SEAREX Newsletter
    4(3), 15–19.
Peltzer, E. T. and R. B. Gagosian, 1983: 'Source marker studies of
    organic compounds found in atmospheric aerosols', SEAREX News-
    letter, 6(1), 13–21.
Peltzer, E. T. and R. B. Gagosian, 1984: 'Naturally derived organic
    compounds in aerosols and rain samples collected at Samoa', SEAREX
    Newsletter, 7(1), 22–26.
Peltzer, E. T., J. B. Alford, and R. B. Gagosian, 1984: 'Methodology
    for sampling and analysis of lipids in aerosols from the remote
    marine atmosphere', Tech. Rept. 84–9, Woods Hole Oceanographic
    Inst., Woods Hole, MA.
Peltzer, E. T., R. B. Gagosian and J. T. Merrill, 1985: 'Naturally
    occurring organic compounds in aerosol samples collected at Ninety
    Mile Beach, New Zealand: Preliminary results', SEAREX Newsletter,
    8(1), 1–7.
Plouvier, V., 1963: 'Distribution of aliphatic polyols and cyclitols',
    In: Chemical Plant Taxonomy, ed. T. Swain, Academic, Orlando, FL.,
    313–336
Rosen, H., T. Novakov, and B. A. Bodhaine, 1981: 'Soot in the Arctic',
    Atmos. Environ., 15, 1371–1374.
Sargent, J. R., 1976: 'The structure, metabolism and function of
    lipids in marine organisms', In: Biochemical and Biophysical Per-
    spectives in Marine Biology, Volume 3, eds. D. C. Malins and J.
    R. Sargent. Academic Press, 149–212.
Sargent, J. R., R. F. Lee and J. C. Nevenzel, 1976: 'Marine Waxes',
    In: Chemistry and Biochemistry of Natural Waxes, ed. P. E. Kolat-
    tukudy. Elsevier, 49–91.
Sargent, J. R., R. R. Gatten and R. McIntosh, 1977: 'Wax esters in
    the marine environment - their occurrence, formation, transforma-
    tion, and ultimate fate', Mar. Chem., 5, 573–584.
Scalen, R. S. and J. E. Smith, 1970: 'An improved measure of the odd-
    even predominance in the normal alkanes of sediment extracts and
    petroleum', Geochim. Cosmochim. Acta, 34, 611–620.
Schneider, J. K., R. B. Gagosian, J. K. Cochran, and T. W. Trull,
    1983: 'Particle size distributions of n-alkanes and ^{210}Pb in
    aerosols off the coast of Peru', Nature, 304, 429–432.

Schneider, J. K. and R. B. Gagosian, 1985: 'Particle size distribu-
    tions of lipids in aerosols off the coast of Peru', J. Geophys.
    Res., 90, 7889-7898.
Settle, D. M. and C. C. Patterson, 1982: 'Magnitudes and sources of
    precipitation and dry deposition fluxes of industrial and natural
    leads to the North Pacific at Enewetak', J. Geophys. Res., 87,
    8857-8869.
Shoreland, R. B., 1963: 'The distribution of fatty acids in plant
    lipids', In: Chemical Plant Taxonomy, ed. T. Swain. Academic,
    Orlando, FL, 253-312.
Simoneit, B. R. T., 1977: 'Organic matter in eolian dust over the
    Atlantic Ocean', Mar. Chem., 5, 443-464.
Simoneit, B. R. T., 1979: 'Biogenic lipids in eolian particulates
    collected over the ocean', In: Proceedings Carbonaceous Particles
    in the Atmosphere, ed. T. Novakov, 233-244 (NSF-LBL).
Simoneit, B. R. T., 1980: 'Eolian particulates from oceanic and rural
    areas-- Their lipids, fulvic and humic acids and residual carbon',
    In: Advances in Organic Geochemistry 1979, eds. A. G. Douglas and
    J. R. Maxwell. Pergamon Press, New York, p. 343-352.
Simoneit, B. R. T. and M. A. Mazurek, 1981: 'Air pollution: The
    organic components', In: Critical Reviews in Environmental
    Control, 11 (3), 219-276, CRC Press.
Simoneit, B. R. T. and M. A. Mazurek, 1982: 'Organic matter of the
    troposphere: II - Natural background of biogenic lipid matter in
    aerosols over the rural western United States', Atmos. Environ.,
    16, 2139-2159.
Simoneit, B. R. T., M. A. Mazurek and W. E. Reed, 1983: 'Characteriza-
    tion of organic matter in aerosols over rural sites: Physto-
    sterols', In: Advances in Organic Geochemistry 1981, eds. M.
    Bjoroy et al. J. Wiley and Sons Ltd., Chichester, 355-361.
Simoneit, B. R. T., 1984a: 'Organic matter of the troposphere: III
    -- Characterization and sources of petroleum and pyrogenic
    residues in aerosols over the Western United States', Atmos.
    Environ., 18, 51-67.
Simoneit, B. R. T., 1984b: 'Application of molecular marker analysis
    to reconcile sources of carbonaceous particulates in tropospheric
    aerosols', In: Proc. 2nd Int. Conf. Carbonaceous Particles in the
    Atmosphere, Science of the Total Environ., 36, 61-72.
Tulloch, A. P., 1976: 'Chemistry of Waxes of Higher Plants', In:
    Chemistry and Biochemistry of Natural Waxes, ed. P. E. Kolat-
    tukudy. Elsevier, 236-252.
Turekian, K. K. and J. K. Cochran, 1981a: '^{210}Pb in surface air at
    Enewetak and the Asian dust flux to the Pacific', Nature, 292,
    522-524.
Turekian, K. K. and J. K. Cochran, 1981b: '^{210}Pb in surface air at
    Enewetak and the Asian dust flux to the Pacific: A correction',
    Nature, 294, 670.
Uematsu, M., R. A. Duce, J. M. Prospero, L. Chen, J. T. Merril and R.
    L. McDonald, 1983: 'Transport of mineral aerosol from Asia over
    the North Pacific Ocean', J. Geophys. Res., 88, 5343-5352.

Uematsu, M., J. Merrill, R. Duce and J. Prospero, 1985: 'Interannual variation of mineral aerosol concentration over the North Pacific', SEAREX Newsletter, 8(1), 38-42.

Van Vaeck, L. V. and K. Van Cauwenberghe, 1978: 'Cascade impactor measurements of the size distribution of the major classes of organic pollutants in atmospheric particulate matter', Atmos. Environ., 12, 2229-2239.

Van Vaeck, L., G. Broddin and K. Van Cauwenberghe, 1979: 'Differences in particle size distributions of major organic pollutants in ambient aerosols in urban, rural and seashore areas', Environ. Sci. Tech., 13, 1494-1502.

Van Vaeck, L., K. Van Cauwenberghe and J. Janssens, 1984: 'The gas-particle distribution of organic aerosol constituents: Measurement of the volatilization artifact in hi-vol cascade impactor sampling', Atmos. Environ., 18, 417-430.

Wakeham, S. G., 1982: 'Organic matter from a sediment trap experiment in the equatorial North Atlantic: wax esters, steryl esters, triacylglycerols, and alkyldiacylglycerols', Geochim. Cosmochim. Acta, 46, 2239-2257.

Wakeham, S. G., J. W. Farrington and J. K. Volkman, 1983: 'Fatty acids, wax esters, triacylglycerols and alkyldiacyclglycerols associated with particles collected in sediment traps in the Peru upwelling', In: Advances in Organic Geochemistry 1981, ed. M. Bjoroy, Wiley, New York, 185-197.

Weschler, C. J., 1981: 'Identification of selected organics in the Arctic aerosol', Atmos. Environ., 15, 1365-1369.

Zafiriou, O. C., R. B. Gagosian, E. T. Peltzer, and J. B. Alford, 1985: 'Air-to-Sea fluxes of lipids at Enewetak Atoll', J. Geophys. Res., 90, 2409-2423.

THE MARINE MINERAL AEROSOL

Roy Chester
Department of Oceanography
The University of Liverpool
P.O. Box 147
LIVERPOOL L69 3BX
United Kingdom

1. INTRODUCTION

In 1972 I wrote an article on what was then a relatively new topic,
i.e. the implications of the presence of a dust veil over the oceans
(Chester, 1972), and in 1985 I put together a state of the art review
on our present knowledge of this marine dust veil (Chester, 1985).
During the period between these two articles there has been a great
leap forward in our understanding of how processes occurring in the
atmosphere can affect many aspects of chemical oceanography, and it is
perhaps only now that we are beginning to grasp the full implications
that arise from the presence of this marine dust veil, particularly
with respect to the effects it has on the major oceanic biogeochemical
cycles.

The suspension of solid and liquid material in a gaseous medium is
usually referred to as an aerosol. The marine aerosol consists of a
variety of components which include salts, acids, volcanic debris,
organic combustion products, biological material, extraterrestrial
material and continental weathering products. Various aspects of the
marine aerosol have been considered elsewhere in this volume and the
present treatment is limited to the mineral aerosol (dust) present over
the oceans, although of course this cannot be described in isolation
from other aerosol components. Nonetheless, it is this mineral dust,
which arises from continental weathering, that will form the basis of
the present chapter.

2. THE CONCEPT OF THE MARINE DUST VEIL

Eolian transport is by no means a newly discovered phenomenon. For as
long as mariners have sailed the oceans it has been known that dusts
from the land areas can be transported for long distances out to sea.
Darwin (1846) was one of the first natural scientists to recognize that
the atmosphere represented an important pathway for the transport of
material to the oceans. The presence of atmospherically-transported
eolian dust in marine sediments was reported by Murray and Renard

P. Buat-Ménard (ed.), The Role of Air-Sea Exchange in Geochemical Cycling, 443–476.
© *1986 by D. Reidel Publishing Company.*

(1891) in deposits off the coasts of Africa and Australia. Later,
Radczewski (1939) identified wind blown Saharan desert material in
deep-sea sediments off West Africa, and many other workers have
subsequently found eolian components in oceanic sediments; for a review
of this topic-see Riley and Chester (1971).   Maps constructed almost
fifty years ago showing the distribution of haze (dry aerosol) over the
oceans also hinted at the scale of this atmospheric transport (see
e.g.; McDonald, 1938).   However, the concept that there is a dust veil
over all oceanic areas is relatively new, and much of our knowledge
concerning its nature has been acquired over the past two decades or
so. As with all scientific advances many threads must be drawn
together to gain a full picture of how our knowledge of the topic
increased. However, if one had to identify a 'catalyst' that sparked
off subsequent investigations, the work reported by Delany et al.
(1967) would form a convenient starting point. Prior to this work, most
investigations into the transport of continental dust to marine regions
had been based on evidence acquired from deep-sea sediments, i.e. the
end point in the oceanic transport cycle. Delany et al. (1967),
however, entered the cycle at what may be regarded as the mid-point
between the mobilization of the dust on the land masses and its
deposition at the base of the water column, and collected material
directly from the marine atmosphere itself. The study was carried out
on the island of Barbados which lies in the path of the Atlantic north
east trades several thousand kilometres to the west of the African
coast. By modern standards the collection technique they used was
crude, involving meshes which were suspended in the atmosphere from a
tower on the coast of the island. The basic problem with these meshes
is that they collect only the soil-sized fraction (> 2µm diameter) of
the aerosol and do not effectively retain the small-size fraction (<
0.5µm) which is extremely important from a chemical point of view.
Despite this disadvantage, Delany et al. (1967) were able to demonstrate
that dust originating in the West African deserts could be transported
for thousands of kilometres across the Atlantic by the north east trade
winds. Following this inital work, many other collections of soil-sized
dusts were made over a variety of marine areas, and it became apparent
that the dust veil was a global feature of the marine atmosphere. It
also became apparent that the concentrations of material in this marine
dust veil varied, over as much as several orders of magnitude, at
different locations and on the basis of the early data Aston et al.
(1973) were able to map the distributions of mineral dust over a
considerable section of the World Ocean. Despite the limitations
imposed by the crude collection techniques it was possible, even on the
basis of this limited data, to broadly identify the factors which
controlled the concentration of material in the marine dust veil.
However, in order to fully assess the effects that the transport of
atmospheric material has on oceanic processes it is necessary to have a
quantitative knowledge of the concentration of material in the dust
veil, and this came later from the use of more efficient aerosol
collection systems which were designed to retain the full spectrum of
atmospheric particles, up to 10µm diameter, either as total population

samples (e.g. using filters) or as sized-fractionated samples (e.g. using cascade impactors); for a full description of aerosol collection techniques-see Liss and Slinn (1983).

As aerosol collection techniques improved, and the number of oceanic regions studied increased, scientists began to develop a much more detailed picture of the kinds and distributions of material in the marine dust veil. The following review of the transport of mineral dust to marine areas, and its effects on oceanic processes, is based on post-1972 data. Thus, it will up-date the early ideas and will attempt to summarize our present concept of the marine dust veil.

3.   SOURCES OF MATERIAL TO THE MARINE ATMOSPHERE

The particulate and gaseous material supplied to the atmosphere can originate from either natural or pollutant sources. We are concerned here mainly with one type particle, i.e. mineral dust, but since it is present in a mixture of other components it is necessary to identify these and to do this it is important to distinguish between high temperature and low temperature generation processes. The reason for this is that the form in which elements are present in the atmosphere, i.e. their speciation, can be strongly dependent on the temperature at which they are mobilized from their parent material.

Natural sources of material to the atmosphere include the earth's crust, the ocean surface, the biosphere, outer space and volcanic activity. Anthropogenic sources arise from a wide variety of processes which include fossil fuel burning, mining and the processing of ore materials, waste incineration, the production of chemicals, agricultural utilization and numerous social and industrial activities. The emission source strengths of some of the particle-generating processes are listed in Table I.

Two general points must be made when the data in Table I are related to the distribution of particulate material in the marine atmosphere. (i) The most important direct sources of particulates to the atmosphere involve low temperature processes associated with both the generation of sea-salts from the ocean surface and the weathering mobilization of crustal material. (ii) A number of the particle sources are localized in nature (e.g. industrial and volcanic activity), whereas others (e.g. sea-salt generation and crustal weathering) have a global importance.

It is apparent, therefore, that on a global basis the predominant direct producers of particulate material supplied to the marine atmosphere are the generation of sea-salts and crustal weathering. The sea-salts are recycled from an internal (i.e. oceanic) source, whereas the crustal weathering products are transported to the oceans from an external (i.e. continental) source. As a result, the continental weathering processes represent the single largest external source for directly-generated particulate material found in the marine atmosphere. Various terms are used to describe the material present in the marine aerosol but, in general, dust usually refers to continental weathering products, together with some volcanic material, which are in the coarse particle (> 1μm) size range. For our purposes, therefore, we will consider the marine dust veil to be made up largely of these continental weathering products and in the following sections their distribution in the marine atmosphere, their composition, and some of their effects on the biogeochemistry of the oceans will be reviewed.

Table I. Estimates of the global emissions of particulate material to the atmosphere.

Source	Global production		
	1	2	3
Man-made			
Direct particle production	30		
Particles formed from gases			
Converted sulphates	200		
Others	50		
Total man-made	280		200
Natural			
Direct particle production			
Forest fires	5	36	
Volcanic emissions	25	10	
Vegetation		75	
Crustal weathering- (mineral dust)	250	500	
Sea salt	500	1000	
Particles formed from gases			
Converted sulphates	335		
Others	135		
Total natural	1250		
Overall total	1530		

* Units, 10^{12} g yr^{-1}
1, Prospero et al. (1983); 2, Nriagu (1979); 3, Lantzy and Mackenzie (1979).

4.   THE DISTRIBUTION OF MATERIAL IN THE MARINE DUST VEIL

4.1 Introduction

Continental weathering products, which form the bulk of the mineral aerosol, are mobilized into the atmosphere by wind erosion. Chester

(1985) has pointed out that the process is strongly dependent on the
nature of the surface cover in the source (or catchment) area, which
itself is dependent on the prevailing geological, weathering and
general climatic regimes. In regions having loose surface deposits,
e.g. deserts and arid lands, there is a readily available reservoir of
particulate material which is susceptible to wind erosion.   In
contrast, surface covers of forest, grassland, snow and ice will
considerably reduce the production rates of atmospheric dusts; for a
detailed treatment of the factors which affect dust emissions-see
Gillette (1979). This surface cover control on the amount of material
mobilized into the atmosphere means that there will be an overall
latitudinal dependence imposed on mineral dust concentrations in the
atmosphere, since the arid and desert areas are concentrated into low
and mid latitude belts. The altitude to which particles are lifted
depends on local conditions (e.g. storm outbreak), but for model
calculations scale heights of   3 - 5km are often employed.   Once
material is mobilized into the atmosphere it is transported by the
major wind systems, i.e. the polar easterlies, the mid-latitude
westerlies and the low latitude trades. It is removed from the
atmosphere to the sea surface by dry or wet deposition processes.
These processes are extremely complex and it is only recently that we
have begun to understand them in detail; for a review of this topic-see
Slinn (1983). In order to construct a dust deposition budget for a
specific oceanic area, it is necessary to have data on the
concentration of the material in the air and also on the rates at which
both the wet and dry removal processes operate. Data on the rates of
these removal processes are often difficult to acquire. However, in
order to give some idea of the time scales involved it is generally
assumed that particles are dispersed into the atmosphere to a height of
between 3 - 5 km, and that they have residence times ranging from 5 -
10 days (see e.g.; Prospero et al., 1983). Mineral dusts are dispersed
by the various wind systems but their relatively short residence times
in the air impose a constraint on the distance to which they can be
transported, and the atmospheric circulation patterns tend to inhibit
even mid-latitude/equatorial exchange, let alone inter-hemispheric
exchange across the Inter-Tropical Convergence Zone (ITCZ), because the
various pressure belts work to restrict transfer between circulation
cells. Inter-hemispheric transfer can, however, take place; for
example, when the ITCZ is shifted across the equator, or when it is not
present at all during the development of some monsoon systems (Chester,
1985). Overall, a combination of their relatively large particle size,
their relatively short tropospheric residence times and the constraints
on their transport imposed by atmospheric circulation patterns, tends
to restrict the dispersion of mineral dust raised from the surfaces of
the continents. However, mineral dusts can undergo very long-range
transport and they need not be confined to the latitudinal zone in
which they originated or to the wind system in which they were
initially injected. For example, Uematsu et al. (1983) have
demonstrated that during the spring, high level westerly winds can
carry Asian dust 'pulses' out over the Pacific to lower latitudes where

the dust can be advected from east to west by the trade winds. In
contrast, during the summer the weakening of the westerlies and the
northward expansion of the trades effectively blocks off this transport
path which, together with a decrease in dust storm outbreaks in Asia,
leads to a transition from a high-dust to a low-dust period across much
of the North Pacific-see below. Thus, the transport of mineral dust,
and other atmospheric components, can be affected by seasonal changes
in wind circulation patterns.

It may be concluded, therefore, that a variety of factors affect
the amount of mineral aerosol transported in the marine atmosphere.
These include the nature of the continental surface regime from which
the particles are generated, the strengths and circulation patterns of
the wind system into which they are mobilized, and the rates at which
they are removed from the air. Sufficient data are now available to
make a general assessment of the extent to which these factors combine
to control the concentrations of material in the marine dust veil.
These data have generally been obtained in one of two ways. (i)
Directly from gravimetric measurement of the dusts, or (ii) indirectly
from the concentrations of particulate Al in the air; the latter being
based on the assumption that the element is exclusively associated with
crust-derived aluminosilicates and that the mineral aerosols will have
the same Al concentrations as their parent soils, i.e. an average of 8%
(see e.g.: Uematsu et al., 1983; Chester et al., 1984a).

The post-1972 data presently available on the distribution of
mineral dust over marine areas has been summarized by Chester (1985).
These data are reviewed below in terms of the major oceans, and are
then used to construct a global map of the marine dust veil.

4.2   The Atlantic Ocean And Surrounding Waters

In many ways the Atlantic Ocean,and its surrounding waters, provide a
classic example of the factors which control the distribution of
mineral aerosols over marine regions. This is because the Atlantic is
a relatively narrow ocean with a large latitudinal extent. As a
result, it covers a variety of wind systems, and has a number of
contrasting climatic regimes and particle catchment areas on its
surrounding land masses. It is therefore potentially rewarding to
assess the latitudinal controls on the distribution of the mineral
aerosol relative to a generalized north to south, i.e. Arctic-
Atlantic-Antarctic, transect.

a) The Arctic. Although the Arctic is a remote polar region its
atmosphere contains a higher concentration of non-ice/snow aerosols
than would be expected on the basis of its distance from the major
particle sources. These aerosols have been termed 'Arctic Haze' (see
e.g.; Mitchell, 1957), and contain particulate material from both
anthropogenic emissions (e.g. sulphates) and natural sources (e.g.
crustal dust). Much of the particulate material is well aged and has a
distant origin; probably in Euroasia, central Europe and the eastern
United States. Concentrations of mineral aerosol over the Arctic show

considerable variation, depending on whether the sampling is carried
out in background air or in a haze episode, and the long range
transport of mineral aerosol to the region has been described by Rahn
et al. (1979). For example, during an Asian dust episode over Alaska
Rahn et al. (1981), reported an average mineral dust concentration of
3.3 ug m^{-3} of air in the haze bands.

b) The North Atlantic; 65°N-40°N. In the open-ocean environment
of this region mineral aerosol concentrations are probably < 0.5µg m^{-3}
of air, rising to around 2 µg m^{-3} of air closer to the continental
margins. This sector of the Atlantic underlies the general area of the
North Atlantic westerlies, and although these winds transport material
originating in the US-European 'pollution belt', the surface cover in
these temperate latitudes prevents the large-scale mobilization of
crustal dust into the atmosphere.

c) The North Atlantic; 40°N-0°N. In this region maximum
concentrations of mineral dust are found in the north east trades,
which transport large quantities of crustal solids raised into the
atmosphere from the North African desert belt. Most westward transport
of this dust occurs above the trade wind inversion in the 'harmattan',
in a layer which extends from ∿1500m to ∿5000m in altitude over the West
African coast converging to ∿1500m to ∿3000m over the Caribbean Sea
(Schutz, 1979); for a detailed description of the origin and composition
of these 'Saharan' aerosols-see the volume edited by Morales (1979).
Some of the highest concentrations of mineral aerosol over the World
Ocean have been reported for the Atlantic north east trades, where they
can range up to as much as 10 µg m^{-3} of air in dust storm outbreaks.
However, the mineral aerosol concentrations can be extremely variable,
and large-scale transport often takes place in the form of these dust
storm-mediated 'pulses'. For example, mineral aerosol concentrations at
Sal Island, off the coast of West Africa, during Saharan air out-breaks
varied over the range < 10 to > 180µg m^{-3} of air during July, August
and September, 1974 (Savoie and Prospero, 1977). Chester et al.
(1984b) also reported extreme variations in the atmospheric
concentrations of Al in a latitudinal section through the Atlantic north
east trades off the coast of West Africa, and related them to dust storm
outbreaks. The average mineral aerosol loading derived from their Al
data was 15µg m^{-3} of air, but individual values extended over the range
< 1µg m^{-3} of air to > 700µg m^{-3} of air for ∿12 hour collection periods.
Hoffman et al. (1974) gave data on the atmospheric concentrations of Fe
on an east-west tropical North Atlantic transect which are equivalent to
mineral aerosol loadings of 1µg m^{-3} of air to 750 µg m^{-3} of air;
i.e. in the same range as those reported by Chester et al. (1984b).
Estimates of the Saharan dust input to the North Atlantic are in the
range 60 - 200 x 10^{12} g yr^{-1} (see e.g.: Schutz et al., 1980; Prospero,
1981), although quite clearly much of this takes place in individual
dust 'pulses'. It may be concluded, therefore, that there is a dust
'envelope' over the North Atlantic underlying the north east trades
between ∿ 30°N and ∿5°N (see e.g.; Chester et al., 1979), although the

limits of the area over which the Saharan dust is transported varies
seasonally, with a southerly shift in the dust 'envelope' in winter
(see e.g.; Prospero, 1968).

d) The South Atlantic. Fewer data are available on the concentrations of
mineral dust over the South Atlantic than over the much more heavily
studied North Atlantic. However, those that have been reported reveal
an overall pattern in which there is a dramatic fall in the amount of
mineral dust present in the air across the equator in the south east
trades (average mineral aerosol, $0.5\mu g$ m^{-3} of air) as the influence of
the North African deserts falls off. It is apparent, therefore, that
the desert regions of southern África do not act as massive particle
reservoirs for the supply of mineral aerosol to the Atlantic south east
trades (see e.g.; Chester et al., 1971). Further south, the decrease in
mineral aerosol concentrations continues, and such data as are found in
the literature suggest that loadings in the south Atlantic westerlies
are about one order of magnitude lower than those in the south east
trades (see e.g.; Chester et al., 1984b).

e) The Antarctic. According to Maenhaut et al. (1979) 'no place
on earth is more remote from the major natural and anthropogenic
sources than the snow-covered Antarctic polar plateau'. In this
pristine air the concentrations of mineral dust can fall to values as
low as $0.002\mu g$ m^{-3} of air. However, even here aerosol concentrations
are subject to seasonal variations in the transport of material from
distant sources (Cunningham and Zoller, 1981).

Two very important conclusions can be drawn from the various data
on the distribution of the mineral aerosol over the Atlantic Ocean and
its surrounding waters. (i) There is a strong latitudinal control on
the distribution of the aerosol, with maximum concentrations found at
low latitudes. The degree of this latitudinal control apparently
restricts much of the mineral aerosol to the zone in which it was
mobilized on the continents, i.e. these are the zonal dusts identified
by Chester and Stoner (1974). Because of this latitudinal control it
should be expected that the composition of the Atlantic mineral aerosol
will reflect that of parent surface deposits from which it was
mobilized, and this question is considered in Section 5.2. (ii) Much
of the transport of mineral aerosol from the major dust reservoir in
the Atlantic, i.e. the Sahara Desert, takes place in the form of
discrete 'pulses'. As a result of this, the deposition of mineral
aerosol to the sea surface from the Atlantic north east trades will be
intermittent, and so can cause local non-steady state conditions which
may impose temporal variations on the effects that the dust has on
marine processes.

4.3 The Mediterranean

The Mediterranean Sea, like the Atlantic, is an extremely interesting
marine region from the point of view of the distribution of mineral
aerosols, although for different reasons. The Mediterranean Sea lies in
a relatively narrow latitudinal band, but there is a marked contrast

between the particle-generation regions on its northern and southern
shores (see e.g.; Chester et al., 1981). It is bounded in the north by
nations having a variety of economies, ranging from industrial and semi-
industrial to agricultural, and in the south by the North African desert
belt. There is also volcanic activity in the region. Together, these
surrounding land areas provide a wide variety of particulate material to
the Mediterranean atmosphere. As a consequence, the concentrations of
mineral aerosol exhibit large variations, depending on the origin of the
air mass in which they are transported. For example, 'pulses' of
Saharan dust can invade the northern shores of the Mediterranean (see
e.g. Chester et al., 1984c), and because they are intermittent and
seasonal in nature they make it difficult to estimate an average mineral
dust concentration for the Mediterranean atmosphere. During out-breaks
of Saharan air, Chester et al. (1984c) found that concentrations of
mineral dust can reach values of 50µg m^{-3} of air. However, the average
concentration on an east-west Mediterranean transect (Chester et al.,
1981; Chester et al., 1984c) was 10µg m^{-3} of air, which is reasonably
close to that of 4µg m^{-3} of air given by Prospero (1979).

4.4 The Pacific Ocean

Until recently no large-scale studies had been carried out on the
distribution of the mineral aerosol over the Pacific, although a number
of studies had been made on individual areas and these have been
summarized by Uematsu et al. (1983). To fill this gap, two studies have
recently been carried out as part of SEAREX (Sea-Air Exchange Program),
one in the North and one in the South Pacific.
    a) The North Pacific. In 1981 the SEAREX Asian Dust Study (SADS)
was initiated. The programme involved the collection of aerosol samples
at seven island locations in the North Pacific. The initial results
were reported by Uematsu et al. (1983), and may be summarized as
follows. A distinct seasonal variability was found in the transport of
mineral dust over much of the central North Pacific, with concentrations for the period February to June (mean, 1.3µg m^{-3} of air) being
higher than those for the period July to January (mean, 0.25µg m^{-3} of
air). The seasonal pattern is generally similar to that of the frequency of dust storms on the Asian mainland. Mean concentrations taken
over both the high and low dust periods show that in the spring,
westerlies carry the dust out into the Pacific to lower latitudes where
it is advected east to west by the trade winds; this is the high dust
season. During the summer the weakening of the westerlies and the
northward expansion of the trades blocks this path, leading to a low
dust season. In addition to this seasonal variation, there is also a
geographical variability in the SADS data, with the setting up of a
concentration gradient away from the Asian mainland. Mean concentrations taken over both the high and low dust periods show that
the highest mineral dust concentrations are found in the mid-latitudes
extending from ~50°N to ~20°N (mean, 0.79µg m^{-3} of air for the three
island stations in these latitudes - Shemya, Midway and Oahu).
Concentrations are lower at ~10°N (mean for the Enewetak Island station,

$11^{\circ}N$, $162^{\circ}E$; $0.26\mu g$ m^3 of air), and reach even smaller values around
the equator (mean for the Fanning Island station, $4^{\circ}N$, $159^{\circ}E$;
$0.07\mu g$ m^{-3} of air). Uematsu et al. (1983) estimated that $20 \times 10^{12} g$
of Asian dust are transported each year to the central North Pacific,
and larger quantities are probably deposited over the western North
Pacific closer to the Asian sources, and total deposition to the North
Pacific maybe as high as 40-$50 \times 10^{12} g$ yr^{-1} (R. A. Duce; Pers. Comm.).
The estimate of $20 \times 10^{12} g$ yr^{-1} for the transport of mineral dust to the
central North Pacific is small compared to that of $60 - 200 \times 10^{12} g$ yr^{-1}
carried to the North Atlantic in the north east trades, but still
represents an important input of land-derived solids to this oceanic
region. Further, as Uematsu et al. (1983) have pointed out, non-
biogenic sedimentation rates in the North Pacific are lower than those
in the North Atlantic, thus enhancing the potential significance of the
mineral dust transported through the atmosphere to the central North
Pacific.

b) The South Pacific. Although a number of individual studies have
been carried out on aerosols from the South Pacific (see e.g.; Maenhaut
et al., 1983), relatively little mineral dust data is available for
much of this large oceanic area. In order, therefore, to evaluate
mineral dust loadings in the South Pacific, the SEAREX South Pacific
Aerosol Network (SPAN) was set-up in 1983, and a number of general
conclusions can be drawn from the data which have emerged from the
programme to date (see Uematsu et al., 1985). (i) The highest
concentrations of mineral aerosol in the South Pacific were found at
Norfolk ·Island (mean $0.79\mu g$ m^{-3} of air). This station is the closest
in the network to the Australian mainland, and the relatively high
loadings probably reflect an input of continental dust from this source
and from higher latitudes. (ii) Intermediate mineral aerosol
concentrations were reported near the equator in the latitudinal band
extending from $0^{\circ}S$ to $\sim 10^{\circ}S$ (mean, $0.23\mu g$ m^{-3} of air).
(iii) The lowest mineral aerosol concentrations were found further
south in the latitudinal band extending from $\sim 14^{\circ}S$ to $\sim 25^{\circ}S$ (mean,
$0.01\mu g$ m^{-3} of air). The lowest concentration reported for any of the
South Pacific stations was at Samoa (mean, $0.018\mu g$ m^{-3} of air), and the
range ($< 0.003 - 0.14\mu g$ m^{-3} of air) included values as low as those
found over Antarctica. The SPAN data are at present being evaluated in
more detail (Uematsu; Pers. Comm).

4.5 The Indian Ocean

The literature contains relatively few data on the concentrations of
mineral aerosol over the open-ocean regions of the Indian Ocean. It is
to be expected that the desert and arid land areas surrounding the
northern Indian Ocean should be highly productive mineral aerosol
sources, and this has been confirmed in several studies. For example,
Chester et al. (1984a) demonstrated that the atmosphere is an important
pathway for the transport of crustal material to the northern Arabian
Sea from the arid regions of the Iran-Makran and Rajasthan. The

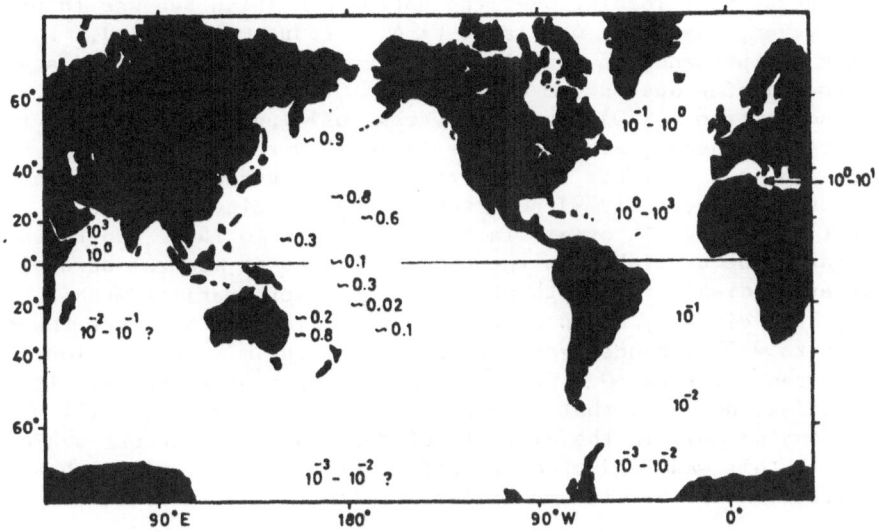

Figure 1. Tropospheric mineral dust concentrations over the World
Ocean. With the exception of the values given for the southern
Indian Ocean, the concentrations are based on filter - collected
data reported since 1972. Some of the data included were originally
presented as concentrations of mineral dust; the remainder have been
obtained from the atmospheric concentrations of Al-see text. For
most regions order of magnitude ranges are given for the mineral
dust concentrations, but mean values are illustrated for the Pacific
Ocean where long term collections have been made on island stations.
The units are in $\mu g\ m^{-3}$ of air.

average mineral dust loading over the northern Arabian Sea was 16 μg m⁻³
of air; however, the range was large ($<$ 4 - 253μg m⁻³ of air),
indicating the presence of dust 'pulses' transported from the adjacent
desert areas during dust storm out-breaks. Over the Bay of Bengal,
Golberg and Griffin (1970) found an average of ~7μg m⁻³ of air for the
total, largely mineral, aerosol; this is still a relatively high
concentration, and reflects inputs from the surrounding land masses.
Further south Prospero (1979) reported an average mineral aerosol
concentration of 5μg m⁻³ of air in the region 7°N to 15°S. No recent
data appear to be available on the concentrations of mineral aerosols in
the southern Indian Ocean south of ~15°S. However, earlier mesh-
collected data given by Aston et al. (1973) indicates that over the area
from~20°S to ~30°S, concentrations range from 1μg m⁻³ of air close
to the African coast to~0.01μg m⁻³ of air in open-ocean areas. It is
to be expected, however, that mineral aerosol concentrations will
increase considerably in the vicinity of the Australian desert sources;
for example, this was predicted by Griffin et al. (1968) on the basis of
the distribution of the clay mineral kaolinite in Indian Ocean deep-sea
sediments.

4.6 Summary

On the basis of the various data discussed above we can now draw a
number of overall conclusions regarding the distribution of the mineral
aerosol over the oceans.
    a) Mineral dust originates on the land surfaces, with the greatest
quantities being generated in the Northern Hemisphere which contains
more of the land masses than does the Southern Hemisphere. This
mineral dust is dispersed over both continental and marine surfaces via
the global atmospheric circulation patterns.
    b) The mineral aerosol can undergo long-range transport to even
the most remote regions, with the result that there is a dust veil
present over all marine areas.
    c) However, as a result of differences in the relative
efficiencies of the processes involved in the mobilization, transport
and removal of the mineral aerosol, the concentrations of material in
the marine dust veil show very large variations; ranging from 10⁻³ μg m⁻³
of air in remote regions, such as the South Pacific and Antarctica,
to $<$ 10³ μg m⁻³ of air off the major deserts, e.g. in the Atlantic north
east trades adjacent to the coast of West Africa.
    d) The most important single factor controlling the amount of
mineral aerosol mobilized into the atmosphere and transported to marine
regions is the nature of the surface deposits in the continental
particle-generation region, with desert areas being the predominant
source reservoirs. Because desert and arid regions tend to be
concentrated into specific belts, one in the Northern and one in the
Southern Hemisphere, the overall net effect is to impose a general
latitudinal control on the concentrations of material in the marine
dust veil. In the Atlantic the highest concentrations of dust are
found at low latitudes (0°N-35°N), reflecting the influence of the North

African deserts to the west. In contrast, in the Pacific the highest
dust concentrations are found at mid latitudes $(20^{0}N - 50^{0}N)$, as a
result of transport from the Asian deserts to the north of the ocean.

e) A significant feature of the supply of material to the marine
atmosphere from the major deserts is that the dust is often carried in
discrete 'pulses' which are intermittent in nature.

The distribution of mineral aerosol over the World Ocean is
illustrated in Figure 1.

5.   THE COMPOSITION OF MATERIAL IN THE MARINE DUST VEIL

5.1 Introduction

The composition of the mineral aerosol is controlled initially by that
of the parent, or source, material, but it can be modified by
fractionation processes which occur across the soil/air interface and
can be further altered in the atmosphere itself. Over recent years a
number of studies have been carried out on both the mineral and
chemical compositions of the material forming the marine dust veil, and
the two topics are considered individually below.

5.2 The Mineral Composition Of The Marine Dust Veil

A wide variety of crust-derived minerals have been identified in
material collected from the marine dust veil. These include the clay
minerals, palygorskite, quartz, feldspars, micas, calcite, dolomite,
amphiboles, pyroxenes, anatase, rutile, haematite, goethite, talc,
serpentine and a number of ferrimagnetic minerals (see e.g.: Delany et
al., 1967; Parkin et al., 1970; Goldberg and Griffin, 1970; Chester et
al., 1971, 1972, 1977, 1984a,c). In addition to the soil-derived
minerals, volcanic products can be ejected into the atmosphere, but
only the finer material (volcanic ash) is transported on a large scale.
This volcanic ash consists of glass shards, together with small amounts
of mineral and crystal fragments. On a global scale, clay minerals,
quartz and feldspars are quantitatively the most important mineral
components of the marine dust veil.

The mineralogy of the surface deposits from which the crust-
derived aerosols originate varies considerably from one area to
another; depending, among other factors, on the surface geology and
overall weathering intensity in a specific particle-generation regime.
However, certain minerals, or more usually suites of minerals, can be
characteristic of particular soil regimes, and as a result different
aerosol populations can have individual mineral signatures. These
minerals therefore have a potential use as tracers for the dispersion
of crust-derived solids within the marine dust veil, and this can be
illustrated with respect to the clay minerals.

The principal clay mineral groups in soils, sediments and
atmospheric dusts are illite, kaolinite, chlorite and montmorillonite.
Studies carried out on the < 2μm fractions of deep-sea sediments have
shown that there are a number of <u>gross</u> trends in the distributions of

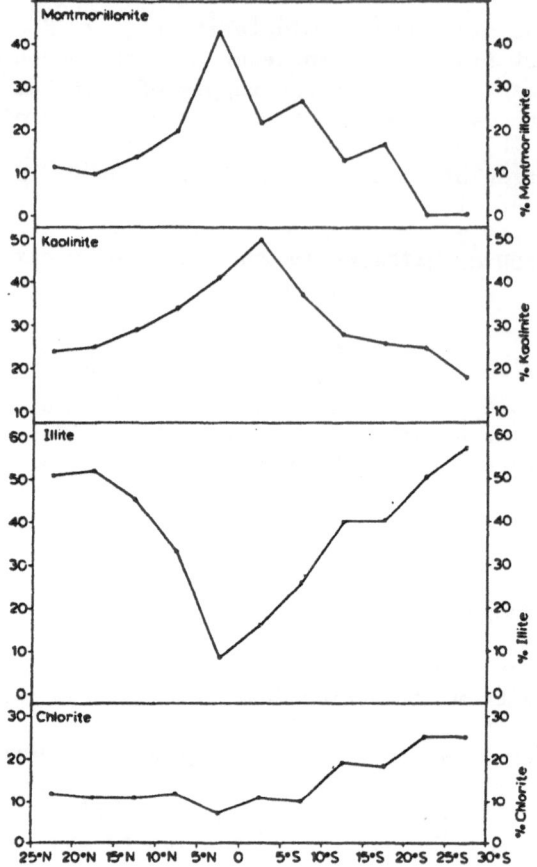

Figure 2.   Latitudinal patterns in the
distributions of the major clay minerals
in atmospheric dusts collected along
the eastern margins of the Atlantic
Ocean; values are averages for each
5° of latitude (from Chester et al.,
1972).

the clay minerals in these deposits (see e.g.: Biscaye, 1965; Griffin
et al., 1968). On the basis of these trends, Griffin et al. (1968)
drew the following conclusions regarding the origins of the clay
minerals in deep-sea sediments. (1) Chlorite is characteristic of high
latidude sediments. In these regions it is supplied from chlorite-rich
polar rocks from which it is released mainly by mechanical weathering.
(2) Kaolinite has its highest concentrations in low latitude sediments.
This mineral can be formed under a variety of weathering conditions,
but it is characteristic of intense tropical and desert weathering
regimes. (3) Illite is a ubiquitious clay mineral and is not
characteristic of any particular weathering regime. It has a mainly
land-derived origin in deep-sea sediments and has its highest
concentrations in deposits of the northern hemisphere, where most of
the world's land mass is found. However, because of its wide spread
occurrence in soils, its overall distribution in the sediments is
controlled largely by the extent to which it is diluted by clays having
more specific sources. (4) Montmorillonite, or smectite, can originate
in several ways but it has its highest concentrations in deep-sea
sediments in regions where there is volcanic activity and a relatively
small input of land-derived solids, e.g. in the South Pacific. Griffin
et al. (1968) concluded, therefore, that much of the montmorillonite in
deep-sea sediments did not have a continental source, but was produced
from the alteration of volcanic material by in situ processes in the
sediment column. Thus, the deep-sea sediment distributions of those
clay minerals which are predominantly land-derived, i.e. chlorite,
kaolinite, and illite, provide information on their dispersal by
fluvial and atmospheric transport from the continents. Since most
crustal atmospheric dusts are generated directly from surface deposits
they represent an early stage in the global sedimentary transport
cycle, and it is therefore of interest to establish whether or not the
clay minerals in the dusts mirror their origins in soils and so have
distinctive clay signatures. In this respect, Chester et al. (1972)
gave data on the distribution patterns of the four major clay minerals
on a north-south Atlantic transect extending from $\sim 25^{\circ}$N to $\sim 30^{\circ}$S. The
results are illustrated in Figure 2, and two principal conclusions can
be drawn from the data. (i) There are strong latitudinal trends in the
distributions of kaolinite and illite in the Atlantic mineral aerosol;
with the concentrations of kaolinite increasing, and those of illite
decreasing, towards low latitudes. A less well-defined trend is found
in the distribution of chlorite, with increasing concentrations towards
higher latitudes in the South Atlantic. These trends are entirely
consistant with the overall distributions of the clays in the adjacent
continental borderland surface soils (see e.g.; Griffin et al., 1968).
(ii) The distribution of montmorillonite in the dusts is rather
complicated. The highest concentrations are found around, and south
of, the equator, which does not appear to follow the patterns found
either in the surface deposits of the surrounding land masses or in the
underlying deep-sea sediments.
        In a more recent study, Chester et al. (1984c) described an
event in which incursions of Saharan dust were identified over the

Tyrrhenian Sea in the Mediterranean. Geostrophic back trajectories
were used to identify potential aerosol particle generation regions and
two distinct dust populations were distinguished; (a) a northern
'European' population, and (b) a Saharan population. The average clay
mineralogies of the two populations are given in Table II.

It is apparent from the data in Table II that there are
differences in the clay mineralogies of the two Tyrrhenian Sea aerosol
populations. The samples originating in northern Europe are
characterised by having relatively high concentrations of illite
(average, 67%) and roughly similar proportions of kaolinite (average
15%) and chlorite (average, 18%). The Saharan samples still have a
relatively high content of illite (average, 56%), but differ from the
northern European population in that the concentration of kaolinite
(average, 35%) considerably exceeds that of chlorite (average, 7%).
Previously, Chester et al. (1977) had identified a 'pure Saharan'
population in the eastern Mediterranean, and the average clay mineral
content of this is given in Table II. Despite the fact that two

Table II. Clay mineralogy of mineral aerosols from the
Mediterranean Sea.

Sample population	I	C	Clay mineral content[1] K	M
Tyrrhenian Sea: average, European population[1]	67	18	15	N.D.
Tyrrhenian Sea: average, Saharan population[2] ;	56	7	35	2
Eastern Mediterranean: average, 'pure' Saharan population[3] ;	40	8	50	5
Tyrrhenian Sea: Saharan end-member sample[2]	34	4	59	3

1 I = illite, C = chlorite, K = Kaolinite, M = Montmorillonite.
The individual clay mineral percentages are expressed in
terms of a 100% clay mineral sample; i.e. (illite +
chlorite + kaolinite + montmorillonite) = 100%. N.D., not
detected.
2   Data from Chester et al. (1984c).
3   Data from Chester et al. (1977)

individual populations could be distinguished, the dusts collected over
the Tyrrenhian Sea are still, at least to some extent, a mixture of
material from Saharan and European sources. However, by using a
combination of mineral and chemical criteria, Chester et al. (1984c)
were able to identify a Saharan end-member sample within the Tyrrhenian
Sea dusts. The clay mineral content of this end-member sample is also
listed in Table II, from which it can be seen that it has a clay mineral
signature which resembles that of the 'pure' Saharan dust. Similar
results for the clay mineralogy of eolian dusts over the Central
Mediterranean were reported by Tomadin et al. (1984), who found that
material collected in southerly winds had more kaolinte and less
chlorite than that in samples taken from northerly winds.

5.2.1 Summary. Mineral aerosols, as defined here, are mobilized
directly from the earth's crust and the spectra of individual minerals
which they contain will therefore be determined by the regions of
lithogenesis, defined on the basis of their weathering regimes, on the
land masses. Some minerals are characteristic of particular weathering
regimes, and so have a potential use as tracers for the dispersion of
mineral aerosols about the planet. This was illustrated with respect
to the clay minerals, and it was shown that these can be used to
characterize certain mineral aerosol populations. Other minerals which
have been used for this purpose include quartz, dolomite, palygorskite
and talc (see e.g.: Delany et al., 1967; Goldberg and Griffin, 1970;
Chester et al., 1972; Buat-Menard et al., 1982).

5.3 Chemical Characteristics Of The Mineral Aerosol.

In this section some aspects of the chemistry of mineral dusts will be
discussed. However, the marine aerosol consists of an intimate mixture
of material from a variety of sources and for this reason the chemistry
of the mineral component cannot be considered in isolation, but rather
must be related to that of the total aerosol population.The present
treatment is restricted to the inorganic chemistry of crustal dusts;
organic aspects of aerosols are discussed elsewhere in this volume and
for a detailed review of the topic-see Duce et al., 1983a).

5.3.1 The identification of the crustal source. One of the most widely
used methods of relating an element in an aerosol to its source is by
employing an indicator which is derived predominantly from one specific
source. In order to assess the enrichment of the element relative to
this source, the 'excess' (or non-source) fraction can be defined in
terms of an enrichment factor (EF). Aluminium is normally used as the
indicator element for the crustal source, and although there are a
number of problems inherent in the selection of a composition for the
source material (see e.g.; Chester, 1985) most authors have used that
of average crustal rocks. The crustal enrichment of an element in
aerosol is then calculated according to the following equation:

$$EF_{crust} = (E/Al)_{air}/(E/Al)_{crust},$$

in which $(E/Al)air$ are the concentrations of an element and Al in the aerosol and $(E/Al)crust$ are their concentrations in average crustal rocks. However, $EFcrust$ values should only be treated as order of magnitude indicators of the crustal source. Thus, values close to unity are taken as an indication that an element has a mainly crustal source, and those > 10 are considered to indicate that a substantial portion of the element has a non-crustal origin. By using crustal EFs, we therefore have a means of making an assessment of the extent to which an element in an aerosol is crust-dominated. Rahn (1976) has tabulated the $EFcrust$ values for 70 elements in over a hundred samples from the World Aerosol, and a summary of his data is given in Table III.

Over half of the elements originally listed by Rahn (1976) have average $EFcrust$ values which are < 10^1 ; these elements are present in the World Aerosol in roughly crustal proportions, and are termed the crustal, or non-enriched, elements (NEE). Nonetheless, it must be remembered that EFs are only order of magnitude indicators of enrichments and so elements can still be 'enriched' at values < 10^1 ; for this reason, lower limit $EFcrust$ values should perhaps be set for truly crustal dusts-see e.g. Table V. The remaining elements have $EFcrust$ values in the range 10^1 - > 10^3 ; these are referred to as the anomalously enriched elements (AEE), and are not usually crust-dominated. This is a very important distinction, but for some elements

Table III. Average $EFcrust$ values for some elements in the World Aerosol (after Rahn, 1976).

$EFcrust$	Element
> 10^3	N,C,Se,Br,Cd,Se,Pb,Sb.
10^2 - 10^3	Sn,Ag,Cl,Au,S,Hg,I,As,Zn,Bi,Mo,Cu.
10^1 - 10^2	In,H,Ge,Ni,B,W,V,Cs.
< 10^1	Cr,Li,Ba,Mn,Rb,Be,Ca,Mg,Ga,Fe,Sr,Zr,Sc,Si.

it must be seriously qualified in one respect. The NEE will retain their overall crustal character in all, or at least almost all, aerosol populations. However, the degree to which an AEE is actually enriched can vary considerably from one aerosol population to another. This can be illustrated with respect to the distributions of Fe and Cu in the Atlantic aerosol-see Table IV.

Table IV. Average EF*crust* values for Fe and Cu in
the Atlantic aerosol (after Chester, 1985).

| Oceanic | | EF*crust* | |
region	Fe		Cu
North Atlantic;			
westerlies	1.2		30
north east trades	1.0		1.2
South Atlantic;			
south east trades	1.2		44
westerlies	1.8		225

The aerosols were collected on a north-south Atlantic transect and
Fe, which is a NEE, has average EF*crust* values which are $< 10^1$ in all
the sample groups. Cu is an AEE in the World Aerosol and average Cu
EF*crust* values in most of the Atlantic sample groups are, in fact, $>$
10^1. However, the Cu EF*crust* values vary considerably and reach a
minimum of around unity in aerosols from the north east trades. These
samples were collected off the coast of West Africa where 'pulses' of
Saharan dust-laden air are common-see Section 3.2. The EF*crust* values
of a range of elements in a Saharan-dominated dust population are
listed in Table V.
    It can be seen from the data in Table V that the average EF*crust*
values for all the elements listed are $< 10^1$ in the Saharan-dominated
aerosol population. This illustrates an important point with respect
to the mineral aerosol, i.e. when crustal components are injected into
atmosphere in sufficient quantities they can dominate the total aerosol
population, with the result that elements which are normally enriched
now behave in a non-enriched manner.

5.3.2 The particle size association of elements in the marine aerosol.
Low temperature, mechanically-generated, crustal (and sea-salt)
particles are usually in the coarse range, i.e. they have diameters $>$
1μm. However, during high temperature processes, such as smelting,
fossil fuel burning and volcanic activity, some trace metals can be
released into the atmosphere in the vapour phase from which they can
undergo condensation and gas-to-particle conversion and can be adsorbed
onto the surfaces of ambient aerosol particles. This process is size-
dependent because, in general, smaller particles will have a relatively
large surface area-volume ratio (Rahn, 1976). As a consequence, many of
the elements released during high temperature processes are associated
with particles having a sub-micron size. The particle size spectrum of
an element in an aerosol should therefore offer an insight into its
source, and this has been demonstrated by a number of authors over the
past few years. Data reported by Duce et al. (1983b) for aerosols

Table V.   EF*crust* values of some elements in a series of seven Saharan
-dominated crustal aerosols collected by a filter technique from the
Atlantic north east trades off the coast of West Africa (data from
Murphy, 1985).

Element	EF*crust* Arithmetic average	range
Al	1.0	Indicator element
Fe	1.0	0.7 - 2.2
Mn	1.0	0.9 - 1.4
Ti	2.7	1.6 - 4.7
Cr	1.4	1.0 - 2.7
V	1.6	1.1 - 2.8
Co	1.1	0.7 - 2.1
Ni	1.3	0.7 - 2.5
Cu	1.2	0.8 - 2.2
Zn	3.8	1.7 - 8.1
Pb	9.1	3.5 - 23
Cd	9.4	3.2 - 17

collected at Enewetak, (North Pacific) can be used to illustrate this
point.   The most outstanding feature in the size distributions of the
crustal elements was that the maximum concentrations were found on
particles having a mass median diameter (MMD) of $\sim$3µm.   Further, more
than 75 of the total mass of Al,Mn,Fe and Sc was found on particles
having MMDs in the range 0.92 - 3 µm.   The size distributions of these
elements was similar in both the dry (high dust) and wet (low dust)
seasons, and their EF*crust* values,indicated that they were present in
crustal proportions in all size fractions.
      Relative to the crustal elements, the sea-salt-associated elements
(Na,Ca,Mg,K) and enriched elements (Cd,Zn,Pb,Sb,Cu)had characteristical-
ly different particle size spectrum; sea-salt elements being mainly on
particles with MMDs in the range $\sim$3 -$\sim$7µm, and enriched elements being
largely on submicron particles (<0.5 µm).

5.3.3 The partitioning of elements in the marine aerosol.   The partitioning of elements among the components of an aerosol, i.e. their
speciation, can reveal information on their sources.   Chester et al.
(1986) applied a sequential leaching scheme to investigate the
partitioning of a series of elements among a number of operationally-
defined fractions in crust-dominated marine aerosols.   For this, the
authors used a set of soil-sized mineral dusts collected from the
Atlantic north east trades during Saharan outbreaks as an example of a
crustal aerosol end-member (CAEM).   The partitioning signatures of the

elements are illustrated in Figure 3, in which they are expressed as
the percentages of the total aerosol concentration in each of the host
associations. It is evident that in this CAEM~80% of the total
concentrations of Al,Fe,Cr,Cu and Zn are held in the residual
fractions. In contrast, only ~30% of the Pb, and ~10% of the Cd, is
residual in character.

Overall, the refractory nature of the elements in the CAEM
decreases in the order; Al, Fe (> 90%) > Cu,Zn (~80%) >> Pb (~25% >>
Cd(~10%). The manner in which elements are partitioned among the host
associations should have a major influence on the fate of the elements
following the deposition of the aerosols to the sea surface. In
general, elements in the refractory fraction will be the most
potentially immobile, and those in the loosely-held association the most
potentially mobile (Chester et al., 1985). This is important since the
extent to which an element is refractory depends, at least for some
elements, on its source. For example, in the low temperature-generated
CAEM around 80% of the total Cu and Zn is refractory in nature; however,
Lum et al. (1982) demonstrated that in high temperature-generated urban
aerosols~ 37% of the total Cu and~ 67% of the total Zn were in loosely-
held associations. Partitioning differences of this kind play an
important role in determing the extent to which an element is soluble in
sea water following the deposition of either crust-dominated or
pollutant-dominated aerosols to the sea surface (see e.g.; Hodge et al.,
1978).

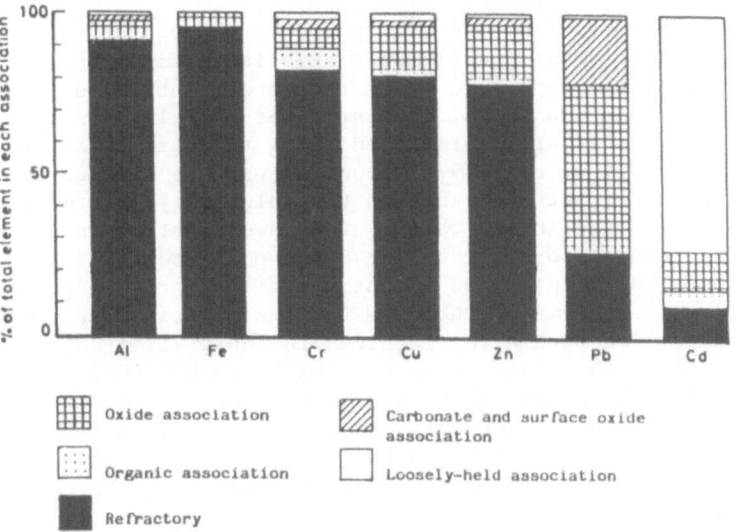

Figure 3. The partitioning of elements in crust-dominated aerosols.
The samples were collected from the Atlantic north east trades during
outbreaks of Saharan dust. The associations refer to individual stages
released during a sequential leaching attack on the dusts (from Chester
et al., 1985).

5.3.4 <u>Summary</u>. The chemical characteristics of the mineral aerosol
end-member component of the marine dust veil can be summarized as
follows.

a) The elements in it are present in crustal proportions (i.e.
their EF*crust* values are generally < 10^1, and often approach unity),
and have most of their total mass associated with particles having MMDs
in the range 1 - 3μm.

b) A relatively large proportion (~80%) of the total
concentrations of some elements, e.g. Al,Fe,Cr,Cu,Zn, is held in the
refractory fractions of the mineral aerosol.

c) When 'pulses' of crustal material are injected into the
atmosphere, e.g. during dust storm outbreaks, the mineral component can
dominate the chemical composition of the total aerosol population,
resulting in elements which are normally enriched in the atmosphere
behaving in a non-enriched manner.

6. THE INFLUENCE OF THE MARINE DUST VEIL ON OCEANIC CYCLES

Atmospherically-transported material deposited at the sea surface can
affect biogeochemical processes which occur in both the water column
and the underlying sediments. Two examples of these effects are
described in the following sections.

6.1 The Water Column.

Aerosols are removed from the atmosphere by wet and dry deposition, and
the subsequent fates of the elements associated with them depends to a
large extent on the degree to which they are solubilized in sea water.
This, in turn, depends on the speciation of the elements, i.e. the
manner in which they are partitioned among the various aerosol
components. Hodge et al. (1978) produced what is perhaps the most
extensive data set yet available on the solubility of atmospherically-
transported elements in sea water. The investigation involved two
contrasting aerosol populations; an anthropogenically-rich population
from Southern California, and a continental dust-rich population from
Baja, California. Some of the data for the solubilities of elements
from the dust-rich population is given in Table VI, from it which it
can be seen that they range from < 1% (Al) to ~80% (Cd).

Clearly, therefore, there are differences in the extent to which
the elements in mineral aerosols are soluble in sea water, and this
will control the degree to which the aerosols affect the dissolved/
particulate distributions of the elements following their deposition to
surface waters.

6.1.1 Particulate elements. There is a strong latitudinal control on
the distribution of mineral aerosols in the Atlantic atmosphere (see
Figure 1) and this is reflected in the surface water distributions of
some particulate elements. In this context, Hoffman et al. (1974)
showed that Fe, V and Mn were enriched in particulate matter from the
micro-layer in the tropical Atlantic off the coast of West Africa, and

Table VI. The sea water solubilities and EF*crust* values of some elements in a dust-rich aerosol population (data after Hodge et al., 1978).

Element	Dust-rich aerosol population (Baja, California)	
	Average sea water solubility, (% total conc.)	Average EF*crust*
Al	<1	1.0
Fe	1.1	0.99
Cr	10	0.99
Cd	80	31
Mn	42	1.2
Pb	14	439
Zn	30	8.3
Cu	16	11

concluded that this was the result of the deposition of aerosols from Saharan 'plumes'. According to Hodge et al. (1978) the Al in mineral aerosols is relatively insoluble in sea water (see Table VI), and it will therefore serve as an example of how atmospheric deposition can affect the distribution of particulate elements in surface waters. Krishnaswami and Sarin (1976) carried out a survey of the concentrations of particulate Al (Alp) in Atlantic surface waters and reported elevated values in polar regions, which they suggested could have resulted from the influx of rock material from glaciers, and off West Africa, which they attributed to an eolian input from the north east trades; here, surface water Alp concentrations reached values as high as 768 ng kg^{-1}, which may be compared to those of < 50 ng kg^{-1} in the open-ocean South Atlantic. A similar trend was also identified by Chester (1982) in Atlantic surface waters; this author showed that although eolian inputs were highest in the area underlying the north east trades, they made a considerable contribution to surface water Alp concentrations even in the open-ocean South Atlantic where mineral aerosol concentrations are relatively low (see Figure 1). Further evidence for the importance of atmospheric deposition on the distribution of Alp in open-ocean waters has been provided by Uematsu et al. (1986), who demonstrated that enhanced Alp concentrations in surface waters of the northwestern Pacific could be correlated with high atmospheric dust concentrations in the area. It is apparent, therefore, that the deposition of mineral

aerosols can provide an important source for Alp in oceanic surface
waters, and the supply of this particulate element to the underlying
sediments is considered in Section **6**.2.

6.1.2 Dissolved elements.   From the data given by Hodge et al. (1978)
it would appear that of the elements they studied Cd and Mn are the
most sea water-soluble from crustal aerosols.   However, even in the
crust-dominated population the average Cd EF*crust* value was ∿30,
indicating that in fact much of the element had a non-crustal source.
For Mn, however, the average EF crust value was around unity, i.e. the
element was almost exclusively crustal in origin.   Thus, if the results
reported by Hodge et al. (1978), which indicate a Mn solubility of 40%-
50% for both crustal and polluted samples, are applied globally they
can be used to estimate the input of dissolved Mn (Mnd) to sea water
from mineral dusts.   Further, at this relatively high solubility,
atmospherically-transported fingerprints of Mnd should be found in the
surface waters of those marine regions which receive large injections,
or 'pulses', of mineral aerosol.   Such 'fingerprints' have, in fact,
been identified in a recent study reported by Kremling (1985).   This
author carried out a survey of the distributions of a number of
elements in unfiltered, acidified, sea waters collected on a north-
south Atlantic transect.   The meridional distribution of total
dissolvable Mn (TDM) showed a characteristic and clear pattern, with
elevated conentrations in the region ∿10°N to ∿30°N - see Figure 4.
This is the region underlying the north east trades which transport
large quantities of Saharan dust westwards across the Atlantic, and
Kremling (1985) concluded that the elevated TDM concentrations in the
surface waters had resulted from an atmospheric input of particulate Mn
which he assumed, following Hodge et al. (1978), was ∿50% soluble.   It
was shown in Section 4.2 that the transport of Saharan dust over the
North Atlantic is extremely variable, and this was reflected in
Kremling's data which showed differences in the surface water TDM
concentrations on two cruises carried out at different times - see
Figure 4.   It is apparent, therefore, that dust 'pulses' can strongly
affect the distribution of dissolved Mn in the surface waters
underlying the Atlantic north east trades.
      Manganese is an element which is relatively soluble from mineral
aerosols and it was shown above that material deposited from the marine
dust veil can clearly influence its distribution is some surface
waters.   Other elements are less soluble from crustal aerosols; for
example, < 1% of the total concentration of Al in this type of aerosol
is soluble in sea water (Hodge et al., 1978).   However, Kremling (1985)
has suggested that because of the relatively large inputs of crustal
material transported by the north east trades in the Atlantic, an
important dissolved atmospheric Al flux cannot be discounted even at
this very low aerosol solubility; see also, Hydes (1979) who concluded
that the dissolution of Al from atmospherically-transported particles
led to its enhanced concentration in surface waters in the North
Atlantic.   Short term variability in the deposition of atmospherically-
transported material will affect the extent to which aerosols influence

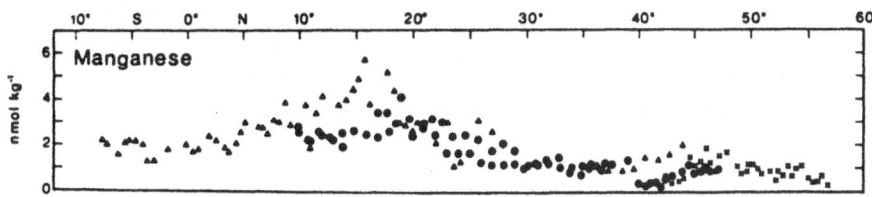

Figure 4.   The meridional distribution of total dissolvable Mn in North
Atlantic surface waters.   Enhanced concentrations between 10°N and 20°N
result from atmospheric transport.   Data points over these latitudes
are from two cruises and reflect short term variations in the input of
mineral dusts (from Kremling, 1985).

surface water concentrations of dissolved Al in the same way as they
affect those of dissolved Mn.   For example, Uematsu et al.   (1985a)
have shown that atmospherically-transported Al in surface waters of the
northwestern Pacific varied by a factor of eight over a two month
period in the summer of 1978.

6.2. The Sediment Column.

Short term variability in the deposition of atmospherically-transported
material to the sea surface, such as that arising from intermittent
dust 'pulses', can be identified in the distributions of some elements
in open-ocean waters.   However, sediments underlying the water column
represent the integrated deposition of material over relatively long
time periods, with the result that any short term variability in
deposition to the sea surface will be smoothed out.   Because of this,
deep-sea sediments offer an oceanic reservoir in which long term trends
in the input of atmospherically-transported material may be 'fingerprinted'.
The 'fingerprints' have often been used in reconstructing paleo-climates.
     Many authors have highlighted the importance of the atmosphere as
a pathway for the transport of material to deep-sea sediments.   Oceanic
regions in which it has been suggested that atmospheric transport plays
a significant role in sedimentation include the following; the North
Atlantic (see e.g.: Radczewski, 1939; Goldberg and Griffin, 1964;
Delany et al., 1967; Prospero, 1979; Chester et al., 1979; Chester,
1981; the North Pacific (see e.g.: Rex and Goldberg, 1958; Duce et al.,
1980; Uematsu et al., 1985a and b); the northern Indian Ocean and
Arabian Sea (see e.g.: Goldberg and Griffin, 1970; Prospero, 1979;
Venkatarathnam et al., 1981; Chester et al., 1984a); and the
Mediterranean Sea (see e.g.: Venkatarathnam and Ryan, 1971; Chester et
al., 1977; Tomadin et al., 1984).
     Qualitative assessments of the role played by the atmosphere in
the transport of material to deep-sea sediments have often been made
using mineral tracers, particularly the clay minerals and quartz.   For
example, over the region of the North Atlantic lying between 05°N and

35°N data provided by Griffin et al. (1968), Chester et al. (1972) and
Behairy et al. (1975) showed that there is a general agreement between
the clay mineral signatures of atmospheric dusts, surface water
particulates and underlying deep-sea sediments-see Table VII.

Table VII. The average distribution of clay minerals in
atmospheric dusts, surface water particulates and deep-sea
sediments from the eastern North Atlantic over the region
05°N to 35°N (after Chester 1981).

Material	Clay mineralogy[1]		
	Montmorillonite	Illite	(Kaolinite + chlorite)
Atmospheric dusts[a]	14	45	41
Surface sea water particulates[b]	15	40	45
Sediments[c]	16	55	30

1 The individual clay mineral percentages are
  expressed in terms of a 100% mineral sample; i.e.
  (montmorillonite + illite + kaolinite + chlorite) = 100%.
a Data from Chester et al. (1972).
b Data from Behairy et al. (1975).
c Data from Griffin et al. (1968).

On the basis of evidence such as that given in Table VII, and the
presence of off-shore concentration gradients in the distributions of
quartz and the clay minerals in surface sediments, Chester et al.
(1979) were able to identify a dust 'envelope' in the tropical North
Atlantic under which atmospheric transport plays a dominant role in
the supply of land-derived material to the ocean floor-see Figure 5.
    Attempts have also been undertaken to make quantitative estimates
of the input of mineral dust to deep-sea sediments. Some of these have
been based on the concentrations of mineral dust in the atmosphere and
its deposition rate to the ocean surface; for example, Delany et al.
(1967) estimated that in the area underlying the Atlantic north east
trades the land-derived fractions of the deep-sea sediments deposited
east of, and on, the Mid-Atlantic Ridge resulted wholly from the
deposition of atmospheric dusts. Dust deposition models for the same
region have also been evaluated by other authors (see e.g.: Prospero,
1979; Jaenicke, 1979). Although the input of mineral aerosols has been

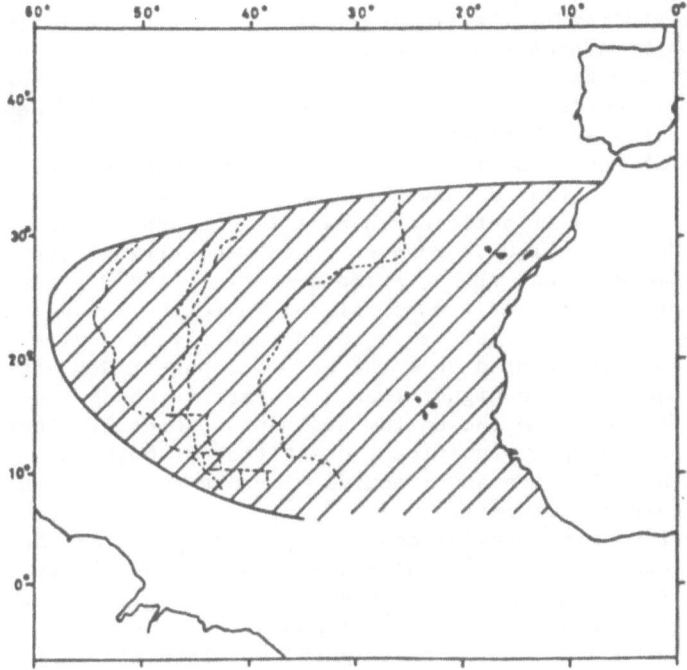

Figure 5.   The 'dust envelope' in North Atlantic deep-sea sediments.
The shaded area indicates the region over which transport in the north
east trades dominates the supply of land-derived material to the
equatorial North Atlantic.   The limits of the envelope were determined
on the basis of the concentrations and atmospheric deposition of the
mineral aerosol, and on a comparision of its mineralogy, biology and
isotopic composition with that of the underlying sediments.   The
positions of the Mid-Atlantic Ridge and ridge flanks within the
envelope are indicated by the dotted lines.   (After Chester et al.
1979).

most extensively studied in the tropical North Atlantic, tentative dust
budgets have also been proposed for other marine areas; these include
the Mediterranean Sea (see e.g.; Ganor and Mamane, 1982), the North
Pacific (see e.g.; Uematsu et al., 1983) and the Northern and Central
North Atlantic (see e.g.; Prospero, 1979). Other estimates have
involved the use of the deposition budgets of individual elements. For
example, Chester et al. (1979) showed that mineral dusts deposited to
the sea surface from the Atlantic north east trades are a significant
source of Al,Fe,Mn,Cr,Co,Ni,Cu,Zn and Pb to the underlying deep-sea
sediments. In a more recent study, Arimoto et al. (1985) estimated
that the net atmospheric deposition of Al and Fe to the sea surface is
the dominant source for these elements to the underlying sediments.
These authors also demonstrated that the net atmospheric deposition of
V, Cu and Zn is equal to, or higher than, the rate at which they are
removed from the system into the bottom sediments. Clearly, therefore,
air-sea exchange can supply significant quantities of both
biogeochemically non-reactive and reactive elements to the oceans, and
eventually to deep-sea sediments.

Evidence derived from several sources (for example from the
similarities between the clay mineral signatures of mineral aerosols,
surface water particulates and deep-sea sediments), implies that there
has been a relatively rapid down-column transport of particles out of
the mixed layer to the sea bed before significant lateral particle
displacement could occur. One mechanism which could result in such
rapid transport is the aggregation of the mineral aerosol particles
into fecal material. Bishop et al. (1977) have estimated that transit
times for this type of aggregate in the equatorial Atlantic are in the
order of $\sim$10 to $\sim$15 days for a 4-km deep water column, which would
result in the preservation of surface water variability in the
underlying sediments. Buat-Menard and Chesselet (1979) took account of
these large-sized fecal aggregates when they constructed a model for
the mass balance between the input of material to the sea surface from
the atmosphere and its removal from the mixed layer by sinking
particles. Chester (1982) used this model to estimate the atmospheric
and down-column fluxes of particulate Al (Alp) in the eastern/North
Atlantic, and added an into-sediment flux to the mass balance
calculations. A summary of the data is given in Table VIII.

The overall conclusion drawn by Chester (1982) was that over the
entire region of the eastern North Atlantic atmospheric sources can be
the most important single contributor to the Alp in surface waters.
However, only in the area directly underlying the north east trades
(i.e. 30°N - 10°N) is the down-column sinking of mineral aerosols of
sufficient magnitude to satisfy the sediment requirement of Alp, and so
impose eolian mineral 'fingerprints' on the bottom deposits. In other
areas, Alp will be supplied to the sediments by processes such as river
input, advective transport at mid-depths from the continental margins,
and bottom movement. However, the study did show that in oceanic
regions which receive large injections of crustal dust, atmospheric
inputs can dominate the supply of land-derived material to the

Table VIII. Fluxes of particulate Al in the eastern North Atlantic
(data from Chester, 1982; units, ng cm^2 yr^{-1})

Oceanic region	Atmospheric deposition	Down-column flux	Into-sediment flux
40°N	3000 - 13 000	23 000	90 000
30°N-10°N	80 000	70 000	90 000
05°S-10°N	13 000	15 000	90 000

underlying sediments.

6.3 Summary.

The deposition of mineral aerosols from the marine dust veil can
influence the particulate and dissolved distributions of some elements
in both surface waters and underlying sediments. The effect on
particulate/dissolved phase distributions will be determined by the
extent to which an element is soluble from the deposited mineral
aerosol, but the magnitude of the atmospheric input to both phases in
surface waters will reflect short term variations in mineral dust
loadings; the distribution of total dissolvable Mn in Atlantic surface
waters was cited as an example of this. In contrast, atmospherically-
derived 'fingerprints' in the underlying sediments will reflect
integrated, i.e. long term, variations in the inputs of mineral dust;
clay mineral and quartz concentration gradients in Atlantic deep-sea
sediments off the coast of West Africa provide an example of this type
of long term eolian input.

ACKNOWLEDGEMENTS

    This paper has benifited from many suggestions made by participants
at the ASI. In particular, I would like to express my thanks to M.
Uematsu and W. Maenhaut who showed a great deal of patience and care in
reviewing the paper; their contributions are gratefully acknowledged.

REFERENCES

Arimoto, R., R.A. Duce, B.J. Ray and C.K. Unni, 1985. Atmospheric
    trace elements at Enewetak Atoll; 2. Transport to the
    ocean by wet and dry deposition. J Geophys. Res., 90,
    2391-2408.

Aston, S.R., R. Chester, L.R. Johnson and R.C. Padgham, 1973.
    Eolian dust from the lower atmosphere of the eastern
    Atlantic and Indian Oceans, China Sea and Sea of Japan.
    Mar. Geology, 14, 15-28.
Behairy, A.K., R. Chester, A.J. Griffiths, L.R. Johnson and J.H.
    Stoner, 1975. The clay mineralogy of particulate material from
    some surface sea waters of the eastern Atlantic Ocean. Mar. Geology,
    18, M45-M56.
Biscaye, P.E., 1965. Mineralogy and sedimentation of recent deep-sea
    clay in the Atlantic Ocean and adjacent seas and oceans.
    Bull. Geol. Soc. Am., 76, 803-832.
Bishop, J.K.B., J.M. Edmonds, D.R. Ketten, M.P. Bacon and W.B.
    Silker, 1977. The chemistry, biology and vertical flux of
    particulate matter from the upper 400 m of the equatorial Atlanic
    Ocean. Deep Sea Res., 24, 511-548.
Buat-Menard, P. and R. Chesselet, 1979. Variable influence of the atmo-
    spheric flux on the trace metal chemistry of oceanic suspended
    matter. Earth Planet. Sc. Lett., 42, 399-411.
Buat-Menard, P., U. Ezat and A. Gaudichet, 1982. Size distribution and
    mineralogy of alumino-silicate dust particles in Tropical Pacific
    air and rain. Proc. 4th . Intern. Conf. on Precipitation Scaveng-
    ing, Dry Deposition and Resuspension, November 29 - December 3,
    Eds. Pruppacher et al., Elsevier, 1259-1269.
Chester, R., 1972. Geological, geochemical and environmental implic-
    ations of the marine dust veil. In: The Changing Chemistry
    of the Sea, Nobel Symposium 20, Eds. D. Dryssen and D. Jagner,
    Almqvist and Wicksell, 291-305.
Chester, R., 1981. Regional trends in the distribution and sources of
    aluminosilicates and trace metals in recent North Atlantic deep-sea
    sediments. Bull. Inst. Geol. Bassin d'Aquitaine. Bordeaux,
    31, 325-335.
Chester, R., 1982. Particulate aluminium fluxes in the eastern Atlantic.
    Mar.Chem., 11, 1-16.
Chester, R., 1985. Marine Geochemistry. Allen and Unwin; In Prep.
Chester, R., H. Elderfield, J.J. Griffin, L.R. Johnson and R.C Padgham,
    1972. Eolian dust along the eastern margins of the Atlantic Ocean.
    Mar.Geol., 13, 91-105.
Chester, R. and J.H. Stoner., 1974. The distribution of Mn, Fe, Cu, Ni,
    Co, Ga, Cr, V, Ba, Sr, Zn and Pb in some soil-sized particulates
    from the lower troposphere over the World Ocean. Mar. Chem.,
    2, 157-188.
Chester, R., H. Elderfield and J.J. Griffin, 1971. Dust transported in
    the northeast and southeast trade winds of the Atlantic Ocean.
    Nature, 233, 474-476.
Chester, R., G.G. Baxter, A.K. Behairy, K. Connor, D. Cross, H.
    Elderfield and R.C. Padgham, 1977. Soil-sized eolian dusts from the
    lower troposphere of the Eastern Mediterranean Sea. Mar. Geology,
    24, 201-217.
Chester, R., A.G. Griffiths and J.M. Hirst, 1979. The influence of soil
    sized atmospheric particulates on the elemental chemistry of deep-

sea sediments of the northeastern Atlantic. Mar. Geology, 32, 141-154.

Chester, R., A.C. Saydam and E.J. Sharples, 1981. An approach to the assessment of local trace metal pollution in the Mediterranean marine atmosphere. Mar. Poll. Bull., 12 426-431.

Chester, R., E. J. Sharples and G.S. Sanders, 1984a. The concentrations of particulate aluminium and clay minerals in aerosols from the Northern Arabian Sea. J. Sed. Petrol., 55, 37-41.

Chester, R., E.J. Sharples and K.J.T. Murphy, 1984b. The distribution of particulate Mo in the Atlantic aerosol. Oceanologica Acta, 2, 441-450.

Chester, R., E.J. Sharples, G.S. Sanders and A.C., Saydam, 1984c. Saharan dust incursion over the Tyrrhenian Sea. Atmos. Environ., 18, 29-935.

Chester, R., K.J.T. Murphy, J. Towner and A. Thomas, 1986. The partitioning of elements in crust-dominated aerosols. Chem. Geology; In press.

Cunningham, W.C. and W.H. Zoller, 1981. The chemical composition of remote area aerosols. J. Aerosol Sci., 12, 367-384.

Darwin, C., 1846. An account of the fine dust which falls on vessels in Atlantic Ocean. Q.J. Geol. Soc. (London), 2, 26.

Delany, A.C., Audrey C. Delany, D.W. Parkin, J.J. Griffin, E.D. Goldberg and B.E.F. Reinmann, 1967. Airborne dust collected at Barbados. Geochim. Cosmochim Acta, 31, 885-909.

Duce, R.A., C.K. Unni and B.J. Ray, 1980. Long-range atmospheric transport of soil dust from Asia to the Tropical North Pacific: temporal variability. Science, 209, 1522-1524.

Duce, R.A., A. Mohnen, P.R. Zimmerman, D. Grosjean, W. Cautreels, R. Chatfield, R. Jaenicke, J.A. Ogren, E.D. Pellizzari and G.T. Wallace, 1983a. Organic material in the global troposphere. Rev. Geophys. Space Phys., 21, 921-952.

Duce, R.A., B. Arimoto, B.J. Ray, C.K. Unni and P.J. Harder, 1983b. Atmospheric trace elements at Enewetak Atoll: 1, concentrations, sources and temporal variability. J. Geophys. Res., 88, 5321-5342.

Ganor, E. and V. Mamane, 1982. Transport of Saharan dust across the eastern Mediterranean. Atmos. Environ., 10, 1079-1084.

Gillette, D.A., 1979. Environmental factors affecting dust emission by wind erosion. In: Saharan Dust. Ed. C. Morales, John Wiley, 71-91.

Goldberg, E.D. and J.J Griffin, 1970. The sediments of the northern Indian Ocean. Deep-Sea Res., 17, 513-537.

Goldberg, E.D and J.J. Griffin, 1964. Sedimentation rates and mineralogy in the South Atlantic. J. Geophys. Res., 69, 4293-4309.

Griffin, J.J. H. Windom and E.D. Goldberg, 1968. The distribution of clay minerals in the world Ocean. Deep-Sea Res., 15, 433-459.

Hodge, V., S.R. Johnson and E.D. Goldberg, 1978. Influence of atmospherically transported aerosols on surface ocean water composition. Geochem.Jour., 12, 7-20.

Hoffman, E.J., G.L. Hoffman and R.A. Duce, 1974. Chemical fractionation of alkali and alkaline earth metals in atmospheric particulate matter over the North Atlantic. J. Rech. Atmos., 8, 675-688.

Hydes, D.J., 1979. Aluminium in sea water: control by inorganic processes. Science, 202, 1260-1262.

Jaenicke, R., 1979. Monitoring and critical review of the estimated source strength of mineral dust from the Sahara. In Saharan Dust. Ed. C. Morales, John Wiley, 233-242.

Kremling, K., 1985. The distribution of cadmium, copper, nickel, manganese, and aluminium in surface waters of the open Atlantic and European shelf area. Deep-Sea Res., 32, 531-555.

Krishnaswami, S. and M.M. Sarin, 1976. Atlantic surface particulates: composition, settling rates and dissolution in the deep-sea. Earth Planet. Sci Let., 32, 430-440.

Lantzy, R.J. and F.T. Mackenzie, 1979. Atmospheric trace metals: global cycles and assessment of man's impact. Geochim. Cosmochim. Acta, 43, 511-525.

Liss, P.S. and W.G.N. Slinn, 1983. Appendix B, Annotated Bibliography of Sampling. In Air-Sea Exchange of Gases and Particles. Eds., P.S.Liss and W.G.N. Slinn, Reidel Publishing Company, 543-549.

Lum, K.R., J.S. Betteridge and R.R. Macdonald, 1982. The potential availability of P,Al,Cd,Co,Cr,Fe,Mn,Ni,Pb and Zn in urban particulate matter. Environ. Tech. Letts., 3, 57-62.

Maenhaut, W., W.M. Zoller, R.A. Duce and G.L. Hoffman, 1979. Concentration and size distribution of particulate trace elements in the South Polar atmosphere. J. Geophys. Res., 84, 2421-2431.

Maenhaut, W.H., Raemdonck, A. Selen, R. Van Grieken and J.W. Winchester, 1983. Characterization of the atmospheric aerosol over the eastern Pacific. J. Geophys. Res., 88, 5353-5364.

McDonald, W.F., 1938. Atlas of Climatic Charts of the Oceans. Dep.of Agr.,U.S. Weather Bur., Washington, D.C.

Mitchell, J.M., 1957. Visual range in the polar regions with particular reference to the Alaskan Arctic. J. Atmos. Terr. Phys., Spec. Suppl., Pt. 1., 195-211.

Morales, C., 1979. Saharan Dust. Ed. C. Morales, John Wiley, 297 pages.

Murphy, K.J.T., 1985. The Trace Metal Chemistry of the Atlantic Aerosol. Unpub. Ph.D. Thesis, University of Liverpool, 380 pages.

Murray, J and A.F. Renard, 1891. Deep-Sea Deposits. Rep. 'Challenger' Exped., III, London, 327 pages.

Nriagu, J.O., 1979. Global inventory of natural and anthropogenic emissions of trace metals to the atmosphere. Nature, 279, 409-411.

Parkin, D.W.., D.R. Phillips, R.A.L. Sullivan and L.R. Johnson, 1970. Airborne dust collections over the North Atlantic. J. Geophys. Res.,75, 1782-1793.

Prospero, J.M., 1968. Atmospheric dust studies on Barbados. Bull. Am. Meteorol. Soc., 49, 645-652.

Prospero, J.M., 1979. Mineral and sea salt aerosol concentrations
    in various ocean regions. J. Geophys. Res., 84, 725-731.
Prospero, J.M., 1981. Eolian transport to the World Ocean. In: The
    Oceanic Lithosphere, vol. 7, The Sea, Ed. C. Emiliani, John
    Wiley, 801-874.
Prospero, J.M., R.J. Charlson, V. Mohnen, R. Jaenicke, A.C. Delany,
    J. Moyers, W. Zoller, and K. Rahn, 1983. The atmospheric aerosol
    system: an overview. Rev. Geophys. Space Phys., 21, 1607-1629.
Radczewski, O.E., 1939. Eolian deposits in marine sediments. In: Recent
    Marine Sediments. Ed. P. Trask. Am. Assoc. Pet. Geol., Tulsa,
    Okla., 496-502.
Rahn, K.A., 1976. The chemical composition of the atmospheric aerosol.
    Tech. Rept., Graduate School of Oceanography, University of Rhode
    Island, Kingston, R.I.
Rahn, K.A., R.D. Borys, G.E. Shaw, L. Shutz and R. Jaenicke, 1979. Long-
    range impact of desert aerosol on atmospheric chemistry: two examples.
    In Saharan Dust. Ed. C.Morales, John Wiley, 243-266.
Rahn, K.A., R.D. Borys and G.E. Shaw, 1981. Asian desert dust
    over Alaska: anatomy of an Arctic haze episode. Geol. Soc. Am.,
    Spec. Paper 186; 37-70.
Rex, R.W. and E.D. Goldberg, 1958. Quartz contents of pelagic sediments
    of the Pacific Ocean. Tellus, 10, 153-159.
Riley, J.P. and R. Chester, 1971. Introduction to Marine Chemistry.
    Academic Press, London, 465 pages.
Savoie, D.L. and J.M. Prospero, 1977. Aerosol concentration statistics
    for the tropical North Atlantic. J. Geophys. Res., 82, 5954-
    5963.
Schutz, L., 1979. Sahara dust transport over the North Atlantic
    Ocean-model calculations and measurements. In: Saharan Dust. Ed. C.
    Morales, John Wiley, 267-277.
Schutz, L., R. Jaenicke and H. Pietrek, 1980. Saharan dust transport
    over the North Atlantic Ocean. Geol. Soc. Amer. Spec. Paper,
    186, 87-100.
Slinn, W.G.N., 1983. Air-to-sea transfer of particles. In Air-Sea
    Exchange of Gases and Particles. Eds., P.S. Liss and W.G.N.
    Slinn, Reidal Publishing Company, 299-396.bTomadin, L., R. Lenaz,
    V. Landuzzi, A. Mazzucotelli and R. Vannucci, 1984. Wind-blown
    dusts over the Central Mediterranean. Oceanologica Acta,
    7, 13-23.
Uematsu, M., R.A. Duce, J.M. Prospero, L. Chen, J.T. Merrill and R.L.
    McDonald, 1983. Transport of mineral aerosol from Asia over
    the North Pacific Ocean. J. Geophys. Res., 88, 5343-5352.
Uematsu, M., R.A., Duce, T. Patterson and J.M. Prospero, 1985a. Spatial
    distribution of mineral aerosol over the Southwest Pacific
    Ocean. SEAREX News., 8, 34-38.
Uematsu, M., R.A. Duce, S. Nakaya and S. Tsunogai. 1986 Short-
    term temporal variability of eolian particles in surface waters of
    the Northwestern North Pacific. J. Geophys. Res., In Press.

Venkatarathnam, K. and W.B.F. Ryan, 1971. Dispersal patterns of
    clay minerals in the sediments of the Eastern Mediterranean Sea.
    Mar. Geology, 11, 261-282.
Venkatarathnam, K.J.A. Kostecki, J.A. Robinson and P.E. Biscaye, 1981.
    Distributions and origins of clay minerals and quartz in surface
    sediments of the Arabian Sea. J. Sed.Petrol., 51, 563-569.
Walsh, P.R. and R.A. Duce, 1976. The solubilization of
    anthropogenic atmospheric vanadium in sea water. Geophys. Res.
    Lett., 3, 375-378.

AIR TO SEA TRANSFER OF ANTHROPOGENIC TRACE METALS

Patrick Buat-Ménard
Centre des Faibles Radioactivités,
Laboratoire mixte CNRS-CEA,
Domaine du CNRS, BP N°1,
91190, Gif sur Yvette,
France

1. INTRODUCTION

There is more and more evidence that, at least in the Northern hemisphere, the atmospheric cycles of some trace metals, especially heavy metals such as Pb, Cd, Sb, As, Sn, are now strongly perturbed by human activities (Andreae et al., 1984). How such a geochemical perturbation in the atmosphere interacts with the oceanic cycles of these elements is a matter of crucial interest since many trace metals are both essential for the development of marine life and potentially toxic for marine life above certain concentration levels. This induced shift of the cycles of some trace metals in surface and deep waters from a steady to a transient state offers also the perspective of using new oceanic tracers for improvements in ocean circulation models. Such a shift is still extremely difficult to model because of the lack of accurate knowledge of a) the temporal and spatial distribution of the atmospheric deposition of these elements b) the scale of anthropogenic perturbations in the atmosphere and c) the biogeochemical processes which drive the fate of trace metals in the ocean.

We know now that the behavior of most trace metals in the ocean is not conservative, i.e. not solely controlled by water mixing processes (Bruland, 1983). Indeed, depending on their chemistries, these elements can undergo varying degrees of internal recycling prior to their ultimate removal to marine sediments. This internal cycle results from a complex series of biogeochemical processes for which a first order understanding now exists. Such an understanding is a prerequisite before attempting to model the air to sea transfer of anthropogenic trace metals. Insight into these processes was first provided by studies in the 1960's of the behavior of artificial and natural radionuclides, primarily brought to ocean waters by atmospheric fall-out (Lowman et al., 1971). In the last 10 years, advances in analytical chemistry together with increasing attention paid to the control of potential sample contamination have enabled a considerable gain in our knowledge of the atmospheric and oceanic chemistry of trace metals. In this chapter, I will first review briefly the major processes which control

P. Buat-Ménard (ed.), The Role of Air-Sea Exchange in Geochemical Cycling, 477–496.
© *1986 by D. Reidel Publishing Company.*

the fate of atmospheric trace metals in sea water. Some insight into the magnitude of air to sea tranfer rates will then be given, with emphasis on the geographical variability of metal fluxes from the atmosphere to the oceans. Finally, when enough information is available, the effects of anthropogenic perturbations from the atmosphere will be assessed from consideration of the horizontal and vertical oceanic distributions of trace elements. Pb will be taken as a case study element since the perturbation of its oceanic cycle due to eolian inputs of anthropogenic origin is rather well documented, thanks primarily to the elegant studies conducted over the last 20 years by C. Patterson and his co-workers (Schaule and Patterson, 1981; Settle and Patterson, 1982; Flegal and Patterson, 1983).

2. THE FATE OF ATMOSPHERIC TRACE METALS IN OCEAN WATERS.

2.1 Physical and chemical forms of metals in the atmosphere.

The first step towards an understanding of the fate of atmospheric trace metals in ocean waters is the knowledge of the physical and chemical form of such elements in the marine atmosphere. Such information is needed both for air and rain. Moreover, since the air to sea flux of many metals consists of a natural and an anthropogenic component, one can expect that the partitioning between the different physical and chemical forms of metals will vary with time and space. Because of experimental difficulties, such data are at present very scarce.
       The form of a metal is also essential in controlling its cycle through the atmosphere. The most abundant form of mercury is gaseous elemental Hg, yet the deposition flux of mercury appears to be controlled by the particulate fraction (Fitzgerald, this issue). There appear to be at least two gaseous forms of selenium. One is derived from high temperature processes, Se° or SeO_2, and the other from the marine and terrestrial biosphere, may be a methylated species (Mosher and Duce, 1983; Andreae et al., 1984). These two forms appear to have quite different residence times. It is therefore essential to identify these species and to study their chemical behavior in the atmosphere. Besides Hg and Se, there is no evidence to date of any significant vapor phase for most metals in the marine atmosphere, although a careful search has yet to be made. In the case of Pb, there is some evidence that a small fraction is present in the vapor phase, at least in urban areas, very likely as tetraalkyllead compounds (Harrison and Laxen, 1978; Rohbock et al., 1980). Evidence suggests that about 10% or less of As in the marine atmosphere is present either in the gas phase or on particles considerably smaller than 0.2 μm (Walsh et al., 1979a). The chemical form of vapor phase As is unknown, but may be a methylated species (Wood 1974). However, this form, if it exists, should have a very short residence time since methylated forms of arsenic in marine rain are almost non existent (Andreae, 1980).
       We know very little about what happens to metals in air and rain after they enter the ocean. To date there is virtually no information from remote marine regions on the fraction of metals in rain which are

in the particulate or dissolved forms. We know nothing yet about reactions that occur during the mixing of rain and sea water. Solubilization, sorption, or flocculation processes involving metals may occur, and such processes would change the physical and chemical forms of the metals in the ocean and may have a significant impact on the cycling of the metals in surface waters. Similar concerns are obvious relative to dry deposition of aerosol particles containing metals. Reactions of these kinds would affect the residence times of the metals in the ocean and their interactions with biological systems.

Walsh and Duce (1976) showed that a major fraction of the anthropogenically derived vanadium present on coastal North Atlantic aerosol particles was soluble in sea water. Similar results were found for several metals during aerosol particles studies along the southern California coast (Hodge et al., 1978), coastal Washington (Crecelius, 1980), and the southeastern U.S. (Mullins, 1978). Results of some of these seawater solubility studies are presented in table 1.

TABLE 1. Seawater soluble fraction of metals on coastal aerosols

Metal	Per cent soluble	Metal	Per Cent soluble
	La Jolla, CA (3hrs)[1]	Quillayute, WA (1 hr)[2]	
Pb	39	As	50
Cd	84	Zn	65
Zn	68		
	Ensenada, Mexico (3 hrs)[1]	Quillayute, WA (24 hrs)[2]	
Pb	13	As	49
Cd	80	Zn	72
Zn	24		
	Narragansett, RI (1 hr)[3]		
Excess V	72		

[1] Hodge et al. (1978)
[2] Crecelius (1980)
[3] Walsh and Duce (1976)

To determine whether particulate material, atmospherically supplied to the ocean surface, may be significantly leached not only by direct interaction with sea water but also as a result of zooplankton grazing, Moore et al. (1984) have leached samples of atmospheric dust at pH values between 5.4 and 8.0. The elements Cu, As and V showed significant increases in the proportion leached as the pH was lowered. Approximately 10% of the iron was leached, independent of pH. An estimate of the atmospheric flux of Fe in a leachable form supports the idea that this source may be the primary contributor to the Fe requirements of the biota. Since the efficiencies with which phytoplankton can grow at low Fe concentrations vary among species, it follows that the meteorological factors that govern the rate of transport of atmospheric dust to the ocean areas could affect plankton ecology and productivity (Duce, this issue).

As will be discussed later, there is direct evidence from water column studies that a considerable fraction of the atmospheric Pb and Sn of anthropogenic origin is soluble in sea water (Schaule and Patterson, 1981; Byrd and Andreae, 1982). Studies of artificial and metallic radionuclides also indicate that a large fraction of the atmospheric input of some metals is soluble in sea water. However, these studies may be limited in their application to the understanding of the cycling of stable trace metals. For example, it is currently accepted that ^{210}Pb can trace the biogeochemical cycling of common lead in sea water because significant percentages of each pass through the dissolved state. However, as pointed by Schaule and Patterson (1981) these two types of lead are supplied through different mechanisms. While the delivery of atmospheric ^{210}Pb to the ocean surface reflects a steady state input process, the common lead input to the ocean has increased significantly over the last 200 years because of anthropogenic effects. Whether the chemical forms of natural ^{210}Pb and pollution-Pb attached to aerosol particles are similar in the marine atmosphere is at present unknown.

2.2 Biogeochemical cycling of trace metals in the ocean.

We have now a good approximation of processes to which anthropogenic inputs of metals from the atmosphere may be subject to throughout the water column (Buat-Ménard, 1983). These processes involve to varying degrees uptake of dissolved trace elements by particulate phases (mainly biogenic) and subsequent regeneration at greater depths as the particulate carrier phase undergoes oxidation and/or dissolution. These uptake or "scavenging" processes can be active or passive. Active uptake of trace elements by phytoplankton in the photic zone can be partly understood in terms of their function as essential micronutrients. Trace elements such as V, Cr, Mn, Fe, Co, Ni, Cu, Zn and Mo play important roles in metal required and metal activated enzyme systems (Bruland, 1983). Collier and Edmond (1984) have conducted an extensive study of the trace element composition of plankton which indicates that the primary carrier for most of the metals studied is the non-skeletal organic fraction of the particulate material. Direct examination of $CaCO_3$ and opal demonstrated that these phases contained only small percentages of the trace elements in the plankton.

In surface waters, most of this plant material is grazed upon by zooplankton and the undigested residues are packaged into fecal materials. In some cases the accumulation of metals is even greater in this detritus, perhaps the result of self-regulation by zooplankton of its metal requirements and detoxification processes (Lambert, 1981). Since such detritus can reach the deep-sea sediments within a month (Brun-Cottan, this issue), one may infer that the benthic fauna is already feeding on material containing enhanced levels of some toxic metals supplied from the atmosphere. The reasons for the accumulation of metals in biogenic detritus are still poorly understood. It seems however that precipitation of metallic minerals, such as barite (Dehairs et al., 1980), can occur within such reducing microenvironments. Surface-adsorption of metals onto the surface of such particles might also be important. The latter process might be the most important geochemically. There is considerable evidence that any particle, immersed in natural sea water, rapidly looses its native surface charge characteristics and becomes negatively charged. This is consistent with the formation of a macromolecular organic surface film. The nature of such organic films is still largly unknown (Hunter and Liss, 1979; Lion and Leckie, 1981). There is some evidence, however, that carboxylic acids and phenolic functional groups are the major ionizable groups in such organic coatings (Hunter, 1981). According to Balistrieri et al. (1981), such coatings should play a key role in controlling adsorption properties of suspended particles in the ocean.

In surface waters, where most of the biological activity takes place, metals can be, like nutrients, recycled many times between the dissolved phase and the particulate phase. At depths greater than about 100 m, biogenic detritus and other particles are less frequently recycled so that they leave the surface water compartment by sedimentation. Thus, two removal processes from surface waters to deep waters are at work: water mixing and transport by settling particles. As a consequence, the residence time of reactive elements in the surface waters must be shorter than that of the water itself ($\simeq$20 years). This is indeed the case (Bruland, 1983). This implies that atmospheric inputs of recent origin, such as anthropogenic inputs, are partially or totally recorded in the flux associated with particles sinking from surface waters. Therefore induced chemical changes in the deep water column will be first recorded in that particulate flux. Also, for these elements for which removal in the particulate form is the dominant sink from surface waters, a simple mass-balance between input via the atmosphere and particulate removal from the surface layers can tell us how important is the atmosphere as a source for these elements in the surface ocean. Such an approach has already been used based either on model calculations of the particulate fluxes (Buat-Ménard and Chesselet, 1979; Wallace et al., 1983; Collier and Edmond, 1984) or on direct measurements by sediment traps (Jickells et al., 1984).

Because of this short residence time in surface waters, the distributions of dissolved (and also particulate) trace elements within this layer are also controlled to a large extent by their dominant input sources (Bruland, 1983). These sources can be external (atmospheric input and advective transport from rivers and/or shelf sediments) or internal (vertical mixing). The rates of inputs from these different sources exhibit different geographical patterns. Therefore a careful consideration of mixed layer distributions of trace metals along oceanic transects can also help to factor out the respective influences of each type of input, as will be discussed in the last section of this chapter.

Once in deep waters, dispersal of atmospheric inputs of trace metals will depend on their internal cycle in that reservoir. As in surface waters, chemically active elements will undergo complex mineralization/solubilization processes. The biological cycling is still at work in deep waters, although it is much slower than in surface waters. Elements involved in the biological cycle will recycle between particulate and soluble phases down to different depths, depending on the chemical behavior and biological involvement of the element. Also, adsorption and desorption reactions on the surface of particles can occur at every depth, as shown recently for thorium isotopes by Bacon and Anderson (1982). A very important distinction has to be made here between small particles and large particles in the ocean. Large particles (>50μm), which form only 10% or less of the mass in suspension, account for 90% of the mass which sediments (Bishop et al., 1978; Chesselet, 1979). Small particles account for the other 10%, although they represent 90% of the standing stock. Large particles (foraminifera, fecal pellets, aggregates) remain in the water column only a few weeks or days. On the other hand, small particles have a mean residence time of 50 to 100 years in the deep water column (Lambert et al., 1981) and are therefore subject to horizontal advection of the water masses in which they occur. As a consequence it has been suggested that the transit of small particles will induce the largest measurable in-situ chemical changes in the dissolved composition of several elements whereas large particles serve as a means of transporting organic matter and inorganic matter directly to the sediment-sea water interface (Lal, 1977; 1980). Moreover, large particles may break or be ingested by pelagic organisms en route, which will also lead to in-situ chemical changes via remobilization of some elements. The relative importance of these various processes is not yet known for most trace elements, with the exception of barium (Dehairs et al., 1980). Finally, due to a variety of biological and physical processes, some trace metals may return to the water column from the water-sediment interface, (Bruland, 1983; Lambert et al., 1984).

Depending on their chemical reactivity, trace metals may pass once or cycle many times in the ocean before being finally buried in sediments. Elements which have a mean oceanic residence time much longer than the mixing time of deep ocean waters (≃1000 years) have their distribution in the deep ocean mostly controlled by their internal cycle (Wollast, this issue). For such elements, which behave like nutrient elements, the response time to recent anthropogenic inputs from the atmosphere is too long to allow, at present time, the detection of changes of their distribution in the deep open ocean. On the other hand, the distribution of elements with very short residence times (<1000 years) is strongly influenced by external sources. For such elements, e.g. Pb, Sn, the deep water column is sensitive to recent changes in input rates. Because there is not sufficient time for mixing, the magnitude of the perturbation should exhibit geographical variations reflecting those of the dominant input source. In the following sections, I will review the available knowledge on the geographical variability of the atmospheric input of trace metals and its effect on the distribution of these elements in surface and deep waters.

3. GEOGRAPHICAL VARIABILITY OF METAL FLUXES FROM THE ATMOSPHERE TO THE OCEAN.

I have already discussed in another chapter of this book the uncertainties involved in the estimation of net air to sea fluxes for materials attached to aerosol particles. In the best case, present estimates are valid within a factor of 2 or 3. These large uncertainties result a) from inaccuracies in the measurements or in the model calculations of the total deposition to the ocean surface and b) from the existence of a "recycled" component of the total flux, i.e. flux associated with the fall-out of sea salt particles (Buat-Ménard, 1983; Arimoto et al., 1985; Buat-Ménard and Duce, 1985).

The net atmospheric input, per unit area of the ocean surface, will depend on the remoteness from continental areas and on the initial load of continental air masses. Table 2 shows that the rate of the net input decreases by many orders of magnitude from coastal areas to the remote regions of the Tropical North Pacific. Flux values shown for the North Sea and the Tropical North Pacific are based on direct measurements. The other values given in this table are based on mean atmospheric concentration data. The fluxes have been derived assuming either that tropospheric aerosols are removed 40 times per year (Duce et al., 1976b), or that the total deposition velocity of continental aerosols in the marine atmosphere is 1 cm s^{-1} (Buat-Ménard and Chesselet, 1979; Arnold et al., 1982) based on the ^{210}Pb total deposition data (Turekian et al., 1977; Turekian and Cochran, 1981). The use of ^{210}Pb data to parameterize atmospheric deposition is probably relevant for elements attached to small particles (<1μm). It is precisely in this size range that most of the mass of pollution-derived material is found in remote regions. One can note from table 2 that for the North Atlantic, there is relatively good agreement between the values calculated by Duce et al. (1976b) from air concentration data obtained during 1973 and 1974 at Bermuda, those calculated by Jickells et al. (1984) from rain concentration data obtained during 1981 and 1982 at Bermuda, and those calculated by Buat-Ménard and Chesselet (1979) from atmospheric samples collected during four cruises in the Tropical North Atlantic in 1974 and 1975. For the Tropical North Pacific the net atmospheric lead input flux calculated by Arimoto et al. (1985), 7 ng cm^{-2} y^{-1} is very close to that calculated by Patterson and his co-authors, 6 ng cm^{-2} yr^{-1} (Settle and Patterson, 1982). This group has found that at American Samoa, in the Tropical South Pacific, the net atmospheric lead input flux is about half of its value over the Tropical North Pacific (Flegal and Patterson, 1983). To my knowledge, the latter data are the only reliable ones published to date for the remote Southern Hemisphere.

With the exception of Pb, it is very difficult at the present time, to assess quantitatively which fraction of the eolian input of trace metals is anthropogenic, especially over remote marine regions. There is strong evidence that in coastal areas and regional seas most of the eolian input of enriched metals is entirely dominated by anthropogenic sources (Arnold, 1985). However, over remote marine regions, the contribution from natural sources through volcanic activity, marine and terrestrial biological processes may still be dominant especially over

Table 2. Estimated mean fluxes of trace metals from the atmosphere to the sea-surface (ng cm^{-2} yr^{-1})

Element	New York Bight (1)	North Sea (2)	Western Mediterranean (3)	South Atlantic Bight (4)	Bermuda (5)	Bermuda (6)	Tropical North Atlantic (7)	Tropical North Pacific (8)
Al	6,000	30,000	5,000	2,900	3,900	-	5,000	1,200
Sc	-	5	1	-	0.6	-	1.1	0.18
V	-	480	-	-	5	-	1.7	7.8
Cr	-	210	49	-	9	-	14	-
Mn	-	920	-	60	45	41	70	9
Fe	5,700	25,500	5,100	5,900	3,000	739	3,200	560
Co+	-	39	3.5	-	1.2	-	2.7	-
Ni+	-	260	-	390	3	32	20	-
Cu+	1,400	1,300	96	220	30	101	25	8.9
Zn+	-	8,950	1,080	750	75	176	130	67
As+	-	280	54	45	3	-	-	-
Se+	-	22	48	-	3	-	14	4.2
Ag+	-	-	3	-	-	-	0.9	-
Cd+	30	43	13	9	4.5	9	5	-
Sb+	-	58	48	-	1.0	-	3.5	-
Au+	-	-	0.05	-	-	-	0.1	-
Hg+	-	-	5	24	-	-	2.1	-
Pb+	3,900	2,650	1,050	660	100	118	310	7.0
Th	-	4	1.2	-	-	-	0.9	0.67

Notes: + denotes elements generally enriched in marine aerosols and which may therefore have an anthropogenic origin (Buat-Ménard, 1984).

(1) Duce et al. (1976a); (2) Cambray et al. (1975); (3) Arnold et al. (1982); (4) Windom (1981); (5) Duce et al. (1976b); (6) Jickells et al. (1984); (7) Buat-Ménard and Chesselet (1979); (8) Arimoto et al. (1985)

the Southern Hemisphere. For example, from mass-balance model calculations, Walsh et al. (1979b) have concluded that arsenic inputs are predominantly anthropogenic in the Northern Hemisphere and predominantly natural in the Southern Hemisphere. It has to be emphasized that the results of such calculations depend heavily on the choice of source strengh estimates. Whereas anthropogenic sources are reasonably well documented, we still know very little about natural sources (Buat-Ménard and Duce, 1985). In recent years some information has emerged on the importance of volcanic activity, and we now have approximate values for the global volcanic source strength, especially for rather volatile metals such as Cd, Hg, Pb and As (Arnold et al., 1981; Settle and Patterson, 1982; Fitzgerald et al., 1983; Zoller, 1984). On the other hand, there is little data for such sources as crustal degassing and the terrestrial biosphere. Metal release from plants during their growth has been suggested as a source of zinc in the atmosphere (Beauford et al., 1977). The importance of terrestrial biological sources for trace elements has yet to be established on a global basis. Little is known about mobilization of trace elements by forest fires or other land-burning practices. Also, the role of the oceans in the atmospheric chemistry of metals cannot be simply restricted to the ejection of sea-salt particles and their associated trace metals into the atmosphere. Microorganisms in the marine environment can produce methylated forms of As, Hg, Se, Sn and other metals which may be released to the air. Evidence for this has been shown for Hg by Fitzgerald et al. (1984) and for Se by Mosher and Duce (1983). Clearly, these processes could be particularly important for the anthropogenic metals.

As an example, it is presently estimated that total arsenic emissions into the atmosphere are of the order of 30,000 T/yr. The uncertainty on this estimate is probably less than a factor of two. On the other hand, according to Chilvers and Peterson (1985), the range for the volcanic source strength is 2,000-20,000 T/yr and the natural release of arsenic through low temperature volatilization (biological methylation) is estimated to vary between 3,000 to 30,000 T/yr. It would seem premature from these numbers to draw any final statement about the scale of the perturbation of the atmospheric As cycle by human activity.

4. EFFECTS OF EOLIAN INPUTS OF ANTHROPOGENIC ORIGIN ON THE DISTRIBUTION OF TRACE METALS IN SURFACE AND DEEP MARINE WATERS.

As mentioned in section 2, a first order estimate of the importance of the atmosphere as a source of trace metals for surface waters can be obtained using a steady-state approach. Such an approach considers the surface water compartment as a "black box" where the input fluxes are mostly balanced by removal associated with sinking particles. Because the residence times of dissolved and particulate trace metals are generally short (<20 years) in surface waters, the assumption of steady-state relative to input via the atmosphere is approximately valid. Such a simple approach can be applied in open ocean regions where a well-developed mixed layer does exist, such as sub-tropical gyres. It

can tell us how sensitive are the concentrations of trace elements in
the surface layer to atmospheric inputs. Until very recently, the amount
of data on the distribution of trace elements in surface waters was
insufficient to warrant any further complexity in the models.

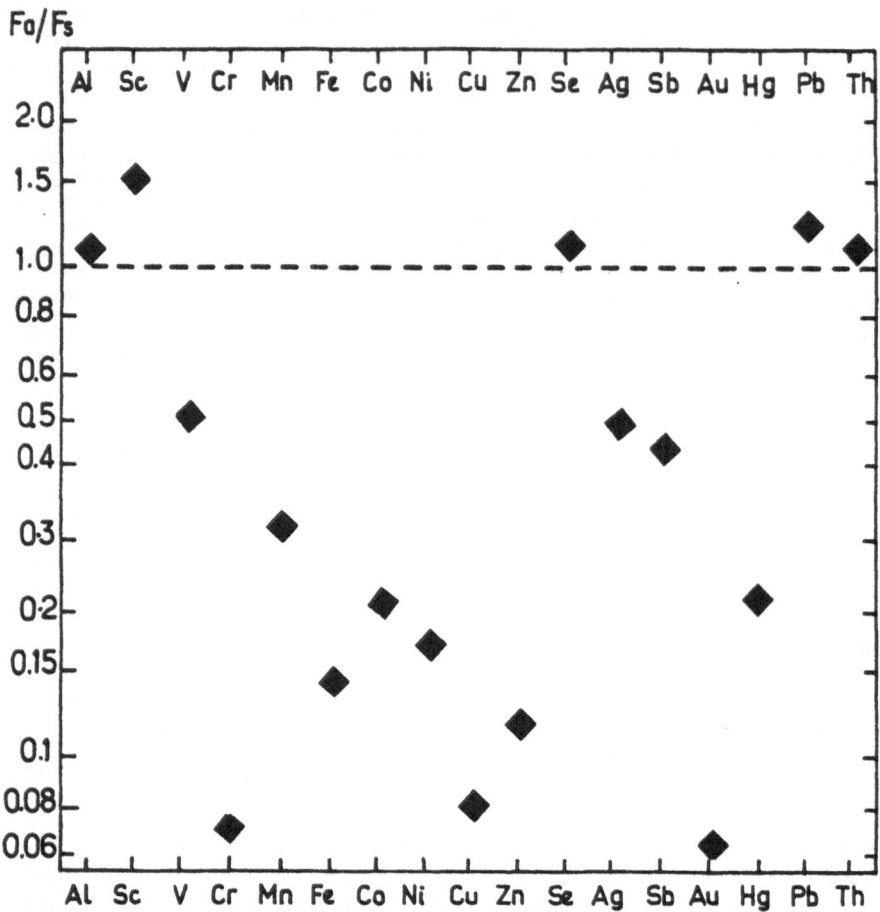

Figure 1. Ratio of flux from atmosphere, Fa, to downward particulate
flux from surface layers, Fs, in the Tropical North Atlantic. Adapted
from Buat-Ménard and Chesselet, 1979).

Buat-Ménard and Chesselet (1979) have compared atmospheric inputs
to the Tropical North Atlantic with downward fluxes of metals in
association with both large and small particles. The major components of
the downward flux was associated with the flux of large particles. The
large particle flux was estimated using the particulate organic C flux
of Wallace et al. (1977) and calculated metal to C ratios for the large
particles. The model revealed that the atmospheric deposition of trace
metals to the ocean is a significant source in the North Atlantic (Fig.
1). Considerable differences were found, however, from one element to

another. The atmospheric flux was found to be the dominant source in the case of lead, for which it was inferred that the particulate flux in the ocean is, at present, primarily anthropogenic. For other trace elements, anomalously enriched in marine aerosols possibly because of an anthropogenic component (Cu, Zn, Cd) the present anthropogenic air-sea flux still appeared to be small relative to natural fluxes in the ocean. The latter fluxes were supposed to result primarily from upwelling and diffusion to surface waters. Wallace et al. (1977, 1983) have used a somewhat similar approach which they have applied to the Sargasso Sea and Gulf Stream. The downward fluxes were calculated from an extensive data set of suspended elemental concentrations of metals and organic C and using a first-order removal rate constant determined from the vertical profile of Al in the upper 100 m (Wallace et al., 1981). Their conclusion for Pb was the same as that of Buat-Ménard and Chesselet (1979). However, different diagnostics were obtained for Cu, and Cd, mostly due to the choice of metal/carbon ratios used in estimating the downward large particles flux. As discussed by Wallace et al. (1983), the downward fluxes calculated by Buat-Ménard and Chesselet (1979) were probably overestimated for some elements by as much as a factor of 4. This demonstrates that, as for the atmospheric input, such an estimate of the vertical flux in the ocean is still model-dependent. This also indicates the need for high quality and extensive data on the trace element composition of marine biogenic particulate matter, especially in the large particle size fraction, such as those obtained by Martin and Knauer (1973) and Collier and Edmond (1984). There is also an urgent need to investigate the spatial and temporal heterogeneities of the biogenic particulate flux and its relationship with primary productivity if we want a better understanding of the fate of metallic pollutants in ocean waters (Bruland, 1983; Deuser et al., 1983; Bacon et al., 1985).

The other approach to assess the importance of the atmospheric input as a net source of trace elements in the ocean is to consider the distributions of dissolved concentrations in sea water. I will use, as an example, the studies undertaken by C. Patterson and his co-workers (Schaule and Patterson, 1981; Flegal and Patterson, 1983) which have established that the upper ocean is enriched in dissolved lead as a result of the atmospheric deposition of anthropogenic lead. Such a result is essentially based on a synthesis of data obtained on eolian inputs, water profiles and sediment outputs of lead.

Lead has a very fast cycle in the ocean. ^{210}Pb based calculations yield at most a few years in surface waters and at most a few centuries in deep waters (Bacon et al., 1976; Nozaki et al., 1976). At present, the atmospheric lead burden (at least in the Northern Hemisphere) is entirely dominated by industrial inputs. Such inputs have increased dramatically, almost by five orders of magnitude, during the last three thousand years, with the major increase occuring since the beginning of this century, as indicated by the Greenland Ice Core study of Murozumi et al. (1969). The present lead cycle in the ocean is therefore not at steady state with respect to atmospheric inputs. Since the residence time of lead in the ocean is much smaller than the turnover time of deep waters, different vertical distributions of lead should be expected from one oceanic region to another, depending on the local atmospheric source

strength. As shown by Figure 2, the lead concentrations in the surface
waters are always higher than in deep waters. Also, the magnitude of the
concentrations in the waters are related to the levels of the
atmospheric input rate. Present atmospheric input rates to the open
North Pacific exceed the prehistoric oceanic output flux of
anthropogenic lead recorded in pelagic sediments about ten-fold. Further
evidence of the industrial character of the present day eolian lead
which now accounts for most of the lead in Pacific and Atlantic surface
waters was obtained from studies of the stable isotope ratios of lead.
Such ratios in surface waters do not match those for ancient natural
authigenic lead in Pleistocene pelagic sediments, but they can be
related to those of ore leads used to manufacture lead alkyls, which is
one of the principal sources of atmospheric lead inputs.

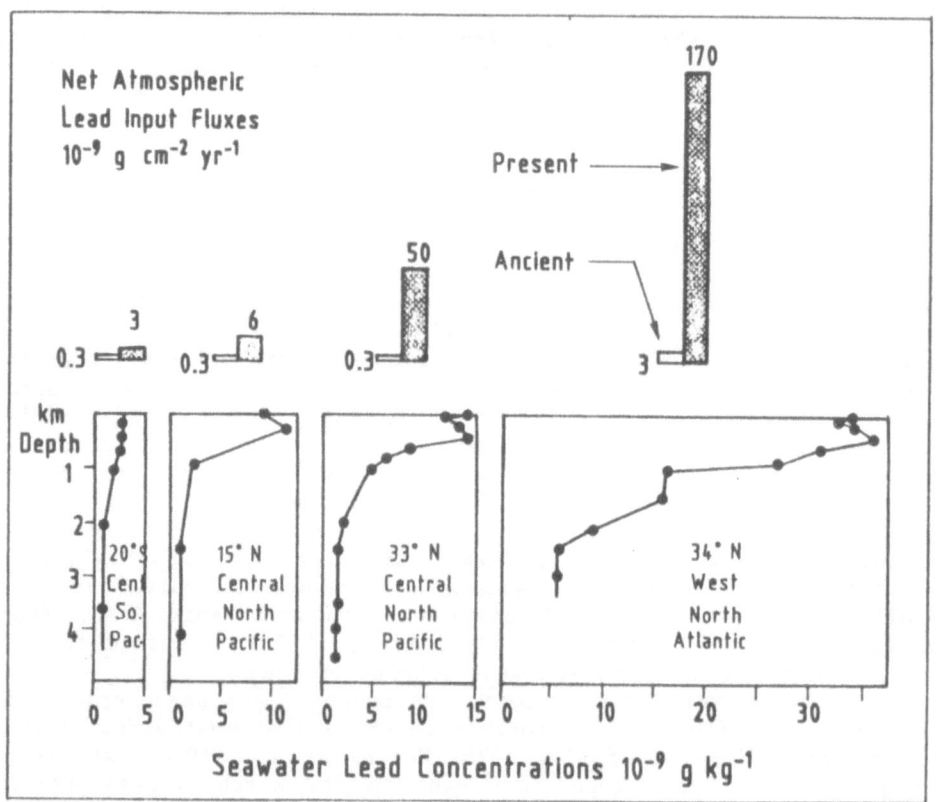

Figure 2. Comparison of eolian lead inputs with concentration profiles
at indicated locations (Adapted from Flegal and Patterson, 1983).
Present eolian data from Settle and Patterson (1982).

    As another example, recent data reported by Byrd and Andreae (1982)
indicate that the vertical distribution of tin in the northwestern
Atlantic is somewhat similar to that of lead and would also reflect the
fact that the present fluxes of tin through the atmosphere and the ocean

are dominated by anthropogenic effects. It should be emphasized however, that if the observation of an enrichment of a given element in the upper part of the water column suggests that the inputs of that element are eolian, other factors must be considered. Indeed, as pointed in the above mentioned works, such enrichments might also result from inputs from the ocean margins carried to the open ocean by lateral diffusive and advective transport, as is the case for Mn (Bruland, 1983). Figure 3 illustrates how the different input mechanisms can be inferred from the study of the spatial distribution of trace elements in surface waters. Between Monterey Bay, California, and Hawaii, Mn, Cu and Ni exhibit the negative horizontal surface gradient expected for metals supplied from rivers, shelf sediments or shallow depths by coastal upwelling, and then transported horizontally to the open ocean by diffusion/advection. Such a process has been shown to explain the surface waters levels of ^{228}Ra introduced at the ocean margins by the decay of ^{232}Th in sediments (Kaufman et al., 1973). For Pb, and also ^{210}Pb, the direction of this gradient is opposite to that exhibited by Cu or Ni. This does not allow diffusion or advection from the continental margins to be the supply route for lead of the open North Pacific. The correlation observed between lead and ^{210}Pb, whose input is mainly eolian (Nozaki et al., 1976), demonstrates that Pb enters the open ocean mainly from the atmosphere. The fact that the concentrations of both forms of lead increase towards the open ocean reflects primarily changes in the removal rates from surface waters. Elevated levels occur within the subtropical gyre where productivity is low, lower levels are observed outside of the gyre as a result of increased scavenging due to higher biological productivity and particle flux. It is likely that the weak concentration increase of Mn between the central gyre and Hawaii (Figure 3) reflects also an eolian input, probably resulting from the solubilization of Mn from eolian soil-dust particles (Chester, this volume). Other studies in the Atlantic have also shown evidence for an eolian input origin of the enrichment of some dissolved trace elements in surface waters, especially in low productivity waters (Elderfield and Greaves, 1982; Measures et al., 1984; Kremling, 1985).

There is strong evidence that, in coastal areas and regional seas, the atmospheric deposition of metallic pollutants is a significant, if not the major source of pollution (Buat-Ménard, 1983). Unfortunately, the time scale of water mixing is generally too short to allow the determination of the natural chemical composition of such waters. For example, in the Western Mediterranean, the average turnover time of deep waters is of the order of 20 years (Merlivat, pers. comm.). As a consequence, direct evidence of the eolian origin of trace metals in such waters cannot easily be inferred solely from the consideration of their vertical distribution in the water column. For example in the Western Mediterranean (Nicolas et al., 1985) have found that Pb and Zn concentrations are higher in surface waters than in deep waters, which clearly suggest an eolian source, likely anthropogenic (Arnold et al., 1982), for these elements. On the other hand, Cu, Cd, Ni exhibit similar levels in surface and deep Western Mediterranean waters (Laumond et al., 1984; Boyle et al., 1985; Nicolas et al., 1985). Boyle et al. (1985) have suggested that the enrichment of Cu, Ni, Cd in Western

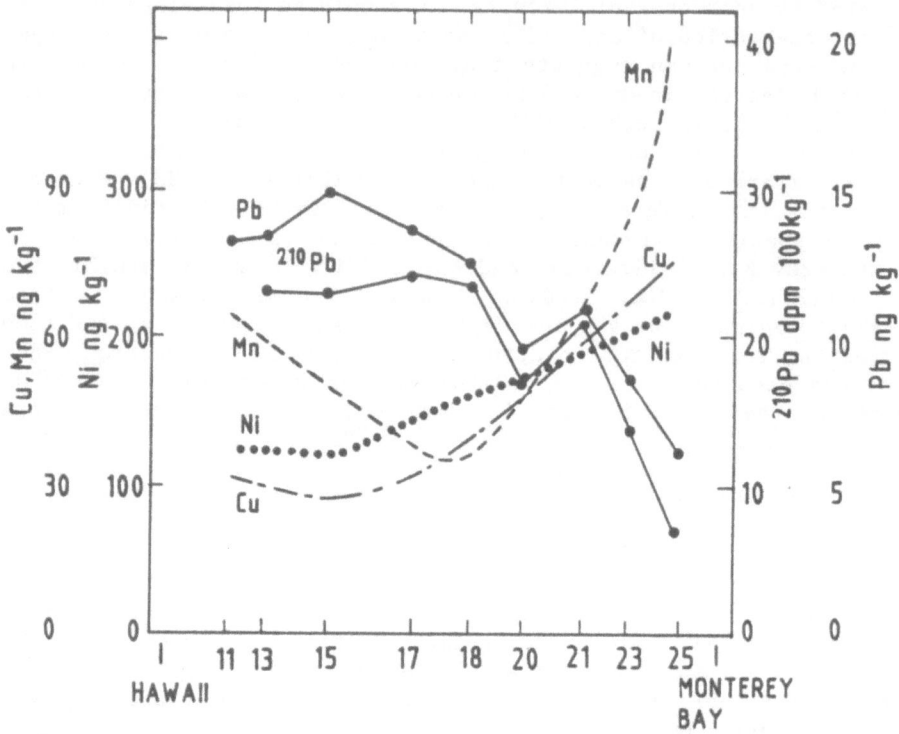

Figure 3. Surface water concentrations of Pb (Schaule and Patterson, 1981), and ²¹⁰Pb (Landing and Bruland, 1980) together with smoothed trends for Mn (Landing and Bruland, 1980), Cu and Ni (Bruland, 1980) along a transect between Hawaii and Monterey, California. Numbers 11 through 25 identify stations along this transect which leaves Hawaii to the northeast, but which has a break at station 17, turning more easterly. At station 15, which is located closest to the North Pacific Gyre (highest salinity along transect), the horizontal metal profile reaches its maximum for Pb and its minimum for Cu and Ni. This figure is adapted from Schaule and Patterson (1981).

Mediterranean surface waters would be expected before human intervention as a consequence of less frequent recycling of nutrients compared to the open ocean. Although this is probably true, mass balance considerations indicate that, with the possible exception of Ni, the present levels of these elements in surface and deep Western Mediterranean waters may reflect primarily eolian inputs of anthropogenic origin. Also, when considering such inputs (Arnold et al., 1982) and the Mediterranean waters concentration data, it appears that most of the anthoropgenic Cu, Zn and Cd delivered to these waters is exported to the Atlantic Ocean within the Mediterranean outflow water. Trace metal anomalies have been observed in this outflow over more than 1,000 km from the Gibraltar Straits (Isley et al, in preparation). For Cu, Zn and Cd, such results would indicate little removal of these elements within the Western Mediterranean Basin, in agreement with the model calculations of Boyle et al. (1985) for nutrient-type elements. On the other hand, only 10% of the pollutant lead entering the Western Mediterranean via atmospheric deposition appears to be exported to the Atlantic Ocean. Most of this lead should therefore be deposited in Western Mediterranean sediments. This conclusion is substantiated by analyses of lead in Mediterranean surficial sediments (Lambert, pers. comm.). This implies that the residence time of lead in Western Mediterranean waters is much shorter (4 years) than that of the turnover time of Western Mediterranean waters (20 years). Similar conclusions can be reached when considering that the relative abundance of Pb to the other "conservative" metals is about 4 times lower in Mediterranean waters than in the atmospheric deposition flux.

5. CONCLUSIONS

Clearly much more work is needed before we can properly evaluate the impact of atmospheric input on the biogeochemical cycling of metals in the ocean and these natural cycles must be understood before we can clearly identify the anthropogenic perturbations. There is a need of carefully coordinated field programs to estimate more accuratly net atmospheric fluxes to the ocean, as well as vertical particulate fluxes in the ocean at the mesoscale (coastal regions, enclosed seas) and at the global scale. Data from the southern hemisphere, where the atmospheric flux, both natural and anthropogenic, is a minimum, should be obtained with high priority to assess natural processes and fluxes in the ocean. Also, the observed temporal variabilities for both the atmospheric flux and the biogenic particulate flux clearly indicate that stready-state approaches are not valid on time scales of the order of one month, or when we consider gradually increasing inputs of pollutants. Sampling programs over several years are probably needed at a given location. Together with atmospheric flux assessments, the sampling strategy in the water column should consider concurrent sampling with sediment traps, large volume in situ filtration units and conventional, small-volume water sampling. In this context, continued intercalibration of collection and analytical methods for trace elements in the marine atmosphere and ocean is mandatory.

Using a few examples, I have attempted to show that it is absolutely necessary to take into account the interplay between input source functions, hydrography and removal for an accurate understanding of the processes controlling the distribution and water column variability of trace elements in the ocean. As accurate data begin to accumulate a fascinating picture of the variable influence of eolian inputs both natural and anthropogenic on the chemistry of marine waters is emerging. The time scales of the physical processes in the ocean are and will be precisely calibrated through the use of the anthropogenic transient tracers (Tritium, Helium-3, Freons, Carbon-14, Krypton-85; Broecker and Peng, 1982) which have a well defined input function from the atmosphere. It should thus be possible in the near future to quantify the mechanisms by which reactive elements are removed from the ocean surface and as a consequence to predict the effects of the pollution. Also, with more accurate source-emission inventory data, we will have in hand the means to monitor and model the future evolution of the geochemical perturbations induced by human activities. For example, the quantity of Pb entering the atmospheric and marine environment has decreased since 1975 and will probably continue to decrease, due to the phasing out of leaded gasoline in U.S.A., and soon in other countries. Model-calculations based on a simple box-model (Boyle, pers. comm.) suggest that the lead concentrations in the surface waters of the ocean should decline with a response time of a few years following the change in input. This offers the perspective for beautiful, large-scale experimental situations in atmospheric and marine geochemistry.

ACKNOWLEDGEMENTS

I wish to thank the people at the ASI who contributed comments and suggestions to this manuscript, especially R. Collier, Q. Espey, C. Jeandel and H. Maring. Thanks are also due to C.E. Lambert for stimulating discussions over these last years. This is contribution C.F.R. n° 729.

REFERENCES

Andreae, M.O., 1980: 'Arsenic in rain and the atmospheric mass balance of arsenic'. J. Geophys. Res., 85, 4513-4518.

Andreae, M.O. et al., 1984: 'Changing Biogeochemical Cycles'. In: Changing Metal Cycles and Human Health. (T.O.Nriagu, ed.), Dahlem Konferenzen, Berlin, Springer-Verlag, 359-374.

Arimoto, R., R.A. Duce, B.J. Ray and C.K. Unni, 1985: 'Atmospheric trace elements at Enewetak Atoll: 2. Transport to the ocean by wet and dry deposition'. J. Geophys. Res., 90, 2391-2408.

Arnold, M., 1985: Géochimie et transport des aérosols métalliques au-dessus de la Méditerranée occidentale. Ph.D. Thesis, University of Paris VII, 226 pp.

Arnold, M., P. Buat-Ménard and R. Chesselet, 1981: An estimate of the input of trace metals to the global atmosphere by volcanic activity. RVEAC Symp. IAMAP, Hamburg, 17-28 August. Abstract.

Arnold, M., A. Seghaier, D. Martin, P. Buat-Ménard and R. Chesselet, 1982: 'Géochimie de l'aérosol marin au-dessus de la Méditerranée occidentale'. Workshop on Pollution of the Mediterranean. Cannes, December 2-4, CIESM, 27-37.

Bacon, M.P. and R.F. Anderson, 1982: 'Distribution of Th isotopes between dissolved and particulate forms in the deep sea'. J. Geophys. Res., 87, 2045-2056.

Bacon, M.P., D.W. Spencer and P.O. Brewer, 1976: '^{2}&?Pb/^{22} Ra and ^{2}&?Po/^{2}&?Pb disequilibrium in seawater and suspended particulate matter'. Earth Planet. Sc. Lett., 32, 277-296.

Bacon, M.P., C. Huh, A.P. Fleer and J.W. Murray, 1981: 'Seasonality in the flux of natural radionuclides and plutonium in the deep

Sargasso Sea'. Deep Sea Res. 32, 273-286.

Balistrieri, L., P.G. Brewer and A.R. Barringer, 1977: 'Scavenging residence times of trace metals and surface chemistry of sinking particles in the deep ocean'. Deep Sea Res. 28A, 101-121.

Beauford, W.J., J. Barlser and A.R. Barringer, 1977: 'Release of particles containing metals from vegetation into the atmosphere'. Science, 195, 571-573.

Bishop, J.K.B., M.R. Ketten and J.M. Edmond, 1978: 'The chemistry, biology and vertical flux of particulate matter from the upper 400 m of the Cape Basin in the Southeast Atlantic Ocean'. Deep Sea Res., 25, 1121-1161.

Boyle, E.A., S.D. Chapnick, X.X. Bai and A. Spivack, 1985: 'Trace metal enrichments in the Mediterranean Sea'. Earth Planet. Sc. Lett., 74, 405-419.

Broecker, W.S. and T.H. Peng, 1982: Tracers in the Sea, Lamont-Doherty Geological Observatory, Palisades, N.Y., 690 pp.

Bruland, K.W., 1980: 'Oceanographic distributions of cadmium, zinc, nickel and copper in the North Pacific'. Earth Planet. Sc. Lett., 47, 176-198.

Bruland, K.W., 1983: 'Trace elements in Sea Water'. In: Chemical Oceanography, vol. 8, (J.P. Riley and R. Chester eds). Academic Press, New-York, 157-220.

Buat-Ménard, P., 1983: 'Particle geochemistry in the atmosphere and oceans'. In: Air-Sea Exchange of Gases and Particles, (P.S. Liss and W.G.N. Slinn, eds), D. Reidel Publishing Company, 455-532.

Buat-Ménard, P., 1984: 'Fluxes of metals through the atmosphere and oceans': In: Changing Cycles of Metals and Human Health (T.O. Nriagu, ed.), Dahlem konferenzen, Springer Verlag, Berlin, 43-69.

Buat-Ménard, P. and R. Chesselet, 1979: 'Variable influence of the atmospheric flux on the trace metal chemistry of oceanic suspended matter'. Earth Planet. Sc. Lett., 42, 399-411.

Buat-Ménard, P. and R.A. Duce, 1985: 'Metal transfer across the air-sea interface: myths and mysteries'. In: Metal cycling in the Environment (T.C. Hutchinson, ed.,), Wiley, New York (in press).

Byrd, J.T. and M.O. Andreae, 1982: 'Tin and Methyllin species in seawater: concentration and fluxes'. Science, 218, 565-569.

Cambray, R.S., D.F. Jeffries and G. Tapping, 1975: An estimate of the input of trace metals to the North Sea and the Clyde Sea (72-73). U.K. A.E.A., Harwell Report, 7733.

Chesselet, R., 1979: 'Modes of settling and organic input to the sediment seawater interface, a review'. Coll. Inter. CNRS. N°293, 27-33.

Chilvers, D.C. and P.J. Peterson, 1985: 'Global cycling of Arsenic'. In: Metal cycling in the Environment (T.C. Hutchinson ed.), Wiley, New York (in press).

Collier, R. and J. Edmond, 1984: 'The trace element geochemistry of marine biogenic particulate matter'. Prog. Oceanog., 13, 113-199.

Crecelius, E., 1980: 'The solubility of coal fly ash and marine aerosols in water'. Mar. Chem., 8, 245-250.

Dehairs, R., R. Chesselet and J. Jedwab, 1980: 'Discrete suspended particles of barite and the barium cycle in the open ocean'. Earth

Planet. Sc. Lett., 49, 528–550.

Deuser, W.G., P.G. Brewer, T.D. Jickells and R.F. Comneau, 1983: 'Biological control of the removal of abiogenic particles from the surface ocean'. Science, 219, 388–391.

Duce, R.A., G.T. Wallace and B.J. Ray, 1976a: Atmospheric trace metals over the New York Bight. NOAA Technical Report. ERL 361-MESA 4.

Duce, R.A., G.L. Hoffman, B.J. Ray, I.S. Fletcher, G.T. Wallace, J.L. Fasching, S.R. Piotrowicz, P.R. Walsh, E.J. Hoffman, J.M. Miller, and J.L. Heffter, 1976b: 'Trace metals in the marine atmosphere: Sources and Fluxes'. In: Marine pollutant transfer (H. Windom and R. Duce, eds.), D.C. Heath and Co., Lexington, Mass., 77–119.

Elderfield, H. and J.M. Greaves, 1982: 'The rare earth elements in sea water'. Nature, 296, 214–219.

Fitzgerald, W.F., G.A. Gill and A.D. Hewitt, 1983: 'Air-sea exchange of mercury'. In: Trace metals in Sea Water (C.S. Wong et al., eds), Plenum Press, New York, 297–315.

Fitzgerald, W.F., G.A. Gill and J.P. Kim, 1984: 'An equatorial Pacific Ocean source of atmospheric mercury'. Science, 224, 597–599.

Flegal, A.R. and C.C. Patterson, 1983: 'Vertical concentration profiles of lead in the Central Pacific at 15°N and 20°S'. Earth Planet. Sc. Lett., 64, 19–32.

Harrison, R.M. and D.H.P. Laxen, 1978: 'Sink processes for tetraalkyllead compounds in the atmosphere'. Environ. Sci. Technol., 12, 1384–1392.

Hodge, V., S.R. Johnson and E.D. Goldberg, 1978: 'Influence of atmospherically transported aerosols on surface ocean water composition'. Geochem. J., 12, 7–20.

Hunter, K.A., 1981: 'Microelectrophoretic properties of natural surface active organic matter in coastal seawater'. Limnol. Oceanogr., 25, 807–822.

Hunter, K.A. and P.S. Liss, 1979: 'The surface charge of suspended particles in estuarine and coastal waters'. Nature, 2821, 823–825.

Jickells, T.D., A.H. Knap and T.M. Church, 1984: 'Trace metals in Bermuda Rainwater'. J. Geophys. Res., 89, 1423–1428.

Kaufman, A., R. Frier and W.S. Broecker, 1973: 'Distribution of ^{228}Ra in the world ocean'. J. Geophys. Res., 73, 8827–8836.

Kremling, K., 1985: 'The distribution of cadmium, copper, nickel, manganese and aluminium in surface waters of the open Atlantic and European shelf area'. Deep Sea Res., 32, 531–555.

Lal, D., 1977: 'The organic microcosm of particles'. Science, 198, 997–1009.

Lal, D., 1980: 'Comments on some aspects of particulate transport in the oceans'. Earth Planet. Sc. Lett., 49, 520–527.

Lambert, C.E., 1981: Le cycle interne du fer et du manganese et leurs interactions avec la matière organique dans l'océan. Ph. D. Thesis, Université de Picardie, 235 pp.

Lambert, C.E., C. Jehanno, J.S. Silverberg, J.C. Brun-Cottan and R. Chesselet, 1981: 'Log-normal distributions of suspended particles in the open ocean'. J. Mar. Res., 39, 77–98.

Lambert, C.E., J.K.B. Bishop, P.E. Biscaye and R. Chesselet, 1984: 'Particulate aluminium, iron and manganese chemistry at the deep

Atlantic boundary layer'. Earth Planet. Sc. Lett., 70, 237-248.

Landing, W. and K.W. Bruland, 1980: 'Manganese in the North Pacific'. Earth Planet. Sc. Lett., 49, 45-56.

Laumond, F., G. Copin-Montegut, P. Coureau and E. Nicolas, 1984: 'Cadmium, copper and lead in the western mediterranean'. Mar. Chem., 15, 251-261.

Lion, L.W. and J.O. Leckie, 1981: 'The biochemistry of the air-sea interface'. Ann. Rev. Earth Planet. Sc., 9, 449-486.

Lowman, F.G., T.R. Rice and F.A. Richards, 1971: 'Accumulation and redistribution of radionuclides by marine organisms'. In: Radioactivity in the Marine Environment, N.A.S., Washington DC, 161-199.

Martin, J.H. and G.A. Knauer, 1973: 'The elemental composition of plankton'. Geochim. Cosmochim. Acta, 37, 1639-1653.

Measures, C.I., B. Grant, M. Khadem, D.S. Lee and J.M. Edmond, 1984: 'Distribution of Be, Al, Se and Bi in the surface waters of the western North Atlantic and Carribean'. Earth Planet. Sc. Lett., 71, 1-12.

Moore, R.M., J.E. Milley and A. Chatt, 1984: 'The potential for biological mobilization of trace elements from aeolian dust in the ocean and its importance in the case of iron'. Oceanol. Acta., 7, 221-228.

Mosher, B.W. and R.A. Duce, 1983: 'Vapor-phase and particulate selenium in the marine atmosphere'. J. Geophys. Res., 88, 6761-6768.

Mullins, B.M., 1978: Geochemical aspects of atmospherically transported trace metals over the Georgia Bight. M.S. Thesis, Georgia Institute of Technology, Atlanta, GA, 66pp.

Murozumi, M., T.J. Chow et C.C. Patterson, 1969: 'Chemical concentrations of pollutant lead aerosols, terrestrial dusts and sea salts in Greenland and Antarctic snow strata'. Geochim. Cosmochim. Acta, 33, 1247-1294.

Nicolas, E., P. Coureau, G. Copin-Montegut and J.P. Bethoux, 1985: 'Trace metals in the Mediterranean'. Terra Cognita, 5, 189.

Nozaki, Y., J. Thomson and K.K. Turekian, 1976: 'The distribution of ^{210}Pb and ^{210}Po in the surface waters of the Pacific Ocean'. Earth Planet. Sc. Lett., 22, 304-312.

Rohbock, E., H.W. Georgii and J. Muller, 1980: 'Measurements of gaseous lead alkyls in polluted atmospheres'. Atmos. Environ., 14, 89-98.

Schaule, B.K. and C.C. Patterson, 1981: 'Lead concentrations in the Northeast Pacific: Evidence for global anthtopogenic perturbations'. Earth Planet. Sc. Lett., 54, 97-116.

Settle, B.K. and C.C. Patterson, 1982: 'Magnitudes and sources of precipitations and dry deposition fluxes of industrial and natural leads to the north Pacific at Enewetak'. J. Geophys. Res., 87, 8857-8869.

Turekian, K.K. and J.K. Cochran, 1981: '^{210}Pb in surface air at Enewetak and the Asian dust flux to the Pacific'. Nature, 292, 522-524. Corrigenda, Nature 294, 670.

Turekian, K.K., Y. Nozaki and L.K. Benninger, 1977: 'Geochemistry of atmospheric radon and radon products'. Ann. Rev. Earth Planet. Sc., 5, 227-255.

Wallace, G.T. Jr., G.L. Hoffman and R.A. Duce, 1977: 'The influence of organic matter and atmospheric deposition on the particulate trace metal concentration of northwest Atlantic surface seawater'. Marine Chem., 5, 143-190.

Wallace, G.T. Jr., O.M. Mahoney, R. Dulmage, F. Storti and N. Dudek, 1981: 'Particulate aluminium in oceanic surface layers- evidence for first order removal'. Nature, 293, 729-731.

Wallace, G.T. Jr., N. Dudek, R. Dulmage and O.M. Mahoney, 1983: 'Trace element distributions in the Gulf Stream adjacent to the Southeastern Atlantic continental shelf- influence of atmospheric and shelf waters inputs'. Cann. J. Fish. Aquat. Sci., 40, 183-191.

Walsh, P.R. and R.A. Duce, 1976: 'The solubilization of anthropogenic atmospheric vanadium in seawater'. Geophys. Res. Lett., 3, 375-378.

Walsh, P.R., R.A. Duce and J. Fasching , 1979a: 'Tropospheric arsenic over marine and continental regions'. J. Geophys. Res., 84, 1710-1718.

Walsh, P.R., R.A. Duce and J.L. Fasching, 1979b: 'Considerations of the enrichment, sources and fluxe of arsenic in the troposphere'. J. Geophys. Res., 84, 1719-1726.

Windom, H.L., 1981: 'Comparison of atmospheric and riverine transport of trace elements to the continental shelf environment'. In: River Inputs to Ocean Systems. UNEP and UNESCO, 360-369.

Wood, J.M., 1974: 'Biological cycles for toxic elements in the environment'. Science, 183, 1049-1052.

Zoller, W.H., 1984: 'Anthropogenic perturbations of metal fluxes into the atmosphere'. In: Changing Metal Cycles and Human Health (T.O. Nriagu, ed.) Dalhem Konferenzen, Springer Verlag, Berlin, 27-41.

THE IMPACT OF ATMOSPHERIC NITROGEN, PHOSPHORUS, AND IRON SPECIES ON MARINE BIOLOGICAL PRODUCTIVITY

Robert A. Duce
Center for Atmospheric Chemistry Studies
Graduate School of Oceanography
University of Rhode Island
Kingston, Rhode Island 02881 U.S.A.

1.   INTRODUCTION

1.1 Atmospheric Deposition: Geochemical and Biological Impact

Marine chemists have long been concerned with the geochemical cycling of trace species in the ocean and with evaluating the sources, sinks, and transport paths of chemicals through the marine system. In recent years there has been a growing realization that rivers - previously believed to be the primary input path for materials entering the ocean - may be equalled and even surpassed in some cases by other transport paths and sources. Recent investigations of hydrothermal vents and submarine volcanism have highlighted the importance of these sources on the ocean floor for a number of substances in the ocean. Over the past ten years it has become apparent that the atmosphere can also be an important transport path for many substances entering the sea. Indeed, one of the primary objectives of this Advanced Study Institute is an evaluation of the importance of air/sea exchange processes in controlling or affecting the concentrations of a wide variety of marine organic and inorganic substances.
 Equally important, of course, is the impact these atmospherically derived substances may have on ocean chemical processes and biological systems. It is the latter topic, an evaluation of the importance of certain atmospherically derived substances to biological productivity, that will be the subject of this chapter. In the past it has often been impossible to evaluate the biological and biochemical impacts of atmospheric inputs since accurate estimates of the input fluxes were not available. This is now changing rapidly as the scope and areal coverage of high quality research programs generating reliable input data have increased. Investigations in both coastal and open ocean regions now provide us with sufficient data that we can begin to assess the potential impact of atmospheric deposition of several substances on the biological mobilization of these materials in the sea. However, I must stress that this is still an area of research in its infancy and with many outstanding questions.
 Once a substance enters the ocean its past history becomes

497

P. Buat-Ménard (ed.), The Role of Air-Sea Exchange in Geochemical Cycling, 497–529.
© *1986 by D. Reidel Publishing Company.*

unimportant, and it is extremely difficult to determine the fraction
of that substance in the ocean that is derived from the atmosphere.
Thus we are left with two options as we try to assess the biological
importance of the atmospheric component:

1.) to attempt to compare the atmospheric input rate of a
substance in surface ocean waters with its input from other sources; or

2. to simply evaluate how a particular substance entering the
ocean from the atmosphere is utilized by biological systems, with no
quantitative information about relative source strengths.

In this paper I will explore option 1 and will concentrate on
only three elements important for marine productivity. I will attempt
to summarize what we know and what we might reasonably expect to be
the impact of atmospheric deposition of these elements in selected
ocean environments. Elements considered include nitrogen, phosphorus,
and iron.

I must point out that I will only be considering the "immediate",
or short term sources for these materials used biologically in the
photic zone. In general this means a comparison of atmospheric input
with mechanisms transporting material vertically from deeper waters.
We will not discuss the more fundamental geochemical question of the
longer time scale sources or transport paths for these materials in
the ocean as a whole, e.g. riverine input, submarine volcanism and hot
vents, and release from sediments, relative to atmospheric transport.

There are many problems associated with making accurate estimates
of the atmospheric input of various chemical species to the ocean by
wet and dry removal processes. Many of these problems are discussed
in more detail in the chapter "The Ocean as a Sink for Atmospheric
Particles" by Buat-Menard in this volume and will not be considered
here. However, one point deserves brief mention. The gross
deposition of a substance to the ocean, as determined from rain
chemistry or by a model applying deposition velocities to atmospheric
concentrations, is composed of a net input (i.e. that component
derived from some non-marine source) as well as a component associated
with either recycled sea salt aerosols or marine-derived gases (see
e.g. Arimoto et al., 1985; Jickells et al., 1984; Settle and
Patterson, 1982). For substances present on aerosol particles, simple
use of the ratio of substance X to Na in seawater to attempt to
correct for the recycled marine component is often not successful.
This is due to the scavenging by rising bubbles of a fraction of
substance X associated with surface active organic material (see e.g.
Weisel et al., 1984; Buat-Menard, 1983; Williams, 1967; MacIntyre,
1974; Duce and Hoffman, 1976). An accurate correction for this
chemical enrichment process is quite difficult.

An additional complication is related to the possible
gas-to-particle conversion of marine derived gases over the ocean.
Some fraction of a substance present an aerosols entering the ocean
may be recycled material if that substance or a precursor was
initially emitted from the sea in a gaseous form. Examples of
elements for which this process is likely quite important include
sulfur (see Andreae chapter in this volume) and selenium (Mosher and
Duce, 1983). Some species can degas from the sea salt aerosol leading

to a vapor phase component which may then reenter the ocean.
Excellent examples are HCl released by the acidification of
atmospheric sea salt (Clegg and Brimblecombe, 1985) and boron released
as boric acid (Fogg and Duce, 1985).

Evaluating the net atmospheric input to the ocean is far from
hopeless, however. There have been remarkable strides in the past few
years in analytical capabilities, in collection methods, in
contamination avoidance techniques, in temporally extended and
geographically extensive sampling programs, and in the development of
more sophisticated modeling approaches. We now can estimate the net
input of a number of substances from the atmosphere to the ocean much
more accurately, even though uncertainties on a global scale are still
often a factor of 3 to 5. However, most importantly, we can perhaps
finally say that at least we are becoming aware of many of the
deficiencies in making these estimates. This is certainly a
necessary, if not sufficient, step toward an accurate evaluation of
the net input of these materials to the ocean.

1.2 Nutrients and Primary Productivity

Before we can evaluate the importance of the atmospheric input of
nitrogen or other species to biological systems in marine surface
waters, it is useful to review very briefly chemical mass balances and
sources of nutrients involved in primary production. This topic is
reviewed in detail by Eppley (1980), Fogg (1975, 1982), Eppley and
Peterson (1979), Dugdale and Goering (1967), and McCarthy and
Carpenter (1983), among others.

Primary production in the ocean depends on the input of nutrients
to the photic zone from fixation (in the case of nitrogen) in surface
waters, from deeper waters, and from the atmosphere (new production)
as well as from nutrient recycling in the surface waters (regenerated
production) (see e.g. Eppley and Peterson, 1979; Dugdale and Goering,
1967). There has been considerable discussion and some controversy
concerning which nutrient limits primary productivity in the ocean.
There now seems to be general agreement that in fresh waters
phosphorus is limiting, while in coastal ocean waters and probably
most regions of the open ocean, at least on short time scales,
nitrogen is limiting (e.g. see Ryther and Dunstan, 1971). At times
phosphorus and other trace substances (e.g. silicate, iron, vitamin
B_{12}, etc.) may also be critical for primary production in the open
ocean (Ryther and Dunstan, 1971; Howarth and Cole, 1985; Thomas, 1969;
Ryther and Guillard, 1959; Fogg, 1975; Thayer, 1974; Eppley et al.,
1979). On longer times scales most geochemists believe phosphorus is
ultimately the limiting nutrient in the ocean because nitrogen
deficits can be met by nitrogen fixation, with nitrogenous nutrients
accumulating until the available phosphorus is utilized (Redfield,
1958; Smith, 1984; Broecker and Peng, 1982).

As pointed out by Eppley and Peterson (1979), new production is
equivalent to that material (carbon) which sinks from the euphotic
zone to the deep ocean, that present in the fish and shellfish catch,
and the deposition of guano on land by birds. The latter two removal

pathways are believed to be small compared with sinking particles. Assuming the entire system is in steady state, the remaining material is termed regenerated production and, in the case of nitrogen, is derived from ammonia, urea, amino acids, and other dissolved organic nitrogen compounds which are produced by animals and heterotrophic microorganism metabolism in the photic layer. Thus new production is supported by nutrients entering the photic layer from all other sources (e.g. the atmosphere, vertical transport from deep water, riverine input in coastal areas, and nitrogen fixation in the surface waters). These processes are illustrated simply in Figure 1 (adapted from Eppley and Peterson, 1979). Obviously in an ideal closed system with no losses from the bottom or top, nutrients could continue to cycle through an enclosed food web indefinitely. In the ocean, however, the loss of sinking particles to deep water, the fish catch, and the guano deposits result in an export of nitrogen and other nutrients which must be balanced by inputs from other sources, i.e. the new nutrients responsible for the new production.

SURFACE OCEAN PRODUCTIVITY

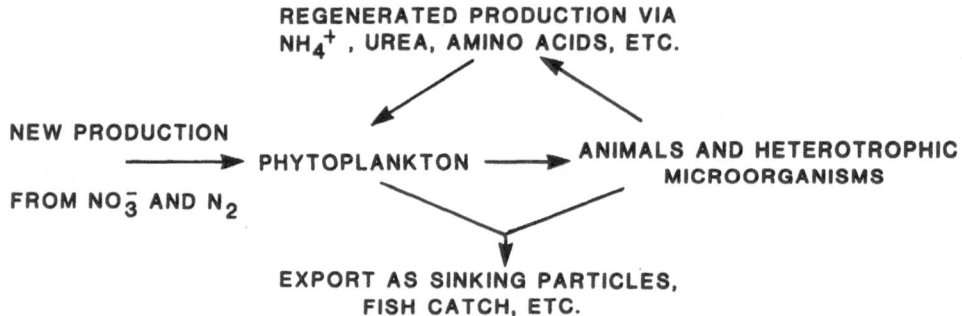

Figure 1. Production system in the surface ocean illustrating new and regenerated productivity. Adapted from Eppley and Peterson (1979).

1.3 A Simple Model of Vertical Transport to the Photic Zone

To evaluate the importance of the atmospheric component of these new nutrient substances in open ocean regions it is necessary to compare the atmospheric fluxes to the other fluxes entering the photic zone, i.e. vertical transport and fixation (in the case of nitrogen). In our discussions we will consider two rather geographically different oceanic regions - the area near Bermuda in the Sargasso Sea in the North Atlantic and the region of the Central North Pacific Gyre near Hawaii. As a first crude approximation, the vertical flux of a substance, X, from deeper waters into the photic zone can be assumed to take place by advection and by eddy diffusion. We will use a very simple one-dimensional model to evaluate the upward flux of material into the photic zone by these two processes With this simple model the vertical flux can be represented by:

$$F = wR + K_z(dX/dz) \qquad (1), \qquad where$$

F is the combined flux of X from deeper water, in mol X m^{-2} d^{-1}
w is the vertical velocity of the water, in m d^{-1},
R is the mean concentration of substance X in the meaningful portion
  of the X gradient below the photic zone, in mol X m^{-3},
K_z is the coefficient of eddy diffusivity, in m^2 d^{-1}, and
dX/dz is the X gradient, in mol X m^{-4}

Values for w in the North Atlantic have been given by Leetma and Bunker (1978). In the area around Bermuda they found w to vary seasonally and geographically. They found about +0.04 m d^{-1} (upward transport) for much of the year, but as low as -0.08 m d^{-1} (downward transport) during the winter. For our calculations of advective input to the photic zone we will use a conservative range of 0 to +0.04 m d^{-1}, recognizing that at some times vertical advection will be downward, not upward. In the Pacific Munk (1966) obtained a value of ~+0.012 m d^{-1}, which we will use in this region.

In the Sargasso Sea region Rooth and Ostlund (1972) obtained a value for K_z of 1.7 m^2 d^{-1} from tritium profiles. A value of 3.4 m^2 d^{-1} was found by Wunsch and Minster (1982) at 36°N in the North Atlantic. In the northeast Pacific a maximum value of 10 m^2 d^{-1} was found by Pritchard et al. (1971) using ^{14}C data. In the same region a value of 1.7 m^2 d^{-1} was determined by Roether et al. (1970) from GEOSECS data. We will use a range of 1.7-10 m^2 d^{-1} in our considerations for both regions, similar to the range of values used by McCarthy and Carpenter (1983) in their consideration of nitrogen fluxes in these regions. Since these values of w and K_z will be used in several sections of this paper, they are summarized in Table 1.

Table 1. Assumed Values for w and K_z

Region	w (m d^{-1})	K_z (m^2 d^{-1})
Sargasso Sea	0 - 0.04	1.7 - 10
North Pacific Gyre	0.012	1.7 - 10

I must emphasize again the simplistic approach being used here. Recent results, to be discussed later, point out clearly that on short timescales strong ocean-atmosphere dynamical coupling may result in very effective turbulent transport of nutrients from deep waters into the photic zone. Because this transport is likely sporadic, as would

be the productivity it stimulates, it may not be easily related to the
non-realistic mean conditions assumed in the model above.

2.   NITRATE AND OTHER NITROGEN SPECIES

2.1. Calculations of Nitrogen Fluxes in Open Ocean Regions

2.1.1. <u>Introduction</u>. Nitrogen species are required nutrients for
biological growth. The upward flux of nitrate from deep water into
the euphotic zone has traditionally been believed to be the primary
source for the new nitrogenous nutrients, with atmospheric input and
nitrogen fixation believed to be considerably smaller sources.
     Until recently our understanding of nitrogen cycling in the
near-surface waters of the ocean could be summarized by arguements
given by McCarthy and Carpenter (1983). These authors point out that
in recent years we have been able to improve substantially the
estimates for nitrogen supply to the euphotic zone from nitrogen
fixation and remineralization. They also point out that the same can
be said for atmospheric washout, although they only consider wet
deposition of ammonium ion in their discussion. These authors believe
that less progress has been made in quantifying the upward transport
of nitrate from deeper waters, which they believe is the dominant
source of new nitrate to the euphotic zone and which they estimate in
some detail in their paper. McCarthy and Carpenter (1979) also point
out that riverine input of nitrogen species to the coastal ocean is
still poorly understood. It is instructive to evaluate carefully the
estimates of nitrogen fluxes derived by McCarthy and Carpenter (1983).

2.1.2. <u>Nitrogen Fluxes from Deep Waters</u>. For their estimates of the
nitrate fluxes in the water column McCarthy and Carpenter use a value
for R of ~0.7 mmol NO_3-N m^{-3} for the Sargasso Sea at a depth of
~100m. A range of 0.5-1.0 mmol NO_3-N m^{-3} is observed in the
GEOSECS (1981) data near Bermuda. For the North Pacific Gyre region
McCarthy and Carpenter (1983) use a range from 0.8 (Gundersen et al.,
1976) to 1.8 mmol NO_3 m^{-3} (Kiefer et al., 1976) observed at
150-175m for R. Similar values are observed from the GEOSECS (1982)
data. With the values of w given previously and using Equation 1,
these values for R result in an advective flux ranging from 0 to +40
μmol NO_3-N m^{-2} d^{-1} in the Sargasso Sea and 10-22 μmol
NO_3-N m^{-2} d^{-1} in the North Pacific.
     For the calculation of the eddy diffusion flux in the Sargasso
Sea region the nitrate gradient, dN/dz, used by McCarthy and Carpenter
was 0.013 mmol NO_3-N m^{-4}. A value of 0.021 mmol NO_3-N
m^{-4} is found from the GEOSECS (1981) data in this region. We will
use a range of 0.013 to 0.021 mmol NO_3-N m^{-4}. In the North
Pacific McCarthy and Carpenter used a range for dN/dz of 0.035 to 0.05
mmol NO_3-N m^{-4} as determined by Gundersen et al. (1976) and
Kiefer et al. (1976) respectively. Somewhat higher values of 0.061 to
0.073 mmol NO_3-N m^{-4} are determined from the GEOSECS (1982)
data. We will use a range of 0.035 to 0.073 mmol NO_3-N m^{-4} for

dN/dz.

Unfortunately at this point McCarthy and Carpenter (1983) made several errors in their paper, and I am indebted to A. Pszenny who first pointed out several of these to me. One typographical error gave the Gundersen et al. (1976) gradient as 0.35 μmol NO_3-N m^{-4} rather than the correct units of 0.035 mmol NO_3-N m^{-4}, although the correct values were apparently used in the calculation. The more serious error involved the conversion of units for K_z. The original data were given in cm^2 s^{-1}, and when these were converted to m^2 h^{-1} in their paper an order of magnitude error resulted. For example, a K_z value of 0.20 cm^2 s^{-1} was converted to 0.72 m^2 h^{-1} instead of the correct value of 0.072 m^2 h^{-1}. (Note, we are using units of m^2 d^{-1}.) The result is that all the eddy diffusion fluxes calculated by McCarthy and Carpenter (1983) are too high by a factor of 10. The eddy diffusion fluxes calculated using the correct range of values of K_z given in Table 1 are 22 to 210 μmol NO_3-N m^{-2} d^{-1} in the Sargasso Sea area and 60 to 730 μmol NO_3-N m^{-2} d^{-1} in the North Pacific (see Table 2). Platt et al. (1982) also estimated a maximum diffusive flux in this region of ~800 μmol N m^{-2} d^{-1}.

Table 2. Comparison of New Nutrient Nitrogen Source Strengths in Surface Waters of the Sargasso Sea and the North Pacific Gyre

Source	Flux (μmol N m^{-2} d^{-1})	
	Sargasso Sea	North Pacific Gyre
Vertical advection	0-40	10-22
Eddy diffusion	22-210	60-730
Nitrogen fixation	0.5-5	0.2-2
Atmospheric input	26-54	8-26
Total	50-310	80-800

2.1.3. Nitrogen Fluxes from Fixation. Measurements and estimates of nitrogen fixation can vary by two to three orders of magnitude in the ocean, depending on season, location, etc. (see e.g. Carpenter, 1983; Platt et al., 1982; Fogg, 1978). Carpenter (1983) assumes Trichodesmium is the primary species responsible for marine nitrogen fixation. From his review we will use a fixation rate range of 0.2 to 2 μmol N m^{-2} d^{-1} for the North Pacific and 0.5 to 5 μmol N m^{-2} d^{-1} for the Sargasso Sea region, although these values are quite uncertain.

2.1.4. Nitrogen Fluxes from the Atmosphere. The atmospheric input
flux of nitrogen at both sites was calculated by McCarthy and
Carpenter (1983) from the data of Menzel and Spaeth (1962b). This was
one of the earliest papers attempting to evaluate the importance of
atmospheric input of nitrogenous species to the ocean. Menzel and
Spaeth (1962b) presented evidence that a significant fraction of the
ammonia observed in surface waters near Bermuda was derived from
rainfall, pointing out that the surface ocean water ammonium
concentrations were correlated with the five day input rates of
rainfall ammonium ion. These authors found that the total NH_4-N
added to the ocean during an annual period (extrapolated here from 10
months of data) was approximately 10 mmol NH_4-N m^{-2}, with monthly
values ranging from 0.19 to 1.4 mmol NH_4-N m^{-2}. The annual value
is equivalent to an average daily value of 27 μmol NH_4-N m^{-2}
d^{-1}. Note that only NH_4-N was considered in this estimate.

Since the paper of Menzel and Spaeth there have been a number of
investigations of the atmospheric concentrations and deposition to the
ocean of a variety of nitrogen species. The species and input
mechanisms which have been considered include aerosol nitrate and
ammonium (wet and dry deposition) and gaseous ammonia, nitric acid,
and NO_x (direct dry deposition). Let us look briefly at some recent
data to make estimates of total atmospheric nitrogen species input to
the ocean in the Sargasso Sea and Central North Pacific Gyre regions.

Table 3 presents available data on the wet input (i.e. via
precipitation) of nitrogen species to the ocean in remote regions.
The reader is referred to the original papers for details on how these
fluxes were determined. Wet deposition rates for nitrate in the
Sargasso Sea region are apparently on the order of 10 - 20 μmol N
m^{-2} d^{-1}. This is a factor of 2 to 3 higher than most estimates of
a global background input rate for nitrate in rain, and this is almost
certainly related to the higher nitrate concentrations over the
western North Atlantic as a result of pollution sources in North
America. Input rates for nitrate in the Central North Pacific are in
the range of 3-6 μmol N m^{-2} d^{-1}. Note that there is some
suggestion that input rates may be lower in remote regions of the
South Pacific.

For the wet deposition of ammonium the input rate to the Sargasso
Sea area is approximately 10-15 μmol N m^{-2} d^{-1}, a factor of 2 to 3
less than that estimated from the data of Menzel and Spaeth (1962b).
There are few data on NH_4^+ over the North Pacific, but ammonium
does not appear to be as variable as nitrate, in agreement with the
lower anthropogenic source strength for NH_3. The wet NH_4-N input
rate in the North Pacific region is likely in the range of 2-10 μmol N
m^{-2} d^{-1}.

Table 4 presents the available data on input of nitrogen species
to the ocean via dry deposition. The individual references should
again be consulted for details on how these values were obtained. We
must reiterate that to date there are no methods for directly
measuring dry deposition of these substances, and all these estimates
have a very large uncertainty. The dry deposition of aerosol nitrate
is apparently higher over the Sargasso Sea region than the North

Table 3. Input of Atmospheric Nitrogen Species to the Ocean by Wet Deposition*

Location	Flux (μmol N m^{-2} d^{-1})		Reference
	NO$_3$	NH$_4^+$	
Atlantic Ocean:			
Bermuda	20	9.8	Galloway (1985)
Bermuda	17	13	Knap et al. (1985)
Atlantic Ocean, ship	14	12	Galloway et al. (1983) and Galloway (1985)
Sargasso Sea	3.2 - 16	-	Savoie (1984)
Pacific Ocean:			
Hawaii	1.0 - 6.3	-	Miller pers. comm. (in Galloway, 1985)
North Pacific Gyre	4.4	-	Savoie (1984)
Ship (R/V Korolev)	5.9	12	Galloway (1985)
Ship (R/V Discoverer)	9.8	12	Galloway (1985)
Samoa	1.0	-	Pszenny et al. (1982)
New Zealand	1.6		Pszenny and Duce (unpub. data)
Tasman Sea	7.8	-	Pszenny and Duce (unpub. data)
Indian Ocean:			
Amsterdam Island	4.1	5.5	Galloway and Gaudry (1984)
Global Estimates:			
	2.0 - 5.9	2.0 - 5.9	Logan (1983), Galloway et al. (1982)
	2.0 - 7.8	3.9 - 14	Soderlund and Svensson (1976)
	3.9	-	Savoie (1984)
	5.1	3.9	Bottger et al. (1984)

*Table largely adapted from Galloway (1985)

Pacific, again due to the proximity of North America. A best guess is that the dry nitrate deposition to the Sargasso Sea is 5-15 μmol N m^{-2} d^{-1} while that to the North Pacific is 2-6 μmol N m^{-2} d^{-1}.

Aerosol ammonium data over remote regions are very sparse, and an estimate of 0.2-1 μmol NH$_4$-N m^{-2} d^{-1} at both locations is very uncertain. Similarly there are few data for HNO$_3$, and the estimate

Table 4. Input of Atmospheric Nitrogen Species to the Ocean
by Dry Deposition*

Location	Flux (μmol N m^{-2} d^{-1})					Reference
	NO_3	NH_4^+	HNO_3	NH_3	NO_x	
Sargasso Sea	4.5-19	–	–	–	–	Savoie (1984)
N. Pacific Gyre	5.5	–	–	–	–	Savoie (1984)
New Zealand	2.0	–	–	–	–	Pszenny and Duce, unpub. data
"Remote Marine" Estimates:	1.2	–	1.2-2.0	–	1.2-2.0	Logan (1983)
	1.0	0.20	1.6	–	0.20	Galloway (1985)
	0.06	0.6-2.0	–	4.0	3.9-9.8	Soderlund and Svensson (1976)
	–	–	0.4	–	0.03	Bottger et al. (1978)
	–	–	–	0.6	–	Georgii and Gravenhorst (1977)

*Table largely adapted from Galloway (1985)

of 0.5-2 μmol HNO_3-N m^{-2} d^{-1} at each site is also extremely
tenuous and perhaps too high (Prospero et al., 1985).

Ammonia measurements are also rare in marine areas, and there is
some uncertainty whether the sea is a source or a sink for atmospheric
ammonia. It is likely the ocean can function as both source and sink,
depending upon oceanic and atmospheric conditions. We will assume
there is no net transfer of ammonia, although this will likely be
variable in time and space.

Only in the past 2-3 years have there been reasonably accurate
measurements of NO in the marine atmosphere. Early estimates of the
input of NO_x, such as that by Soderlund and Svensson (1976), are
probably high because of the high values for NO_x they used in their
model. The range of input estimates is still large, even if this
value is discarded, and we will use an estimated range of 0.05-1 μmol
NO_x-N m^{-2} d^{-1} at both sites.

Our best estimates of these atmospheric inputs of nitrogen
species are presented in Table 5. If these individual estimates are
added, the total atmospheric input rate to the Sargasso Sea ranges
from 26-54 μmol N m^{-2} d^{-1} while that in the North Pacific ranges
from 8-26 μmol N m^{-2} d^{-1}. Note that apparently wet deposition
exceeds dry deposition, a generally accepted fact for aerosol species,
and that the aerosol nitrogen flux apparently dominates the gaseous
flux, although the source of the nitrate in rain, i.e. gaseous

Table 5.   Estimates of the Atmospheric Input of Nitrogen Species
to the Sargasso Sea and Central North Pacific Gyre

Flux (μmol N m^{-2} d^{-1})

Sargasso Sea

Wet NO$_3^-$	Wet NH$_4^+$	Dry NO$_3^-$	Dry NH$_4^+$	Dry HNO$_3$	Dry NH$_3$	Dry NO$_x$	Total
10-20	10-15	5-15	0.2-1.0	0.5-2	0	0.05-1	26-54

Central North Pacific Gyre

3-6	2-10	2-6	0.2-1.0	0.5-2	0	0.05-1	8-26

(HNO$_3$) or particulate nitrate, is not clear.

We must consider the possibility of a recycled component of the atmospheric input of nitrogen species. No relationship was found between aerosol nitrate and atmospheric sea salt at Samoa (Duce, 1983) and this has also been found for aerosol samples from a network of island sites in the North Pacific (Prospero and Duce, unpublished results). This suggests most of the aerosol nitrate is not derived from the ocean on sea salt. Logan (1983) and Zafiriou and McFarland (1981) suggest that the ocean could be a significant but not major source of gaseous precursors for nitrate (NO$_x$), but non-marine sources are believed to predominate. Ammonium is found on submicrometer aerosol particles over the ocean, while atmospheric sea salt is primarily on particles greater than 1 μm in radius, suggesting most of the NH$_4^+$ is not derived directly from the ocean on sea salt. It is possible, however, that in regions where the ocean is a source for atmospheric NH$_3$, some of this NH$_3$ may be returned to the ocean in the form of NH$_4^+$ on aerosols and in rain. However, it is likely that most of the atmospheric deposition of nitrogen species represents a true net input to the ocean.

One major question remains, however, and that concerns organic nitrogen. There are few data on organic nitrogen in rain over the ocean. In rain samples collected off the California coast and near Samoa, Williams (1967) found 33% and 40% respectively of the total dissolved nitrogen to be organic. Knap et al. (1985) suggest that significant quantities of N in rain near Bermuda are organically associated. Williams (1967) suggested that much of the organic nitrogen could be recycled from the ocean surface. Clearly this is a fraction of nitrogen in the atmosphere which requires considerable additional study.

2.1.5. Comparison of Nitrogen Sources and a Major Question. In Table 2 the magnitudes of the four sources or transport paths of new nitrogen in the photic zone are compared. According to these data, atmospheric input might account for from 8 to 70% of the new nitrogen input to the Sargasso Sea region and from 1 to 27% to the North Pacific Gyre. These are significant percentages and would suggest that atmospheric input might be quite important as a source of N for marine organisms.

However, Knap et al. (1985) have recently questioned the importance of atmospheric input of fixed nitrogen to primary productivity in the Sargasso Sea area. These authors point out that Jenkins and Goldman (1985) have suggested the one dimensional eddy diffusion/advection model used by McCarthy and Carpenter (1983) to estimate the vertical flux of nitrate may grossly underestimate the flux of nitrate across the thermocline. On the basis of an oxygen budget analysis Jenkins and Goldman (1985) estimate that new productivity in the photic zone is on the order of 50 g C m^{-2} y^{-1} in the Sargasso Sea area, approximately a factor of 10 larger than the new production estimates based on ^{14}C primary productivity which have been previously accepted. This quantity of new productivity would require ~0.65 mol new N m^{-2} y^{-1}, which is equivalent to ~1800 μmol N m^{-2} d^{-1}. As can be seen from Table 2, this is approximately a factor of 10 more nitrogen than that calculated to be available for new productivity in the Sargasso Sea.

Jenkins and Goldman (1985) point out that the model of Klein and Coste (1984) predicts sufficient nitrogen transport from deep water to support their higher level of new productivity. This model is a one-dimensional, time-dependent, second-order, turbulence closure model. The model of Klein and Coste (1984) evaluates the temporal evolution of the physical entrainment of nutrients brought about by the erosion of the thermocline by turbulence, which is induced primarily by wind stress. The model shows that there can be short term bursts of nutrient transport from deep waters into the photic zone. In a simulation run in the western Mediterranean, Klein and Coste (1984) found pulses occurring approximately every 25 hours, with a time duration of about 5-10 hours. Jenkins and Goldman (1985) integrated these short term fluxes in the Mediterranean and calculated an annual nitrate flux of ~0.9 mol NO$_3$-N m^{-2} y^{-1}, equivalent to a daily flux of ~2500 μmol NO$_3$-N m^{-2} d^{-1}. If this flux were also representative of the Sargasso Sea region, this mechanism would produce sufficient nitrate to support the high new productivity value (50 g C m^{-2} y^{-1}) predicted by Jenkins and Goldman (1985).

If this model is representative of turbulent nutrient fluxes into the photic zone in open ocean regions, this source of new nitrogenous nutrients dominates all other sources presented in Table 2. Under these conditions atmospheric input of new nitrogen is obviously unimportant in most open ocean regions. Clearly this is a subject requiring further work. In particular, the model of Klein and Coste (1984) should be applied to other regions of the ocean, and the high new productivity value proposed by Jenkins and Goldman (1985) must be verified.

2.2 Atmospheric Nitrogen Input to Coastal Waters

While the previous discussion highlights the present uncertainties
relative to the importance of atmospheric input to the new nitrogen
utilized in primary productivity in open ocean regions, a recent paper
by Paerl (1985) suggests that in certain coastal and near-urban
regions atmospheric nitrogen input may be quite important. Paerl has
investigated the impact of rainfall on chlorophyll a production and
$^{14}CO_2$ assimilation in the waters of Bogue Sound on the east coast
of North Carolina. Figure 2 presents the results he obtained
following the addition of 10 and 20% v/v distilled water, 10 and
20% v/v rainwater (each from two separate rainfalls), 100 and 200
ppb nitrate and 50 and 100 ppb phosphate to samples of Bogue Sound
water. The pH of Rain A was 5.85, which Paerl (1985) suggests is
typical of marine rain (although much lower pH values have been
observed for marine rains in remote regions, see e.g. Galloway,
1985). The pH of Rain B was 4.05, which Paerl (1985) suggests is
representative of continental rain with considerable anthropogenic
components. The strongly acidic rainfall resulted in increased
chlorophyll a production relative to the more neutral rain. Paerl
relates this increased production to the higher nitrate/nitrite
concentration in the strongly acidic rain. Parallel nitrate and
phosphate additions (see Fig. 2) showed exclusive nitrate stimulation
of productivity. Both rains stimulated productivity for several days,
but the more acidic rain resulted in stimulation lasting 6-7 days
while that of the more neutral rain only continued for 2-3 days. Note
that the 10-20% additions of rain to seawater in the laboratory

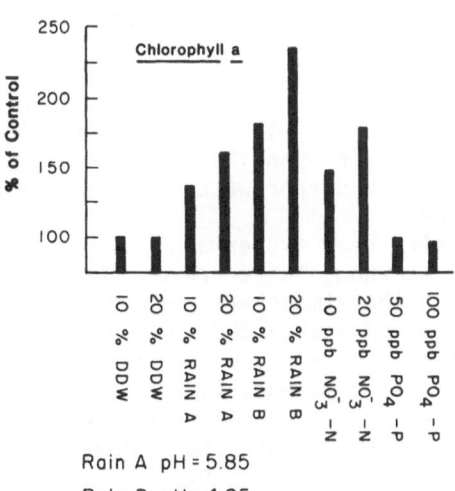

Figure 2. Effect of rainwater on chlorophyll a production in Bogue
Sound, North Carolina. All treatments were compared with 'no
addition' controls. Adapted from Paerl (1985).

experiments appear to be rather large relative to the effect one might expect from several cm of rain falling on the ambient ocean surface.

Paerl (1985) determined that the acidity itself was not responsible for the enhanced production by showing that rain neutralized with NaOH yielded the same enhancement as untreated rain. The addition of the rain to the seawater did not alter the pH of the seawater significantly. Independent bioassays showed that productivity was nitrogen-limited throughout the spring and summer of 1984 in Bogue Sound. Paerl (1985) pointed out that acidic rainfall with its associated high nitrate concentrations may well be important in affecting both the patterns and magnitudes of primary productivity along coastal regions in proximity to extensive urban or industrial activities. In mid-latitudes the eastern coasts of continents may be particularly susceptible to the input of anthropogenic NO_3.

Paerl's paper highlights another aspect of the atmospheric input of nitrogen (and other substances as well) to the ocean - the episodic nature of the input. Precipitation input is discontinuous in time and space and is notably patchy. In addition, source strengths and transport paths for many substances may vary annually, seasonally, or on even shorter time scales. Since in most coastal and open ocean waters nitrogen is the nutrient limiting productivity, large pulses of atmospheric input may be quite important on short time scales. These factors must be considered carefully before the importance of the atmospheric input of nitrogen species to the ocean can be adequately evaluated.

3. PHOSPHORUS

3.1. Atmospheric Phosphorus and the Phosphorus Cycle

Phosphorus is also essential for biological growth in the ocean. There has been relatively little concern about the atmosphere as a transport path for marine phosphate, with most investigators assuming that riverine input is primarily responsible for phosphate in the sea and that vertical transport from deeper waters is responsible for the phosphate in the photic zone.

The atmospheric chemistry of P is certainly much less complex than that of N. As far as we know, there are no significant vapor phase forms of phosphorus in the atmosphere with any appreciable lifetime, although phosphine and methylated forms of phosphine have been suggested. Indeed, these forms may be produced in certain reducing environments, especially swamps, tidal flats, etc. However, it is very likely that their atmospheric lifetime in this form would be extremely short, perhaps only seconds. Certain pesticides containing phosphorus are also undoubtedly present in the vapor phase in small concentrations, but there are virtually no data on these substances. Thus at the present time we can only consider aerosol phosphorus and its transport to the ocean.

There has been relatively little research on the atmospheric phosphorus cycle. In fact data on phosphorus in atmospheric particles

and rain in marine areas are sparse, with most measurements in the
past 10-15 years related to the rainfall input of phosphorus to local
terrestrial ecosystems. Most published global phosphorus cycles omit
consideration of the atmosphere. Only Pierrou (1976) has presented a
phosphorus cycle with an atmospheric component, and even he only
devotes a single page to atmospheric phosphorus. Perhaps the most
comprehensive study of atmospheric phosphorus is that of Graham and
Duce (1979). These authors developed a global atmospheric cycle of
phosphorus and estimated that the net atmospheric transport of
phosphorus into the global ocean was approximately 3.2×10^{10} mol
y^{-1}. This is approximately 5-7% of the estimated riverine input
of phosphorus to the ocean (e.g. see Emery et al., 1955 or Lerman et
al., 1975) of 5.6 to 6.5×10^{10} mol y^{-1} (dissolved) and 45 to 65 $\times$
10^{10} mol y^{-1} (particulate).

Phosphorus concentrations in the marine aerosol have been
measured by Graham and Duce (1979), by Graham et al. (1979), and by
Chen et al. (1985). Aerosol measurements in the South American rain
forest have been made by Lawson and Winchester (1979). These data are
summarized in Table 6. The effect of continents, particularly the
Sahara dust plume, is clearly evident in the data. Concentrations
over the mid-Pacific appear to be ~0.015 nmol m^{-3}, but considerably
higher concentrations may be observed north of the equator during
Asian dust outbreaks (Uematsu et al., 1983). Concentrations north of
~45°N over the Atlantic are similar to those over the North Pacific,
but between the equator and 45°N, and particularly near North America,
concentrations over the Atlantic are typically 0.15-0.30 nmol m^{-3},
with 1.5 nmol m^{-3} common in the Sahara dust plume. Graham and Duce
(1981) and Chen et al. (1985) found the major mass of total phosphorus
in the marine aerosol to be on particles with radii of ~1-3 μm, with
perhaps 10-20% of the mass on submicrometer particles.

3.2 Calculations of Phosphorus Fluxes in Open Ocean Regions

3.2.1. **Phosphorus Fluxes from Deep Waters**. We can again estimate
vertical fluxes of P in the water column by advection and eddy
diffusion using Equation 1. The midpoint concentration of the
PO_4^{3-} gradient in the range of 150 to 175 meters depth in the
Pacific region gives us a value for R of 0.2-0.4 mmol PO_4-P m^{-3}
(GEOSECS, 1982 and Gundersen et al., 1976). In the Sargasso Sea
region, R is ~0.05-0.1 mmol PO_4-P m^{-3} (Carpenter and McCarthy,
1975 and GEOSECS, 1981) in the depth range of 100-150 m. With the
values for w given in Table 1, we can calculate an advective flux of
2.4-4.8 μmol P m^{-2} d^{-1} in the North Pacific and 0-4 μmol P m^{-2}
d^{-1} in the Sargasso Sea. These values are presented in Table 7.

From Gundersen et al. (1976) and GEOSECS data (GEOSECS, 1982)
from stations not far from Hawaii we can estimate a P gradient (dP/dz)
in the Hawaii region of 0.003 to 0.005 mmol P m^{-4}. Similarly, from
GEOSECS data in the vicinity of Bermuda (GEOSECS, 1981) we can obtain
a P gradient of ~0.001 mmol P m^{-4}. With the eddy diffusion
coefficients given previously for these regions we obtain eddy
diffusion fluxes of 5-50 μmol P m^{-2} d^{-1} in the Pacific and 1.7-10

Table 6.   Aerosol Phosphorus Concentrations in the Marine Atmosphere*

Location	Concentration Range (nmol P m^{-3})
Western North Atlantic (30°-45°N)	0.03 - 0.50
Bermuda	0.02 - 0.30
Near Shore North America - Atlantic	0.5 - 2.0
North Atlantic trades	0.1 - 2.6
Central North Atlantic (45°-60°N)	0.02 - 0.50
West coast of South America	0.2 - 0.65
Hawaii	0.013 - 0.030
Samoa	0.013 - 0.026
Central Pacific (20°N-15°S)	0.0013- 0.030
New Zealand coast	0.06 - 0.31
South American Rain Forests	0.87 - 2.5

*All data from Graham and Duce (1979) except New Zealand, which
are from Chen et al. (1985), and South America, which are from
Lawson and Winchester (1979).

μmol P m^{-2} d^{-1} for the Sargasso Sea region. These values are also
given in Table 7. Note that there is no analog to nitrogen fixation
in the phosphorus cycle.

3.2.2. <u>Phosphorus Fluxes from the Atmosphere</u>. Graham and Duce (1982)
attempted to evaluate the importance of atmospheric phosphorus input
to a restricted area of the ocean - a roughly triangular region
bounded by the North American coastline, 25°N, and 65°W. This area
includes Bermuda and much of the Sargasso Sea region. Over 30 aerosol
samples were collected from ships in this region and from the island
of Bermuda. Several forms of phosphorus were determined in these
samples: total phosphorus, reactive (or water soluble) phosphorus, and
the difference between these two forms, which represents organic plus
acid soluble phosphorus (although it is generally referred to as
organic phosphorus). The original data are available in Graham and
Duce (1982). Total phosphorus concentrations were somewhat higher in
the region north of a line between Cape Hatteras and Bermuda. This is
illustrated in Figure 3, where total phosphorus concentration is
plotted against the approximate distance of the sample from the North
American coast. The analytical results have been separated into two
groups: those samples north of the line mentioned above (solid
squares) and those south of this line (open circles). Samples from
Bermuda (triangles) appear to fall into two groups which fit well with
the northern and southern sample sets. A simple first order equation
has been fitted to the two sample sets illustrated in Figure 3:

$$C = C_0 e^{-kD} \qquad (2), \quad \text{where}$$

Table 7. Comparison of New Nutrient Phosphorus Source Strengths
in Surface Waters of the Sargasso Sea and the
North Pacific Gyre

Source	Flux (μmol P m^{-2} d^{-1})	
	Sargasso Sea	North Pacific Gyre
Vertical advection	0 - 4	2.4 - 4.8
Eddy diffusion	1.7 - 10	5 - 50
Atmospheric deposition	0.033 - 0.080	0.005 - 0.012
Total	1.7 - 14	7 - 55

C is the concentration of phosphorus at any distance D, in nmol m^{-3},
C_o is the concentration of phosphorus at the coast, in nmol m^{-3},
D is the distance from the coast, in km, and
k is a constant, in km^{-1}.
   As can be seen from Figure 3, C_o for the northern samples is
~0.65 nmol m^{-3}, while C_o for the southern samples is ~0.3 nmol
m^{-3}, or about half that for the northern samples, suggesting the
importance of anthropogenic sources in the northeastern section of
North America.
   Graham and Duce (1982) also estimated a total atmospheric
phosphorus input into this area of the North Atlantic. They used
three different approaches to make this estimate, and the details are
given in the original paper. The first estimate is made by using the
C_o values from Figure 3 and assuming a certain vertical distribution
and an average transport velocity of 3.8 m s^{-1}, determined from air
mass transport times to Bermuda. This estimate simply gives the
amount of atmospheric phosphorus transported eastward across the North
American coast. The second estimate is based on determination of
total deposition velocities (i.e. wet plus dry removal) from data in
Figure 3, assuming the decrease in concentration as one moves away
from the coast is due entirely to deposition to the ocean. These
total deposition velocities are combined with mean concentrations in
the northern and southern sectors to yield fluxes. The third approach
is to use total deposition rates for phosphorus actually measured by
Graham and Duce (1982) using bucket collectors at Bermuda and to apply
them to the entire region. Graham and Duce (1982) found that 36±15%
of the total phosphorus present on aerosols collected on the U.S. east
coast was soluble in seawater. I will assume that one third of the
atmospheric phosphorus entering the ocean is soluble, resulting in the
soluble P fluxes presented in Table 8. Note that the agreement among

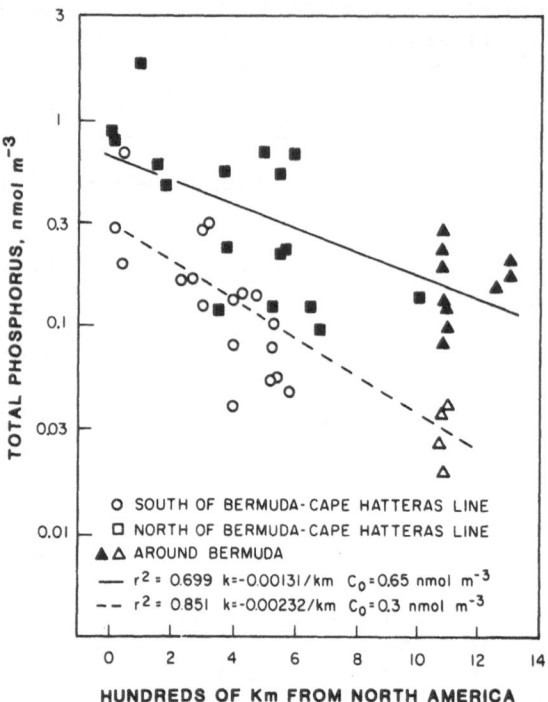

Figure 3. Total atmospheric phosphorus concentrations as a function
of distance from the eastern North American coastline. See text for
details. Adapted from Graham and Duce (1982).

the three approaches is within a factor of 2.5, and considering the
necessary approximations involved in these calculations, this
agreement is reasonably good.
     The total deposition samples collected in Bermuda (estimate No.
3) may include substances from the Sahara Desert along with material
from North America, and thus this estimate of the flux from North
America may be too high. Method No. 2 may also overestimate the P
flux, as no allowance was made for the dilution of the continental air
masses by marine air masses of lower P content. However, the latter
two estimates do bracket estimate No. 1, the P transport across the
North American coast, suggesting that the bulk of the material leaving
North America may be deposited in the western part of the North
Atlantic. The estimates in Table 8 can be assumed to be net input
rates of P to the ocean. Estimates No. 1 and 2 should represent
continentally derived P. Estimate No. 3 was corrected by subtracting
30% of the total P flux, since multivariate regression analysis
showed that approximately 30% of the explicable variance of total P
could be accounted for by the regression with sea salt Na (Graham and
Duce, 1982), suggesting that that fraction was associated with
recycled sea salt. This phosphorus was found to be largely associated
with organic material.

Table 8. Estimates of the Atmospheric Input of Soluble Phosphorus to the Sargasso Sea Region*

Model	Flux $(10^9$ mol P $y^{-1})$	Flux $(\mu mol$ P m^{-2} $d^{-1})$
Vertical distribution/ transit time$^\alpha$	0.08	0.060
Concentration/total deposition velocity	0.04	0.033
Measured deposition at Bermuda	0.11	0.080

*All data from Graham and Duce (1982).
$^\alpha$Quantity transported across the North American coast.

We can make an estimate of the net input of atmospheric P to the Central North Pacific Gyre region by comparing the atmospheric P concentrations in the Pacific region with those in the Bermuda area, using the P fluxes calculated above, and assuming that net atmospheric deposition is proportional to atmospheric concentration. From Table 6 we will use an atmospheric P content in the Pacific region of 0.015-0.03 nmol m^{-3}. From Figure 3 we can see that the concentration at Bermuda varies depending upon the air mass source region in North America, but we can assume a mean value over the Sargasso Sea of ~0.1-0.2 nmol m^{-3}. Combining this information with the atmospheric soluble P deposition estimate in the Sargasso Sea region of 0.033 -0.080 μmol P m^{-2} d^{-1} (see Table 8) results in an estimate of net soluble P input to the Pacific region of 0.005-0.012 μmol m^{-2} d^{-1} (Table 7).

3.2.3. Comparison of Phosphorus Sources. We can see from Table 7 that the atmospheric input of phosphorus is apparently of only limited importance in the Sargasso Sea region, accounting for perhaps as much as 5% or as little as 0.2% of the new phosphorus in the photic zone. In the Central North Pacific Gyre region atmospheric input is even less important, accounting for less than 1% of the new phosphorus in the surface waters. Note that these percentages would be even less if the new models of nutrient input to the photic zone by Jenkins and Goldman (1985) and Klein and Coste (1984), which consider nutrient transport by bursts of turbulence, are included.
It is possible to look at atmospheric transport of P into the ocean from a different perspective, however, and that is to compare the original sources of P in the water. Consider the Sargasso Sea. The two primary sources of P in the water are atmospheric input and transport via rivers. A total atmospheric soluble P input of

0.04-0.11×10^9 mol y^{-1} was found in the region of the Sargasso Sea discussed by Graham and Duce (1982). Riverine input from the east coast of North America between 25° and 45°N has been estimated to be $\sim 5 \times 10^{14}$ 1 y^{-1} (Judson and Ritter, 1964). Using an average total P content of ~ 2 μmol 1^{-1} for P in rivers (Stumm, 1973), 1×10^9 mol y^{-1} of P can be calculated to enter the western North Atlantic each year from rivers, assuming all the riverine P makes its way through the estuaries into the open sea. Thus the atmospheric input of nutrient phosphorus amounts to about 10% of the riverine input.

3.2.4. <u>Temporal Variability of Atmospheric Phosphorus Input</u>. Uematsu et al. (1983, 1985) have found that most of the mineral aerosol entering the North Pacific does so in three or four short duration (3-5 days) bursts associated with major dust storms originating over the deserts of China in the spring. Reactive phosphorus (PO_4^{3-}) in the marine aerosol is largely associated with mineral particles (Graham et al., 1979). A similar seasonal dependence of dust input to the Sargasso Sea from the Sahara Desert has been observed (Chen and Duce, 1983). The mean reactive P/Al molar ratio observed in aerosols at Hawaii was 0.011 ± 0.004 (Graham and Duce, 1981). Atmospheric Al concentrations can occasionally reach 11 nmol m^{-3} in Hawaii, resulting in an atmospheric reactive P concentration of ~ 0.12 nmol m^{-3}, about a factor of 4-8 times higher than that used above to estimate the atmospheric input of P to the North Pacific Gyre. Thus, it is possible that short term daily input rates of reactive P may be considerably higher than the mean values presented in Table 6. Similar short term bursts undoubtedly take place in the Sargasso Sea region as well. The impact of this intense short term atmospheric input of P on marine biological activity, particularly in the spring, is unknown.

4.   IRON

4.1   Introduction

It has been suggested by several authors that the availability of certain trace metals in the ocean may regulate some aspects of biological productivity. For example, this has been suggested for iron (Morel and Hudson, 1985; Lewin and Chen, 1971; Menzel and Ryther, 1961), cobalt, as part of vitamin B_{12} (Menzel and Spaeth, 1962a), copper (Knauer and Martin, 1983); molybdenum (Howarth and Cole, 1985), and others. I shall only consider the biological mobilization of atmospherically derived iron. Iron is important in the production of chlorophyll and in photosynthesis. Murphy et al. (1984) investigated the importance of iron limitation in certain diatoms and suggested that iron limitation may be quite important in open ocean regions, with species utilizing iron efficiently having a competitive advantage. The input of atmospheric trace metals to the ocean and their impact on oceanic trace metal distributions is discussed in some detail in the chapter "Air to Sea Transfer of Anthropogenic Trace

Metals" by Buat-Menard in this volume.

The question of the availability to biological systems, particularly in marine regions, of atmospherically derived trace metals has received only minimal research to date. While there have been several studies of the solubilization of metals present on aerosol particles in distilled water or acidic solutions (see e.g. Lindberg and Harriss, 1983; Perry et al., 1984; Walker and Wechsler, 1980; Ochs and Gatz, 1980; and Tanaka et al., 1981), there have been relatively few studies of the solubility of these metals in seawater (Hodge et al., 1978; Crecelius, 1980; Hardy and Crecelius, 1981; Walsh and Duce, 1976; and Moore et al., 1984). In most of the studies above, urban or near urban aerosol particles were investigated. The only study of the solubility in seawater of metals present on aerosol particles from remote marine areas is that of Maring (1985) for lead, copper and aluminum.

4.2 Solubility of Atmospheric Iron

Crecelius (1980) investigated the solubility in seawater of iron present on fly ash particles. He found that less than 1% of the iron was soluble. Hardy and Crecelius (1981) investigated the seawater solubility of iron on aerosols collected in Seattle and St. Louis as well as at a rural site (Quillayute) 5 km from the Washington coastline. They found iron solubilities ranging from 1 to 8%. Hodge et al. (1980) investigated the seawater solubility of metals on aerosol particles collected from the Scripps Pier at LaJolla, California and from a coastal site in Baja California, Mexico. These authors found $1.3 \pm 0.8\%$ and $0.3 \pm 0.3\%$ of the iron to be soluble respectively at these two sites.

Moore et al. (1984) have presented the most extensive discussion of the potential importance of atmospheric iron on marine productivity. These authors point out that simple measurements of the seawater solubility may not be a good indication of the potential for trace metals to become involved in biological systems. Since particles containing iron, including mineral aerosols, are ingested by zooplankton, the chemical environment within the animal's gut (e.g. lower pH) may mobilize a greater fraction of the metal. Thus the study of Moore et al. investigated the release of iron (and other metals) in seawater at several pH's. Aerosol samples for this study were collected at a coastal site near Halifax, Canada. Moore et al. (1984) determined the solubility of metals relative to that of aluminum, assuming aluminum was largely insoluble. Table 9 presents the results of the relative solubility of iron compared with aluminum for unleached aerosols and for aerosols leached at different pH's. It appears that about 10-12% more of the iron, relative to aluminum, was soluble in seawater. While the solubility of other metals was affected by pH, there was no measurable affect on the iron solubility in the pH range 5.4 to 8.0.

Maring (1985) has investigated the seawater solubility of aluminum present in marine aerosols collected at a mid-Pacific site, Enewetak Atoll. Considering such factors as dust loading, pH, time of

Table 9.   Solubility of Fe Relative to Al in Atmospheric Particles*

pH	Relative Fe/Al Ratio
Unleached	1.0
5.39	0.90
5.60	0.88
6.16	0.90
6.40	0.86
7.99	0.86

*Adapted from data in Moore et al. (1984)

contact with seawater, concentration of organic matter in the seawater, and presence of sunlight, Maring (1985) found that ~5% of the crustal aerosol aluminum was solubilized. Thus, considering the solubilization studies of Moore et al. (1984) above, their determination that ~10% of crustal Fe in aerosols dissolves in seawater is perhaps a minimum value.

Moore et al. (1984) used an approach similar to that used here for N and P to estimate the relative importance of the various sources of Fe in the photic layer. Although the fluxes were not quantified in the form we have used, their estimates indicated the atmospheric input of biologically useable Fe considerably exceeded that entering the photic zone from deeper waters.

4.3 Calculations of Iron Fluxes in Open ocean Regions

4.3.1. Iron Fluxes from Deep Waters.   Equation 1 can again be used to estimate vertical fluxes of iron in the water column. However, in the case of Fe there are no vertical profiles in the Sargasso Sea area or in the North Pacific Gyre. Gordon et al. (1982) measured dissolved iron off the coast of Mexico, with the most remote site only several hundred km west of the coastline. Very low values (0.1-0.2 μmol Fe m^{-3}) were observed in surface waters, increasing to 2-2.5 μmol Fe m^{-3} at depths of 70 to 150 m. Roughly similar values were found by Landing and Bruland (1981) 160 km west of Santa Cruz, California. From these data we can estimate R as 1-1.5 μmol Fe m^{-3}, but these are likely maximum values for the North Pacific Gyre due to the proximity of North America.

In the Atlantic the only dissolved Fe data are also near North America. Hong (1984) measured dissolved Fe at a slope water station at 37°17'N, 71°05'W and found surface concentrations of 0.8-1 μmol Fe m^{-3} increasing to about 2.5 μmol Fe m^{-3} at 60m. Concentrations of dissolvable Fe (i.e. that iron soluble when the seawater sample was acidified to a pH of 2) measured by Wallace et al. (1983) in the Gulf

Stream at the shelf boundary were slightly higher (2-4 μmol m^{-3}). We will again estimate R as 1-1.5 μmol Fe m^{-3}, but these may also be maximum concentrations. Using the values of w given in Table 1, the vertical advection fluxes calculated are 0.012 to 0.018 μmol Fe m^{-2} d^{-1} in the North Pacific and 0 to 0.06 μmol Fe m^{-2} d^{-1} in the Sargasso Sea region. These values are given in Table 10.

The Fe gradients, d(Fe)/dz, calculated from these studies are 0.02-0.04 μmol Fe m^{-4} in the North Pacific and the North Atlantic. Using the values for K_z given in Table 1, we can calculate eddy diffusion fluxes of 0.034 to 0.4 μmol Fe m^{-2} d^{-1} in the North Pacific and the same rates in the Sargasso Sea region. These values are also presented in Table 10.

Table 10. Comparison of Iron Source Strengths in Surface Waters of the Sargasso Sea and the North Pacific Gyre

Source	Flux (μmol Fe m^{-2} d^{-1})	
	Sargasso Sea	North Pacific Gyre
Vertical advection	0 - 0.06	0.012 - 0.018
Eddy diffusion	0.034 - 0.4	0.034 - 0.4
Atmospheric deposition	0.2 - 0.8	0.08 - 0.16
Total	0.2 - 1.3	0.13 - 0.58

4.3.2. Iron Fluxes from the Atmosphere. Atmospheric iron fluxes may be more accurately estimated than water column fluxes because there are data on mineral particles and iron in the atmosphere, in rain, and in total deposition samples over the Central North Pacific and the Sargasso Sea. Uematsu et al. (1985) have measured the annual deposition of mineral aerosol particles at several island sites in the North Pacific, including Oahu, Hawaii. Their data indicate that on an annual basis the mean mineral aerosol deposition at Oahu is 0.12 μg cm^{-2} d^{-1}. This ranges from a mean of 0.10 μg cm^{-2} d^{-1} during the "clean" period from July-January to 0.14 μg cm^{-2} d^{-1} during the dusty season of February-June. During the dusty period daily fluxes can be much higher than the mean value indicated above due to the short term episodic nature of the Asian dust storms responsible for the dust at Oahu (see 3.2.4). The annual mean dust deposition converts to an iron flux of 1.2 μmol Fe m^{-2} d^{-1}, assuming 5.6% of the dust is iron (Taylor, 1964). Assuming further that 10% of the atmospheric Fe is soluble in seawater results in an atmospheric soluble Fe flux of 0.12 μmol Fe m^{-2} d^{-1}, with a range from 0.08 to 0.16 μmol Fe m^{-2} d^{-1} probably most appropriate in the central

North Pacific.

In the Sargasso Sea region Church et al. (1984) and Jickells et al. (1984) have measured iron in rain samples from Bermuda. Church et al. (1984) found that the acid soluble (below pH 2) iron wet deposition rate at Bermuda ranged from 0.22 to 0.37 μmol Fe m^{-2} d^{-1}. While this acid treatment will certainly result in a higher fraction of the Fe solubilizing than it would in seawater, a major fraction of the iron will likely remain insoluble. Thus while these values are likely somewhat high relative to the quantity of Fe that will be biologically available in the ocean, they will be used. Uematsu et al. (1985) found that approximately 80% of the mineral aerosol deposition to the ocean near Hawaii was by wet deposition and 20% was by dry deposition. Using this percentage, the total Fe deposition to the Sargasso Sea would be 0.28 to 0.46 μmol Fe m^{-2} d^{-1}.

Extensive data on atmospheric particulate Fe were obtained by Duce et al. (1976) at Bermuda. These authors found an annual mean atmospheric Fe concentration of 1.7 nmol m^{-3}. Studies by Uematsu et al. (1985) found that the rain scavenging ratio, W, for mineral particles over the Pacific was ~1000. W is defined as:

$$W = \rho C_r / C_a \qquad (3), \quad \text{where}$$

W is the scavenging ratio (dimensionless),
ρ is the density of air at standard conditions (1.2 kg m^{-3}),
C_r is the concentration of Fe in rain, in mol kg^{-1}, and
C_a is the concentration of Fe in air, in mol m^{-3}.

Using the values for W, ρ, and C_a given above combined with a mean annual rainfall at Bermuda of ~1.4 m y^{-1} leads to a wet deposition estimate of 5.5 μmol Fe m^{-2} d^{-1} to the ocean. As discussed above, I will make the conservative assumption that 10% of this Fe is soluble in seawater, leading to a dissolved Fe input rate of 0.55 μmol Fe m^{-2} d^{-1}. If wet deposition accounts for 80% of the total input, the total deposition to the ocean at Bermuda would then be 0.69 μmol Fe m^{-2} d^{-1}.

These fluxes can be compared with other estimates of the atmospheric input of Fe to the Sargasso Sea region, assuming 10% of the atmospheric Fe is soluble. These estimates are all presented in Table 11. Note that not all the estimates in Table 11 are independent, as several of them rely on the atmospheric Fe concentrations reported by Duce et al. (1976). The estimates in Table 11 range from approximately 0.2 to 0.8 μmol Fe m^{-2} d^{-1}, and this range is presented in Table 10.

Weisel et al. (1984) determined that considerably less than 1% of the atmospheric Fe over the North Atlantic was derived from the ocean. Church et al. (1984) similarly determined that less than 10% of the iron they measured in rain collected at Bermuda was derived from the ocean. Thus it is likely that the atmospheric deposition rates given above represent net inputs to the ocean, i.e. recycled iron from seawater is probably unimportant.

Table 11. Estimates of the Atmospheric Input of Iron
to the North Atlantic

Location	Deposition (μmol Fe m^{-2} d^{-1})	Reference
Bermuda	0.69	This paper
Bermuda	0.28-0.46	Church et al. (1984)
Bermuda	0.39	Duce et al. (1976)
Bermuda	0.83	Duce et al. (1976)
North Atlantic	0.50	Moore et al. (1983)
Bermuda	0.15	Wallace et al. (1977)
Tropical North Atlantic	0.16	Buat-Menard and Chesselet (1979)
Gulf Stream, Shelf Break	0.20	Wallace et al. (1983)

4.3.3. <u>Comparison of Iron Sources</u>. The comparison of the vertical advection, eddy diffusion, and atmospheric deposition fluxes in Table 10 suggests atmospheric deposition of Fe may be quite important in both regions. According to the data in Table 10, atmospheric input might account for from 30% to 96% of the iron input to the Sargasso Sea region and from 16% to 76% to the North Pacific Gyre. Note again that these may be minimum percentages since the eddy diffusion flux estimates may be too high. These results are in general agreement with those of Moore et al. (1984), Wallace et al. (1983), and Buat-Menard and Chesselet (1979), who point out the importance of the atmospheric input of iron to the iron observed in the surface waters of the North Atlantic. Note also, however, that if the models of Jenkins and Goldman (1985) and Klein and Coste (1984) are correct, even for iron the atmospheric input may be overwhelmed by short term turbulent inputs from deep waters.

5. GENERAL DISCUSSION AND CONCLUSIONS

If nothing else, this exercise has served to alert us to just how little we really know about the importance of atmospherically derived material to biological systems in the ocean. Unfortunately, it has also pointed out that we are no better off, and in fact often in worse condition, when we try to estimate the importance of other sources and transport paths for the new nutrients necessary for life in the sea. From our considerations of the three elements in this paper, it appears that for the photic zone as a whole the atmosphere may be most important for iron and likely of little importance for phosphorus on a broad geographic scale, with nitrogen somewhere in between.

   It is also interesting to compare briefly the N/P molar ratio for atmospheric material entering the ocean with the N/P ratio observed in the soft tissue of plankton, i.e. the Redfield ratio. At both the Sargasso Sea and North Pacific Gyre sites the atmospheric N/P ratio ranges from a few hundred to a few thousand, while the Redfield ratio is ~16. This supports our conclusion that the impact of material from the atmosphere is potentially much more important to primary production for nitrogen than for phosphorus in the regions examined.

Let us take one final look at these three elements from a slightly different perspective. We can compare the daily atmospheric input per square meter to the ocean for nitrate (not total nitrogen), phosphorus, and iron with the quantities of these substances already present in the very surface waters of the photic zone. We can make this comparison by changing the atmospheric input rate (see Tables 5, 7, and 10) into an equivalent depth of water already containing that same mass of each element in the surface water. For surface water concentrations at both locations we will use the following:

nitrate: 70 μmol m^{-3};
phosphorus: 50 μmol m^{-3}
iron: 1 μmol m^{-3}

The results of this simple calculation are presented in Table 12. Note that the quantities of both nitrate and iron equivalent to that in roughly 0.5 m of surface seawater enter the ocean from the atmosphere each day. This is clearly a significant input to these often nutrient depleted waters.

Table 12. Atmospheric Input: Surface Water Depth Equivalent

Substance	Sargasso Sea (m d^{-1})	North Pacific Gyre (m d^{-1})
Nitrate	0.37 to 0.53	0.08 to 0.20
Phosphorus	0.002 to 0.005	0.0003 to 0.0008
Iron	0.2 to 0.8	0.4 to 0.8

Finally, we must remember two very important caveats relative to the crude calculations made in this paper. First, if the recently proposed mechanisms and rates of input of nutrients from deep waters to the photic zone as a result of bursts of wind driven turbulence are correct, it is likely that the atmospheric input of all three of these substances is relatively unimportant to productivity on the short time scale. Clearly this mechanism and the proposed model must be verified. However, even if this is the case, the input of nutrients from the atmosphere into the surface waters may have a more important effect than our previous discussion would indicate. Jenkins and Goldman (1985) suggested that the euphotic zone may be considered as two very distinct regions, with the upper portion essentially devoid of new nutrients and completely dependent upon nutrient regeneration for productivity. They used the analogy of a "spinning wheel" for this region. The lower portion of the euphotic zone would be dependent primarily on new nutrients from deeper waters. Under these conditions the only new nutrients entering the upper portion would be from the atmosphere, and thus atmospheric input may be particularly critical for new productivity in the near surface waters.

Second, the input of atmospherically derived materials can vary by at least two orders of magnitude temporally at a given site, and it can vary by at least that much geographically as well. (Similar variability in the fluxes from fixation and the water column may also be observed.) Thus the impact of short term pulses of material, for example from a major dust storm reaching thousands of miles from shore and lasting for 3-4 days, is completely unknown. Can the large number of atmospheric particles themselves during these events be important in enhancing or suppressing productivity? Could the desert dust pulses and their associated chemical materials entering the ocean in large quantities during the spring, just as blooms are beginning in the North Pacific, play any role in initiating or sustaining production? And what might be the effect of such a large input of such material in the oligotrophic waters in the central ocean gyres? Finally, we have not even considered the concentration of chemicals, including synthetic organics, in the ocean surface microlayer from atmospheric deposition and the possible role these substances might play in the cycle of organisms that live there. All these questions remain unanswered.

ACKNOWLEDGEMENTS

I wish to thank the many people at the ASI who contributed comments and suggestions to this manuscript. I especially appreciated discussion and comments from A. Pszenny, J.D. Smith, M. Fontugne, G. Ayers, R. Gagosian, and P. Buat-Menard. Thanks are also due to the National Science Foundation, Divisions of Ocean and Atmospheric Science for their generous support over the years.

REFERENCES

Arimoto, R., R.A. Duce, B.J. Ray, and C.K. Unni, 1985: 'Atmospheric trace elements at Enewetak Atoll: 2. Transport to the ocean by wet and dry deposition', J. Geophys. Res., 90, 2391-2408.

Bottger, A., D.H. Ehhalt, and G. Gravenhorst, 1978: 'Atmospharische Kreislauf von Stickstoffoxiden und Ammoniak, Berichte der Kernforschungsanlage Julich, JUL 1558, ISSN 0366-0885.

Broecker, W.S. and T.H. Peng, 1982: Tracers in the Sea, Eldigo Press, Palisades, NY, 690 pp.

Buat-Menard, P., 1983: 'Particle geochemistry in the atmosphere and oceans', in Air-Sea Exchange of Gases and Particles, P.S. Liss and W.G.N. Slinn, eds., Reidel, Dordrecht, 455-532.

Buat-Menard, P. and R. Chesselet, 1979: 'Variable influence of the atmospheric flux on the trace metal chemistry of oceanic suspended matter', Earth Planet. Sci. Lett., 42, 399-411.

Carpenter, E.J., 1983: 'Nitrogen fixation by marine Oscillatoria (Trichodesmium) in the world's oceans', in Nitrogen in the Marine Environment, E.J. Carpenter and D.G. Capone, eds., Academic Press, New York, 65-103.

Carpenter, E.J. and J.J. McCarthy, 1975: 'Nitrogen fixation and uptake of combined nitrogenous nutrients by Oscillatoria (Trichodesmium) thiebautii in the western Sargasso Sea', Limnol. Oceanogr., 20, 389-401.

Chen, L. and R.A. Duce, 1983: 'The sources of sulfate, vanadium, and mineral matter in aerosol particles over Bermuda', Atmos. Environ., 17, 2055-2064.

Chen, L., R. Arimoto, and R.A. Duce, 1985: 'The sources and forms of phosphorus in marine aerosol particles and rain from northern New Zealand', Atmos. Environ., 19, 779-787.

Church, T.M., J.M. Tramontano, J.R. Scudlark, T.D. Jickells, J.J. Tokos, A.H. Knap, and J.N. Galloway, 1984: 'The wet deposition of trace metals to the western Atlantic Ocean at the mid-Atlantic coast and on Bermuda', Atmos. Environ., 18, 2657-2664.

Clegg, S.L. and P. Brimblecombe, 1985: 'Potential degassing of hydrogen chloride from acifified sodium chloride droplets', Atmos. Environ., 19, 465-470.

Crecelius, E.A., 1980: 'The solubility of coal fly ash and marine aerosols in seawater', Mar. Chem., 8, 245-250.

Duce, R.A., 1983: 'Biogeochemical cycles and the air/sea exchange of aerosols', Chapter 16 in The Major Biogeochemical Cycles and Their Interactions, SCOPE 21, B. Bolin and R.B. Cook, eds., Wiley, Chichester, 427-456.

Duce, R.A. and E.J. Hoffman, 1976: 'Chemical fractionation at the air/sea interface', Ann. Rev. Earth and Planet. Sci., 4, 187-228.

Duce, R.A., G.L. Hoffman, B.J. Ray, I.S. Fletcher, G.T. Wallace, J.L. Fasching, S.R. Piotrowicz, P.R. Walsh, E.J. Hoffman, J.M. Miller, and J.L. Heffter, 1976: 'Trace metals in the marine atmosphere: sources and fluxes', in Marine Pollutant Transfer, H. Windom and R. Duce, eds., D.C. Heath, Lexington, 77-119.

Dugdale, R.C. and J.J. Goering, 1967: 'Uptake of new and regenerated forms of nitrogen in primary productivity', Limnol. Oceanogr., 12, 196-206.

Emery, K.O., W.L. Orr, and S.C. Rittenberg, 1955: 'Nutrient budgets in the ocean', in Essays in the Natural Sciences in Honor of Captain Allen Hancock, University of Southern California, Los Angeles, 299-309.

Eppley, R.W., 1980: 'Estimating phytoplankton growth rates in the central oligotrophic oceans', in Primary Productivity in the Sea, P.G. Falkowski, ed., Plenum, New York, 231-242.

Eppley, R.W. and B.J. Peterson, 1979: 'Particulate organic matter flux and planktonic new production in the deep ocean', Nature, 282, 677-680.

Eppley, R.W., E.H. Renger, and W.G. Harrison, 1979: 'Nitrate and phytoplankton production in southern California coastal waters', Limnol. Oceanogr., 24, 483-494.

Fogg, G.E., 1975: 'Primary productivity', Chapter 14 in Chemical Oceanography, 2, J.P. Riley and G. Skirrow, eds., Academic Press, London, 385-453.

Fogg, G.E., 1978: 'Nitrogen fixation in the oceans', Environmental Role of Nitrogen-fixing Blue-green Algae and Asymbiotic Bacteria, Ecol. Bull., Stockholm, 26, 11-19.

Fogg, G.E., 1982: 'Nitrogen cycling in sea waters', Phil. Trans. Roy. Soc. London, B296, 511-520.

Fogg, T.R. and R.A. Duce, 1985: 'Boron in the troposphere: distribution and fluxes', J. Geophys. Res., 90, 3781-3796.

Galloway, J.N., 1985: 'Atmospheric deposition of S and N to remote areas', in Atmospheric Cycling of S and N in Remote Atmospheres, J. Galloway, R. Charlson, M. Andreae, and H. Rodhe, eds., 1985: Report of a NATO Advanced Research Workshop, Reidel, Dordrecht, In press.

Galloway, J.N. and A. Gaudry, 1984: 'The composition of precipitation on Amsterdam Island, Indian Ocean', Atmos. Environ., 2649-2656.

Galloway, J.N., A.H. Knap, and T.M. Church, 1983: 'The composition of western Atlantic precipitation using shipboard collectors', J. Geophys. Res., 88, 10,859-10,864.

Galloway, J.N., G.E. Likens, W.C. Keene, and J.M. Miller, 1982: 'The composition of precipitation in remote areas of the world', J. Geophys. Res., 87, 8771-8786.

Georgii, H.W. and Gravenhorst, G., 1977: 'The ocean as a source or sink of reactive trace gases', Pure Appl. Geophys., 115, 503-511.

GEOSECS, 1981: GEOSECS Atlantic Expedition, Vol. 1, Hydrographic Data, NSF, Washington, D.C., 121pp,.

GEOSECS, 1982: GEOSECS Pacific Expedition, Vol. 3, Hydrographic Data, NSF, Washington, D.C., 137pp,.

Gordon, R.M., J.H. Martin, and G.A. Knauer, 1982: 'Iron in north-east Pacific waters', Nature, 299, 611-612.

Graham, W.F. and R.A. Duce, 1979: 'Atmospheric pathways of the phosphorus cycle', Geochim. Cosmochim. Acta, 43, 1195-1208.

Graham, W.F. and R.A. Duce, 1981: 'Atmospheric input of phosphorus to remote tropical islands', Pacific Science, 35, 241-255.

Graham, W.F. and R.A. Duce, 1982: 'The atmospheric transport of phosphorus to the western North Atlantic', Atmos. Environ., 16, 1089-1097.

Graham, W.F., S.R. Piotrowicz, and R.A. Duce, 1979: 'The sea as a source of atmospheric phosphorus', Mar. Chem., 7, 325-342.

Gundersen, K.R., J.S. Corbin, C.L. Hanson, M.L. Hanson, R.B. Hanson, D.J. Russell, A. Stollar, and O. Yamada, 1976: 'Structure and biological dynamics of the oligotrophic ocean photic zone off the Hawaiian Islands', Pac. Sci., 30, 45-68.

Hardy, J.T. and E.A. Crecelius, 1981: 'Is atmospheric particulate matter inhibiting marine primary productivity?', Environ. Sci. Tech., 15, 1103-1105.

Hodge, V., S.R. Johnson, and E.D. Goldberg, 1978: 'Influence of atmospherically transported aerosols on surface ocean water composition', Geochem. J., 12, 7-20.

Hong, H., 1984: Chemistry of Iron in Different Marine Environments and the Binding of Iron, Copper, Manganese and Aluminum with Particles in a Microcosm System, PhD dissertation, University of Rhode Island, Kingston, 240 pp.

Howarth, R.W. and J.J. Cole, 1985: 'Molybdenum availability, nitrogen limitation, and phytoplankton growth in natural waters', Science, 229, 653-655.

Jenkins, W.J., and J.C. Goldman, 1985: 'Seasonal oxygen cycling and primary production in the Sargasso Sea', J. Mar. Res., 43, 465-491.

Jickells, T.D., A.H. Knap, and T.M. Church, 1984: 'Trace metals in Bermuda rainwater', J. Geophys. Res., 89, 1423-1428.

Judson, S. and D.F. Ritter, 1964: 'Rates of regional denudation in the United States', J. Geophys. Res., 69, 3395-3401.

Kiefer, D.A., R.J. Olson, and O. Holm-Hansen, 1976: 'Another look at the nitrite and chlorophyll maxima in the central North Pacific', Deep-Sea Res., 23, 1199-1208.

Klein, P. and B. Coste, 1984: 'Effects of wind stress variability on nutrient transport into the mixed layer', Deep-Sea Res., 31, 21-37.

Knap, A.H., T.D. Jickells, A. Pszenny, and J. Galloway, 1985: 'The significance of atmospheric derived fixed nitrogen on the productivity of the Sargasso Sea', Submitted to Nature.

Knauer, G.A. and J.H. Martin, 1983: 'Trace elements and primary production: problems, effects, and solutions', in Trace Metals in Sea Water, C. Wong et al., eds., Plenum, New York, 823-840.

Landing, W.M. and K.W. Bruland, 1981: 'The vertical distribution of iron in the northeast Pacific', abstract, EOS, 62, 906.

Lawson, D.R. and J.W. Winchester, 1979: 'Sulfur, potassium, and phosphorus associations in aerosols from South American tropical rain forests', J. Geophys. Res., 84, 3723-3727.

Leetma, A. and A.F. Bunker, 1978: 'Updated charts of the mean wind stress, convergences in the Ekman layers, and Sverdrup transports in the North Atlantic', J. Mar. Res., 36, 311-321.

Lerman, A., F.T. MacKenzie, and R.M. Garrels, 1975: 'Modeling of geochemical cycles: phosphorus as an example', Geol. Soc. Am. Mem., 142, 205-218.

Lewin, J. and C.H. Chen, 1971: 'Available iron: a limiting factor for marine phytoplankton', Limnol. Oceanogr., 16, 670-675.

Lindberg, S.E. and R.C. Harriss, 1983: 'Water and acid soluble trace metals in atmospheric particles', J. Geophys. Res., 88, 5091-5100.

Logan, J.A., 1983: 'Nitrogen oxides in the troposphere: global and regional budgets', J. Geophys. Res., 88, 10,785-10,807.

MacIntyre, F., 1974: 'Chemical fractionation and sea-surface microlayer processes', in The Sea, 5, Marine Chemistry, Wiley and Sons, New York, 245-299.

Maring, H.B., 1985: The Impact of Atmospheric Aerosols on Trace Metal Chemistry in Open Ocean Surface Seawater, PhD dissertation, Univ. of Rhode Island, Kingston, 138 pp.

McCarthy, J.J. and E.J. Carpenter, 1979: 'Oscillatoria (Trichodesmium) theibautii (Cyanophyta) in the central North Atlantic Ocean', J. Phycol., 15, 75-82.

McCarthy, J.J. and E.J. Carpenter, 1983: 'Nitrogen cycling in near-surface waters of the open ocean', Chapter 14 in Nitrogen in the Marine Environment, E.J. Carpenter and D.G. Capone, eds., Academic Press, New York, 487-512.

Menzel, D.W. and J.H. Ryther, 1961: 'Nutrients limiting the production of phytoplankton in the Sargasso Sea, with special reference to iron', Deep-Sea Res., 7, 276-281.

Menzel, D.W. and J.P. Spaeth, 1962a: 'Occurrence of vitamin B_{12} in the Sargasso Sea', Limnol. Oceanogr., 7, 151-154.

Menzel, D.W. and J.P. Spaeth, 1962b: 'Occurrence of ammonia in Sargasso Sea waters and in rain water at Bermuda', Limnol. Oceanogr., 7, 159-162.

Moore, R.M., J.E. Milley, and A. Chatt, 1984: 'The potential for biological mobilization of trace elements from aeolian dust in the ocean and its importance in the case of iron', Oceanologica Acta, 7, 221-228.

Morel, F.M. and R.J.M. Hudson, 1985: 'The geobiological cycle of trace elements in aquatic systems: Redfield revisited', Chemical Processes in Lakes, W. Stumm, ed., J. Wiley, New York, 251-281.

Mosher, B.W. and R.A. Duce, 1983: 'Vapor phase and particulate selenium in the marine atmosphere', J. Geophys. Res., 88, 6761-6768.

Munk, W.H., 1966: 'Abyssal recipes', Deep-Sea Res., 13, 707-730.

Murphy, L.S., R.R.L. Guillard, and J.F. Brown, 1984: 'The effects of iron and manganese on copper sensitivity in diatoms: differences in the responses of closely related neritic and oceanic species', Biol. Oceanogr., 3, 187-201.

Ochs, H.T. and D.F. Gatz, 1980: 'Water solubility of atmospheric aerosols', Atmos. Environ., 14, 615-616.

Paerl, H.W., 1985: 'Enhancement of marine primary production by nitrogen-enriched rain', Nature, 315, 747-749.

Perry, M.E., A.W. Elzerman, and T.J. Overcamp, 1984: 'Solubility of atmospheric particulate matter', Chapter 6 in Chemistry of Particles, Fogs, and Rain, J.L. Durham, ed., Butterworth, Boston, 237-257.

Pierrou, 1976: 'The global phosphorus cycle', in Nitrogen, Phosphorus, and Sulfur - Global Cycles, SCOPE Report 7, B.H. Svensson and R. Soderlund, eds., 75-88.

Platt, T. M. Lewis, and R. Geider, 1982: 'Thermodynamics of the pelagic ecosystem: elemental closure conditions for biological production in the open ocean', in Flows of Energy and Materials in the Marine Ecosystem: Theory and Practice, M.J. Fasham, ed., Plenum Press, New York, 49-84.

Pritchard, D.W., R.O. Reid, A. Okubo, and H.H. Carter, 1971: 'Physical processes of water movement and mixing', Radioact. Environ., 90-136.

Prospero, J.M., D. Savoie, R.T. Nees, R.A. Duce, and J. Merrill, 1985: 'Particulate sulfate and nitrate in the boundary layer over the North Pacific Ocean', J. Geophys. Res., 90, 10,586-10,596.

Pszenny, A.A.P., F. MacIntyre, and R.A. Duce, 1982: 'Seasalt and the acidity of marine rain on the windward coast of Samoa', Geophys. Res. Lett., 9, 751-754.

Redfield, A.C., 1958: 'The biological control of chemical factors in the environment', Amer. Sci., 46, 205-222.

Rooth, C. and H.G. Ostlund, 1972: 'Penetration of tritium into the Atlantic thermocline', Deep-Sea Res., 19, 481-492.

Ryther, J.H. and W.M. Dunstan, 1971: 'Nitrogen, phosphorus, and eutrophication in the coastal marine environment', Science, 171, 1008-1013.

Ryther, J.H. and R.R.L. Guillard, 1959: 'Enrichment experiments as a means of studying nutrients limiting to phytoplankton production', Deep-Sea Res., 6, 65-69.

Savoie, D.L., 1984: Nitrate and Non-Sea-Salt Sulfate Aerosols Over the World Ocean: Concentrations, Sources, and Fluxes, PhD Dissertation, Univ. of Miami, Miami, FL, 432 pp.

Settle, D.M. and C.C. Patterson, 1982: 'Magnitudes and sources of precipitation and dry deposition fluxes of industrial and natural leads to the North Pacific at Enewetak', J. Geophys. Res., 87, 8857-8869.

Smith, S.V., 1984: 'Phosphorus vs nitrogen limitation in the marine environment', Limnol. Oceanogr., 29, 1149-1160.

Soderlund, R. and B.H. Svensson, 1976: 'The global nitrogen cycle', in Nitrogen, Phosphorus, and Sulfur - Global Cycles, SCOPE Report 7, B.H. Svensson and R. Soderlund, eds., 89-134.

Stumm, W. 1973: 'The acceleration of the hydrogeochemical cycle of phosphorus', Wat. Res., 7, 131-144.

Tanaka, S., M. Darzi, and J.W. Winchester, 1981: 'Elemental analysis of soluble and insoluble fractions of rain and surface waters by particle-induced x-ray emission', Environ. Sci. Technol., 15, 354-357.

Taylor, S.R., 1964: 'Abundance of chemical elements in the continental crust: A new table', Geochim. Cosmochim. Acta, 28, 1273-1285.

Thayer, G.W., 1974: 'Identity and regulation of nutrients limiting phytoplankton production in the shallow estuaries near Beaufort, N.C.', Oecologia, 14, 75-92.

Thomas, W.H., 1969: 'Phytoplankton nutrient enrichment experiments off Baja California and in the eastern equatorial Pacific Ocean', J. Fish. Res. Bd. Can., 26, 1133-1145.

Uematsu, M., R.A. Duce, J.M. Prospero, L. Chen, J.T. Merrill, and R.L. McDonald, 1983: 'Transport of mineral aerosol from Asia over the North Pacific Ocean', J. Geophys. Res., 88, 5343-5352.

Uematsu, M., R.A. Duce, and J.M. Prospero, 1985: 'Deposition of atmospheric mineral particles to the North Pacific Ocean', J. Atmos. Chem., 3, 123-138.

Walker, M.V. and C.J. Wechsler, 1980: 'Water soluble components of size=fractionated aerosols collected after hours in a modern office building', Environ. Sci. Technol., 14, 594-597.

Wallace, G.T., N. Dudek, R. Dulmage, and O. Mahoney, 1983: 'Trace element distributions in the Gulf Stream adjacent to the southeastern Atlantic continental shelf - influence of atmospheric and shelf water inputs', Can. J. Fish. Aquat. Sci., 40, 183-191.

Wallace, G.T., G.L. Hoffman, and R.A. Duce, 1977: 'The influence of organic matter and atmospheric deposition on the particulate trace metal concentration in northwest Atlantic surface seawater', Mar. Chem., 5, 143-170.

Walsh, P.R. and R.A. Duce, 1976: 'The solubilization of anthropogenic atmospheric vanadium in sea water', Geophys. Res. Lett., 3, 375-378.

Weisel, C.P., R.A. Duce, J.L. Fasching, and R.W. Heaton, 1984: 'Estimates of the transport of trace metals from the ocean to the atmosphere', J. Geophys. Res., 89, 11,607-11,618.

Williams, P.M., 1967: 'Sea surface chemistry: organic carbon and inorganic nitrogen and phosphorus in surface films and subsurface waters', Deep-Sea Res., 14, 791-800.

Wunsch, C. and J.F. Minster, 1982: 'Methods for box models and open circulation tracers: Mathematical programming and non-linear inverse theory', J. Geophys. Res., 87, 5647-5662.

Zafiriou, O.C. and M. MacFarland, 1981: 'Nitric oxide from nitrite photolysis from the central Equatorial Pacific', J. Geophys. Res., 86, 3173-3182.

ANDREAE, M.O.
Dept of Oceanography,
Florida State University,
Tallahasee, FL 32306, U.S.A.

ANDRIE, C.
Laboratoire d'Océanographie dynamique et de Climatologie CEN Saclay,
91191 Gif sur Yvette, Cedex, France

ARIMOTO, R.
Graduate School of oceanography
University of Rhode Island
Kingston, RI 028881, U.S.A.

ATLAS, E.
Dept of Oceanography,
Texas A & M University,
College Station, Texas 77843, U.S.A.

AUSTIN, S.L.
Dept of marine sciences,
Plymouth Polytechnic, Drake circus,
Plymouth, Devon PL4 8AA, U.K.

AYERS, G.
CSIRO, Division of atmospheric Res.
Private bag n°1
Mordialloc, VIC 3195, Australia.

BARNOLA, J.M.
Laboratoire de Glaciologie et
Géophysique de l'environnement
Domaine Universitaire, BP 96
38402 St Martin d'Hères, France.

BELVISO, S.
Centre des Faibles Radioactivités
Laboratoire mixte CNRS CEA
91190 Gif sur Yvette, France

BERGAMETTI, G.
Laboratoire de Chimie minérale des
milieux naturels, Univ. Paris VII
2 PLace Jussieu,
75251 Paris Cedex 05, France

BINGEMER, H.
Max Planck Institut für Chemie
Saarstrasse 23
D-6500 Mainz, Germany

BLANCHARD, D.C.
Atmospheric Science Research Cent.
ES 324, S.U.N.Y. at Albany
Albany, NY 12222, U.S.A.

BONSANG, B.
Centre des Faibles Radioactivités
Laboratoire mixte CNRS-CEA
91190 Gif sur Yvette, France

BOWYER, P.A.
Dept. of Oceanography
University College
Galway, Ireland.

BRUN-COTTAN, J.C.
Laboratoire de Physique et chimie
marines, Université P. et M. Curie
4 Place Jussieu,
75230 Paris Cedex 05. France

BRUYNSEELS, F.
University of Antwerp (UIA)
Dept. of Chemistry
Universiteitsplein
B-2610 Antwerp-Wilrijk, Belgium.

BUAT-MENARD, P.
Centre des Faibles Radioactivités
Laboratoire mixte CNRS-CEA
91190 Gif sur Yvette, France.

CABON, S.
Laboratoire de Physique et Chimie
marines, Université P. et M. Curie
4 Place Jussieu
75230 Paris Cedex 05, France

CACHIER, H.
Centre des Faibles Radioactivités
Laboratoire mixte CNRS-CEA
91190 Gif sur Yvette, France

CAMOES, M.F.
Dpto Quimica. Univ. de Lisboa,
Faculdade de Ciencias
Rua da escola politecnica
1200 Lisboa, Portugal.

CARLIER, P.
Laboratoire de Physico-chimie
Université Paris VII,
2 Place Jussieu, 75251 Paris
Cedex 05. France

CHESSELET, R.
C. N. R. S.
15 quai Anatole France
75700 Paris, France

CHESTER, R.
Dpt of Oceanography,
University of Liverpool
Bedford Street, North box 147
Liverpool L69 3BX, U.K.

CIESLIK, S.
Institut d'astronomie et Géophysique
Université catholique de Louvain
3 chemin du Cyclotron,
B. 1348, Louvain la Neuve, Belgium

CIPRIANO, R.J.
Atmospheric Science Research Center
S.U.N.Y. at Albany, 100 Fuller Road
Albany NY 12205, U.S.A.

CLEGG, S.L.
School of environmental sciences
University of East Anglia
Norwich, NR4 7TJ, U.K.

COLLIER, R.W.
College of Oceanography,
Oregon State University
Corvallis, OR 97331, U.S.A.

COLOM-ALTES, M.A.
Calle Andres, Coll. 8 Soller,
Mallorca
Spain.

CONRAD, R.
Max Planck Institut für Chemie
Saarstrasse 23, D-6500 Mainz,
Germany

DE MORA, S.J.
Chemistry Department
The University of Auckland,
Private bag
Auckland, Nex Zealand.

DUCE, R.A.
Graduate School of Oceanography
University of Rhode Island
Kingston, RI 02881, U.S.A.

DUPLESSY, J.C.
Centre des Faibles Rdaioactivités
Laboratoire mixte CNRS-CEA
91190 Gif sur Yvette, France

ELBAZ, F.
Laboratoire de Géologie de l'Ecole
Normale Supérieure
46 rue d'Ulm,
75038, Paris Cedex 5, France

ERICKSON, D.J.
Graduate School of Oceanography
University of Rhode Island
Kingston, RI 02881, U.S.A.

EPSEY, Q.I.T.
Dept of Oceanography
The University
Southampton S09 5NH, U.K.

EWALD, M.
Laboratoire de Chimie Physique A
Université de Bordeaux I
351 Cours de la Liberation
33405 Talence Cedex, France

FITZGERALD, W.F.
Marine Science Dpt.
University of Connecticut
Groton, CT, 06340, U.S.A.

FLAMENT, P.
Laboratoire de Chimie Marine
Université de Lille I, Cité Scien-
tifique, bâtiment C, 59655
Villeneuve d'Asq, Cedex, France

FOGG, T.R.
Science Applications International
Corps, 476 Prospect Street,
La Jolla, 92038, U.S.A.

FONTUGNE, M.
Centre des Faibles Radioactivités
Laboratoire mixte CNRS-CEA
91190 Gif sur Yvette, France

FRANKIGNOULLE, M.
Laboratoire d'Océanologie
Institut de Chimie B6
4000 Sart Tilman, Liège, Belgium.

GAGOSIAN, R.B.
Dept of Chemistry, Woods Hole
Oceanographic Institution,
Woods Hole, MA 02543, U.S.A.

GAO YUAN
Graduate School of oceanography
University of Rhode Island
Kingston, RI, 02881, U.S.A.

GARCON, V.
Laboratoire de Physique et Chimie
de l'hydrosphere, IPG,
4 Place Jussieu
75230, Paris Cedex 05, France

GARRIGUE, P.
Laboratoire de Chimie Physique A
Université de Bordeaux I
351 Cours de la Libération
33405 Talence Cedex, France

GOYET, C.
Laboratoire de Physique et Chimie
marine, Université P. et M. Curie
4 Place Jussieu,
75230 Paris Cedex 05, France

HARITONIDIS, S.
Botanical Laboratory,
University of Thessaloniki
54006 Greece.

HUNTER, K.A.
Dept. of Chemistry
University of Otago
Box 56, Dunedin, New Zealand

JEANDEL, C.
C.N.E.S./G.R.G.S.
18 Avenue L. Belin
31055, Toulouse Cedex, France

KIM, J.
Marine Sciences Institute Avery
point, University of Connecticut
Groton, CT 06340, U.S.A.

LABEYRIE, L.D.
Centre des Faibles Radioactivités
Laboratoire mixte CNRS-CEA
91190, Gif sur Yvette, France.

LALOU, C.
Centre des Faibles Radioactivités
Laboratoire mixte CNRS-CEA
91190, Gif sur Yvette, France.

LANG, R.F.
Environmental Research Laboratory
Atlantic Oceanographic & meteoro-
logical Laboratory, OCLB NOAA
Rickenbacker Causeway Miami
FL 33149, U.S.A.

LECK, C.
Dept of Meteorology-Arrhenius Lab.
University of Stockholm
10691 Stockholm, Sweeden

LEGRAND, M.
Laboratoire de Glaciologie et
Géophysique de l'environnement
38402 St Martin d'Hères, France

LI QI CHEN
3rd Institute of Oceanography
National bureau of oceanography
POB 70 Xiamen Fujian. The people's
Republic of China.

LISS, P.
School of environmental sciences
University of East Anglia
Norwich NR4, U.K.

Mc KAY, W.A.
Environmental & medical Science Div.
AERE Harwell, Oxfordshire
OX11, ORA, U.K.

MAENHAUT, W.
Institut voor nucleaire witten-
schappen, Proeftuinstraat 86
B-9000 Gent, Belgium

MANTISI, F.
Laboratoire de Physique et Chimie
marines, Université P. et M. Curie,
4 Place Jussieu,
75230 Paris Cedex 05, France.

MARING, H.
Graduate School of Oceanography,
University of Rhode Island
Kingston, RI 02881, U.S.A.

MERLIVAT, L.
Laboratoire d'Océanographie dynamique et de Climatologie, CEN Saclay
91191 Gif sur Yvette, France

MERRILL, L.T.
Graduate School of oceanography
University of Rhode Island
Kingston, RI 02881, U.S.A.

MEYRHAN, H.
Max Planck Institute für Chemie
Saarstrasse 23
6500 Mainz, Germany

MINSTER, J.F.
C.N.E.S./G.R.G.S.
18 Avenue E. Belin
31055 Toulouse, France.

MONAHAN, E.C.
Dept of Oceanography
University College
Galway, Ireland

MONFRAY, P.
Centre des Faibles Radioactivités
Laboratoire mixte CNRS-CEA
91190, Gif sur Yvette, France.

NICOLAS, E.
Laboratoire de Physique et Chimie
Marine, BP 8
06230 Villefranche sur mer, France

NGUYEN, B.C.
Centre des Faibles Radioactivités
Laboratoire mixte CNRS-CEA
91190, Gif sur Yvette, France

NURSE, C.
Institute of ocean sciences
9860 West Saanich Rd.
Sidney BC V8L 4B2 Canada

PARUNGO, F.
Environmental Research Laboratory
R/E 2 N.O.A.A.
Boulder CO 80303, U.S.A.

PATTERSON, T.
Graduate School of oceanography
University of Rhode Island
Kingston, RI, 028881, U.S.A.

POISSON, A.
Laboratoire de Physique et Chimie
marines, Université P. et M. Curie
4 Place Jussieu
75230 Paris Cedex 05, France

PRADO-FIEDLER, P.
Institut für Meeres kunde,
Dusternbrooker Weg 20
2300 Kiel, Germany

PSZENNY, A.
Dept of environmentral Sciences
Clarck Hall, University of Virginia
Charlottesville, V. 22903 U.S.A.

RADWAN, J.
Laboratoire d'Océanographie dynamique et de Climatologie, CEN Saclay
91191 Gif sur Yvette, France

ROETHER, W.
Institut für Umweltphysik des
Universitat Heidelberg
IM Neuenheimerfeld 366,
69 Heidelberg, Germany

ROSS, H.B.
Dept of meteorology, Arrhenius Lab.
Stockholm University
10961 Stockholm, Sweden.

SARMIENTO, J.
Princeton University. Geophysical
fluid dynamics Program,
James Forrestal campus, Box 308
Princeton, NJ 08540, U.S.A.

SAYDAM, A.C.
Marine Science Institute
P.K. 28 Erdemli Icel, Turkey.

SCHNEIDER, B.
Institut für Meereskunde,
Dusternbrooker Weg 20
2300, Kiel, Germany.

SEILER, W.
Max Planck Institut für Chemie,
Saarstrasse 23
6500 Mainz, Germany.

SICRE, M.A.
Laboratoire de Physique et Chimie
Marines, Université P. et M. Curie
4 Place Jussieu,
75230 Paris Cedex 05, France.

SIECENTHALER, U.
Universityat Bern, Physikalisches
Institut, Silerstrasse 5
3012, Bern, Switzerland

SMITH, J.D.
Marine chemistry Laboratory
School of chemistry, University of
Melbourne Parkville 3052 Australia.

SMITH, M.H.
Physics Department
UMIST
Manchester M60 1QD, U.K.

SPITZY, A.
Geologisch-Paleontologisch Inst.
Universitat Hamburg
Bundestrasse 55
2000 Hamburg 13, Germany.

STORMS, H.
UIA, Dpt Scheikunde
Universiteitsplen 1,
2610 Wilryk-Antwerpen, Belgium.

STRAMSKA, M.
Institute of oceanology
Polish Academy of sciences
Powstancow W-WY55 Poland.

TOKOS, J.
Graduate School of oceanography
University of Rhode Island
Kingston, RI 02881, U.S.A.

UEMATSU, M.
Graduate School of oceanography
University of Rhode Island
Kingston, RI 02881, U.S.A.

VAN GEEN
M.I.T. E 34 174
Cambridge
MA 02139 U.S.A.

VAN NESTE, A.
Graduate School of oceanography
University of Rhode Island
Kingston, RI 028881, U.S.A.

WALKER, M.
University of East Anglia,
School of environmental Sciences
Norwich, NR4 7TJ U.K.

WANNINKHOF, R.
Lamont Doherty Geological
Observatory
Palisades, NY 10964, U.S.A.

WATTS, S.
University of East Anglia
School of environmental Sciences
Norwich, NR4 7TJ U.K.

WEBER, H.
Institut für Meereskund abt maritime
meteorologie, Dusternbrooker Weg 20,
2300, Kiel, Germany

WOLLAST, R.
Laboratoire d'Océanographie
Faculté des Sciences,
Université libre de Bruxelles
Avenue F.D. Roosevelt, 50
1050 Bruxelles, Belgium

WOOLF, D.K.
Department of Oceanography
University College
Galway, Ireland.

ZAFIRIOPOULOS, D.
Environment Research & Studies
Anapiron Polemou
11521, Athens, Greece.

ZAFIRIOU, O.
Department of Chemistry
Woods Hole Oceanographic Inst.
Woods Hole, MA 02543, U.S.A.

INDEX

Accoustic cross section 131
Acidic rainfall 509
Adiabatic 43, 44, 47
Adsorption 311, 313
Advection 67
Advection of air 38, 46, 47, 56
Advective 36
Aerosol, 192, 443
      budget 143
      dry deposition 165
      generation models 129, 139
      indirect production 157
      nitrogen flux 506
      particle concentration 131
      particles 165
      particle size 461
      partitioning of elements in 462
      production 165
      sources 165
      speciation of elements 464
      spectra 142, 155
      spectrometer probe 131
      wet deposition 165
Air borne fraction of CO_2 223, 225, 227, 228
Air mass trajectory 50, 51, 53, 54, 428, 429
Air-sea exchange, fluxes 185
Air-sea flux 314
Air-sea gas exchange 113, 114, 198, 199
Air-sea interface 201, 368
Air-water partition coefficient 309
Albedo 237
Algae 286,338
Aliphatic hydrocarbons 411
Alkalinity 69, 219, 230
Alkyl halides 284
Aluminium, 517
      particulate in seawater 465
Aluminosilicate particles 175
Aminoacids 500
Ammonia 500, 504, 506,
Ammonium, 507
      oxidizers 276

Anthropogenic emissions 401, 402
Anticyclones 42, 43
Arsenic (As) 478, 485
Artic Haze 448
Asian dust 424
Asparagosis 288, 289
Association constants 312
Atmosphere,
        circulation 36, 39, 42, 44, 47
        general circulation 35, 57
Atmospheric,
            boundary layer 36, 37, 38, 40, 43, 46, 53, 54
            chemistry 186
            circulation models 58, 59
            deposition 497
            particle scavenging 310
            particle size 431
            particulate matter 192
            pCO_2, 69
            stability 147

Bacteria 276, 368
Baroclinic 43
Beaufort velocity 150
Beryllium-10 234, 235
Biogenic sources 280
Biogeochemical cycling 480
Biological,
          mobilization 497, 498
          processes 77
          transformations 322
Biomass burning 284
Biospheric release of CO_2 229
Boron 499
Boundary layers 200
Box models, 67, 69, 70, 80
            geochemical 66
Breaking waves 116, 130
Bromoform 288
Bubbles, 116, 118, 203
            cloud 130
            spectrum 135, 141
            terminal rise velocity 130
Buffer factor 219

Carbon,
      reservoirs 211
      residence time 210, 211
      transport by particles in the ocean 220, 226
Carbon-13 212, 233, 239

Carbonate sediment dissolution 230
Carbon cycle, 209, 222
Carbon dioxide (CO_2), 209
  air-sea exchange 213, 249
  air-sea transfer velocity 214, 215, 216, 217
  anthropogenic increase 223
  concentration 214
  equatorial upwelling of 241
  future concentrations 231
  glacial interglacial changes 242
  natural variations 229
  partial pressure 215, 216, 217, 218
  pre-industrial 214, 225
  seasonal variations 230, 231, 240
  uptake by ocean 225
Carbondisulfide 352
Carbontetrachloride 289, 290
Carbon monoxide (CO),
  air-sea exchange 269
  in Tropical Atlantic ocean 272
  transfer velocity 227
  zonal distribution 272
Carbonyl sulfide 331, 352
Cascade impactor 175
Chemical,
 enhancement factor 115, 116
 reaction 114
Chlorinated hydrocarbons 296
Chlorite 457
Chlorofluoromethanes 186
Chloroform 287
Chlorophyll 274, 516
Chlorophyll a 387, 388, 393, 397, 509
Chromophore 195, 196
Clay mineral signatures 470
Climate models 237
Climatic effects of CO_2 234
Clouds 39, 41, 52, 193
Cloudwater oxydation 194
Cobalt 516
Conservation equation 66, 78
Constant flux layer 38, 173
Cosmic radiation 234
Coulter counter 86
Crustal degassing 401
Cyanobacteria 276
Cyclones 42, 43

Deep-sea circulation 249
Deep-sea sediments 467
Deep water formation 220

Deep waters 481
Deforestation 223, 229
deposition flux 315
Deposition layer 173
Di-2 ethylhexyl phtalate 296
Diffusion 66
Diffusivity 41, 391, 392
Dimethyldisulfide 356
Dimethylsulfide (DMS) 331, 388, 393
Dimethyl sulfonium ions 285, 286, 287
Dimethyl sulfonium propionate 338
Dimethyl sulfoxide 338
Di-n butyl phtalate 296
Dissolved gases in sea water 119
Dissolved mercury chloride gas 390
Dissolved organic carbon (DOC) 313
Distribution constants 313
Droplet photochemistry 193
Dust flux 180
Dust storms 516, 519
Dry deposition, 202, 498, 504, 506, 520
                of aerosol particles 172, 177
                velocity 145, 310
Dry removal velocity 74

Eddy, 47
     correlation 119
     diffusion 226, 500, 502, 503,   511, 513, 519,
     diffusion-advection model 508
     flux technique 175
     heat viscosity 40
Electrostatic space charge 141
El Niño 240

Energy balance of the earth 236
Energy input, anthropogenic 237
Enrichment factor 459
Enzimatic reactions 10
Equatorial currents 385
Eukaryotes 276
Eulerian model 55,56
Euphotic layer 272
Euphotic zone 498, 500, 522, 523
Evaporation 46, 47
Excess sulfate 351
Exchange processes 321
Extraction technique 270

Fatty acids, 413, 420, 421
            esters 410
            salts 425, 428

Fatty alcohols 411, 420, 421, 425, 428
Feedback, climatic 237
Fertilization by CO_2 230
Film droplets 138
Films 116
First order decay reaction 3
Fluvial input 394
Flux 40, 41, 47
Fly ash particles 517
Fossil fuels 211, 222, 223
Freon-11 289
Friction velocity 116, 149, 173
Front 39, 43, 44
Frontal 38, 40
Fugacity 313
Fungi 284

'Gaia' 287
Gas,
     flux 114, 125
    -particle partitioning 434
     scavenging 311
    -to particle conversion 498
     transfer velocity 114, 123, 125
Gelbstoff 276
General circulation models 237
Geochemical cycling, 284
                of bromine 288
Geochemical modelling 1
Geomagnetic field 234
Geostrophic approximation 78
Glacial times 249
Gradient technique 175
Greenhouse effect 234
Gross deposition   178

Halocarbon gases 282
Haloforms 287
Harmattan wind 449
Henry's law constant 114, 282, 289, 306, 390, 391
Heterogeneous catalysis 10
Heterogeneous photochemistry 191
Hexachlorocyclohexanes, 296
                Air-sea flux 319
High winds 156
Horizontal transport in the ocean 83
Humic acids 312
Hydroxyl radical (OH) 187, 346, 369

Hydrogen (H$_2$),
        air-sea exchange 269
        chloride 499
        sulfide 331
Hydrothermal vents 497

Illite 457
Infrared radiation 236
Inter halogen compounds 289
Inter tropical convergence zone (ITCZ) 47, 48, 381, 383, 447
Inverse diagnostic techniques 78, 79
Iodine, 282
    geochemical cycle 287
Iron (Fe), 497, 498, 516, 522
    atmospheric deposition 519, 521
Isentropic, 45, 46, 54
    trajectories 376

Jet droplets 138

Kaolinite 457
Kinematic viscosity 173
Kinetic dissolution coefficient 104
Kinetic geochemical cycles 2

Land breeze 38, 39
Laminar film thickness 278
Laminaria, 286
    digitata 287
Lead (Pb), 478
    eolian input to the ocean 488
    oceanic cycle 487
Lead-210 431
Light field 196
Light intensity 269
Limiting nutrients 221, 231, 242
Log normal distribution 84
Long range transport 428

Macro algae 289
Manganese 466
Marine aerosols, 165, 203, 416, 443, 511
        lipids in 416
Marine biological productivity 221, 242, 497
Marine carbonate chemistry 219
Marine dust Veil, 443
        mineral composition 455
Mass-transfer rate 309
Mauna Loa 223

Mean oceanic residence time 482
Mercury (Hg), 363, 478
        air-sea exchange 363, 398, 402
        analyses 370
        atmospheric cycle 363, 365, 385
        atmospheric cycling 397
        atmospheric residence time 366
        degassing 366
        depositional flux 393, 396
        depositional velocity 396
        dissolved gaseous 372, 387, 397
        divalent 367
        efflux 392
        elemental 366, 390, 393,
        geochemical cycle 365
        global cycle 364, 385, 398, 400
        longitudinal distribution 380, 381
        methyl compounds 367
        oceanic evasion 382
        particulate fraction 376
        pollution 364, 401
        pre-man cycle 364
        reactive 386, 390
        total gaseous 370, 378
        volatile species 369, 393
        volcanic contribution 365
Methanesulfonic acid 347, 351
Methodology, 413
        analytical 413
        sampling 413
Methyl bromide 285
Methyl chloride 186, 284
Methyl iodide, 284, 286
        role in atmospheric chemistry 287
Methylmercaptan 331, 356
Microbial activity 269
Micrometeorological techniques 119
Micro organisms 276
Mineral aerosol 179, 443, 516, 519
Missing CO_2 sink 229
Mixed layer thickness 73, 76
Mixing 67
Modeling,
   oceanic transport of dissolved constituants 65
Models,
     advection diffusion 71, 72, 78, 80
     boundary layer 115, 116, 118
     box diffusion 226, 227, 228, 232, 238
     diagnostic 65, 70
     film 115

   first order 3
   first order production 9
   laminar film 269
   perturbation 225, 226
   physico-chemical 367, 397
   prognostic 65
   steady state 1
   surface renewal 115
   tracer 226
Molecular diffusivity 115, 278
Molybdenum 516
Montmorillonite 457

N-alkane, 420, 421, 425, 428
   anthropogenic 431
Natural water photochemistry 186
New nutrients 521, 523
New production 499, 500, 508
Nitrate 502, 504, 509
Nitrate radical (NO_3) 346, 190
Nitric acid 504
Nitrogen, 497, 498, 502, 522
   atmospheric input of 502, 505, 507, 509
   fixation 502
Nitrous oxide 186
NO flux 200
Non steady state sedimentation 102
North African desert belt 449
NO_x 504,506
Nutrient,
   concentrations 387
   regeneration 523
Nutrients 499

O (^{1}D) atoms 187
Ocean,
   heat capacity of 238
   models 66
   settling velocity 92
Oceanic,
   carbon cycle 220
   flow 66
   Hg evasion 379, 386
   mixed layer 397
   salinity 66
   suspended matter 83
   temperature 66
OH radical 189
Oligotropic 385, 393
Organic material 151
Organic nitrogen 507

Organo-Hg-complexes 368, 371, 372, 379, 382
Organosulfur compounds 352
Osmolytes 338
Outcrop of deepwater 227, 228, 232
Oxygen,
 cycle 221, 222
 in sea water 68, 76
 isotopic composition 222
Ozone 187, 200, 287, 289
Ozone fluxes 119

Particle,
 counter 136
 generating processes 445
Particulate organic matter in the atmosphere,
 composition 409
 fluxes 409
 sources 409,415
 transport pathways 409,415
Particulate transformations 196
Particles in the ocean,
   coagulation 84
   density 83
   disagregation 84
   dissolution process 99
   geochemical cycles 83
   mass concentration 84,90
   regeneration rate 104
   settling velocity 84, 90
   standing crop 104
   surface area concentration 90
   vertical flux 95
   vertical transport 83
Partitioning 311
Peat bog 400
Periodic fluctuations 22
pH 509, 517
Phosphate (PO_4), 509, 510
  oceanic distribution 76, 78
  regeneration 76,77
Phosphine, 510
 methylated forms 510
Phosphorus (P), 497, 498, 510, 522
  atmospheric,
    cycle 510, 512
    deposition 513, 515
    riverine 516
    sources 513, 515
    species 512

Photochemical,
            models 355
            oxidation 369, 397
Photochemistry 185
Photometabolism 274
Photo oxidation 276
Photoreaction types 197
Photosynthesis 221, 222, 239, 516
Photoxidative destruction 388
Phtalic acid esters 296
Phytoplankton, 274, 338, 368, 393, 480
            activity 388
Planetary temperature 236
Plant respiration 222
Pluvial flux 394, 395
Polychlorobiphenyls, 296
                        air-sea flux 319
Potential temperature 44,54
Power law function 83, 149
Precipitation, 46,47, 202
            concentrations 314
            scavenging 319
Preformed nutrient 69
Primary production, 222, 392, 499
                gross 212
                net 211
Primary productivity 341, 487, 499
Primitive equation model 79
Profile method 119
Productivity 338

Radical oxidation 187
Radiocarbon (^{14}C), 70, 212, 213
            bomb produced 214, 226, 228
            in the ocean 220
            natural 120
            primary productivity 508, 509
            variations 233
Radon deficiency method 120, 214
Rain,
    chemical composition 167, 169
    collection 167
    contamination 167
    scavenging ratio 520
Rainfall, 372, 394
        scavenging 395
Rainout-washout deposition 201
Rain water 373
Recycled sea-salt 498
Recycled sea-spray 178
Red algae 288

Redfield ratio 70, 221, 242, 522
Regenerated production 499, 500
Residence time 5, 83
Resistance 114
Response time 5
Reversible reactions 12
Riverine input 319, 498

Sahara desert 516
Saharan air out-breaks 449
Soil photochemistry 198
Salinity 306
Saturation 269
Scavenging,
       by bubbles 498
       by sea-spray 176
       ratio 169
Schmidt number 116, 118, 124, 125, 174
Sea breeze 38, 39
Sea-salt,
     flux 152
     particles 129, 176
     removal 155
Sea spray 131
Sea spray sulfate 332
Sea surface microlayer 343, 434
Sea-water,
     photochemistry 194, 199
     saturation state 314
     solubility 464
Sea-weed 289
Second order reactions 25
Secondary aerosols 202
Sedimentation, 320
       flux 83
Sediment trap 77
Selenium (Se) 478, 498
Short range transport 429
Silica in the ocean 6
Southern ocean 242
Southern oscillation 240
Spume drops 156
^{90}Sr, 72
   atmospheric flux 73
   in rainfall 73, 74
   ocean concentration 73, 75
Stagnant films 308
Steady-state 370, 402
Steroidal alcohols 411
Stokes number 174

Stokes settling velocity 92
Stratosphere 285, 289
Submarine volcanism 497
Suess effect 233
Sulfate reduction, 334
                    assimilatory 335, 336
                    dissimilatory 335
Sulfur, 498
     dioxide (SO_2) 351
     hexafluoride 290
Sunlight 518
Supersaturation 391
Surface,
        temperature 236
        water 269
        waters 481
Synthetic organic compounds 295

Temperature anomalies 238
Terrestrial,
            organic compounds 429
            plant waxes 425
            source marker compound 429
Tin (Sn) 488
Total carbon 69
Total CO_2 210, 219
Trace gases, 190, 191, 270
            infrared-active 239
Trace metals, 477, 516
            air to sea flux 483
            air to sea transfer 477
            anthropogenic 477
            atmospheric input 516
            chemical forms in the atmosphere 478
            oceanic distributions 516
            physical forms in the atmosphere 478
            solubilization in seawater 517
Trace organics,
            analytical aspects 296
            atmospheric concentrations 300
            depth profiles 299
            dry deposition 303, 309
            gas exchange flux 305
            in surface ocean waters 299
            particle vapor partitioning 302
            wet deposition 303
Tracers, 71, 121
     freon as 290
Transfer velocity 391
Transient tracer 72, 491
Transport by settling particles 481

Tree-rings 233, 234
Triglycerides 410
Tritium, 72
      penetration depth 226, 227
Tritium helium-3   78
Turbulence 508, 521
Turbulence closure model 508
Turbulent nutrient fluxes 508, 521
Two-film resistance model 306

Undersaturation 391
Upwelling 340, 385, 387

Vertical advection 500, 502, 503, 510, 511, 513
Vertical mixing in the ocean 83
Vertical transport in the ocean 500
Viscosity 151

Warning, CO_2 induced 232, 234, 238
Washout factor 169
Water fall 136
Water,
    mixing 477, 481
    vapour fluxes 119
    wind tunnel 175
Wave field 121
Waves 116, 118
Wax esters 410, 421
Weir 135
Wet deposition, 498, 504, 505, 520
          assessment 167
Wet removal velocity 74
White cap, 124, 129,
      atlas 153
      bubbles 133
      decay of 130
      foam patch 133
      simulation tank 131
Wind speed, 115, 120, 269
      near surface 146
Wind stress 508

Zooplankton 517